PRACTICAL HANDBOOK OF

Estuarine and Marine Pollution

Marine Science Series

The CRC Marine Science Series is dedicated to providing state-of-the-art coverage of important topics in marine biology, marine chemistry, marine geology, and physical oceanography. The Series includes volumes that focus on the synthesis of recent advances in marine science.

CRC MARINE SCIENCE SERIES

SERIES EDITORS

Michael J. Kennish, Ph.D.
Peter L. Lutz, Ph.D.

PUBLISHED TITLES

Chemical Oceanography, 2nd Edition, Frank J. Millero
Ecology of Estuaries: Anthropogenic Effects, Michael J. Kennish
Ecology of Marine Invertebrate Larvae, Larry McEdward
Ecology of Marine Bivalves: An Ecosystem Approach, Richard F. Dame
Morphodynamics of Inner Continental Shelves, L. Donelson Wright
Ocean Pollution: Effects on Living Resources and Humans, Carl J. Sindermann
Physical Oceanographic Processes of the Great Barrier Reef, Eric Wolanski
The Physiology of Fishes, David H. Evan
Practical Handbook of Estuarine and Marine Pollution, Michael J. Kennish
Practical Handbook of Marine Science, 2nd Edition, Michael J. Kennish

---

FORTHCOMING TITLES

Benthic Microbial Ecology, Paul F. Kemp
The Biology of Sea Turtles, Peter L. Lutz and John A. Musick
Chemosynthetic Communities, James M. Brooks and Charles R. Fisher
Coastal Ecosystem Processes, Daniel M. Alongi
Environmental Oceanography, 2nd Edition, Tom Beer
The Physiology of Fishes, 2nd Edition, David H. Evans
Seabed Instability, M. Shamim Rahman
Sediment Studies of River Mouths, Tidal Flats, and Coastal Lagoons, Doeke Eisma

PRACTICAL HANDBOOK OF

Estuarine and Marine Pollution

Michael J. Kennish

Institute of Marine and Coastal Sciences
Rutgers University
New Brunswick, New Jersey

CRC Press
Boca Raton New York London Tokyo

Senior Editor:	Paul Petralia
Editorial Assistant:	Cindy Carelli
Project Editor:	Sarah Fortener
Assistant Managing Editor:	Gerry Jaffe
Marketing Manager:	Susie Carlisle
Direct Marketing Manager:	Becky McEldowney
Cover design:	Dawn Boyd
PrePress:	Carlos Esser
Manufacturing:	Sheri Schwartz

Library of Congress Cataloging-in-Publication Data

Kennish, Michael J.
Practical handbook of estuarine and marine pollution / by Michael J. Kennish.
p. cm. -- (CRC Press marine science series)
Includes bibliographical references and index.
ISBN 0-8493-8424-9
1. Estuarine pollution. 2. Marine pollution. I. Title.
II. Series.
GC1085.K46 1996
628.1′68′09162--dc20 96-14030
CIP

International Standard Book Number 0-8493-8424-9
Library of Congress Card Number 96-14030
Printed in the United States of America 1 2 3 4 5 6 7 8 9 0
Printed on acid-free paper

Abstract

This handbook provides comprehensive coverage of significant developments in estuarine and marine pollution. It is designed to serve the multidisciplinary research needs of contemporary marine biologists, chemists, geologists, and oceanographers and, as such, represents a valuable reference for students and professionals in all disciplines of marine science. The book examines in detail anthropogenic effects on estuarine and marine ecosystems from local, regional, and global perspectives. Each chapter consists of expository, tabular, and illustrative material that will benefit marine scientists in academia, government, and industry. A truly international collection of data is presented in an organized framework on a wide range of subject areas, including eutrophication, organic loading, oil pollution, polycyclic aromatic hydrocarbons, halogenated hydrocarbons, trace metals, radioactive waste, dredging and dredged-spoil disposal, and effects of electric generating stations. This work will be of particular value to administrators dealing with the management of estuarine and marine pollution problems.

Abstract

Preface

Anthropogenic waste in estuarine and oceanic environments continues to be a major societal issue. While estuaries and coastal marine waters are the systems subjected to the greatest pollution impacts, the open ocean also receives considerable volumes of waste from a variety of human activities. The proximal deep ocean, for example, is being used to dispose of dredged spoils, sewage sludge, pharmaceutical and industrial wastes, and low-level radioactive waste. As tighter regulatory controls are placed on the discharge and dumping of waste in coastal environments, increasing pressure may be expected for future use of the deep sea as a final repository of unwanted and perhaps even dangerous substances. A waste management problem clearly exists not only in the U.S., but also in countries worldwide. Hence, it is critical to formulate more effective regional and global policies involving many nations that will ensure long-term legal protection against hazardous waste disposal in the deep sea.

Practical Handbook of Estuarine and Marine Pollution provides the most up-to-date and comprehensive reference material on estuarine and marine pollution. It is designed to serve the multidisciplinary research needs of contemporary marine scientists, but also should be of value to administrators and other professionals dealing in some way with the management of estuarine and marine pollution problems. Its primary objective is to amass significant data that will have appeal and utility for practitioners and students of marine science who are engaged in some aspect of marine pollution studies.

The editor is well aware of possible gaps and omissions in this volume. The significant limitations of space inherent in a single volume dealing with such a broad range of environmental problems preclude equal coverage of the subject material comprising this important field of study. Thus, constructive criticisms and suggestions for improving future editions of the handbook are urgently requested from the readership.

Author

Michael J. Kennish, Ph.D., is a research marine scientist in the Institute of Marine and Coastal Sciences and a member of the graduate faculty at Rutgers University, New Brunswick, NJ.

He graduated in 1972 from Rutgers University, Camden, NJ, with a B.A. degree in geology and obtained his M.S. and Ph.D. degrees in the same discipline from Rutgers University, New Brunswick, in 1974 and 1977, respectively.

Dr. Kennish's professional affiliations include the American Fisheries Society (Mid-Atlantic Chapter), American Geophysical Union, American Institute of Physics, New England Estuarine Research Society, Atlantic Estuarine Research Society, Southeastern Estuarine Research Society, Gulf Estuarine Research Society, Pacific Estuarine Research Society, New Jersey Academy of Science, and Sigma Xi. He is also a member of the Advisory Board of the Fisheries and Aquaculture TEX Center of Rutgers University, overseeing the development of fisheries and shellfisheries in estuarine and marine waters of New Jersey.

Although maintaining research interests in broad areas of marine ecology and marine geology, Dr. Kennish has been most actively involved in studies of anthropogenic effects on estuarine and coastal marine ecosystems and with investigations of deep-sea hydrothermal vents and seafloor spreading centers. He is the author or editor of six other books dealing with various aspects of estuarine and marine science. In addition to these books, Dr. Kennish has published more than 70 research articles and presented papers at numerous conferences. Currently, he is the marine science editor of the journal, *Bulletin of the New Jersey Academy of Science,* and series editor of Marine Science books for CRC Press. His biographical profile appears in *Who's Who in Frontiers of Science and Technology* and *Who's Who Among Rising Young Americans.*

Acknowledgments

I am deeply appreciative of my colleagues, family members, and friends who encouraged and inspired me during the production of this work. In the Institute of Marine and Coastal Sciences at Rutgers University, I thank J. F. Grassle, N. P. Psuty, and R. A. Lutz. Special acknowledgment is given to R. A. Lutz for innumerable exchanges of ideas on marine science. I am grateful to personnel in the Editorial Department of CRC Press, especially Paul Petralia and Cindy Carelli, for their efficiency and guidance during the publication process. Finally, I express my love and devotion to my wife, Jo-Ann, and sons, Shawn and Michael, for their patience and understanding during the preparation of the book.

Acknowledgments

Contents

Chapter 6 Heavy Metals

Chapter 7 National Monitoring Surveys

1 Introduction

MARINE POLLUTION DEFINED

The world's estuaries and oceans are the ultimate repository for a vast array of substances discharged deliberately or accidentally via human activities. The immediate and most acute impacts of these activities occur in the coastal zone where population growth has increased dramatically over the years. Concomitant with this growth have been conspicuous changes at the land-sea interface associated with construction of industrial installations, maintenance of harbors and other waterways, domestic development of the coastline, demands of tourism, and other uses of coastal space. While the coastal zone is clearly at greatest risk from various anthropogenic impacts, the open ocean is also not immune to pollution. For example, the input of toxic chemicals from atmospheric transport and deposition, as well as from shipping operations beyond the continental shelf, can adversely affect open ocean waters. Contaminant inputs from atmospheric fallout alone can be delineated in all components of the marine environment — seawater, sediments, and biota. However, because of the great volume of all the oceans (137×10 km^3) and their great dilution capacity, the concentrations of these contaminant inputs usually are insufficient to cause detectable problems in deep-sea environments.

In contrast to conditions in the open ocean, shallow estuarine and nearshore marine waters continue to be extensively degraded by both point and nonpoint sources of pollution. Systems characterized by a slow rate of exchange relative to their volume (e.g., semi-enclosed estuaries and embayments) are most susceptible to contaminant inputs. These systems typically have a very limited assimilative capacity for pollutants. Consequently, certain unassimilated materials, such as synthetic toxic organic compounds, can accumulate and persist for long periods of time, posing a potential long-term danger to marine food webs.

The Joint Group of Experts on the Scientific Aspects of Marine Pollution (GESAMP), an IMO/FAO/UNESCO/WHO/IAEA/United Nations/UNEP-sponsored advisory body, has played a leading role in assessing global marine pollution problems. Established in 1969, GESAMP provides authoritative information on marine pollution problems to its sponsors, to the Intergovernmental Oceanographic Commission, to other organizations of the United Nations system, and to state members of the United Nations.[1] This international advisory body prepares periodic reports on the health of the marine environment, identifying specific problem areas (e.g., coastal development, oil pollution, organochlorine contaminants, thermal loading). The last global report on the state of the marine environment by GESAMP was issued in 1990.[2] This report is widely accepted as an authoritative reference on the subject.[1]

GESAMP[3] defines marine pollution as "the introduction by man, directly or indirectly, of substances or energy into the marine environment, resulting in such deleterious effects as harm to living resources, hazards to human health, hindrance to marine activities including fisheries, impairment of quality for use of seawater, and reduction of amenities." According to Clark,[4] contamination occurs "when a man-made input increases the concentration of a substance in seawater, sediments, or organisms above the natural background level for that area and for the organisms." Contaminants enter estuarine and oceanic waters via five primary pathways: (1) riverine input, (2) nonpoint source runoff from land, (3) direct pipeline discharges, (4) discharges and dumping from ships, and (5) atmospheric deposition.[5-7] The most common anthropogenic wastes found in estuarine and coastal marine environments worldwide are dredged spoils, sewage, and industrial and municipal discharges.[4,8] These wastes generally contain a wide range of pollutants, notably heavy metals, petroleum hydrocarbons, polycyclic aromatic hydrocarbons, chlorinated hydrocarbons, and other substances.

Systems exhibiting the greatest pollution impacts remain those in close proximity to population centers. In U.S. coastal waters, examples include Boston Harbor, the New York Bight Apex, and Baltimore Harbor on the east coast, and San Diego Harbor, San Francisco Bay, and Commencement Bay on the west coast. Many of the pollution problems encountered in these systems are largely attributable to overpopulation, overdevelopment, and poor coastal resource management.

TYPES OF CONTAMINANTS

In heavily impacted areas, such as the aforementioned U.S. coastal systems, the total contaminant burden derives from many land-based sources. Chief among the contaminants affecting these waters are (1) organic carbon enrichment related to elevated nutrient inputs, particularly nitrogen and phosphorus; (2) heavy metals associated with sewage effluents and sewage sludges; (3) organochlorine compounds originating from widespread domestic and agricultural use of herbicides and pesticides, as well as various industrial wastes; (4) petroleum hydrocarbons from oil spills, sewage, and nonpoint source runoff; and (5) polycyclic aromatic hydrocarbons from industrial effluents, pyrolysis of organic matter, and other sources. Domestic, industrial, and municipal wastes have accumulated for years in some coastal waters bordering metropolitan centers. However, because of the enactment of stringent regulations to control the input of these wastes, the degree of pollution in the U.S. coastal marine environment appears to be declining in many areas.[9]

From a global perspective, five classes of contaminants are considered to be critical to the environmental health of the ocean:[10]

1. Petroleum hydrocarbons (crude oil and its refined products)
2. Halogenated hydrocarbons
3. Heavy metals (particularly mercury, cadmium, and lead)
4. Radionuclides (especially cesium-137, strontium-90, and plutonium-239, 240)
5. Litter

There is evidence for a decline in concentrations of at least certain constituents of the first four classes of these critical contaminants. The amount of litter, however, seems to be on the increase around the world.[10] Perhaps most alarming is the occurrence of persistent plastics which has been on the rise in oceanic waters for a number of years.

McIntyre[7] identified three pollutant categories of priority concern today in coastal marine environments (i.e., sewage, nutrients, and synthetic organic compounds). Sewage input, eutrophication from excessive nutrients, and the toxic effects of persistent organic compounds in estuaries and nearshore oceanic waters pose significant public health risks. Three other pollutant categories (i.e., heavy metals, radionuclides, and oil) are seen as less threatening in the sea. As recounted by McIntyre,[7] more general impacts of heavy metals on marine communities are evident only in the immediate vicinity of metal-rich discharges or mine-tailing effluents where concentrations are very high. Aside from nuclear accidents, marine inputs of radionuclides are now restricted to the relatively small number of discharges from nuclear power stations and reprocessing plants, which are rigorously controlled by various national or international agencies. After enactment of the Partial Test Ban Treaty of 1963, oceanic input of radionuclides from atmospheric testing of nuclear weapons decreased sharply, considerably reducing the atmospheric influx of anthropogenic radionuclides. Damage from oil pollution at sea, while potentially severe, is localized in space and time, with the worst impacts arising from oil moving ashore and contaminating coastal wetlands or becoming buried in sandy beaches. The main global concern regarding oil pollution appears to be operational discharges from large tankers which circulate on ocean currents and affect beach amenity far from the original source. Used oils from land-based sources may pose a potential risk to the coastal zone worldwide.

Marine wastes may be organized into several distinct categories: degradable wastes, fertilizers, dissipating wastes, particulates, and conservative wastes.[4] Degradable wastes, composed of organic material, are subject to bacterial attack. They constitute by far the largest volume of anthropogenic wastes added to estuaries and coastal marine waters. Included under this category are sewage wastes, oil spills, agricultural wastes, chemical industry wastes, food processing wastes, brewing and distillery wastes, and paper pulp mill wastes. Fertilizers release substantial amounts of nitrogen and phosphorus to waters bordering land areas, and may promote eutrophication of susceptible systems at certain times of the year. Some industrial wastes rapidly lose their damaging properties after they enter an outfall area at a point of discharge. Examples of such dissipating wastes are the release of heated waters from coastal power plants and cyanide from metallurgical industries. Particulates encompass a group of diverse inert materials, as well as matter which may be chemically contaminated. Dredged spoils, fly (powdered) ash from coal-fired power plants, china clay waste, colliery waste, and plastics represent common types of particulate wastes found in estuarine and marine environments. Conservative wastes (e.g., heavy metals, halogenated hydrocarbons, and radioactivity) are substances potentially reactive with marine organisms and occasionally harmful to them.

LEGISLATIVE CONTROL OF MARINE POLLUTION

UNITED STATES

The U.S. has provided legal protection of the marine environment to various degrees over the last century, although some of the most significant legislation has been enacted during the past 25 years. In response to the great public outcry toward environmental pollution problems in the 1960s and 1970s, federal and state legislation was enacted often as a result of crisis management. At times, overlapping regulations were formulated to control point sources of contaminant input to marine waters.[11] Nevertheless, Congress has played a crucial role in developing and implementing effective U.S. policy measures on waste disposal at sea since 1972.

The River and Harbor Act of 1890 prohibited any obstruction to the navigable capacity of U.S. waters and led to a compilation of all navigation laws in Sections 9 through 20 of the River and Harbor Appropriations Act of 1899.[12] Section 10 requires authorization from the Secretary of the Army, acting through the Corps of Engineers, for the excavation from or deposition of material in navigable waters of the U.S., the construction of any structure in such waters, and any obstruction or alteration therein. Not only does the law apply to dredging and dredged-spoil disposal activities, but also to excavation, filling, rechannelization, construction, and any other modification of navigable waters (defined as those subject to the ebb and flow of the tide shoreward to the mean high water mark and/or are presently used to transport interstate or foreign commerce). The building of any structure in or over navigable waters, such as jetties, groins, wharfs, weirs, bank protection (e.g., bulkhead, revetment, and riprap), moored structures (e.g., pilings), moored floating vessels, boat ramps, and recreational docks, requires compliance with Section 10 of the act. A major intent of this federal law is to regulate and limit adverse environmental effects of dredging, dredged-spoil disposal, construction, or any other modification of navigable waters and wetlands.[13]

The Federal Water Pollution Control Act of 1948, Federal Water Pollution Control Act Amendments of 1972, the Clean Water Act of 1977, and the Clean Water Act Amendments of 1987 deal with point sources of municipal and industrial wastes, oil spills, and releases of hazardous materials, as well as other types of pollutants. The Federal Water Pollution Control Act regulates the discharge of effluents into freshwater and marine waters. Section 404 of the Clean Water Act is the principal tool used by the U.S. Environmental Protection Agency and the U.S. Army Corps of Engineers to regulate the discharge of dredged or fill material into wetlands and waters of the U.S.[14] The objective of the Clean Water Act is to "restore and maintain the chemical, physical, and biological integrity of the nation's waters." Hence, its provisions provide broad protection for the control of both point and nonpoint sources of pollution into natural waters.[15] While the 1977 Clean Water Act established a technology-based approach to regulate individual point source discharges through National Pollutant Discharge Elimination System (NPDES) permits, the 1987 Clean Water Act Amendments identified the remaining serious pollution problems, including nonpoint source impacts associated with eutrophication, hydrologic modification, accumulation of toxic pollutants, sedimentation, and increased turbidity.[16] Much of the legislative authority of the U.S. Environmental Protection Agency to regulate and protect the quality of surface waters derives from the Clean Water Act.

Passage of the National Environmental Policy Act in 1969 ushered in an era of more aggressive long-term management of marine wastes. For example, the Marine Protection, Research, and Sanctuaries Act of 1972 (commonly called the Ocean Dumping Act) was passed to strictly regulate the dumping of all types of materials into marine waters that would adversely affect human health, welfare, the marine environment, or economic potentialities (e.g., sewage sludges, dredged spoils, industrial wastes). It regulates the transportation and dumping of substances in the ocean seaward of the territorial sea baseline.[17] Under the Marine Protection, Research, and Sanctuaries Act, the oceanic disposal of certain hazardous materials, such as high-level radioactive wastes and biological, chemical, and radiological warfare agents, was banned. In addition, any dumping of wastes in the ocean required a permit from the U.S. Environmental Protection Agency. The dumping of dredged spoils from navigable waters was placed under regulatory control of the U.S. Army Corps of Engineers.[18]

Congress passed amendments to the Ocean Dumping Act in 1977 which called for an end to sewage sludge and industrial waste disposal at sea as soon as possible. The Water Resources Development Act of 1986 moved the site for dumping of sewage sludge from the New York Bight Apex to the edge of the continental shelf 190 km east of Atlantic City, NJ. The Ocean Dumping Ban Act of 1988 stated that all ocean dumping of sewage sludge and industrial waste would cease after December 31, 1991, whether or not it unreasonably degraded the marine environment.

Passage of the Toxic Substances Act of 1976 also had significant implications for improving the health of the oceans. This legislation enabled the U.S. Environmental Protection Agency to ban production of PCBs in the U.S. As a result, the concentrations of polychlorinated biphenyls (PCBs) in water, sediments, and biota from a number of coastal sites nationwide, especially some of those in proximity to known industrial sources and other "hot spot" areas, have substantially declined during the past 20 years.[6,9,19,20] However, there are a few "hot spot" areas (e.g., New Bedford Harbor and St. Lawrence River) where, in spite of trends toward a decline, the concentrations of PCBs are such that mitigative actions are being considered.[21]

The Low-Level Radioactive Waste Policy Act of 1980 and the Nuclear Waste Policy Act of 1982 provided a framework for resolving many of the management problems associated with low- and high-level radioactive waste. The Low-Level Radioactive Waste Policy Act established two major national policies: (1) each state must assure adequate disposal capacity for the low-level radioactive waste generation within its boundaries, except for the waste produced by federal defense or research and development; and (2) a regional grouping of states can be allied through interstate agreements to provide the required disposal facilities. The Nuclear Waste Policy Act establishes the following: (1) a schedule for the siting, construction, and operation of high-level radioactive waste depositories; (2) the working and decision-making relationships between the federal government, state governments, and Indian tribes; (3) the federal policy and responsibility for nuclear waste management; and (4) a fund to cover the costs of nuclear waste disposal. Despite this comprehensive radioactive waste legislation, considerable scientific and technical uncertainty exists with regard to the proper waste disposal design required to safeguard the public and the environment in future years.

The Food, Agriculture, Conservation, and Trade Act of 1990 established the Wetlands Reserve Program (Section 1438) which is designed to set aside more than 1 million hectares of land for restoring and protecting wetlands by the mid-1990s. The Wetlands Reserve Program supports demonstration studies to obtain information on: (1) the costs of constructing wetlands, (2) the development of a wetlands restoration program, (3) the establishment of wetlands in different kinds of agricultural landscapes and the acceptability of this approach in different regions of the country, and (4) the effectiveness of wetlands as sinks for contaminants from agricultural runoff. Major technical issues on wetlands restoration that must be addressed are the effects of contaminants (especially sediments and pesticides) on wetland ecosystem composition, structure, and function; the fate of organic contaminants in wetlands; and the development of site selection criteria and design criteria for restored wetlands.[22]

England

Water pollution legislation in England that applies to discharges in inland, estuarine, and tidal waters (Water Act of 1989) differs from that which applies to discharges in the marine environment (Food and Environment Protection Act of 1985). The principal regulatory mechanism for controlling polluting discharges in these waters is a consent or licensing system. Regional water authorities grant consents for industrial discharges, whereas the Department of the Environment grants consents for effluent from the sewage works of water authorities.

Sections 32 and 33 of the Control of Pollution Act of 1974, later replaced by similar terms under Section 107 of the Water Act of 1989, make it an offense to "cause or knowingly permit" the discharge into inland waters of: (1) noxious, poisonous, or polluting matter and (2) any trade or sewage effluent whether or not it is potentially harmful. Controls rest on a consent system combined with criminal offenses. The central government possesses the authority to establish water quality objectives.[23]

INTERNATIONAL INITIATIVES

Marine pollution problems generally are controlled by coastal nations with their own legislative programs. To control marine pollution both regionally and globally, however, international treaties have been developed periodically and signed by member states. Most notable in this respect is the Convention on the Prevention of Marine Pollution by Dumping of Wastes and Other Matter of 1972 (commonly known as the London Dumping Convention) which was brought into force on August 30, 1975. National legislation and regulations of states contracting to the London Dumping Convention provide the principal legal framework for controlling waste dumping at sea. A number of regional conventions also protect against ocean dumping of waste in marginal and semi-enclosed seas. Included among these regional conventions are the Oslo Convention for the North Sea, the Helsinki Convention for the Baltic Sea, and the Barcelona Convention for the Mediterranean Sea.

A second international agreement of major significance is the Partial Test Ban Treaty of 1963, which prohibits the testing of nuclear weapons in the atmosphere by nations signatory to the treaty. Between 1945 and 1963, numerous atmospheric nuclear explosions resulted in elevated levels of radionuclides in the open ocean. Deep-sea sediments have served as a repository for substantial quantities of these radionuclides, although concentrations declined markedly after enactment of the treaty.

A third global treaty of note is the International Convention for the Prevention of Pollution from Ships of 1973, which largely deals with oil pollution from ships. Amended by a Protocol in 1978 (MARPOL 1973/78), the main part of this treaty entered into force in 1983. Annex V of this convention places restrictions on the dumping of garbage at sea. One of the most important provisions of Annex V is the ban placed on the dumping of all plastics in the ocean.

Agenda 21 of the 1992 United Nations Conference on Environment and Development emphasizes the rights and obligations of states and provides the international basis upon which to pursue the protection and sustainable development of marine and coastal environments and their resources.[1] Agenda 21 addresses the major marine environment and development priorities for the international community that will lead nations into the 21st century.[24] New approaches to coastal and oceanic management and development must be pursued not only at the global level but also at the national, subregional, and regional levels.

The approach and general framework of action of the United Nations Environment Program (UNEP) for protecting coastal and open ocean areas against marine pollution involve three closely linked elements dealing with the global marine environment, regional seas, and living marine resources. Globally coordinated marine pollution monitoring is being pursued through the Global Ocean Observing System developed in cooperation between the Intergovernmental Oceanographic Commission, the World Meteorological Organization, and UNEP. An integral component of the planned global monitoring system will be surveys conducted at the regional level. The oceans and coastal areas program of UNEP is supported by intensive cooperation of many international, regional, and intergovernmental organizations, as well as numerous national institutions.[1]

Countries can no longer view marine pollution as strictly a national problem. Hence, many governments are responding to environmental degradation of their coastal marine waters through the development of broadly-based interagency coordinating bodies. Such an integrated and well coordinated management approach is necessary to mitigate the growing environmental deterioration and resource-use conflicts in the coastal zone, to curb the pursuit of unsustainable coastal development, to correct narrowly focused conservation and protection strategies, and to formulate effective marine pollution prevention programs.[25] Sectoral approaches to marine policy and management will have only limited success in mollifying global ocean problems of the 21st century.[26]

PLAN OF THE VOLUME

The *Practical Handbook of Estuarine and Marine Pollution* is a comprehensive treatment of estuarine and marine pollution presented in an expository, illustrative, and tabular format. Chapter 1 provides an overview of primary pollutants found in coastal and oceanic environments and the legal initiatives undertaken to control marine pollution problems. Chapter 2 focuses on enrichment of nutrients and organic carbon in estuaries and coastal marine waters. Large concentrations of nutrient elements enter these waters from point sources (e.g., industrial and municipal wastewaters, dredged material, sewage wastes), as well as nonpoint sources (e.g., agricultural, rural, and urban runoff; atmospheric deposition) of pollution. Excessive nutrient input often fosters high carbon production which can culminate in serious oxygen deficiencies within susceptible bodies of water. These effects are manifested most conspicuously in anoxia or hypoxia of bottom waters in shallow estuaries and enclosed seas possessing poor circulation, where oxygen depleted waters cannot be effectively reoxygenated. Concurrent influx of allochthonous organic carbon can exacerbate these conditions by elevating the biochemical oxygen demand. Persistent and recurrent episodes of anoxia or hypoxia in bottom waters generally endanger benthic communities and compromise the overall health of an ecosystem. In extreme cases, mass mortalities of benthic fauna result in depleted marine life resources in an area, including recreationally and commercially important fin- and shellfish.

Chapter 3 examines estuarine and marine oil pollution. While major oil spills from supertankers and accidents at offshore oil platforms are commonly perceived by the public as the principal causes of oil pollution in the coastal zone, other anthropogenic sources actually release greater quantities of oil to the sea. For example, more than 65% of the oil occurring in estuarine and marine environments derives from municipal and industrial wastes, urban and river runoff, ocean dumping, and atmospheric fallout. Approximately 26% of the anthropogenic oil in the sea originates from transportation of the oil (e.g., deballasting, dry docking, tanker accidents), and only about 9% from fixed installations (e.g., coastal refineries, offshore production facilities, marine terminals). The estimated amount of oil entering estuarine and marine waters from man's activities is about 2.12 million tons (mt)/yr, compared to approximately 0.25 mt/yr from natural seeps.

Crude oil is composed of many thousands of gaseous, liquid, and solid organic compounds. More than 75% by weight of most crude oils consists of hydrocarbon compounds. Among the toxic components are benzene, toluene, xylene, and other low-molecular-weight aromatics, carboxylic acids, phenols, and sulfur compounds.

Several physical-chemical processes change the composition of oil in seawater. Chief among these processes are evaporation, photochemical oxidation, emulsification, and dissolution. Evaporative loss of volatile hydrocarbons removes the toxic lower-molecular-weight components during the first 24 to 48 hours of an oil spill. The loss of these volatile components substantially lowers the overall toxicity of the oil to communities of organisms. Microbial degradation of hydrocarbon compounds, mainly by bacteria, is the principal biotic pathway by which the oil is broken down. Several factors influence biodegradation rates, most notably water temperature, nutrient availability, dissolved oxygen concentration, and salinity. As time passes, the oil increases in density, with the hydrocarbons tending to sorb to particulate matter such as clay, silt, sand, and shell fragments. These processes promote sedimentation of the oil.

Estuarine and coastal marine waters rank high among the most sensitive ecosystems to the damaging effects of oil pollution. Salt marshes, mangroves, seagrasses, mudflats, and other low-energy habitats tend to trap oil, and the toxic components readily destroy the vegetation. Animal inhabitants are impacted either directly via the toxic components of the oil or indirectly via the degradation of critical habitat areas. In severe cases, oil pollution can decimate entire communities of organisms and adversely affect sensitive habitats for years. The application of chemical dispersants, solvents, and other agents during oil cleanup operations commonly exacerbates biotic impacts.

Chapter 4 discusses polycyclic aromatic hydrocarbons (PAHs), a group of ubiquitous compounds widely distributed in estuarine and nearshore oceanic environments, especially those in proximity to urban industrialized centers. PAHs are compounds consisting of hydrogen and carbon arranged in the form of two or more fused benzene rings in linear, angular, or cluster arrangements with substituted groups possibly attached to one or more rings. They are a cause of concern because of their potential carcinogenicity, mutagenicity, and teratogenicity to organisms. These compounds enter aquatic environments through multiple pathways, especially sewage and industrial effluents, oil spills, creosote oil, fossil fuel combustion, waste incineration, and forest and brush fires. Atmospheric deposition of PAHs generated by the pyrolysis of fossil fuels appears to be the primary delivery system to the world's oceans.

Polycyclic aromatic hydrocarbons readily sorb to particulate matter and consequently accumulate in seafloor sediments. Benthic organisms inhabiting heavily contaminated areas of the sea floor are often exposed to high concentrations of PAHs. These compounds undergo bioaccumulation, biotransformation, and biodegradation in many organisms. As the PAH compounds are metabolized, some of them become carcinogenic, mutagenic, or both. Certain biota (e.g., bivalves) lack MFO (enzyme) activity and, therefore, concentrate PAHs. The high-molecular-weight PAHs tend to be carcinogenic, mutagenic, and teratogenic to a large number of organisms; however, they are significantly less toxic than the unsubstituted lower-molecular-weight PAH compounds, which are noncarcinogenic. Aside from the biotic transformation of PAHs by microbes and aquatic animals, photooxidation and chemical oxidation are key abiotic degradative processes acting on PAHs in estuarine and marine environments.

The effects of halogenated hydrocarbons are recounted in Chapter 5. This group of ubiquitous, toxic hydrocarbon compounds characteristically contain chlorine, bromine, fluorine, or iodine. The higher-molecular-weight halogens represent the greatest threat to estuarine and marine ecosystems. These compounds typically accumulate in the lipid-rich tissues of animals, and many of them are highly toxic. Organochlorine compounds in pesticides and certain industrial chemicals (e.g., chlorinated aromatics, chlorinated paraffins) provide examples. Specific chlorinated hydrocarbon compounds (e.g., DDT, PCBs, chlordane) resist breakdown in aquatic environments. They are chemically stable, persistent, highly mobile, and toxic, thereby generating many acute as well as insidious pollution problems. In addition, because of their bioaccumulation and biomagnification effects, they pose a chemical hazard to man.

Many of the chlorinated hydrocarbon pesticides were banned in the U.S. during the 1970s and 1980s due to their extreme toxicity to nontarget organisms. Most of these compounds belong to several chemical groups: DDT and its analogues, the cyclodienes, toxaphene and related chemicals, and the caged structures mirex and chlordecone. These compounds continue to be detected in sediment, water, and biotic samples collected from numerous estuarine and coastal marine sites in the U.S. despite long-term bans on their use. They are not distributed uniformly in environmental samples, but occur in highest concentrations in "hot-spot" areas.

Polychlorinated biphenyls comprise one of the most well known classes of halogenated hydrocarbons. These synthetic compounds are deleterious to marine life, particularly organisms at higher trophic levels (e.g., fish and mammals). They also have been linked to acute and chronic health effects in man, such as cancer, liver damage, skin lesions, and reproductive disorders. Hence, their occurrence in fin- and shellfish suitable for human consumption is a cause for concern. PCBs are ubiquitous in the world's oceans. Although U.S. production of PCBs ended in 1977, their contamination has persisted in many estuarine and nearshore oceanic areas.

Two other classes of halogenated hydrocarbon compounds that are highly toxic to aquatic organisms include chlorinated dibenzo-*p*-dioxins (CDDs) and chlorinated dibenzofurans (CDFs). They occur as trace contaminants in industrial chemicals (e.g., phenoxy herbicides, chlorinated phenols, PCBs). Estuarine and marine fauna exposed to CDDs and CDFs often experience progressive weight loss, reproductive impairment, immunosuppression, impaired liver function, cardiovascular changes, developmental abnormalities, histopathological alterations, and other problems. Similar to PCBs, CDDs and CDFs are globally distributed, highly lipophilic, and environmentally persistent compounds. However, toxicological testing of CCDs and CDFs on estuarine and marine organisms has not been comprehensive, and more investigations are required to assess the factors that alter the bioconcentration, toxicokinetics, and metabolism of these compounds.

Chapter 6 details heavy metal pollution. Heavy metals comprise a group of elements that are potentially toxic to estuarine and marine organisms above a threshold availability. At elevated levels, they act as enzyme inhibitors and can be lethal. Two subgroups are recognized: transitional metals (e.g., cobalt, copper, iron, manganese) and metalloids (e.g., arsenic, cadmium, lead, mercury, selenium, tin). While the weathering of rocks, leaching of soils, eruption of volcanoes, and emission of fluids from deep-sea hydrothermal vents release large concentrations of heavy metals to the sea, anthropogenic inputs from a wide range of sources are also considerable. For example, automobile emissions and wastes from a host of industries (e.g., mining, smelting, refining, and electroplating operations) can create "hot spots" of heavy metal pollution potentially hazardous to aquatic communities.

Important delivery systems of heavy metals to estuarine and coastal marine environments include riverine inflow, atmospheric fallout, and anthropogenic point sources. Municipal and industrial waste discharges in urbanized/industrialized regions often result in significant contamination of neighboring systems (e.g., Boston Harbor, Newark Bay, Commencement Bay). Because metals rapidly sorb to particulate matter, they tend to accumulate in bottom sediments of estuaries and shallow coastal marine systems. Hence, benthic organisms in these regions are at times exposed to high levels of metal contamination.

Many estuarine organisms accumulate heavy metals from seawater, bottom sediments, interstitial waters, or their food supply. Bioaccumulation depends largely on metal speciation and bioavailability. Metal toxicity may be a function of the free metal ionic activity. Some estuarine organisms, particularly mollusks such as mussels and oysters, sequester heavy metals and, therefore, are useful as bioindicators of heavy metal pollution.

Chapter 7 reviews national monitoring surveys of U.S. coastal waters. The most comprehensive monitoring survey is that of the National Oceanic and Atmospheric Administration's (NOAA) National Status and Trends Program which consists of four major components: (1) the Benthic Surveillance Project, (2) the Mussel Watch Project, (3) Biological Effects Surveys and Research, and (4) Historical Trends Assessment. The Benthic Surveillance Project measures concentrations of chemical contaminants (aromatic hydrocarbons, chlorinated hydrocarbons, trace metals) in sediments and bottom-dwelling finfish at selected sites in urban and nonurban embayments. The Mussel Watch Project determines levels of chemical contamination in sediments and bivalve mollusks from more than 240 sampling sites nationwide. The focus of Biological Effects Surveys and Research is to investigate more thoroughly those regions where laboratory analyses of samples indicate a potential for substantial environmental degradation and biological impacts of the contaminants. Historical Trends Assessment involves closer examination of the environmental conditions in different regions of the U.S.

The National Status and Trends Program assesses the levels of more than 70 contaminants in biota and bottom sediments. The average distance between sampling sites in this program is about 20 km in estuaries and 100 km along the open coast. Results of monitoring surveys to date indicate that the highest levels of chemical contamination occur in urbanized estuaries and coastal waters. Between 1986 and 1993, there were many more decreases than increases in contaminant concentrations in coastal regions. Thus, the concentrations of most contaminants measured in the National Status and Trends Program may be declining.

Chapter 8 addresses the subject of radioactive waste. The principal anthropogenic sources of radioactivity in estuarine and marine environments are the nuclear fuel cycle and wastes from various agricultural, industrial, medical, and scientific applications. The detonation of nuclear explosives in the atmosphere contributed significant concentrations of radionuclides to these environments between 1945 and 1980. Oceanic input of radioactive materials from human activity has varied substantially during the past 30 years owing to reductions in fallout from nuclear explosions after enactment of the Second Nuclear Test Ban Treaty in 1963, the increase in the number of nuclear power plants online after 1970, and accelerated usage of radioactive substances in agriculture, industry, and medicine after 1975.

There are six categories of radioactive wastes: (1) high-level wastes, (2) transuranic wastes, (3) low-level wastes, (4) uranium and mill tailings, (5) decontamination and decommissioning wastes from nuclear reactors, and (6) gaseous effluents. Provisions of the London Dumping Convention of 1972 and the U.S. Marine Protection, Research, and Sanctuaries Act of 1972 prohibit the dumping of high-level radioactive wastes in the sea. In the U.S., the Low-Level Radioactive Waste Policy Act of 1980 and its amendments (1985) regulate the disposal of low-level radioactive wastes.

Estuarine and marine organisms take up radionuclides from water, sediments, or other organisms. Bioaccumulation of radionuclides appears to be greatest in areas nearby nuclear fuel processing plants and facilities producing nuclear explosives. Radiation exposure of these organisms can result in physiological and genetic changes, aberrant growth and development, and disease and death. The observed effects of ionizing radiation are a function of the total dose, dose rate, type of radiation, and exposure period.

Chapter 9 covers dredging and dredged-spoil disposal. Adverse effects of these activities on estuarine and coastal marine systems are associated primarily with the destruction of benthic habitat, the impairment of water quality, and direct or indirect impacts on organisms. Beneficial effects include: (1) increased nutrients that can enhance productivity of a system, (2) improved circulation in estuaries and embayments, (3) increased recreational and commercial usage of a waterbody, and (4) increased sediment supply for beach nourishment, landfill projects, and soil improvement.

The removal of bottom sediments during dredging operations frequently causes mass mortality of benthic organisms. These operations usually disrupt the entire benthic community for a period of months to several years. The rate of recovery of the dredged area is temporally and spatially variable and site specific. Recolonization varies considerably with geographical location, sediment composition, and types of organisms inhabiting the area. Apart from the direct impacts of the dredge itself, increased turbidity and the release of toxic substances from the sediments can adversely affect areas beyond the immediate dredging site. However, these effects are temporary.

The chief impacts of dredged-spoil dumping at open-water sites are increased turbidity, burial of benthic organisms, and changes in sediment properties. Recolonization of biota often takes place rapidly at these sites. The open-water disposal of contaminated spoils usually requires some type of containment of the material (e.g., a protective cap). Nevertheless, subaqueous dumping of dredged spoils may be the preferred method of disposal rather than subaerial dumping.

Chapter 10 describes coastal power plant impacts. Large electric generating stations located in the coastal zone adversely affect estuarine and marine communities by the discharge of heated effluents, the release of chemical substances, the impingement of organisms on intake screens, and the entrainment of various life forms in cooling water systems. Thermal discharges from these power plants often alter the water quality and aquatic communities in near-field regions. The physiological and behavioral responses of organisms exposed to the heated effluent typically involve an alteration of metabolic rates leading to a diminution of growth, as well as behavioral adjustment manifested in avoidance or attraction reactions to the heated water. Thermal death may ensue due to the failure of smooth muscle peristalsis, denaturation of proteins in the cells, increased lactic acid in the blood, and oxygen deficit related to increased respiratory activity. Sublethal effects that can be detrimental include a decrease in reproduction or hatching success of eggs and an inhibition in development of larvae. Thermal loading frequently lowers the productivity in outfall waters.

The use of biocides in coastal power plants to control biofouling of the condenser circuits also kills many nontarget organisms. Chlorine is injected periodically into cooling water systems to destroy bacterial slimes and macrofouling organisms, such as mussels or clams, which can lower plant operating efficiencies. Other organisms entrained in the cooling water (e.g., phytoplankton, zooplankton, meroplankton, and ichthyoplankton) are highly susceptible to the chlorine, and mortality typically is high. Exposure of organisms to residual chlorine in outfall waters also can be detrimental, often contributing to decreased productivity of these waters.

The greatest biological impact of coastal power plants appears to be due to impingement of larger organisms on intake screens and entrainment of plankton, microinvertebrates, and small juvenile fish passively drawn into cooling water condenser systems. Annual impingement losses at some plants occasionally exceed a million individuals, and annual in-plant mortality of entrained organisms commonly exceeds a billion individuals. While the absolute mortality caused by impingement and entrainment of organisms at coastal power plants is very high, there are no documented cases of long-term, system-wide biotic problems in coastal waters attributable to any single power plant unit.

REFERENCES

1. Gerges, M. A., Marine pollution monitoring, assessment and control: UNEP's approach and strategy, *Mar. Pollut. Bull.*, 28, 199, 1994.
2. GESAMP, State of the Marine Environment, GESAMP Reports and Studies No. 39, United Nations Environment Program, Nairobi, 1990.
3. GESAMP, Scientific Criteria for the Selection of Waste Disposal Sites at Sea. Reports and Studies No. 16, Inter-Governmental Maritime Consultative Organization, London, 1982.
4. Clark, R. B., *Marine Pollution*, 3rd ed., Clarendon Press, Oxford, 1992.
5. Capuzzo, J. M. and Kester, D. R. (Eds.), *Oceanic Processes in Marine Pollution*, Vol. 1, *Biological Processes and Wastes in the Ocean*, Robert E. Krieger Publishing, Malabar, FL, 1987.
6. Kennish, M. J., *Ecology of Estuaries: Anthropogenic Effects*, CRC Press, Boca Raton, FL, 1992.
7. McIntyre, A. D., The current state of the oceans, *Mar. Pollut. Bull.*, 25, 1, 1992.

8. Capuzzo, J. M., Burt, W. V., Duedall, I. W., Park, P. K., and Kester, D. R., Future strategies for nearshore waste disposal, in *Wastes in the Ocean*, Vol. 6, *Nearshore Waste Disposal*, Ketchum, B. H., Capuzzo, J. M., Burt, W. V., Duedall, I. W., Park, P. K., and Kester, D. R. (Eds.), John Wiley & Sons, New York, 1985, 491.
9. NOAA, Recent Trends in Coastal Environmental Quality: Results from the Mussel Watch Project, Tech. Rep., Department of Commerce, Rockville, MD, 1995.
10. Waldichuk, M., The state of pollution in the marine environment, *Mar. Pollut. Bull.,* 20, 598, 1989.
11. Spencer, D. W., The ocean and waste management, *Oceanus,* 33, 5, 1990.
12. Engler, R. M. and Mathis, D. B., Dredged-material disposal strategies, in *Oceanic Processes in Marine Pollution*, Vol. 3, *Marine Waste Management: Science and Policy*, Camp, M. A. and Park, P. K. (Eds.), Robert E. Krieger Publishing, Malabar, FL, 1989, 53.
13. U.S. Army Corps of Engineers Philadelphia District, Are You Planning Work in a Waterway or Wetland?, Unpublished Tech. Rep., U.S. Army Corps of Engineers, Philadelphia, 1990.
14. Tammi, C. E., Offsite identification of wetlands, in *Applied Wetlands Science and Technology*, Kent, D. M. (Ed.), Lewis Publishers, Boca Raton, FL, 1994, 13.
15. Fields, S., Regulations and policies relating to the use of wetlands for nonpoint source pollution control, in *Created and Natural Wetlands for Controlling Nonpoint Source Pollution*, Olson, R. K. (Ed.), C. K. Smoley, Boca Raton, FL, 1993, 151.
16. Robb, D. M., The role of wetland water quality standards in nonpoint source pollution strategies, in *Created and Natural Wetlands for Controlling Nonpoint Source Pollution*, Olson, R. K. (Ed.), C. K. Smoley, Boca Raton, FL, 1993, 159.
17. Bascom, W. N., Disposal of sewage sludge via ocean outfalls, in *Oceanic Processes in Marine Pollution*, Vol. 3, *Marine Waste Management: Science and Policy*, Camp, M. A. and Park, P. K. (Eds.), Robert E. Krieger Publishing, Malabar, FL, 1989, 25.
18. Kitsos, T. R. and Bondareff, J. M., Congress and waste disposal at sea, *Oceanus,* 33, 23, 1990.
19. O'Connor, T. P., *Recent Trends in Coastal Environmental Quality: Results from the First Five Years of NOAA Mussel Watch Project,* NOAA Office of Ocean Resources Conservation and Assessment, Rockville, MD, 1992.
20. O'Connor, T. P., Cantillo, A. Y., and Lauenstein, G. G., Monitoring of temporal trends in chemical contamination by the NOAA National Status and Trends Mussel Watch Project, in *Biomonitoring of Coastal Waters and Estuaries,* Kramer, K. J. M. (Ed.), CRC Press, Boca Raton, FL, 1994, 29.
21. Rice, C. P. and O'Keefe, P., Sources, pathways, and effects of PCBs, dioxins, and dibenzofurans, in *Handbook of Ecotoxicology*, Hoffman, D. J., Rattner, B. A., Burton, G. A., Jr., and Cairns, J., Jr. (Eds.), Lewis Publishers, Boca Raton, FL, 1995, 424.
22. van der Valk, A. G. and Jolly, R. W., Recommendations for research to develop guidelines for the use of wetlands to control rural nonpoint source pollution, in *Created and Natural Wetlands for Controlling Nonpoint Source Pollution*, Olson, R. K. (Ed.), C. K. Smoley, Boca Raton, FL, 1993, 167.
23. Macrory, R., The legal control of pollution, in *Pollution: Causes, Effects, and Control*, 2nd ed., Harrison, R. M. (Ed.), Royal Society of Chemistry, Cambridge, 1990, 277.
24. Nollkaemper, A., Marine pollution from land-based sources, *Mar. Pollut. Bull.,* 24, 8, 1992.
25. Hildebrand, L. P. and Norrena, E. J., Approaches and progress toward effected integrated coastal zone management, *Mar. Pollut. Bull.,* 25, 94, 1992.
26. Brewers, J. M. and Wells, P. G., Challenges for improved marine environmental protection, *Mar. Pollut. Bull.,* 25, 112, 1992.

2 Eutrophication and Organic Loading

INTRODUCTION

Excessive nutrient enrichment (eutrophication) and organic loading problems occur in many coastal ecosystems worldwide and invariably are coupled to a variety of anthropogenic activities. Estuarine and coastal marine waters often receive large amounts of nutrients from both point and nonpoint sources, including industrial and municipal wastewaters, dredged material, sewage sludge, agricultural and urban runoff, ground water seepage, and atmospheric deposition. In the U.S., widespread eutrophication from excess nutrient input or organic matter enrichment is apparent in many estuaries of the Atlantic and Gulf coasts.[1-3] On a global scale, marine environments receive approximately equal proportions of nutrient inputs from atmospheric transport and riverine discharges.[4,5] The flux of nutrients in rivers attributable to anthropogenic activities is probably greater than the natural flux. Clearly, most of the atmospheric nitrogen influx to coastal regions is of anthropogenic origin. Regardless of nutrient source, the ocean margins trap nearly all of the nutrients delivered by rivers and 90% transported by the atmosphere.[6]

The influx of large concentrations of nutrient elements to the sea is perceived to be a priority problem.[1] It occasionally contributes to the development of toxic phytoplankton blooms termed red tides which can cause mass mortality of invertebrates and fish. Overenrichment also accelerates primary production which may lead to anoxic or hypoxic conditions in bottom waters as decaying plant remains accumulate on the seabed and foster high rates of respiration of organisms, including animals, plants, and the decomposing activity of microorganisms.[1,7]

OXYGEN DEPLETION

Anoxic waters are defined as those with 0 mg O_2 per liter, and hypoxic waters are those with <2 mg O_2 per liter. Coastal systems impacted by severe oxygen deficits exhibit structural and functional changes. For example, increased water column and benthic respiration relative to production signals a major functional change in eutrophied coastal systems.[8] Eutrophic conditions typically culminate in shifts in the structure of animal communities to those dominated by pollution tolerant species. While overenrichment is acknowledged as an increasing problem in many coastal areas, the quantitative relationship between nutrient inputs and the onset of anoxia and hypoxia is poorly understood.[9] This is so because there is little reliable quantitative information on the magnitude and frequency of nitrogen and phosphorus inputs to coastal ecosystems and how these inputs are processed within watersheds and estuaries.[10]

Eutrophication also can create unfavorable conditions for submerged aquatic vegetation. The reduction of some seagrass beds (such as upper Chesapeake Bay), for example, has been ascribed to the shading effects of abundant phytoplankton in the overlying waters and to enhanced epiphytic and macroalgal growth.[11] Such dieback of seagrass beds has potentially dangerous consequences for animal communities, since the beds form important habitats and nursery grounds for many invertebrate and finfish populations. Excessive nutrient enrichment may likewise adversely affect attached macroalgae, as is evident in some rocky intertidal areas, by favoring the proliferation of opportunistic species with rapid growth rates (e.g., *Cladophora* spp., *Enteromorpha* spp.) which lowers the overall floral diversity.[12]

Anoxia and hypoxia arise most commonly in bottom waters of estuaries and enclosed embayments, although both phenonmena also have been reported in open seas (e.g., New York Bight, northern Gulf of Mexico, North Sea, Baltic Sea).[2,9] Jaworski[13] described three types of estuaries based on the level of eutrophication.

1. Hypereutrophic systems having very excessive nuisance conditions, anoxia, and "undesirable" biological communities
2. Eutrophic systems having excessive nuisance conditions, low dissolved oxygen concentrations, and "undesirable" biological communities
3. Noneutrophic systems having biologically healthy and productive components with "desirable" biological communities

Shallow coastal seas that are poorly flushed appear to be the systems most susceptible to dissolved oxygen depletion.[14,15] However, larger, stratified estuaries with deep channels, such as Chesapeake Bay, Long Island Sound, and parts of Puget Sound, also display protracted periods of bottom water anoxia or hypoxia.[16-18] During the past decade, there has been an apparent increase in anoxic and hypoxic conditions in many coastal regions of the world owing to eutrophication. This change has been accompanied by greater frequency, intensity, and geographical distribution of nuisance algal blooms.[19,20]

High inputs of organic solids (e.g., sewage sludge) can exacerbate anoxia and hypoxia by elevating the biochemical oxygen demand (BOD, the oxygen consumed during the microbial decomposition of organic matter) and the chemical oxygen demand (COD, the oxygen consumed through the oxidation of ammonium and other inorganic reduced compounds).[1] Although the bottom waters of these impacted areas may be anoxic, rapid growth of phytoplankton in the overlying, nutrient-enriched surface waters commonly generates substantial concentrations of oxygen. In addition, oxygen levels in the upper waters increase by diffusion and surface entrapment from the atmosphere. Point sources of organic solids often create localized problems in coastal areas; however, the decomposition of phytoplankton remains in eutrophied systems typically affects broader expanses of estuarine and coastal marine environments through extensive oxygen consumption in bottom waters.

NUTRIENTS

In estuarine and coastal marine waters of the temperate zone, nitrogen is the element usually limiting to primary production. However, primary production appears to be phosphorus limited in many tropical estuarine and coastal systems. Hence, increased inputs of nitrogen and, in some cases, phosphorus usually result in higher primary production that can cause eutrophication problems. Elements other than nitrogen and phosphorus are also required by autotrophs, but their availability usually is not limiting to growth. These include major elements (e.g., calcium, carbon, magnesium, oxygen, and potassium), minor and trace elements (e.g., cobalt, copper, iron, molybdenum, vanadium, and zinc), and organic nutrients (e.g., biotin, thiamine, and vitamin B_{12}). Aside from nutrient input, light availability also limits primary production in some coastal water bodies (e.g., the Hudson-Raritan estuary).

Ammonia (NH_3), nitrite (NO_2^-), and nitrate (NO_3^-) are the three principal dissolved inorganic forms of nitrogen in coastal systems, with nitrate occurring in highest concentrations. Orthophosphates constitute the major fraction of dissolved inorganic phosphorus in these systems. Nitrate levels in seawater usually amount to 1 μg-atom/l or less and rarely exceed 25 μg-atom/l. Phosphate concentrations generally range from 0 to 3 μg-atom/l. Dissolved organic nitrogen (e.g., urea, amino acids, peptides) and dissolved organic phosphorus also may be valuable nutrient sources for autotrophic growth in the sea. The total dissolved organic nitrogen and dissolved organic phosphorus concentrations are frequently several times greater than the dissolved inorganic concentrations, especially in marine waters.[19] Most temperate estuaries exhibit seasonal variations in the concentrations of nitrogen and phosphorus compounds; lowest levels typically occur in the spring and summer when autotrophic production is highest, and highest levels occur in the fall and winter when autotrophic production is lowest.

Several processes operating in the water column and bottom sediments influence the biogeochemical cycling of nitrogen. In the water column, these processes include uptake, remineralization, and oxidation; in bottom sediments, they involve burial, remineralization, biological uptake, oxidation, reduction, nitrous oxide production, and denitrification. Benthic-pelagic coupling of nutrients appears to be very important in the primary production of nearshore environments.

The exchange of nutrients between the water column and bottom sediments is a critical pathway for nutrient cycling. Benthic nutrient fluxes depend on temperature, rate of organic deposition, composition of organic matter integrating both surface and subsurface mineralization, denitrification, inorganic exchange/solution processes occurring above and below the oxycline, and burial. These fluxes tend to be greater in temperate systems than tropical ones, primarily due to higher rates of primary production and organic deposition in the mid-latitudes.[21]

Bottom sediments represent important sources of ammonium and phosphate for the water column, especially in summer. Conversely, sediments can be sinks for nitrate through denitrification. Hence, they may provide an important pathway for the removal of nitrogen inputs from anthropogenic activities.[22]

ORGANIC LOADING

The input of organic carbon has been linked to eutrophication problems and decreased dissolved oxygen levels in estuarine and coastal marine waters. The highest concentrations of organic carbon occur in waters receiving domestic and industrial sewage wastes, where both dissolved and particulate organic organic carbon levels occasionally exceed 100 mg/l. Dissolved organic carbon concentrations in estuarine and coastal marine waters typically range from about 1 to 5 mg/l. Particulate organic carbon levels generally range from approximately 0.5 to 5 mg/l in estuaries and 0.1 to 1 mg/l in coastal seawater.[23]

In many cases, large carbon inputs from municipal and industrial wastewaters, sewage sludge, dredged spoils, and other point sources have contributed significantly to oxygen depletion and added substantially to BOD levels. Sewage wastes, in particular, have caused the eutrophication and dissolved oxygen depletion of coastal waters in many regions of the world, especially those in proximity to large metropolitan centers, such as New York, Boston, Los Angeles, London (England), and elsewhere.[24]

Considerable quantities of oxidizable carbon may have entered these receiving waters via piped outfalls, sewage sludge dumping, or riverborne discharges. Of these three pathways of organic carbon loading, sewage sludge dumping usually produces the most immediate effects by altering the seabed and directly impacting the benthic community. More insidious effects are associated with changes in water quality above the sewage sludge dumpsites which generally are enriched in nitrogen, phosphorus, and other elements.

Sewage consists of a mixture of solids (up to 10% by weight) and liquids with a wide range of composition. Many contaminants (e.g., heavy metals, chlorinated hydrocarbons, PCBs) readily sorb to particulate organic matter and, therefore, are common constituents of sewage sludge. The amount of dry solids in the sludge depends on a number of factors such as treatment works capacity, thickening methods, and rainfall.[25] Pathogenic microorganisms — bacteria, viruses, protozoa, and helminths — enter the sludge during the processing of human and animal wastes and pose a health risk to humans who either ingest contaminated seafood products or swim in contaminated waters. Consumption of sewage-contaminated seafood can cause serious illnesses (e.g., viral hepatitis and cholera). Pathogens and contaminants are also responsible for diseased macroinvertebrates and fish inhabiting areas impacted by sewage sludge. For example, fin rot disease in finfish and exoskeletal disease in shellfish (crabs, lobsters) are common maladies observed in animals exposed to sewage sludge.[26]

Chronic organic enrichment at some sewage sludge dumpsites degrades benthic communities, as is often manifested by changes in species richness, relative abundance, and biomass of macrofaunal inhabitants.[25] For example, a sewage sludge dumpsite located 22 km southeast of the Hudson-Raritan estuary in the New York Bight Apex received sewage-sludge-rich inorganic matter and chemical contaminants for more than 60 years (1924 to 1987). Sewage sludge dumping radically altered the benthic habitat and adversely affected benthic communities over a 10- to 15-km^2 region.[27-29] As a result, an area with a radius of 11 km around the sewage sludge dumpsite was closed to shellfishing in May 1970.[30]

Disposal of sewage sludge at the 22-km dumpsite was terminated in December 1987. In order to reduce the potential transport of organic wastes to the coast and to prevent contamination of commercial fishing grounds in inshore waters, sewage sludge disposal was shifted to a 270-km^2 location approximately 196 km southeast of New York Harbor known as the 106-Mile Site. Sewage sludge disposal at the deep-ocean dumpsite caused measurable changes in the concentration of sludge indicators in seafloor sediments — notably linear alkylbenzenes, coprostanol, and spores of the bacterium *Clostridium perfringens.* Increased deposition of utilizable food in the form of sludge-derived organic matter also led to a greater abundance of two species of polychaete worms at the dumpsite. While the density of these populations has increased due to the additional organic carbon, the species diversity of the benthic community has declined. Approximately 36 million metric wet tons of sludge were dumped at the deep-sea site by the end of 1991.[31] Dumping at this site was terminated in 1992.

The National Oceanic and Atmospheric Administration (NOAA) has documented changes in living marine resources and habitats during and following cessation of sewage sludge dumping at the 22-km dumpsite in the New York Bight Apex. Intense sampling at three sites between 1986 and 1989 showed that the numbers of crustacean, molluscan, and total species increased subsequent to the closure of the dumpsite, while the pollution indicator polychaete, *Capitella* sp., decreased during the interval.[29] Changes in benthic habitats during this period included reduced total organic carbon and metal contamination in sediments and increased bottom water dissolved oxygen in summer.[32] The use of coastal waters for the disposal of sewage waste has been a long established practice in many countries. While this disposal method is acknowledged to be the cheapest financial option, it has resulted in many environmental problems. Consequently, the U.S. and some European nations have initiated steps to pursue alternate methods of disposal. Sewage sludge dumping at sea in the U.S. is under strict federal regulations.[33]

European countries intend to stop sewage sludge disposal at sea by the end of 1998.[25] However, other countries outside the U.S. and European communities will continue to dispose of sewage wastes in coastal waters in the forseeable future, thereby continuing to threaten marine communities and sensitive habitats in nearshore areas.

In many estuaries and coastal waters, most of the organic carbon is derived from natural sources. For example, decaying submerged vascular plants (saltmarsh grasses, seagrasses, mangroves), benthic macroalgae, and biodeposits (feces and pseudofeces) are the principal contributors to particulate organic carbon pools in numerous detritus-based estuaries. Detritus concentrations in these systems range from 0.1 to more than 125 mg/l, with most detritus (90%) being deposited in bottom sediments or transported out to sea.

Detrital carbon pools can play an integral role in the development of anoxia/hypoxia in coastal waters. For instance, the mass mortality of shellfish over an 8600-km^2 area of the continental shelf off New Jersey in the summer of 1976 was initially thought to be due to sewage sludge and other waste materials discharged into the New York Bight Apex. Later studies showed, however, that the mortality was caused by anoxia associated with the decomposition of a detrital carbon pool comprised of the remains of a subsurface bloom of the dinoflagellate *Ceratium tripos*.[34] While this anoxic event was the most extensive one recorded in the New York Bight, nearshore areas in this region generally have been subject to periodic episodes of oxygen depletion over smaller expanses of the sea floor.

CASE STUDIES

Research on eutrophication of U.S. coastal waters has centered on Chesapeake Bay and its subestuaries. Eutrophication is considered to be the dominant anthropogenic factor contributing to anoxia and hypoxia in the bay. Although anoxia has been documented in Chesapeake Bay since the 1930s,[35] it has become more prolonged and widespread in recent years.[15,16,36] Oxygen depletion in Chesapeake Bay has been related to a variety of biological, chemical, and physical factors. The severity of dissolved oxygen depletion is contingent upon a combination of factors, but most importantly the vertical density stratification, nutrient and organic carbon inputs, and morphometry of the estuary. Climatic factors can exacerbate hypoxia in bottom waters. Seasonal increases in rainfall or snowmelt, for example, strengthen the density stratification of the estuary. The spring freshet, in particular, increases water column stability and minimizes advective transport of oxygen from surface waters to the deep layer. Increased water column stability in the estuary between February and May occurs concomitantly with higher riverine flow. Anoxic conditions commonly spread along the bottom of the channel of the bay in waters deeper than 10 m during the period from June to September.[37] Organic matter accumulating from the previous year (summer and fall) and settling into the deep water during winter accounts for most of the oxygen demand. Some new primary production in spring and summer, although less of a driving force in the oxygen dissipation of the bay, contributes somewhat to the summer deep-water oxygen depletion. Anoxia has periodically caused severe ecological damage in the estuary.[38]

Hypoxia also has been frequently observed in major tributaries on the western shore of Chesapeake Bay (e.g., Patuxent, Rappahannock, and York rivers). These tributaries are partially mixed coastal plain estuaries characterized by deep basins near their mouths. Systems with stronger gravitational circulation (e.g., James River) rarely exhibit anoxia or hypoxia even though they may receive larger wastewater loadings. Benthic and water column oxygen demand, together with vertical mixing, have a pronounced effect on the severity of anoxia/hypoxia in the bottom waters of the mainstem of Chesapeake Bay and its subestuaries. Where the gravitational circulation is reduced, such as in the lower reaches of the Rappahannock River, anoxia/hypoxia in bottom waters may persist throughout the summer months.[39]

Changes in external nutrient loadings also may influence the development of anoxia/hypoxia in these systems. For instance, investigations of nutrient loadings received by the upper mainstem of Chesapeake Bay, the Patuxent estuary, and the Potomac estuary indicate substantially higher nitrogen to phosphorus ratios during the 1985 to 1989 period than documented from the late 1960s to the late 1970s.[40] These changes may have important implications regarding the role of external nutrients in regulating water quality as well as phytoplankton growth and productivity in the three estuarine systems. Correll et al.[41] have shown that nitrogen and phosphorus discharges into the Rhode River on the western shore of Chesapeake Bay, although a small fraction of the total loadings to the watershed, are large enough to cause overenrichment of waters in the upper estuary.

Periods of anoxia and hypoxia and related water quality problems are a growing threat to other estuarine systems. For example, Portnoy[42] documented summer oxygen depletion in the Herring River estuary, a diked system near Wellfleet, MA. The increase in oxygen demand of organic matter released by diked salt marsh deposits and the occurrence of heavy rains which increase the runoff of wetland organic matter have been correlated with periods of hypoxia and anoxia, respectively. The estuary experiences regular summer hypoxia and 1- to 3-week periods of mainstream anoxia often accompanied by fishkills.

Anoxia appears to be a recent phenomenon in Waquoit Bay, a shallow estuarine lagoon on Cape Cod, MA.[43,44] Recent anoxic events in this system are linked to eutrophication primarily resulting from nitrogen loading to groundwater via septic systems.[44] As a consequence of excessive nutrient loading of the bay, pervasive changes have been delineated throughout the food web of the estuary. In addition to greater primary production by phytoplankton, a thick canopy of macroalgae covers the bay bottom. The high macroalgal biomass dominates the bay ecosystem by altering the nutrient status of the water column and increasing the frequency of anoxic events. The composition of the benthic fauna has shifted significantly due to the change in bottom habitat, with the abundance and species richness of invertebrates being lower in parts of the bay where macroalgal biomass is higher. Valiela et al.[43] have recorded repeated kills of invertebrates and fish following anoxic events in the bay during the summer.

A similar situation to Waquoit Bay occurs along the inner-shelf waters of the Florida Keys. Elevated nutrient concentrations in the Florida Keys are attributed to widespread use of septic tanks that increase nutrients in groundwaters discharging into shallow nearshore waters.[45] The groundwater influx of nutrients results in coastal eutrophication and frequent hypoxic events especially during the summer.

Chronic summer hypoxia also has been discerned in the Pamlico River estuary in North Carolina. Hypoxia occurs in this estuary only when water column stratification develops and water temperatures exceed 15°C.[46] As a result of the reduced dissolved oxygen levels, macrobenthos in the deeper waters of the estuary experience massive kills. Species diversity and density of the macrobenthos also decline during the summer months.

Moderate hypoxia has been chronicled in the central basin of Long Island Sound. More frequent episodes of hypoxia take place in the western basin, and severe, persistent, and recurrent hypoxia is observed in the East River and Eastern Narrows. This general east-to-west gradient of decreasing dissolved oxygen in the sound is ascribed to heavy urbanization and nutrient enrichment exacerbated by sewage effluent from the East River.[47] Oxygen depletion is a cumulative process through the summer corresponding to a period of thermally-controlled stratification in the estuary which promotes the development of hypoxic conditions.[48]

Anoxia and, more frequently, hypoxia occur annually in the northern Gulf of Mexico.[49] Oxygen depletion is evident in nearshore regions, although the severity and spatial extent of the reduced oxygen levels vary and are controlled by physical, chemical, and biological processes that are not completely understood. Hypoxia has been related to nutrient input from discharges of the Mississippi and Atchafalaya Rivers.

Eutrophication leading to anoxic/hypoxic events has now been documented in the coastal waters of many other countries. For example, increased nutrient loading of estuaries and coastal marine waters, often coupled to anthropogenic activities, has produced eutrophic conditions in Australia (e.g., Great Barrier Reef Lagoon), Japan (e.g., Hiroshima Bay, Osaka Bay), the Adriatic Sea, Norwegian coastal waters (e.g., the Skagerrak, the Kattegat), the Netherlands (e.g., western Wadden Sea, Veerse Meer), the German Bight, and many other regions.[50-57] These eutrophicated waters have resulted in a multitude of impacts on biotic communities.

REFERENCES

1. National Research Council, *Managing Wastewater in Coastal Urban Areas*, National Academy Press, Washington, D.C., 1993.
2. Kennish, M. J., Pollution in estuaries and coastal marine waters, *J. Coastal Res.*, Spec. Issue No. 12: Coastal Hazards, 27, 1994.
3. Ortner, P. B. and Dagg, M. J., Nutrient-enhanced coastal ocean productivity explored in the Gulf of Mexico, *Eos Trans AGU*, 76, 97, 1995.
4. GESAMP, Land/sea boundary flux of contaminants: contributions from rivers, Reports and Studies No. 32, Joint Group of Experts on the Scientific Aspects of Marine Pollution, Paris, UNESCO, 1987.
5. GESAMP, Atmospheric input of trace species to the world ocean, Reports and Studies No. 38, Joint Group of Experts on the Scientific Aspects of Marine Pollution, Geneva, 1989.
6. Windom, H. L., Contamination of the marine environment from land-based sources, *Mar. Pollut. Bull.*, 25, 1, 1992.
7. Clark, R. B., *Marine Pollution*, 3rd ed., Clarendon Press, Oxford, 1992.
8. Wassmann, P., Calculating the load of organic carbon to the aphotic zone in eutrophicated coastal waters, *Mar. Pollut. Bull.*, 21, 183, 1990.
9. Wenzel, L. and Scavia, D., NOAA's coastal ocean program, *Oceanus*, 36, 85, 1993.

10. Jaworski, N. A., Groffman, P. M., Keller, A. A., and Prager, J. C., A watershed nitrogen and phosphorus balance: the upper Potomac River basin, *Estuaries,* 15, 83, 1992.
11. Valiela, I., Costa, J., Foreman, K., Teal, J. M., Howes, B., and Aubrey, B., Transport of groundwater-borne nutrients from watersheds and their effects on coastal waters, *Biogeochemistry,* 10, 177, 1990.
12. Baden, S. P., Loo, L. O., Pihl, L., and Rosenberg, R., Effects of eutrophication on benthic communities including fish: Swedish west coast, *Ambio,* 19, 113, 1990.
13. Jaworski, N. A., Sources of nutrients and the scale of eutrophication problems in estuaries, in *Estuaries and Nutrients*, Neilson, B. J. and Cronin, L. E. (Eds.), Humana Press, Clifton, N.J., 1981, 83.
14. Reyes, E. and Merino, M., Diel dissolved oxygen dynamics and eutrophication in a shallow, well-mixed tropical lagoon (Cancun, Mexico), *Estuaries,* 14, 372, 1991.
15. Kennish, M. J., *Ecology of Estuaries: Anthropogenic Effects*, CRC Press, Boca Raton, FL, 1992.
16. Stanley, D. W. and Nixon, S. W., Stratification and bottom-water hypoxia in the Pamlico River estuary, *Estuaries,* 15, 270, 1992.
17. Cooper, S. R. and Brush, G. S., Long-term history of Chesapeake Bay anoxia, *Science,* 254, 992, 1991.
18. Cooper, S. R. and Brush, G. S., A 2500-year history of anoxia and eutrophication in Chesapeake Bay, *Estuaries,* 16, 617, 1993.
19. Wells, M. L., Mayer, L. M., and Guillard, R. R. L., Evaluation of iron as a triggering factor for red tide blooms, *Mar. Ecol. Prog. Ser.,* 69, 93, 1991.
20. Smayda, T. J., A phantom of the ocean, *Nature,* 358, 374, 1992.
21. Furnas, M. J., The behavior of nutrients in tropical aquatic ecosystems, in *Pollution in Tropical Aquatic Systems*, Connell, D. W. and Hawker, D. W. (Eds.), CRC Press, Boca Raton, FL, 1992, 29.
22. Rizzo, W. M., Nutrient exchanges between the water column and a subtidal benthic macroalgal community, *Estuaries,* 13, 219, 1990.
23. Kennish, M. J., *Ecology of Estuaries,* Vol. 1, *Physical and Chemical Aspects*, CRC Press, Boca Raton, FL, 1986.
24. Windom, H. L., Contamination of the marine environment from land-based sources, *Mar. Pollut. Bull.,* 25, 1, 1992.
25. Costello, M. J. and Read, P., Toxicity of sewage sludge to marine organisms: a review, *Mar. Environ. Res.,* 37, 23, 1994.
26. Sindermann, K., *Ocean Pollution: Effects on Living Resources and Humans*, CRC Press, Boca Raton, FL, 1995.
27. Mayer, G. F. (Ed.), *Ecological Stress and the New York Bight: Science and Management*, Estuarine Research Federation, Columbia, SC, 1982.
28. Swanson, R. L., Champ., M. A., O'Connor, T., Park, P. K., O'Connor, J., Mayer, G. F., Stanford, H. M., and Erdheim, E., Sewage-sludge dumping in the New York Bight apex: a comparison with other proposed ocean dumpsites, in *Wastes in the Ocean,* Vol. 6, *Nearshore Waste Disposal*, Ketchum, B. H., Capuzzo, J. M., Burt, W. V., Duedall, I. W., Park, P. K., and Kester, D. R., John Wiley & Sons, New York, 1985, 461
29. Studholme, A. L., Ingham, M. C., and Pacheco, A. (Eds.), Response of the Habitat and Biota of the Inner New York Bight to Abatement of Sewage Sludge Dumping, Third Annual Progress Report — 1989, NOAA Tech. Mem. NMFS- F/NEC-82, NOAA, Woods Hole, MA, 1991.
30. O'Connor, J. S. and Segar, D. A., Pollution in the New York Bight: a case history, in *Impact of Man on the Coastal Environment,* Duke, T. W. (Ed.), EPA-600/8-82-021, U.S. Environmental Protection Agency, Washington, D.C., 1982, 47.
31. Bothner, M. H., Takada, H., Knight, I. T., Hill, R. T., Butman, B., Farrington, J. W., Colwell, R. R., and Grassle, J. F., Sewage contamination in sediments beneath a deep-ocean dump site off New York, *Mar. Environ. Res.,* 38, 43, 1994.
32. Steimle, F. W., Sewage sludge disposal and water flounder, red hake, and American lobster feeding in the New York Bight, *Mar. Environ. Res.,* 37, 233, 1994.
33. Swanson, R. L. and Mayer, G. F., Ocean dumping of municipal and industrial wastes in the United States, in *Oceanic Processes in Marine Pollution*, Vol. 3, *Marine Waste Management: Science and Policy*, Camp, M. A. and Park, P. K. (Eds.), Robert E. Krieger Publishing, Malabar, FL, 1989, 35.
34. Stoddard, A. and Walsh, J. J., Modeling oxygen depletion in the New York Bight: the water quality impact of a potential increase of waste inputs, in *Oceanic Processes in Marine Pollution*, Vol. 5, *Urban Wastes in Coastal Marine Environments*, Wolfe, D. A. and O'Connor, T. P. (Eds.), Robert E. Krieger Publishing, Malabar, FL, 1988, 92.

35. Newcombe, C. L. and Horne, W. A., Oxygen poor waters of the Chesapeake Bay, *Science,* 88, 80, 1938.
36. Kuo, A. Y. and Neilson, B. J., Hypoxia and salinity in Virginia estuaries, *Estuaries,* 10, 277, 1987.
37. Taft, J. L., Taylor, W. R., Hartwig, E. O., and Loftus, R., Seasonal oxygen depletion in Chesapeake Bay, *Estuaries,* 3, 242, 1980.
38. Seliger, H. H., Boggs, J. A., and Biggley, J. A., Catastrophic anoxia in the Chesapeake Bay in 1984, *Science,* 228, 70, 1985.
39. Kuo, A. Y., Park, K. L., and Moustafa, M. Z., Spatial and temporal variabilties of hypoxia in the Rappahannock River, Virginia, *Estuaries,* 14, 113, 1991.
40. Magnien, R. E., Summers, R. M., and Sellner, K. G., External nutrient sources, internal nutrient pools, and phytoplankton production in Chesapeake Bay, *Estuaries,* 15, 497, 1992.
41. Correll, D. L., Jordan, T. E., and Weller, D. E., Nutrient flux in a landscape: effects of coastal land use and terrestrial community mosaic on nutrient transport to coastal waters, *Estuaries,* 15, 431, 1992.
42. Portnoy, J. W., Summer oxygen depletion in a diked New England estuary, *Estuaries,* 14, 122, 1991.
43. Valiela, I., Foreman, K., LaMontagne, M., Hersh, D., Costa, J., Peckol, P., DeMeo-Anderson, B., D'Avanzo, C., Babione, M., Sham, C., Brawley, J., and Lajtha, K., Couplings of watersheds and coastal waters: sources and consequences of nutrient enrichment in Waquoit Bay, *Estuaries,* 15, 443, 1992.
44. D'Avanzo, C. and Kremer, J. N., Diel oxygen dynamics and anoxic events in an eutrophic estuary of Waquoit Bay, Massachusetts, *Estuaries,* 17, 131, 1994.
45. LaPointe, B. E. and Clark, M. W., Nutrient inputs from the watershed and coastal eutrophication in the Florida Keys, *Estuaries,* 15, 465, 1992.
46. Stanley, D. W. and Nixon, S. W., Stratification and bottom-water hypoxia in the Pamlico River estuary, *Estuaries,* 3, 270, 1992.
47. Parker, C. A. and O'Reilly, J. E., Oxygen depletion in Long Island Sound: a historical perspective, *Estuaries,* 14, 248, 1991.
48. Welsh, B. L. and Eller, F. C., Mechanisms controlling summertime oxygen depletion in western Long Island Sound, *Estuaries,* 14, 265, 1991.
49. Pokryfki, L. and Randall, R. E., Nearshore hypoxia in the bottom water of the northwestern Gulf of Mexico from 1981 to 1984, *Mar. Environ. Res.,* 22, 75, 1987.
50. Bell, P. R. F., Status of eutrophication in the Great Barrier Reef Lagoon, *Mar. Pollut. Bull.,* 23, 89, 1991.
51. Hallegraeff, G. M., Harmful algal blooms in the Australian region, *Mar. Pollut. Bull.,* 25, 186, 1992.
52. Seiki, T., Date, E., and Izawa, H., Eutrophication in Hiroshima Bay, *Mar. Pollut. Bull.,* 23, 95, 1991.
53. Doi, T. and Nitta, A., Ecological modeling at Osaka Bay related to long-term eutrophication, *Mar. Pollut. Bull.,* 23, 247, 1991.
54. Crema, R., Castelli, A., and Prevedelli, D., Long-term eutrophication effects on macrofaunal communities in northern Adriatic Sea, *Mar. Pollut. Bull.,* 22, 503, 1991.
55. Stabell, O. B., Pedersen, K., and Aune, T., Detection and separation of toxins accumulated by mussels during the 1988 bloom of *Chrysochromulina polylepis* in Norwegian coastal waters, *Mar. Environ. Res.,* 36, 185, 1993.
56. Nienhuis, P. H., Eutrophication, water management, and the functioning of Dutch estuaries and coastal lagoons, *Estuaries,* 15, 538, 1992.
57. Radach, G., Ecosystem functioning in the German Bight under continental nutrient inputs by rivers, *Estuaries,* 15, 477, 1992.

APPENDIX 1. NUTRIENTS

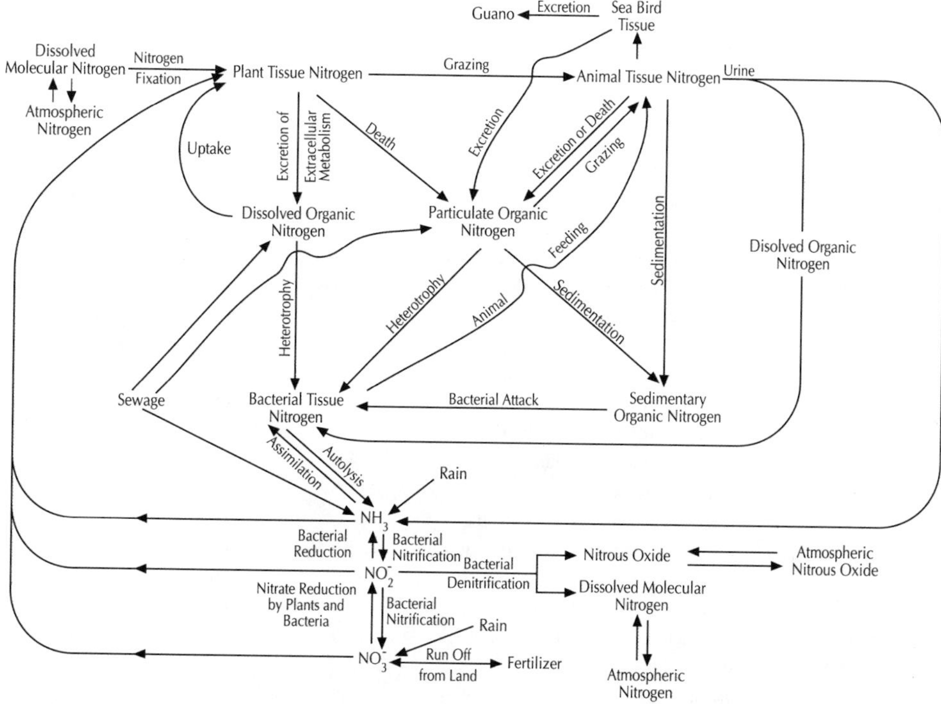

FIGURE 1.1. The nitrogen cycle in ocean waters. (From Millero, F. J. and Sohn, M. L., *Chemical Oceanography,* CRC Press, Boca Raton, FL, 1992, 336. With permission.)

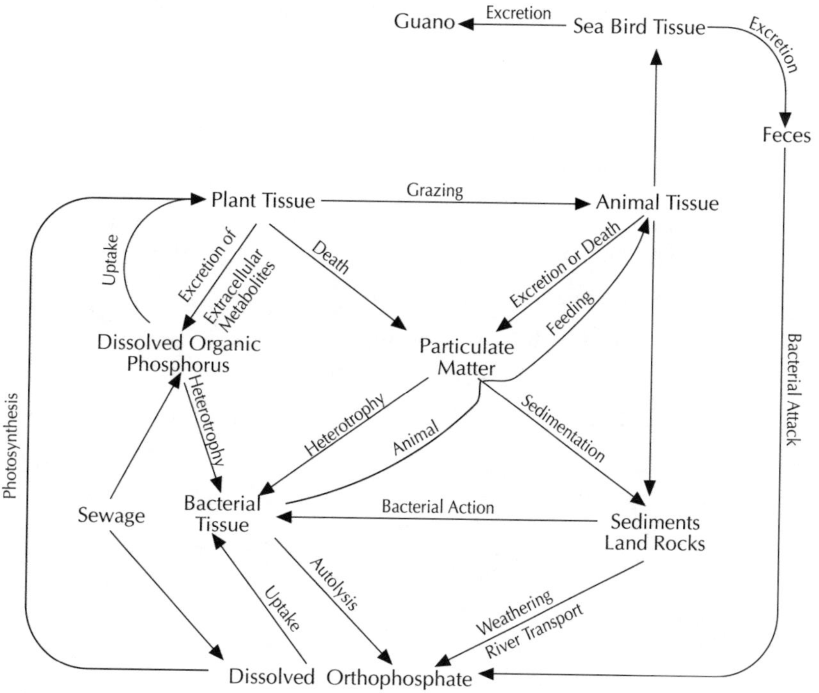

FIGURE 1.2. The phosphate cycle in ocean waters. (From Millero, F. J. and Sohn, M. L., *Chemical Oceanography,* CRC Press, Boca Raton, FL, 1992, 331. With permission.)

TABLE 1.1
Concentrations of Nitrogen and Phosphorus in 28 Estuarine Ecosystems[a]

Estuary	Nutrient Concentration	
	Nitrogen	Phosphorus
River dominated		
Pamlico River, North Carolina	1.5	8.0
Narragansett Bay, Rhode Island	0.6	1.6
Western Wadden Sea, Netherlands	3.0	2.0
Eastern Wadden Sea, Netherlands	4.0	2.5
Mid-Patuxent River, Maryland	4.2	2.3
Upper Patuxent River, Maryland	10.0	2.0
Long Island Sound, New York	1.5	0.5
Lower San Francisco Bay, California	20.6	3.8
Upper San Francisco Bay, California	11.5	2.0
Barataria Bay, Louisiana	4.6	0.8
Victoria Harbor, British Columbia	11.5	2.0
Mid-Chesapeake Bay, Maryland	4.5	0.6
Upper Chesapeake Bay, Maryland	5.0	6.0
Duwamish River, Washington	60.0	3.0
Hudson River, New York	5.0	0.16
Apalachicola Bay, Florida	10.0	0.1
Embayments		
Roskeeda Bay, Ireland	0.4	2.2
Bedford Basin, Nova Scotia	0.6	0.5
Central Kaneohe Bay, Hawaii	0.8	0.3
S.E. Kaneohe Bay, Hawaii	1.0	0.5
St. Margarets Bay, Nova Scotia	1.1	0.5
Vostok Bay, Russia	1.0	0.05
Lagoons		
Beaufort Sound, North Carolina	0.5	0.5
Chincoteague Bay, Maryland	3.2	2.5
Peconic Bay, New York	1.9	1.3
High Venice Lagoon, Italy	2.4	0.05
Fjords		
Baltic Sea	1.3	0.1
Loch Etive, Scotland	1.1	0.06

[a] Nutrient concentrations in μg-atom/l.

Source: Boynton, W. R., Kemp, W. M., and Keefe, C. W., in *Estuarine Comparison,* Kennedy, V. S. (Ed.), Academic Press, New York, 1982, 69. With permission.

TABLE 1.2
Various Oxidation States of Nitrogen

Oxidation State	Compound
+5	NO_3^-, N_2O_5
+4	NO_2
+3	HONO,[a] NO_2^-, N_2O_3
+2	HONNOH,[b] $HO_2N_2^-$, $N_2O_2^{2-}$
+1	N_2O
0	N_2
–1	H_2NOH, HN_3, N_3^-, NH_2OH
–2	H_2NNH_2
–3	RNH_4, NH_3,[c] NH_4^+[c]

[a] pK = 3.35.
[b] $pK_1 = 7.05$, $pK_2 = 11.0$
[c] $pK_B = 4.75$, $pK_A = 9.48$.

Source: Millero, F. J. and Sohn, M. L., *Chemical Oceanography,* CRC Press, Boca Raton, FL, 1992, 332. With permission.

TABLE 1.3
Transformation of Nitrogen Forms in the Nitrogen Cycle of Estuaries

Process	Transformation
A. Nitrogen fixation (oxygen sensitive)	$N_2 \rightarrow NH_3$
B. Dissimilatory reduction (oxygen sensitive)	
1. Respiratory reduction (nitrate respiration)	$NO_3^- \rightarrow NO_2^-$
2. Denitrification	NO_3^-, NO_2^-, $N_2O \rightarrow$ gaseous products (N_2O, N_2, NH_3)
C. Assimilatory reduction (ammonia sensitive)	NO_3^-, etc. $\rightarrow NH_3$
D. Nitrification	$NH_3 \rightarrow NO_3^-$
E. Ammonification	"R-NH_2" $\rightarrow NH_3$

Source: Webb, K. L., in *Estuaries and Nutrients,* Neilson, B. J. and Cronin, L. E. (Eds.), Humana Press, Clifton, NJ, 1981, 25. With permission.

TABLE 1.4
Dissolved and Particulate Nutrient Concentrations in Tropical Aquatic Ecosystems

	NH_4	NO_2–NO_3	DON	PON (μmol/l)	PO_4	DOP	POP	Si
Oceanic (near surface)								
Gulf of Mexico	0.0–0.7	0.0–0.1		0.1–0.6	0.0–0.5			0–3
Subtropical North Pacific		0.1–0.1		0.2–0.3	0.1–0.1			2–3
Equatorial Pacific			6.2–13.8					
Equatorial Atlantic	0.1–0.1	0.0–2.0		–0.3			0.0–0.1	
South Pacific (Moorea)		–0.1			–0.4			–2
Coastal (near surface)								
Campeche Bank	0.0–2.7	0.0–0.3		0.1–5.1	0.0–0.4		0.0–0.1	0–3
Gulf of Papua/Torres Strait	0.0–3.2	0.0–3.1	0.9–13.7		0.3–0.6	0.0–0.2		1–30
Central GBR (18–20S)	0.0–0.5	0.0–0.5	2.4–14.8	1.0–3.8	0.0–0.3	0.0–0.8	0.0–0.2	0–2
Barbados	0.5–2.7	0.4–5.1			0.1–0.2			
Upwelling								
Peru	0.0–3+	0.0–20+		3.0–14+	0.0–2.5+			0–25
Arabian Sea		0.0–20+			0.0–2.0+			0–16
Estuary (near surface)								
Cochin Backwater (India)		0.0–20.3		0.0–150.0				
Missionary Bay (Australia)	0.1–0.4	2.0–7.0		0.1–0.4	0.2–0.6			
Coral reefs, oceanic and atolls								
Canton Atoll	0.1–1.3	0.0–2.4			0.0–0.5			2–3
Enewetak Atoll	0.2–0.3	0.1–0.3	1.7–2.3		–0.2	–0.2		
Tonga Lagoon	0.1–0.7	0.1–1.0	1–23.0	3.0–10.0	0.1–0.9			17–91
Gilbert Islands	0.3–0.5	0.0–2.6	3.8–5.6		0.0–0.4			
Takapoto Atoll	–0.1	–0.2			–0.1			–0
Moorea		–0.1			–0.5			–2
Coral reefs, shelf and fringing								
Kaneohe Bay	0.4–2.4	0.1–2.6	3.4–7.5		0.2–1.0	–0.4		
Jamaica	0.1–3.8				0.0–0.7			
Lizard Island (GBR)	0.1–0.2	0.2–1.0	3–5.0		0.2–0.4			1–2

Davies and Old Reef lagoons (GBR)	–0.2	–0.4			–0.2			–1
Abrolhos Islands	0.1–11.0	0.8–5.2			0.2–2.9	0.0–4.9		1–7
Rivers								
Amazon		–8.5			–0.2			–128
Ganges	0.7–31.4	35.7–770.0			1.9–652.0			21–2869
Maroni (S. America)		5.0–9.0			–0.2			–191
Niger	–1.0	–7.6						–250
Bermejo (S. America)	0.8–35.7	5.5–52.1			0.8–3.3	0.1–0.8	1.5–230.0	
Orinoco					–0.2			–91
Zaire	–0.4	6.7–7.0			0.1–0.8			165–342
Streams								
Savannah Rivers (Uganda)	<0.4–3.9	0.0–61.0			1.1–30.3			87–603
Wet forest streams (Uganda)	<0.4–9.7	0.0–171.0			0.1–8.2			118–825
Lesser Antilles	<2.8–62.8	<8.1–179.0			0.1–1.5			42–533
Lakes								
Lake Waigani (PNG)	–4.4	–21.6	–723.0		–42.6	–22.0		
Lake Victoria (Uganda)	–28.6	–0.4			–0.3			
Lake Nabugado (Kenya)	–<7				–21.0			
Lake Kyoga (Kenya)	–14.0	14.0			–16.8			–326
Lake Chilwa (Malawi)		2180.0–5114.0			0.0–229.0			
Lagartijo Reservoir		–4.3			–0.3			–478
Lakes Robertson/McIlwaine (Zimbabwe)	0.4–8.8	1.3–28.2			0.0–11.8			
Lake Calado (Brazil)	0.0–1.4	0.0–3.2			0.1–0.3			
Lake La Plata (Puerto Rico)	0.0–3.5	0.0–53.0	–46.0		1.5–6.8			
Swamps and Wetlands								
Kawaga Swamp (Uganda)	–8.6	–0.3			–0.9			
North Swamp (Kenya)	–3.2	–4.6	–63.0	–871.0		–1.9	–30.6	
Tasek Bara (Malaysia)	0.0–54.7	0.7–20.7	4–109.0		0.0–3.4	1.1–25.2		–52
Papyrus swamps (Uganda)	–<0.4	0.0–30.3			0.2–1.3			34–558

Note: 0.0 = nondetectable; DON = dissolved organic nitrogen; PON = particulate organic nitrogen; DOP = dissolved organic phosphorus; POP = particulate organic phosphorus.

Source: Furnas, M. J., in *Pollution in Tropical Aquatic Systems*, Connell, D. W. and Hawker, D. W. (Eds.), CRC Press, Boca Raton, FL, 1992, 29. With permission.

TABLE 1.5
Scale of Eutrophication in Chesapeake Bay and Its Tributary Estuaries

Ecosystem	Years	Ecological Description	Surface Area (m^2) (10^6)	Vol. (m^3) (10^6)	Avg. Depth (m)	Retention Time (years)
Patuxent	1963	Noneutrophic	137	660	4.8	1.70
	1969–1971	Eutrophic	137	660	4.8	1.70
	1978	Eutrophic	137	660	4.8	1.70
York	1969–1971	Noneutrophic	210	910	4.3	0.72
Rappahannock	1969–1971	Noneutrophic	400	1780	4.5	1.27
James	1969–1971	Eutrophic	600	2400	3.6	0.39
Potomac	1913	Noneutrophic	1250	7150	5.8	1.07
	1954	Eutrophic	1250	7150	5.8	1.07
	1969–1971	Hypereutrophic	1250	7150	5.8	1.07
	1977–1978	Eutrophic	1250	7150	5.8	1.07
Chesapeake Bay	1969–1971	Localized eutrophic				
(Including tribs)		conditions	11,500	74,000	6.5	1.16
(Excluding tribs)			6500	52,000	8.4	1.32

Source: Jaworski, N. A., in *Estuaries and Nutrients,* Neilson, B. J. and Cronin, L. E. (Eds.), Humana Press, Clifton, NJ, 1981, 37. With permission.

TABLE 1.6
External Nutrient Loadings for Chesapeake Bay and Its Tributary Estuaries

		External Phosphorus			External Nitrogen			At N:P
Ecosystem	Years	(g/yr) (10^6)	(g/m²/yr)	(g/m³/yr)	(g/yr) (10^6)	(g/m²/yr)	(g/m³/yr)	Ratio of Load
Patuxent	1963	170	1.24	0.26	930	6.7	1.4	12
	1969–1971	250	1.82	0.38	1110	8.1	1.7	10
	1978	420	3.06	0.64	1500	11.4	2.4	8
York	1969–1971	160	0.76	0.18	1190	5.6	1.3	17
Rappahannock	1969–1971	180	0.45	0.10	1500	3.8	0.8	19
James	1969–1971	1780	2.70	0.70	10,300	15.6	4.2	13
Potomac	1913	910	0.73	0.13	18,600	14.8	2.6	46
	1954	2000	1.63	0.28	22,600	18.1	3.1	26
	1969–1971	5380	4.30	0.80	25,200	20.2	3.5	11
	1977–1978	2520	2.01	0.35	32,800	26.2	4.6	30
Chesapeake Bay	1969–1971							
(Including tribs)		15,000	1.30	0.20	109,100	9.5	1.5	16
(Excluding tribs)		7350	1.10	0.10	70,160	10.8	1.3	22

Source: Jaworski, N. A., in *Estuaries and Nutrients,* Neilson, B. J. and Cronin, L. E. (Eds.), Humana Press, Clifton, NJ, 1981, 37. With permission.

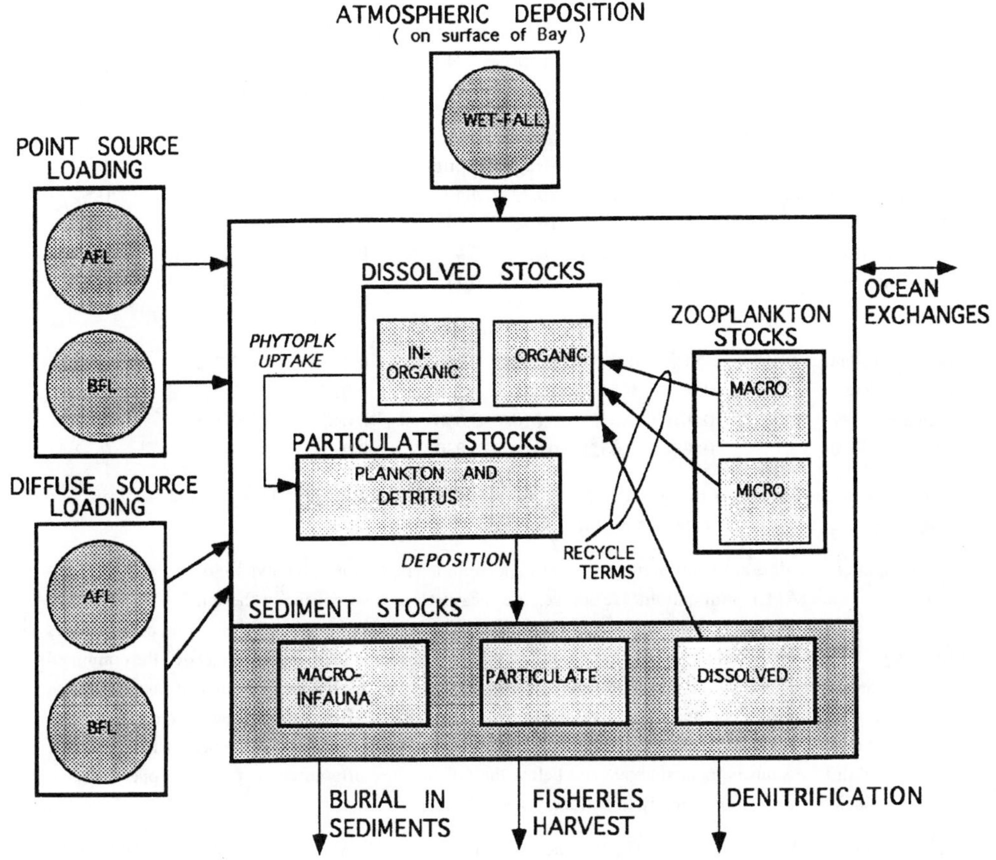

FIGURE 1.3. A schematic diagram of the Chesapeake Bay nutrient budget. Nutrient sources, storages, recycle pathways, internal losses, and exchanges across the seaward boundary are indicated. (From Boynton, W. R., Garber, J. H., Summers, R., and Kemp, W. M., *Estuaries,* 18, 285, 1995. With permission.)

TABLE 1.7
Summary of Annual Loadings of TN and TP from Terrestrial and Atmospheric Sources for the Chesapeake Region[d]

Location	Point Sources TN[a] AFL[b]	Point Sources TN[a] BFL[b]	Diffuse Sources TN[a] AFL[b]	Diffuse Sources TN[a] BFL[b]	Atmospheric Sources TN[c] (wet-fall)	Total Load (kg yr^{-1})	Areal Load (g m^{-2} yr^{-1})
Nitrogen Inputs							
Maryland Mainstem Bay	9.48	9.67	50.72	4.49	6.24	80.60	20.54
Potomac River	2.64	9.30	19.76	1.87	1.92	35.49	29.33
Patuxent River	0.61	0.22	0.21	0.47	0.22	1.73	12.63
Choptank River	0.00	0.14	0.12	0.71	0.57	1.54	4.27
Phosphorus Inputs							
Maryland Mainstem Bay	ND	1.041	2.100	0.360	0.251	3.752	0.96
Potomac River	0.620	0.140	1.880	0.210	0.077	2.927	2.42
Patuxent River	0.070	0.046	0.010	0.060	0.090	0.195	1.42
Choptank River	0.000	0.052	0.010	0.030	0.023	0.115	0.32

Note: All entries have units of kg × 10^6 TN or TP yr^{-1}. TN = total nitrogen; TP = total phosphorus. See original source for specific references for these findings.

[a] Point and diffuse data are from Summers (1989) and were averaged for 1985 and 1986.

[b] Above fall-line (AFL) point and diffuse sources are measured as a composite at the fall-line. The relative contributions of point and diffuse sources were estimated by subtracting known above fall-line point sources from the total load measured at the fall-line. In one case the point source load slightly exceeded the combined load presumably because of TP losses during transport in the river. In all cases the sum of above fall-line point and diffuse sources is equal to the load measured at the fall-line. BFL refers to below the fall-line.

[c] Atmospheric deposition data are from Smullen et al. (1982) and were averaged for the period 1976 to 1981.

[d] Point and diffuse sources located above and below the fall-line are differentiated. The atmospheric source (wet-fall) includes deposition directly to surface waters.

Source: Boynton, W. R., Garber, J. H., Summers, R., and Kemp, W. M., *Estuaries,* 18, 285, 1995. With permission.

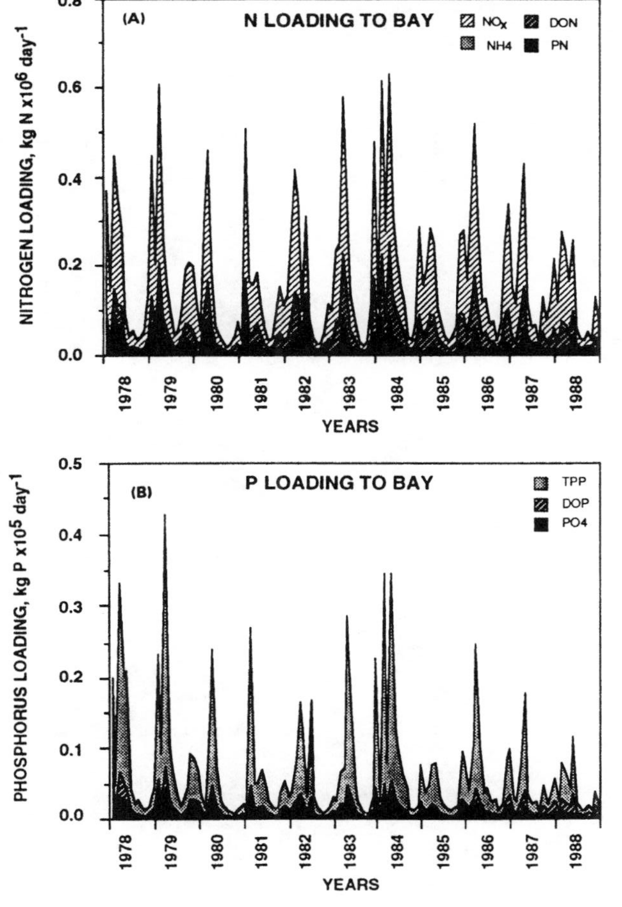

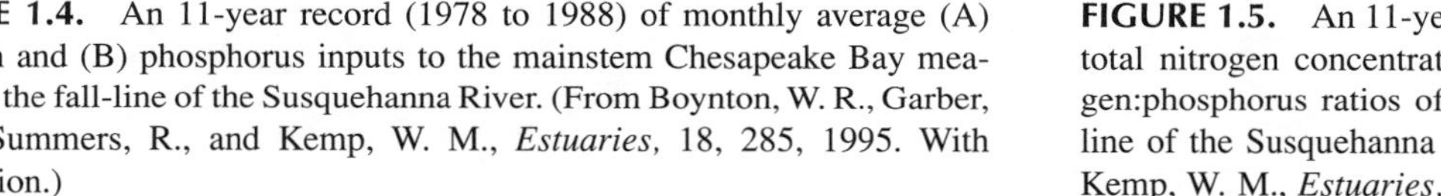

FIGURE 1.4. An 11-year record (1978 to 1988) of monthly average (A) nitrogen and (B) phosphorus inputs to the mainstem Chesapeake Bay measured at the fall-line of the Susquehanna River. (From Boynton, W. R., Garber, J. H., Summers, R., and Kemp, W. M., *Estuaries,* 18, 285, 1995. With permission.)

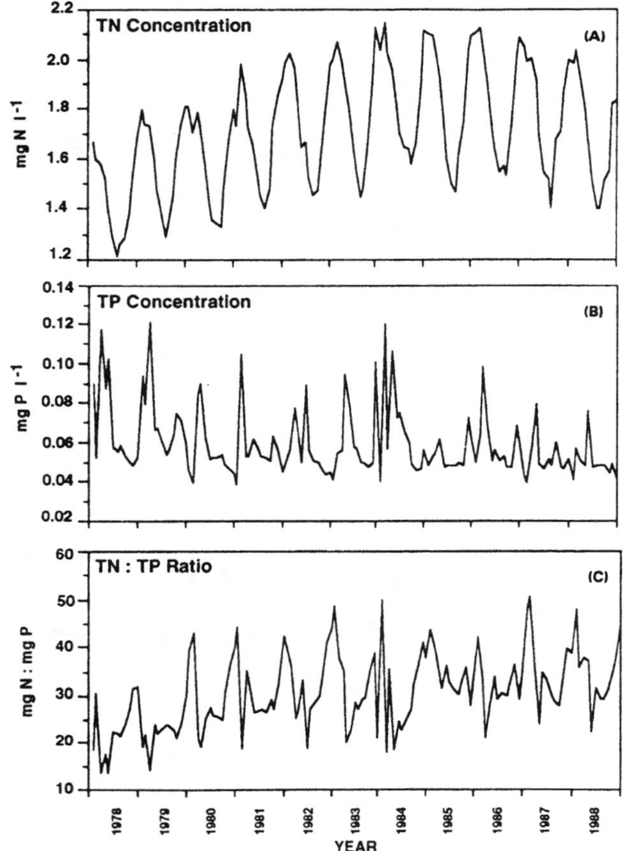

FIGURE 1.5. An 11-year record (1978 to 1988) showing monthly average values of (A) total nitrogen concentrations, (B) total phosphorus concentrations, and (C) total nitrogen:phosphorus ratios of inputs to the mainstem Chesapeake Bay measured at the fall-line of the Susquehanna River. (From Boynton, W. R., Garber, J. H., Summers, R., and Kemp, W. M., *Estuaries,* 18, 285, 1995. With permission.)

TABLE 1.8
A Comparison of Estimates of Annual TN and TP Inputs to Several Chesapeake Bay Tributaries and the Chesapeake Bay System

		Annual Nutrient Loading	
Location	Study Period	Total Nitrogen Load (kg N × 10^6 yr^{-1})	Total Phosphorus Load (kg P × 10^6 yr^{-1})
Patuxent River	1963	0.91	0.17
	1969–1971	1.11	0.25
	1978	1.55	0.42
	1985–1986	1.73	0.21
Potomac River	1913	18.6	0.91
	1954	22.6	2.04
	1969–1971	25.2	5.38
	1977–1978	32.8	2.51
	1985–1986	32.1	3.35
	1985–1986	35.5	2.93
Choptank River	1976–1979	1.81	0.29
	1980–1987	1.32	0.08
	1985–1986	1.54	0.12
Chesapeake system	1979–1981	123	10.30
	1985–1986	152	11.25

Note: TN = total nitrogen; TP = total phosphorus.

Source: Boynton, W. R., Garber, J. H., Summers, R., and Kemp, W. M., *Estuaries,* 18, 285, 1995. With permission.

TABLE 1.9
Summary of Average Annual Stocks (1985 and 1986) of Particulate and Dissolved Nitrogen and Phosphorus in the Water Column, Sediments, and Biota for Selected Areas of Chesapeake Bay

		Maryland Mainstem Bay		Potomac River		Patuxent River		Choptank River	
	Nutrient Stocks	N	P	N	P	N	P	N	P
Water column[a]	Dissolved								
	Inorganic	9.11	0.333	2.76	0.159	0.13	0.018	0.26	0.028
	Organic	7.51	0.159	2.92	0.201	0.26	0.024	0.56	0.052
	Particulate	3.26	0.606	0.95	0.295	0.16	0.031	0.29	0.037
	Total	19.73	1.098	6.63	0.655	0.55	0.073	1.21	0.117
Sediment[b]	Dissolved	0.52	0.014	0.39	0.004	0.01	0.003	0.04	0.016
	Particulate	214.01	36.370	57.88	11.830	7.38	2.120	18.80	3.660
	Total	214.53	36.384	58.27	11.834	7.39	2.123	18.84	3.676
Biota	Macrozooplankton[c]	0.17	0.020	0.03	0.003	0.00	0.001	0.02	0.002
	Benthic	3.34	0.406	0.68	0.083	0.10	0.012	0.10	0.012
	Total macrofauna[d]	3.51	0.426	0.71	0.086	0.10	0.013	0.12	0.012

Note: All entries have units of kg × 10^6 N or P. See original source for specific references for these findings.

[a] Water-column data are from Magnien et al. (1990).

[b] Sediment data are from Boynton and Kemp (1985), Boynton et al. (1990), and J. Cornwell (personal communication).

[c] Macrozooplankton data are from Jacobs (1989).

[d] Benthic macrofauna data are from Holland et al. (1989).

Source: Boynton, W. R., Garber, J. H., Summers, R., and Kemp, W. M., *Estuaries,* 18, 285, 1995. With permission.

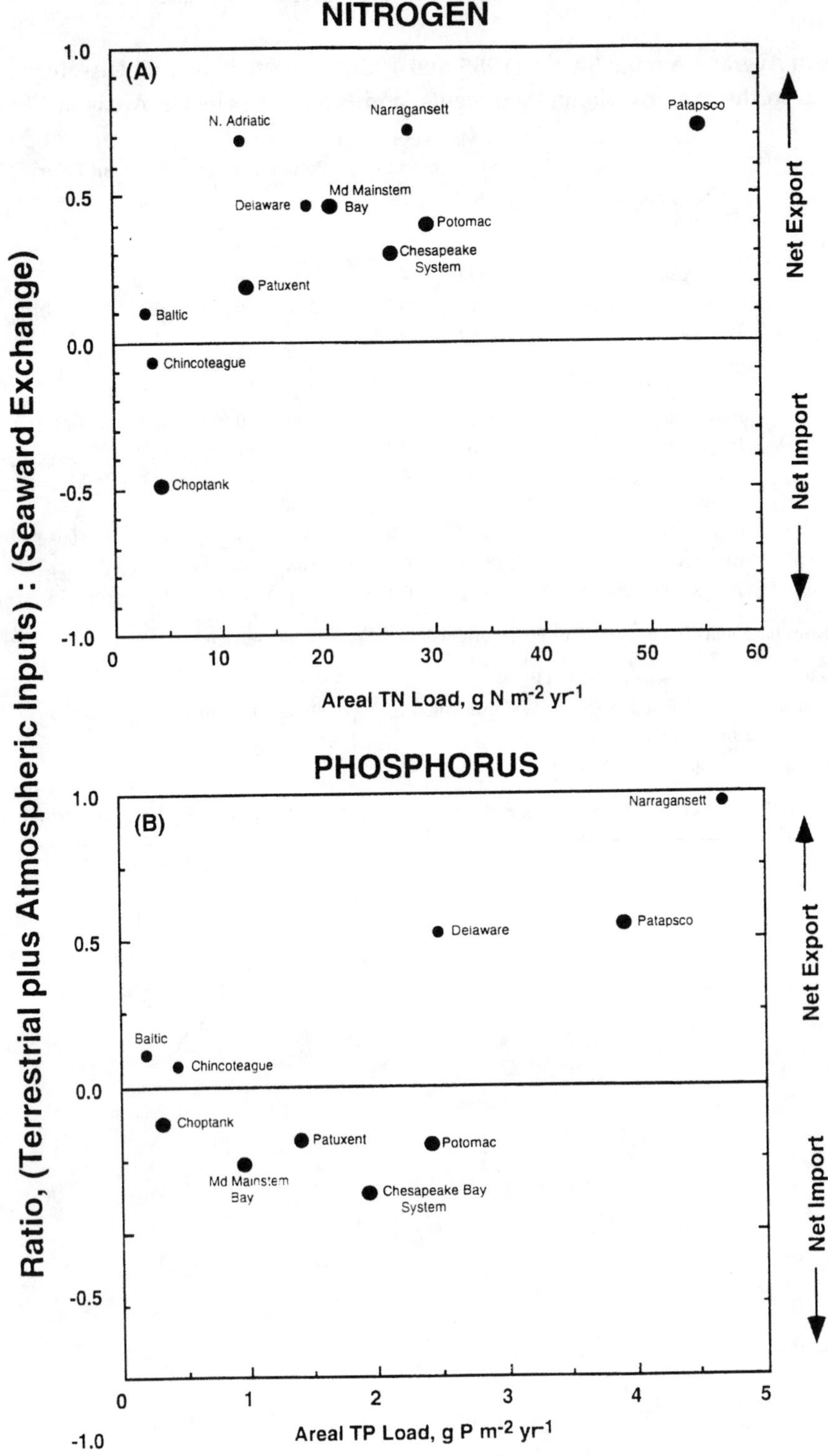

FIGURE 1.6. Scatter plots of annual areal loading rates of (A) nitrogen and (B) phosphorus vs. annual areal export:import ratio for Chesapeake Bay sites and several other locations where appropriate information is available. (From Boynton, W. R., Garber, J. H., Summers, R., and Kemp, W. M., *Estuaries,* 18, 285, 1995. With permission.)

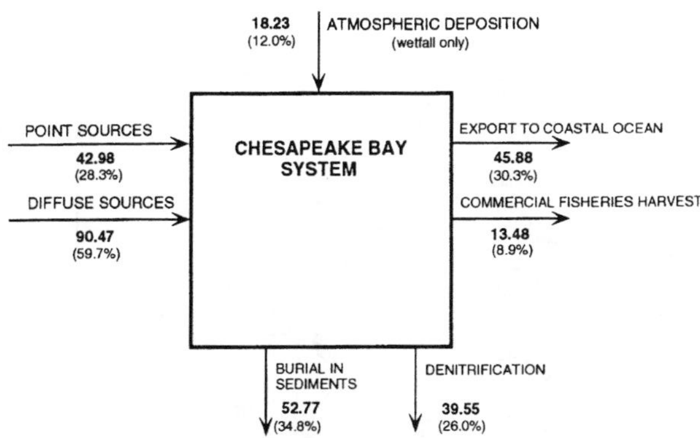

PHOSPHORUS

(kg P x 10^6 yr^{-1})

0.73
(6.5%)
ATMOSPHERIC DEPOSITION
(wetfall only)

POINT SOURCES
3.96
(35.2%)

DIFFUSE SOURCES
6.56
(58.3%)

CHESAPEAKE BAY SYSTEM

IMPORT FROM COASTAL OCEAN
4.11
(36.58%)

COMMERCIAL FISHERIES HARVEST
0.56
(5%)

14.48
(128.7%)
BURIAL IN SEDIMENTS

FIGURE 1.7. Simplified annual total nitrogen (TN) and total phosphorus (TP) budgets for the entire Chesapeake Bay system. (From Boynton, W. R., Garber, J. H., Summers, R., and Kemp, W. M., *Estuaries,* 18, 285, 1995. With permission.)

TABLE 1.10
External Nitrogen (TN) and Phosphorus (TP) Loadings (10^3 kg/yr), Total Nitrogen to Phosphorus Loading Ratios (Atomic), Watershed and Estuarine Areas (km^2), Watershed to Estuarine Area Ratios, and Watershed and Estuarine Area Normalized Loading ($kg/km^2/yr$) for the Upper Chesapeake Bay, Patuxent Estuary, and Potomac Estuary during 1985–1989

	Total Load			Head of Estuary Load			Head of Estuary Load as a Percent of the Total Load		Watershed	Estuarine	Ratio of Watershed to Estuarine	Watershed Area Normalized Load		Estuarine Area Normalized Load	
System	TN	TP	TN:TP	TN	TP	TN:TP	TN	TP	Area	Area	Area	TN	TP	TN	TP
Mainstem	84,100	3638	51	59,511	2167	61	71	60	86,979	3200	27	967	41.8	26,281	1137
Patuxent	1981	152.6	29	1182	88	30	60	58	2358	141	17	840	64.7	14,050	1082
Potomac	37,357	2344	35	32,008	2007	35	86	86	36,563	1251	29	1022	64.1	29,862	1874

Source: Magnien, R. E., Summers, R. M., and Sellner, K. G., *Estuaries,* 15, 497, 1992. With permission.

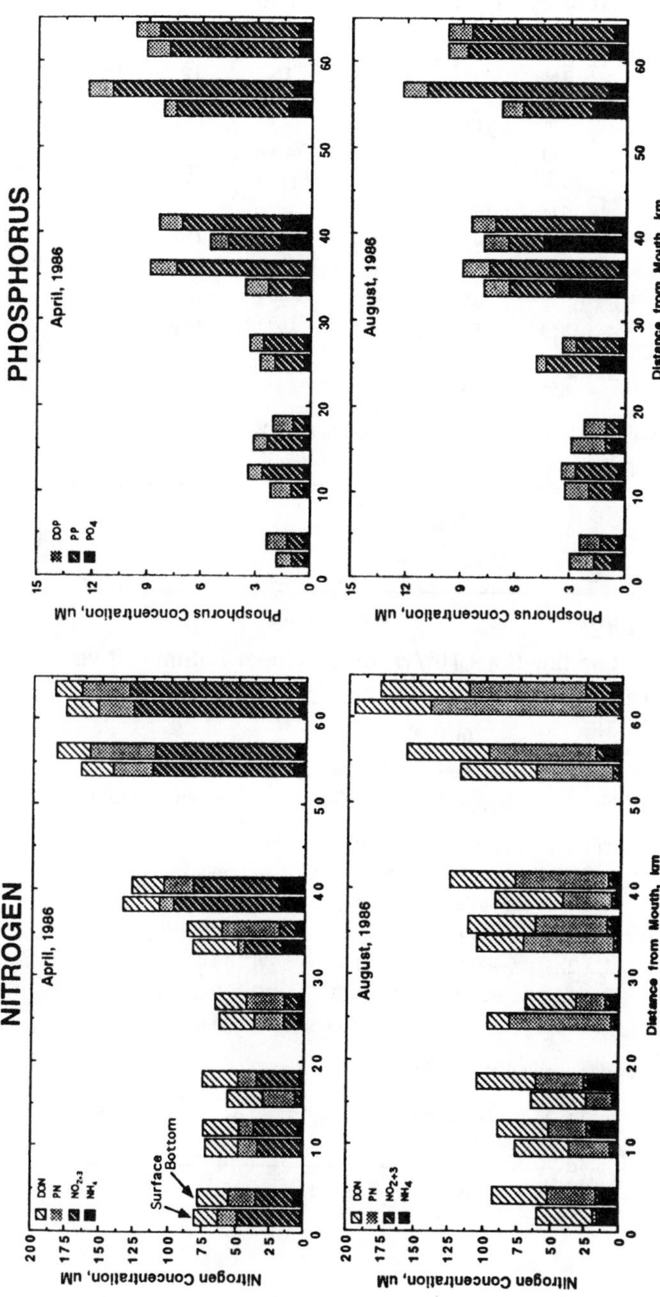

FIGURE 1.8. Longitudinal plots of surface and bottom water nitrogen and phosphorus concentrations along the channel of the Patuxent River during April and August 1986. (From Boynton, W. R., Garber, J. H., Summers, R., and Kemp, W. M., *Estuaries*, 18, 285, 1995. With permission.)

TABLE 1.11
Nitrogen (TN) and Phosphorus (TP) Loadings (10^3 kg/yr) of Wastewater Treatment Plants (WWTPs) in the Patuxent River Watershed and the Estimated Total External Load Delivered to the Patuxent Estuary from 1963 to 1989

	WWTP Load			Total Load		
Year	TN	TP	TN:TP	TN	TP	TN:TP
1963	74	19	9	930	170	12
1967	300	78	9	—	—	—
1969–1971	—	—	—	1110	250	10
1973	624	169	8	—	—	—
1978	798	193	9	1500	420	8
1985	815	138	13	1771	187	21
1986	842	65	29	1745	140	27
1987	859	57	33	1972	143	30
1988	866	50	38	1987	116	37
1989	855	50	38	2441	166	32

Source: Magnien, R. E., Summers, R. M., and Sellner, K. G., *Estuaries,* 15, 497, 1992. With permission.

TABLE 1.12
Nutrient Loading (kg × 10^6/yr) of the Upper Potomac River Basin from 1983 to 1988

		Total			
Year	River Flow[a]	Suspended Solids	Total Phosphorus	Oxidized Nitrogen	Total Nitrogen
1983	413.6	1441.4	2.50	18.0	29.4
1984	463.1	1604.5	3.07	20.6	35.2
1985	324.5	2950.4	3.82	12.7	27.4
1986	228.6	769.1	1.25	9.8	16.2
1987	321.2	960.4	1.91	13.9	23.4
1988	245.1	977.2	1.79	9.2	16.9
Average	332.6	1450.5	2.39	14.0	24.8

[a] Flow = average daily discharge in m^3 sec^{-1}.

Source: Jaworski, N. A., Groffman, P. M., Keller, A. A., and Prager, J. C., *Estuaries,* 15, 83, 1992. With permission.

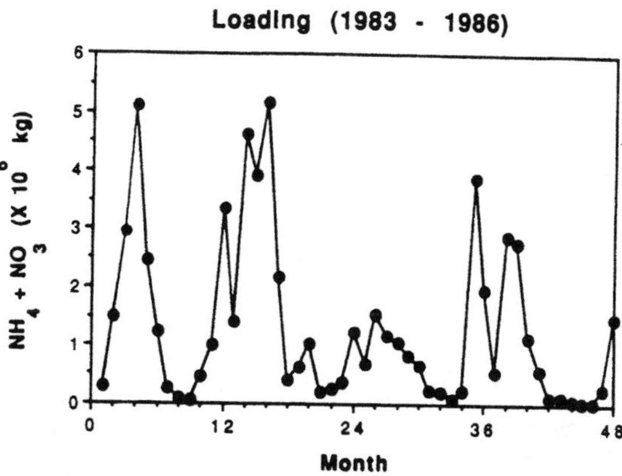

FIGURE 1.9. Monthly river ammonia- and nitrate-nitrogen loading (kg × 10^6/mo) for the Potomac River (1983 to 1986). (From Jaworski N. A., Groffman, P. M., Keller, A. A., and Prager, J. C., *Estuaries,* 15, 83, 1992. With permission.)

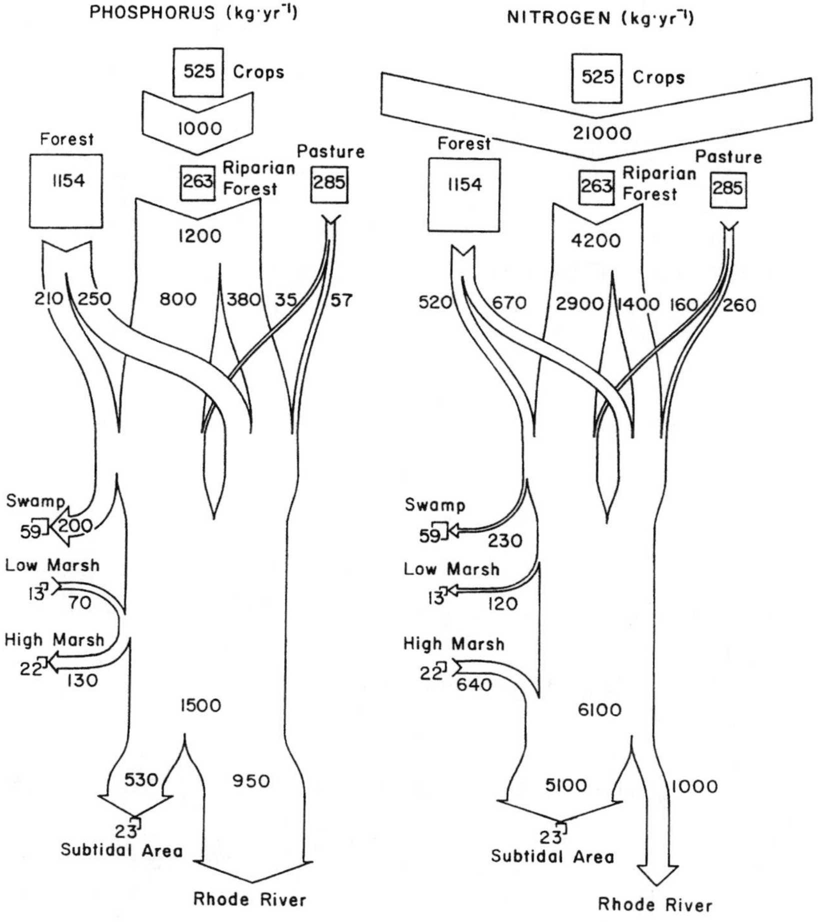

FIGURE 1.10. Net annual nitrogen and phosphorus fluxes (kg/yr) through the component ecosystems of the Rhode River landscape. Arrows indicate watershed discharges from upland systems, net uptake by freshwater swamp, and net tidal exchanges by the marshes and subtidal area. Widths of arrows are proportional to the fluxes given by the numbers in arrows. Numbers in boxes give the area of each ecosystem type (ha). (From Correll, D. L., Jordan, T. E., and Weller, D. E., *Estuaries,* 15, 431, 1992. With permission.)

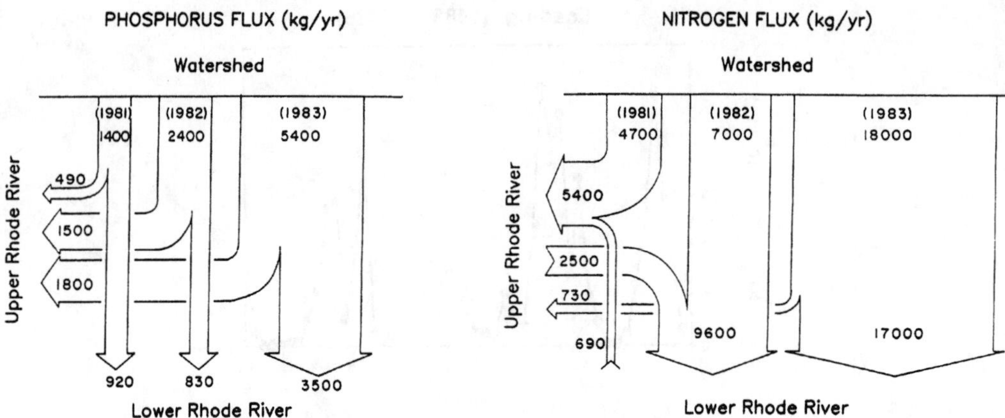

FIGURE 1.11. Variability of annual phosphorus and nitrogen fluxes into and out of the upper Rhode River estuary. (From Correll, D. L., Jordan, T. E., and Weller, D. E., *Estuaries,* 15, 431, 1992. With permission.)

APPENDIX 2. ORGANIC CARBON

TABLE 2.1
Concentrations of Organic Carbon in Natural Waters

Conc. of Organic Carbon (mg/l)	River	Estuary	Coastal Sea	Open Sea Surface	Open Sea Deep	Sewage
Dissolved	10–20 (50)	1–5 (20)	1–5 (20)	1–1.5	0.5–0.8	100
Particulate	5–10	0.5–5	0.1–1.0	0.01–1.0	0.003–0.01	200
Total	15–30 (60)	1–10 (25)	1–6 (21)	1–2.5	0.5–0.8	300

Note: Numbers in parentheses represent extreme values.

Source: Head, P. C., in *Estuarine Chemistry,* Burton, J. D. and Liss, P. S. (Eds.), Academic Press, London, 1976, 53. Copyright: 1976 Academic Press, Inc. (London) Ltd. With permission.

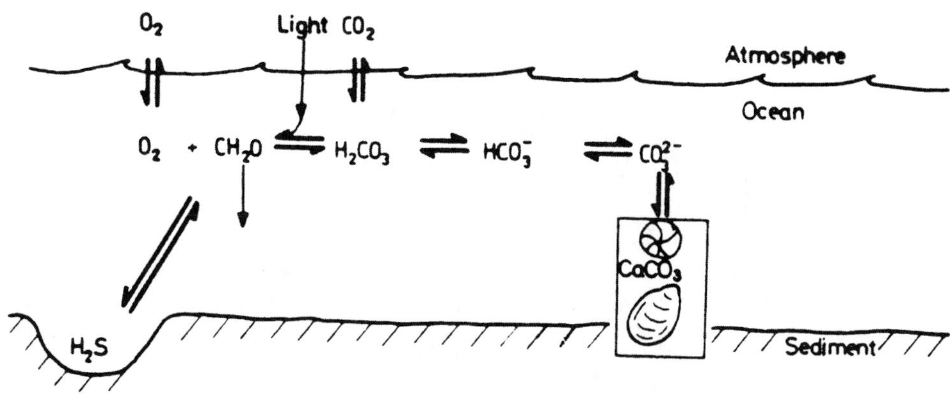

FIGURE 2.1. The CO_2-$CaCO_3$ cycle. (From Olausson, E., in *Chemistry and Biogeochemistry of Estuaries,* Olausson, E. and Cato, I. (Eds.), John Wiley & Sons, Chichester, 1980, 297. Reproduced by permission of copyright owner, John Wiley & Sons, Ltd.)

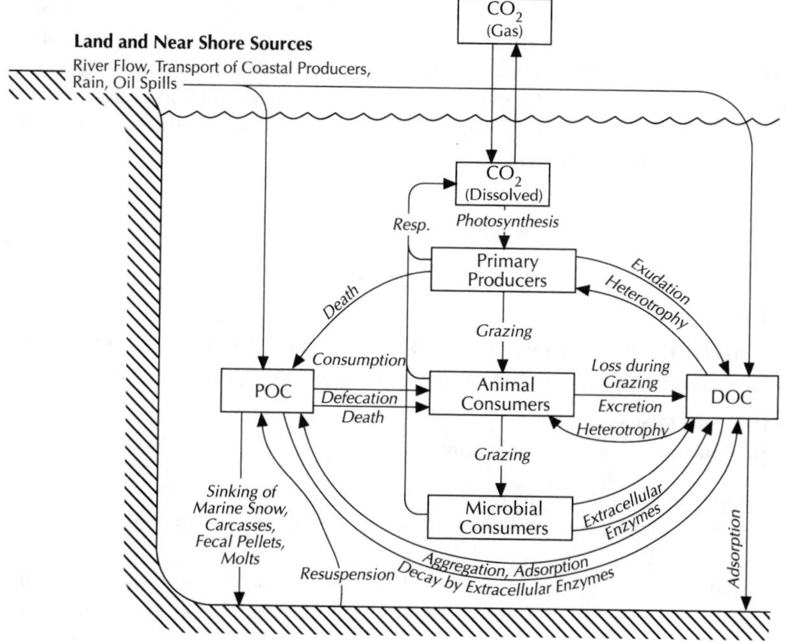

FIGURE 2.2. Carbon transfer in aerobic marine environments. The boxes represent pools and the arrows, processes. The inorganic parts of the cycle are simplified. Organic aggregates and debris comprise marine snow. Some resuspension of dissolved organic carbon from sediments into the overlying water is not shown in the diagram. (From Valiela, I., *Marine Ecological Process,* 2nd ed., Springer-Verlag, New York, 1995. With permission.)

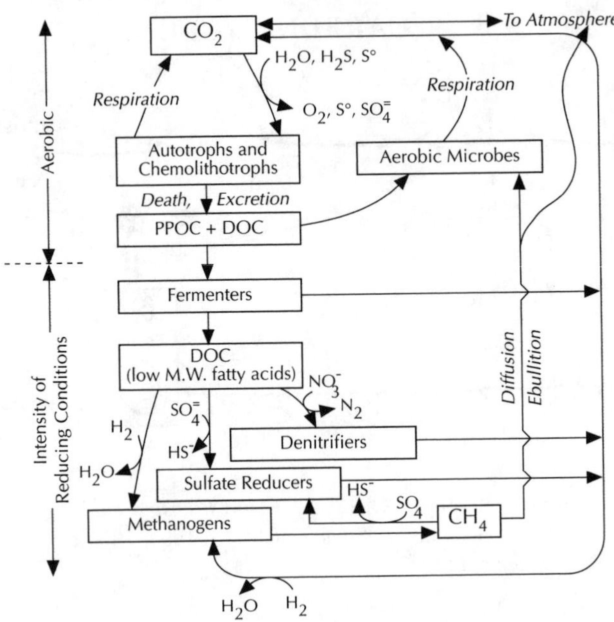

FIGURE 2.3. Carbon transformations in the transition from aerobic to anaerobic situations. The gradient from aerobic to anaerobic can be thought of as representing a sediment profile, with increased reduction and different microbial processes deeper in the sediment. Boxes represent pools or operations that carry out processes; arrows are processes that can be biochemical transformations or physical transport. Elements other than carbon are shown, where relevant, to indicate the couplings to other nutrient cycles. Some arrows indicate oxidizing and some reducing pathways. (From Valiela, I., *Marine Ecological Processes,* 2nd ed., Springer-Verlag, New York, 1995. With permission.)

TABLE 2.2
Concentration Ranges (μg/l) of the Major Identified Groups of Dissolved Organic Substances in Natural Waters

Dissolved Organic Substances	Rain	Groundwater	River	Lake	Sea
Volatile fatty acids	10	40	100	100	40
Nonvolatile fatty acids	5–17	5–50	50–500	50–200	5–50
Amino acids	—	20–350	50–1000	30–6000	20–250
Carbohydrates	—	65–125	100–2000	100–3000	100–1000
Aldehydes	—	—	~0.1	—	.01–0.1

Source: Wotton, R. S. (Ed.), *The Biology of Particles in Aquatic Systems,* CRC Press, Boca Raton, FL, 1990, 119. With permission.

TABLE 2.3
Seasonal Maxima and Minima of Dissolved Organic Matter in Estuaries and Coastal Marine Waters

Location	Measurement	Maximum (Time, conc.)	Minimum (Time, conc.)	Remarks
Estuary, Netherlands	DOC	Summer 4–5 mg l^{-1}	Fall 2 mg l^{-1}	Near mouth
Estuary, Netherlands	Amino acids	March 0.8 mg C l^{-1}	Sept. 0.1 mg C l^{-1}	Near mouth
Estuary, Netherlands	DOC	No pattern 13 mg l^{-1}	No pattern 8 mg l^{-1}	
Estuary, Netherlands	DOC	June and July 2 mg l^{-1}	Mar. and Nov. 0.3–0.5 mg l^{-1}	Autochthonous
Estuary, Netherlands	Amino acids	July 0.6–0.8 mg C l^{-1}	Feb.–Mar. 0.1 mg C l^{-1}	Highest near mouth
Nile estuary	DOM (O_2 demand)	August 9.05 mg O l^{-1}	December 2.37 mg O l^{-1}	Temperature-dependent changes
Maine estuary	DOC	Summer 5–7 mg l^{-1}	Winter 1–2 mg l^{-1}	Up to 12.6 mg l^{-1}
Maine estuary	DFAA	No pattern 300 n*M* l^{-1}	No pattern 100 n*M* l^{-1}	Some summer low values
English Channel	DOC	Mar., Jul., Aug., Oct. 1.8–2.2 mg l^{-1}	Mar., Apr., June 0.8–1.2 mg l^{-1}	Lagged Chl *a*
English Channel	DON	Late summer 7 μmol N l^{-1}	Winter undetectable	Summarized Station E1
English Channel	DFAA	Winter 3–4 μmol N l^{-1}	Summer 1 μmol N l^{-1}	
English Channel	DOC	June and Sept. .88 and .96 mg l^{-1}	March .56 mg l^{-1}	
English Channel	DON	Apr. and Aug. 5–6 μg-atN l^{-1}	October 2.5 μg-atN l^{-1}	
Southern North Sea	DOC	Spring 1.7 mg l^{-1}	Winter 0.4 mg l^{-1}	Summer-up, autumn-down
Dutch Wadden Sea	DOC	May–June 4–5 mg l^{-1}	Winter 1–2 mg l^{-1}	Tidal inlet
Maine coast	DFAA	May 88 μg C l^{-1}	December 10 μg C l^{-1}	July and Oct. peaks
Louisiana coast	DOC	October 3 mg l^{-1}	January 1.5 mg l^{-1}	<Full year
Gulf of Naples	DOC	No pattern 31.4 mg l^{-1}	No pattern 0.4 mg l^{-1}	*Posidonia* sea grass bed
Strait of Georgia, B.C.	DOC	Summer 3 mg l^{-1}	November 1 mg l^{-1}	
Coastal pond, Cape Cod	Carbohydrate	April 2.9 mg l^{-1}	February 1.3 mg l^{-1}	Inverse to chlorophyll
Menai Strait, England	DOC	Autumn 3–4 mg l^{-1}	Winter 1 mg l^{-1}	Spring–fall increase
Irish Sea	DFAA	May and Sept. 31 and 25 μg l^{-1}	Feb. and Aug. 5–10 μg l^{-1}	Late bloom maximum
Irish Sea	Total AA	Jan. and July 120 and 111 μg l^{-1}	February 2 μg l^{-1}	Little pattern

Note: DOC = dissolved organic carbon; DOM = dissolved organic material; DFAA = dissolved free amino acid; DON = dissolved organic nitrogen; AA = amino acid.

Source: Wotton, R. S. (Ed.), *The Biology of Particles in Aquatic Systems,* CRC Press, Boca Raton, FL, 1990, 95. With permission.

TABLE 2.4
Seasonal Maxima and Minima of Particulate Organic Matter in Estuarine and Coastal Marine Waters

Location	Measurement	Maximum (Time, conc.)	Minimum (Time, conc.)	Remarks
Estuary, Netherlands	Amino acids	May 1.5–1.6 mg C l^{-1}	February Near zero	Lower estuary
Estuary, Netherlands	Amino acids	July 4.2 mg C l^{-1}	February .1–.2 mg C l^{-1}	Upper estuary
Estuary, Netherlands	POC	Summer 3 mg l^{-1}	Sept.–Feb. 1 mg l^{-1}	Lower estuary
Estuary, Netherlands	POC	No pattern 8–9 mg l^{-1}	No pattern Near zero	Upper estuary
Estuary, Netherlands	Amino acids	May 1.7 mg C l^{-1}	October <0.1 mg C l^{-1}	Lower estuary
Estuary, Netherlands	POC	Aug.–Oct. 10–20 mg l^{-1}	Nov.–July 2–8 mg l^{-1}	
Arabian Sea estuary	POC	June and Aug. 5.24 mg l^{-1}	Oct.–Jan. 0.28 mg l^{-1}	Max. during monsoons
Estuary, Brazil	TSM	Summer 78.2 mg l^{-1}	Winter 6.9 mg l^{-1}	TSM about 60% organic
Estuary, Brazil	POC	Jan. and Nov. 2.1 mg l^{-1}	July and Oct. .6 and .3 mg l^{-1}	Near mouth
Estuary, Brazil	POC	February 3.2 mg l^{-1}	July 0.5 mg l^{-1}	Upper estuary
Dutch Wadden Sea	POC	May 4–6 mg l^{-1}	Winter 0.5–1.0 mg l^{-1}	Tidal inlet
English Channel	POC	Apr. and July 386 and 320 μg l^{-1}	Sept.–May 120–144 μg l^{-1}	
English Channel	PON	Apr. and July 50 and 37 μg N l^{-1}	Sept.–Mar. 8–17 μg N l^{-1}	
Galveston Bay, TX	Protein	May–June and Aug. 1.7 mg l^{-1}	January 0.2 mg l^{-1}	0.6 mg l^{-1} mid-August
Galveston Bay, TX	Carbohydrate	September 530 μg l^{-1}	Dec. and Aug. Near zero	Weak seasonal pattern
Galveston Bay, TX	Lipid	September 0.8 mg l^{-1}	February 0.1 mg l^{-1}	Similar to CHO pattern
Gulf of Naples	POM dry weight	No pattern 30 mg l^{-1}	No pattern 3.5 mg l^{-1}	*Posidonia* beds
Coastal Arabian Sea	POC	Nov. (30 m) 2.51 mg l^{-1}	May (surface) 0.52 mg l^{-1}	
Puget Sound	POC	July 40 μg l^{-1}	Sep.–Mar. 50–100 μg l^{-1}	Surface
N. Dawes Inlet, AK	POC	July 1.22 mg l^{-1}	November 0.24 mg l^{-1}	
Funka Bay, Japan	POC	July 11.5 g C m^{-2}	Early October 3.5 g C m^{-2}	Integrated upper 50 m
Funka Bay, Japan	Protein	February 3 g C m^{-2}	Early October 1.5 g C m^{-2}	Integrated upper 50 m
Tripoli Harbor	POC	August 1.3 mg l^{-1}	Nov.–Apr. .9–1.1 mg l^{-1}	Surface

TABLE 2.4 (continued)
Seasonal Maxima and Minima of Particulate Organic Matter in Estuarine and Coastal Marine Waters

Location	Measurement	Maximum (Time, conc.)	Minimum (Time, conc.)	Remarks
Strait of Georgia, B.C.	POC	Summer 400 μg l^{-1}	November 170 μg l^{-1}	Seasonal means
Coastal S. California	POC	Feb. and Mar. 351 and 235 μg l^{-1}	September 191 μg l^{-1}	Coastal stations

Note: POC = particulate organic carbon; TSM = total suspended material; PON = particulate organic nitrogen; POM = particulate organic matter.

Source: Wotton, R. S. (Ed.), *The Biology of Particles in Aquatic Systems,* CRC Press, Boca Raton, FL, 1990, 97. With permission.

TABLE 2.5
Seasonal Maxima and Minima of Dissolved Organic Matter in Saltmarsh Waters

Location	Measurement	Maximum (Time, conc.)	Minimum (Time, conc.)	Remarks
Texas marsh	DOC	July 37 mg l^{-1}	Winter 11–12 mg l^{-1}	
Louisiana bays	DOC	Early January 11.5 mg l^{-1}	Late Jan.–Nov. 5 mg l^{-1}	Minor changes after January
Estuary, SC	DOC	February 10.7 mg l^{-1}	July 3.2 mg l^{-1}	Flow-related variability
Estuary, SC	DON	Jan. and Aug. 32–38 μg-atN l^{-1}	Sep., Mar., May 18–19 μg-atN l^{-1}	
Estuary, SC	DOP	July 0.7 μg-atP l^{-1}	Nov.–Feb., Sep. Undetected	
Marsh, SC	DOC	Feb.–May 18.6 mg l^{-1}	December 3.1 mg l^{-1}	Runoff-related maximum
Marsh, VA	DON	Summer 1 mg l^{-1}	Winter 0.1 mg l^{-1}	
Tidal creek, SC	DOC	Fall 3.9 mg l^{-1}	Summer 2.8 mg l^{-1}	Seasonal means
Marsh, VA	DOP	July 0.25 mg l^{-1}	January <0.05 mg l^{-1}	
Duplin R., GA	DOC	Summer 9.6 and 9.2 mg l^{-1}	Winters 4.9 and 3.6 mg l^{-1}	Surface and bottom
Flax Pond, NY	DOC	July 2.5 mg l^{-1}	Jan.–Mar. 1.5 mg l^{-1}	Extreme values

Note: DOC = dissolved organic carbon; DON = dissolved organic nitrogen; DOP = dissolved organic particles.

Source: Wotton, R. S. (Ed.), *The Biology of Particles in Aquatic Systems,* CRC Press, Boca Raton, FL, 1990, 92. With permission.

TABLE 2.6
Seasonal Maxima and Minima of Particulate Organic Matter in Saltmarsh Waters

Location	Measurement	Maximum (Time, conc.)	Minimum (Time, conc.)	Remarks
Texas marsh	POC	July 5–7 mg l^{-1}	Winter 3–4 mg l^{-1}	Usual values, not extremes
Louisiana bays	POC	Feb.–Mar. 5–6 mg l^{-1}	Rest of year Negligible	
Marsh, SC	POC	Summer 4.6 mg l^{-1}	December 0.7 mg l^{-1}	
Marsh, SC	POC	Summer 2–3 mg l^{-1}	Winter ca. 1 mg l^{-1}	
Marsh, SC	Particulate N	July 17 μg-atN l^{-1}	January 2 μg-atN l^{-1}	
Marsh, SC	Particulate P	Sept. and Aug. 1.2–1.3 μg-atP l^{-1}	January 0.1 μg-atP l^{-1}	
Duplin R., GA	POC	Summer 5.9 and 4.3 mg l^{-1}	Winter 1.8 and 1.5 mg l^{-1}	Surface and bottom
Flax Pond, NY	POC	Mar. and May 2.0 and 2.1 mg l^{-1}	Jan. and Dec. 0.2 mg l^{-1}	
Dill Creek, SC	POM	August 58.5 mg l^{-1}	Sept. and Dec. 0.4 and 0.5 mg l^{-1}	
Marsh, MA	POC	Mar. and Sept. 1.0 and 1.6 mg l^{-1}	Apr.–Aug., Nov. .03–0.6 mg l^{-1}	Monthly averages

Note: POC = particulate organic carbon; POM = particulate organic matter.

Source: Wotton, R. S. (Ed.), *The Biology of Particles in Aquatic Systems,* CRC Press, Boca Raton, FL, 1990, 93. With permission.

TABLE 2.7
Accumulation of Organic Matter in Marine Sediments

	Percentage of Primary Production Accreting in Sediments	Sedimentation Rate (mm yr^{-1})
Oceanic sediments		
Abyssal plain, average of several sites	0.03–0.04	0.001
Central Pacific	<0.01	0.001–0.006
Off northwest Africa, Oregon, and Argentina	0.1–2	0.02–0.7
Peru upwelling, Baltic	11–18	1.4
Saltmarsh sediments		
Cape Cod, U.S.	5.3	1
Long Island, U.S.	37	2–6.3

Source: Valiela, I., *Marine Ecological Processes,* 2nd ed., Springer-Verlag, New York, 1995. With permission.

TABLE 2.8
Accumulation of Organic Matter in *Spartina* Marshes

Area	TNPP Accumulating in Sediments (%)	Sedimentation Rate (mm year^{-1})
Massachusetts	5.3	1
New York	37	2–6.3
New York	0	–9.5–37
North Carolina	75	>1.2
South Carolina	0	1.3

Source: Dame, R. F., *Rev. Aquat. Sci.*, 1, 639, 1989. With permission.

TABLE 2.9
Organic Carbon Budget for the Dollard Estuary

Import/Production		Export/Utilization	
Particulate C from North Sea and River Ems	37.1	Dissolved C to North Sea	?
From potato flour mill	33.0	Utilization in water	7.2
From salt marshes	0.5	Utilization in sediment	18.2
Phytoplankton production	0.7	Buried in sediment	9.9
Benthic algae production	9.3	Bird feeding	0.26
Total	80.6	Total	35.56

Note: Units are $\times 10^6$ kg C/year for the entire area of nearly 100 km^2.

Source: Van Es, F. B., *Helgol. Wiss. Meeresunters.*, 30, 283, 1977. With permission.

TABLE 2.10
Seasonal Maxima and Minima of Dissolved Organic Matter in Rivers

Location	Measurement	Maximum (Time, conc.)	Minimum (Time, conc.)	Remarks
Ganges R., Bangladesh	DOC	July 9.3 mg l^{-1}	June 1.3 mg l^{-1}	Max near crest
Ganges R., Bangladesh	Carbohydrate	June 1120 μg l^{-1}	October 141 μg l^{-1}	
Ganges R., Bangladesh	Amino acids	July 616 μg l^{-1}	March 150 μg l^{-1}	
Brahmaputra R.	DOC	July 6.5 mg l^{-1}	Rest of year 1.3–2.6 mg l^{-1}	Rising water maximum
Brahmaputra R.	Amino acids	August 262 μg l^{-1}	March 79 μg l^{-1}	
Brahmaputra R.	Carbohydrate	July 985 μg l^{-1}	October 155 μg l^{-1}	
Indus R.	DOC	Aug.–Sept. 22 mg l^{-1}	Low-flow period 1.2 mg l^{-1} min	Maximum range
Amazon R.	DOC	May–June 6.5 mg l^{-1}	Feb.–March 4.2 mg l^{-1}	
Orinoco R., Venezuela	DOC	May and Dec. 5 and 4 mg l^{-1}	Jun.–Nov. 2–3 mg l^{-1}	
Gambia R., West Africa	DOC	September 3.7 mg l^{-1}	December 1.3 mg l^{-1}	
Columbia River	TOC	Spring–summer 3.2 mg l^{-1}	Late fall 1.8 mg l^{-1}	ca. 89% dissolved
Tigris R., Iraq	DOM	April 13.7–21.5 mg O l^{-1}	October 0.3–1.6 mg O l^{-1}	As O_2 demand
Caroni R., Venezuela	DOC	Aug. and Jan. 8 mg l^{-1}	November 4 mg l^{-1}	Humic rich
Ems River, Netherlands	DOC	February 12 mg l^{-1}	Aug.–Sept. 4–5 mg l^{-1}	Increase in fall–winter
Guatemalan rivers	DOM	June–July 4–36 mg l^{-1}	June or Oct. 3–5 mg l^{-1}	Peak discharge July–August
Alaskan rivers	DOC	Variable 4–6 mg l^{-1}	Aug.–Sept. 1–2 mg l^{-1}	
Shetucket R., CT	DOC	May and Sept. 6.2–10 mg l^{-1}	Jan.–Apr. 2–4 mg l^{-1}	Max. 26.4 mg l^{-1} in runoff
Ogeechee R., GA	DOC	Jan.–May 12–15 mg l^{-1}	Aug.–Dec. 6–8 mg l^{-1}	To 17 mg l^{-1} July storm
Westerwoldse Aa, Neth.	DOC	December 48 mg l^{-1}	July 13 mg l^{-1}	Winter pollution
Black Creek, GA	DOC	Jan.–May 31–38 mg l^{-1}	June–Dec. 14–28 mg l^{-1}	To 42 mg l^{-1} in storm
N. Carolina stream	DOC	July and Oct. 1.1 and 1.3 mg l^{-1}	April 0.4 mg l^{-1}	Undisturbed watershed
N. Carolina stream	DOC	July and Oct. 0.5 and 0.6 mg l^{-1}	Apr.–May 0.3 mg l^{-1}	Clear-cut watershed
Little Miami R., OH	DOC	No pattern 12.5 mg l^{-1}	No pattern 2.5 mg l^{-1}	Pollution related
Loire R., France	DOC	Jan.–Apr., Dec. 5–6 mg l^{-1}	May 2.5 mg l^{-1}	
White Clay Cr., PA	DOC	Summer–fall 2–2.5 mg l^{-1}	Winter 1–1.5 mg l^{-1}	Higher DOC downstream

TABLE 2.10 (continued)
Seasonal Maxima and Minima of Dissolved Organic Matter in Rivers

Location	Measurement	Maximum (Time, conc.)	Minimum (Time, conc.)	Remarks
White Clay Cr., PA	DOC	Autumn 9–12 mg l^{-1}	Mid-winter 2–4 mg l^{-1}	To 18 mg l^{-1} fall peak
Hubbard Br., NH	DOC	No pattern 2 mg l^{-1}	No pattern <0.1 mg l^{-1}	Max. and min. in fall
Moorland Stream, U.K.	DOM	August up to 30 mg l^{-1}	February 0–3 mg l^{-1}	Peak flow in August

Note: DOC = dissolved organic carbon; DOM = dissolved organic material; TOC = total organic carbon.

Source: Wotton, R. S. (Ed.), *The Biology of Particles in Aquatic Systems,* CRC Press, Boca Raton, FL, 1990, 85. With permission.

TABLE 2.11
Seasonal Maxima and Minima of Particulate Organic Matter in Rivers

Location	Measurement	Maximum (Time, conc.)	Minimum (Time, conc.)	Remarks
Ganges R., Bangladesh	Amino acids	July 2395 μg l^{-1}	March 24 μg l^{-1}	
Ganges R., Bangladesh	Carbohydrate	July 1672 μg l^{-1}	March 46 μg l^{-1}	
Indus R., Pakistan	POC	Aug. (1981–82) up to 16 mg l^{-1}	Feb. or Apr. 0.3 mg l^{-1}	Jun. and Sept. peaks (1983)
Indus R., Pakistan	Amino acids	Aug. or Sept. 659–2009 μg l^{-1}	Oct. or Apr. 127–277 μg l^{-1}	Irregular sampling
Indus R., Pakistan	Carbohydrate	Aug. or Sept. 412–1105 μg l^{-1}	Feb., Apr., May 58–122 μg l^{-1}	Irregular sampling
Yangtze R., China	POC	July 17 mg l^{-1}	Jan.–Feb. 3 mg l^{-1}	Averages
Amazon R., near mouth	POC	Feb.–Mar. 8.2 mg l^{-1}	May–June 1–2 mg l^{-1}	
Amazon R., upstream	POC	Feb.–Mar. 15–20 mg l^{-1}	May–June 3.7 mg l^{-1}	
Guatemalan rivers	POC	June 7 mg l^{-1}	Nov.–Apr. Undetected	Tributaries to 15 mg l^{-1}
Orinoco R., Venezuela	POC	No pattern 2.5 mg l^{-1}	No pattern 1 mg l^{-1}	
Gambia R., West Africa	POC	Early Sept. 2 mg l^{-1}	Oct.–Nov. 0.3 mg l^{-1}	No relation to discharge
Columbia River	POC	Summer 860 μg l^{-1}	Winter 2 μg l^{-1}	Highly variable
St. Lawrence River	POC (means)	May and Nov. .67 and .69 μg l^{-1}	February .24 mg l^{-1}	May, terrigenous; Nov., *in situ*
Erriff R., Ireland	POC	No pattern 4 mg l^{-1}	No pattern 0.25 mg l^{-1}	Runoff-related peaks
Bundorragha R., Ireland	POC	No pattern 0.9 mg l^{-1}	No pattern 0.3 mg l^{-1}	Steeper basin than Erriff
Loire R., France	POC	May–Aug. 6 mg l^{-1}	Nov. and Mar. 2–3 mg l^{-1}	Increased at low water
Chalk Stream, England	POM	Fall–winter 10–12 mg l^{-1}	Summer 1–2 mg l^{-1}	Dry weight

Note: POC = particulate organic carbon; POM = particulate organic matter.

Source: Wotton, R. S. (Ed.), *The Biology of Particles in Aquatic Systems,* CRC Press, Boca Raton, FL, 1990, 88. With permission.

TABLE 2.12
Predominant Organic Compounds in Some Canadian Rivers

Sample Location	DOC (mg/l)	Rel. Color (col. units)	Humic Acid (mg/l)	Fulvic Acid (mg/l)	Tannins and Lignins (mg/l)	Phenols (mg/l)	Carbohydrate (mg/l)	Amino Acids (μg/l)	Total Pesticides (μg/l)
Thompson River at Savona (B.C.)	6	17	N.D.	2.1	0.4	N.D.	0.03	172	0.006
Saskatchewan River above Carrot River	13	10	1.7	8.0	1.3	N.D.	0.1	High	0.006
Yukon River at Dawson (Yukon)	5		N.D.	2.7	0.6	N.D.	0.2	676	0.002
Qu'Appelle River at Welby (Sask.)	12	20	0.8	12.0	1.4	N.D.	0.07	Low	0.004
Moose River at Mouth (Ont.)	18		7.1	11.3	4.0		N.D.	291	0.017
Richelieu River at St. Helaire (Que.)	8	10	N.D.	7.7	0.3	N.D.	0.09	189	0.006
St. Lawrence River at Levis (Que.)		10	0.3	6.8	0.7	0.003	0.08	Low	0.004
Annapolis River at Wilmot (NS)	7		0.6	5.2	1.2	N.D.		253	N.D.
St. John River at Woodstock (NB)	14	30	N.D.	12.6	4.0	N.D.	0.1	203	0.001
Exploits River at Millertown (Newf.)	7		0.6	5.5	1.4	N.D.	0.11	100	0.004

Note: DOC = dissolved organic carbon; N.D. = not detected.

Source: Afghan, B. K. and Chau, A. S. Y., *Analysis of Trace Organics in the Aquatic Environment,* CRC Press, Boca Raton, FL, 1989, 316. With permission.

TABLE 2.13
Net Annual Fluxes of Particulate and Dissolved Carbon, Nitrogen, and Phosphorus from the Coral Creek System (Northeastern Australia) via Tidal Transport

Component	Net Annual Exchange ($g\ m^{-2}\ year^{-1}$)	Proportion of Forest Primary Production Requirements (%)
Particulate matter (mainly intact mangrove plant detritus)		
Particulate organic C	–327	–35.9
Particulate organic N	–3.7	–13.4
Particulate organic P	–0.25	–12.2
Dissolved materials		
Dissolved organic C	7.3	0.8
Dissolved organic N	1.3	4.7
Dissolved organic P	0.37	17.9
NH_4	0.15	0.6
$NO_3 + NO_2$	–0.03	–0.1
PO_4	0.13	6.3
Total dissolved N	1.45	5.4
Total dissolved P	0.50	24.2

Note: The exchange rates are given in terms of forest area contained within the Coral Creek basin and in terms of net forest primary production requirements. A negative sign denotes net export.

Source: Connell, D. W. and Hawker, D. W. (Eds.), *Pollution in Tropical Aquatic Systems*, CRC Press, Boca Raton, FL, 1992, 135. With permission.

TABLE 2.14
Flux Estimates from a Number of Marsh-Estuarine Systems Dominated by *Spartina alterniflora*

System	Type	POC	DOC	TOC	NH_4	NN	DON	PN	TN	PO_4	TP
Crommet Creek	I				2.1	0.3				0.6	3.2
Flax Pond	I	68	–8	53	–2.0	1.0				–1.4	–0.3
Canary Creek	I	–62	–38	–100	0.7	1.9	–0.9	–2.9	–1.2	–0.1	
Bly Creek	I	38	–260	–222	–0.8	0.3	–14.6	2.8	–12.5	–0.2	–0.7
Sippewisset	II	–76			–4.2	–3.9	–9.8	–6.7	–24.6	–0.6	
Gott's Marsh	II	–7			–0.4	–0.9	–2.1	–0.3	–3.7		–0.3
Rhode River	II					–3800			–8400		–2200
Ware Creek	II	–35	–80	–115	–2.9	2.2	–2.3		–2.8	–0.1	0.7
North Inlet	II	–128	–328	–456	–6.3	–0.9			–42.7		
Dill Creek	II	–303									–6.4
Sapelo Island	II	–208	–108	–316							
Carter Creek	III	–116	–25	–141	–0.3	0.3	–9.2	4.6	–4.0	–0.6	

Note: All values are in grams per square meter per year of marsh.

Source: Dame, R. F., *Rev. Aquat. Sci.*, 1, 639, 1989. With permission.

TABLE 2.15
Carbon Budget from the Major Organic Decomposition Processes in the Sediments of a New England Salt Marsh and Short Below-Ground *Spartina* Production at the Same Site

Process	Carbon Fluxes ($g\ m^{-2}\ year^{-1}$)	BNPP (%)
Decomposition		
Aerobic respiration	361	31.7
Fermentation + sulfate reduction	432	37.9
Nitrate reduction	5	0.4
Methanogenesis	6	0.5
Burial	89	7.8
Export (as DOC)	36	3.2
Total losses	929	
Below-ground production	1140	

Note: Decomposition was measured in terms of carbon dioxide. DOC = dissolved organic carbon. BNPP = below-ground net primary production.

Source: Dame, R. F., *Rev. Aquat. Sci.*, 1, 639, 1989. With permission.

TABLE 2.16
Yearly Carbon, Nitrogen, and Phosphorus Flux in North Inlet on an Area of Marsh Basis

	Carbon	Nitrogen	Phosphorus
POC	128		
DOC	328		
NH_4		6.3	
$NO_2 + NO_3$		0.9	
TN		42.7	
PO_4			1.7
TP			3.1
ATP	22	3.3	0.2
CH1-a	1.8	0.27	0.016
Zooplankton	–1.2	–0.26	–0.010
Macrodetritus	0.9	0.01	0.001
Birds	0.05	0.01	0.001

Note: C, N, and P fluxes in $g\ m^{-2}\ year^{-1}$; drainage area = 2380 ha; import = –; export = + (sign not shown).

Source: Dame, R. F., *Rev. Aquat. Sci.*, 1, 639, 1989. With permission.

APPENDIX 3. SEWAGE WASTE

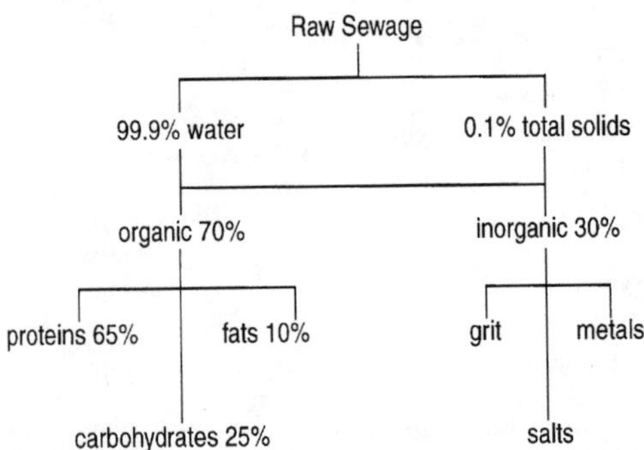

FIGURE 3.1. Composition of a typical raw sewage. (Adapted from Harrison, R. M. (Ed.), *Pollution: Causes, Effects, and Control,* 2nd ed., Royal Society of Chemistry, Cambridge, 1990. With permission.)

TABLE 3.1
Physical Methods of Pretreatment

Process	Aim	Examples
Screening	Removal of coarse solids	Vegetable canneries, paper mills
Centrifuging	Concentration of solids	Sludge dewatering in chemical industry
Filtration	Concentration of fine solids	Final polishing and sludge dewatering in chemical and metal processing
Sedimentation	Removal of settleable solids	Separation of inorganic solids in ore extraction, coal, and clay production
Flotation	Removal of low-specific-gravity solids and liquids	Separation of oil, grease, and solids in chemical and food industry
Freezing	Concentration of liquids and sludges	Recovery of pickle liquor and nonferrous metals
Solvent extraction	Recovery of valuable materials	Coal carbonizing and plastics manufacture
Ion exchange	Separation and concentration	Metal processing
Reverse osmosis	Separation of dissolved solids	Desalination of process and wash water
Adsorption	Concentration and removal of trace impurities	Pesticide manufacture, dyestuffs removal

Source: Harrison, R. M. (Ed.), *Pollution: Causes, Effects, and Control,* 2nd ed., Royal Society of Chemistry, Cambridge, 1990. With permission.

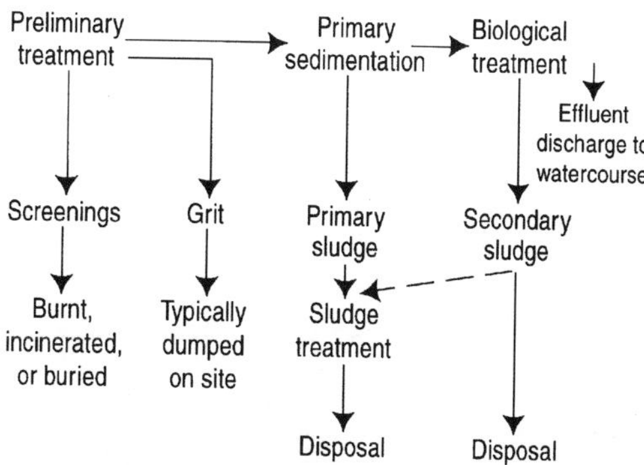

FIGURE 3.2. Flow diagram of a conventional sewage treatment works. (From Harrison, R. M. (Ed.), *Pollution: Causes, Effects, and Control,* 2nd ed., Royal Society of Chemistry, Cambridge, 1990. With permission.)

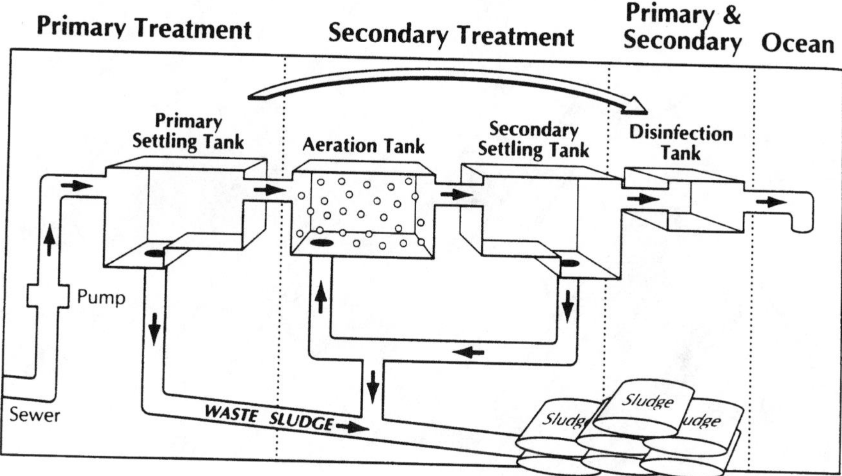

FIGURE 3.3 Scheme of primary and secondary treatment of sewage effluent and sludge. (From Duedall, I. W., *Oceanus,* 33, 29, 1990. With permission.)

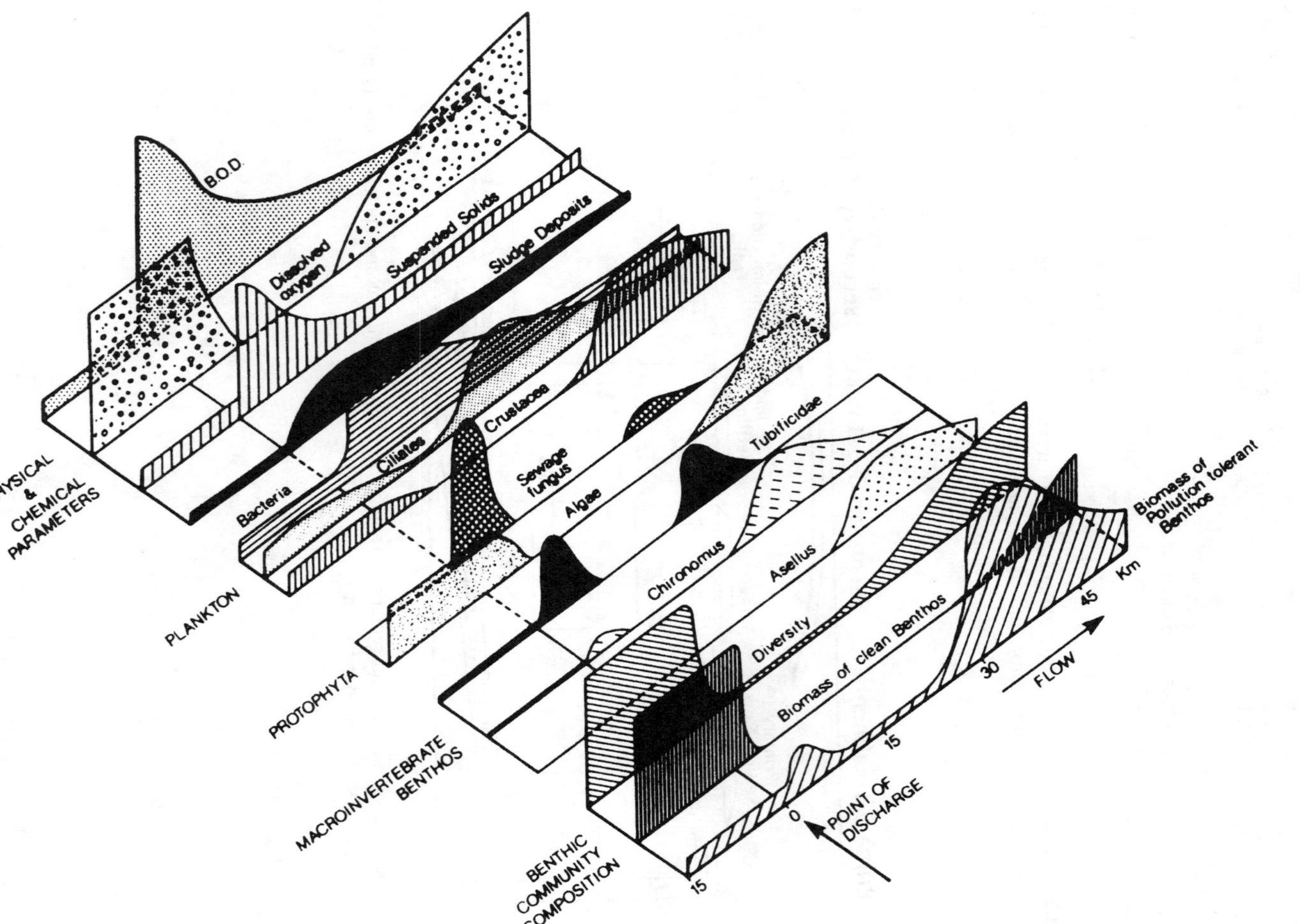

FIGURE 3.4. Spatial variation of physical, chemical, and biological consequences of the continuous discharge of a severe organic load into flowing water. (From Boudou, A. and Ribeyre, F. (Eds.), *Aquatic Ecotoxicology: Fundamental Concepts and Methodologies*, CRC Press, Boca Raton, FL, 1989. With permission.)

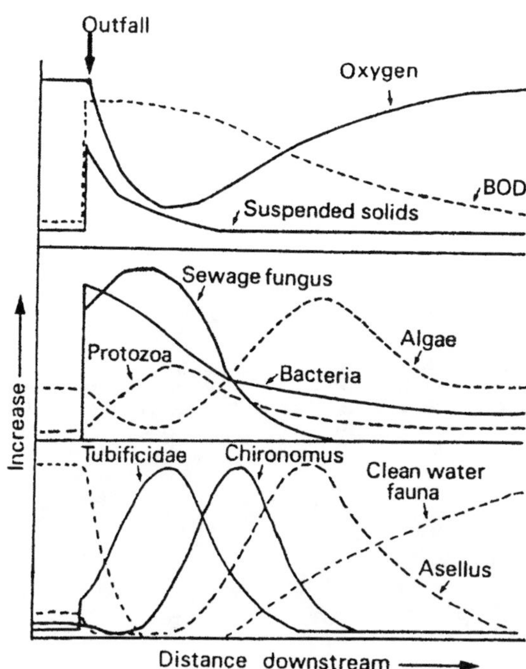

FIGURE 3.5. Changes in water quality and populations of organisms in a river below a discharge of an organic effluent. (From Harrison, R. M. (Ed.), *Pollution: Causes, Effects, and Control,* 2nd ed., Royal Society of Chemistry, Cambridge, 1990. With permission.)

TABLE 3.2
Amounts of Heavy Metal Ions Removed From Sewage by Sludges

	Primary Sedimentation		Percolating Filter Treatment		Activated-Sludge Process	
Heavy Metal Ion	Metal Concentration in Crude Sewage (mg l^{-1})	Proportion Removed by Treatment (%)	Metal Concentration in Settled Sewage (mg l^{-1})	Proportion Removed by Treatment (%)	Metal Concentration in Settled Sewage (mg l^{-1})	Proportion Removed by Treatment (%)
Copper	Up to 0.8	45	Up to 0.44	20	0.4	54
			(as Cr)		Up to 0.44	80
					0.4–25	50–79
	Up to 5 (as Cr)	12			28 (as Cr)	90–93
Dichromate	Up to 1.2	28	Up to 0.86	32	Up to 0.86	67–70
					4.0	6.3
					0.5–2	~100
					5	50
					50	10
Iron (ferric)	3–9	40	1.8–5.4	Nil	1.8–5.4	80
Lead	0.3–0.9	40	0.18–0.54	30	0.18–0.54	90
Nickel	0.1–0.3	20	0.08–10	40	0.08–0.24	30
					2.0	31
					2.5–10	30
Zinc	0.7–1.6	40	0.4–1.0	30	0.4–1.0	60
					2.5	90
					2.5	95
	Up to 5	12			7.5	100
					15	78
					20	74

Source: Harrison, R. M. (Ed.), *Pollution: Causes, Effects, and Control,* 2nd ed., Royal Society of Chemistry, Cambridge, 1990. With permission.

TABLE 3.3
Pathogens Likely To Be Associated with Sewage Sludge

Bacteria	Viruses	Protozoa	Helminths
Salmonella spp.	Poliovirus	*Entamoeba histolytica*	*Echinococcus granulosus*
Shigella spp.	Coxsackie A and B	*Acanthamoeba* spp.	*Hymenolepis nana*
Vibrio spp.	Echovirus	*Giardia* spp.	*Taenia saginata*
Mycobacterium spp.	Adenovirus		*Fasciola hepatica*
Bacillus anthracis	Reovirus		*Ascaris lumbricoides*
Clostridium perfringens	Parvovirus		*Enterobius vermicularis*
Yersinia spp.	Rotavirus		*Strongyloides* spp.
Campylobacter spp.	Hepatitis A		*Trichuris trichiura*
Pseudomonas spp.	Norwalk and related gastroenteric viruses		*Toxocara canis*
			Trichostrongylus spp.
Leptospira spp.			
Listeria monocytogenes			
Escherichia coli			
Clostridium botulinum			

Source: Alderslade, R., in *Recent Experience in the United Kingdom: Characterization, Treatment, and Use of Sewage Sludge,* Hermite, P. L. and Ott, H. (Eds.), Reidel Publishing, Dordrecht, 1981, 372. With permission.

TABLE 3.4
Human Pathogenic Microorganisms Potentially Waterborne

Pathogen	Clinical Syndrome
Bacteria	
Aeromonas hydrophila	Acute diarrhea
Campylobacter spp.	Acute enteritis
Enterotox. *Clostridium perfringens*	Diarrhea
Enterotox. *E. coli*	Diarrhea
Francisella tularensis	Mild or influenzal, febrile, typhoidal illness
Klebsiella pneumoniae	Enteritis (occas.)
Plesiomonas shigelloides	Diarrhea
Pseudomonas aeruginosa	Gastroenteritis (occas.)
Salmonella typhi	Typhoid fever
Other salmonellae	Gastroenteritis
Shigella spp.	Shigellosis ("bacillary dysentery")
Vibrio cholerae	Cholera dysentery (01 serovars) or cholera-like infection (non-01)
V. fluvialis	Gastroenteritis
Lactose-positive *Vibrio*	Pneumonia and septicemia
V. parahaemolyticus	Gastroenteritis
Yersinia enterocolitica	Enteritis, ileitis
Cyanobacteria	
Cylindrospermopsis spp.	Hepatoenteritis
Viruses	
Enteroviruses	Aseptic meningitis, respiratory infection, rash, fever
Poliovirus	Paralysis, encephalitis
Coxsackie virus A	Herpangina, paralysis
Coxsackie virus B	Myocarditis, pericarditis, encephalitis, epidemic pleurodynia, transient paralysis
Echovirus	Meningitis, enteritis
Types 68–71	Encephalitis, acute hemorrhagic conjunctivitis
Hepatitis A	Infectious hepatitis type A
Hepatitis non-A, non-B	Hepatitis type non-A, non-B
Influenza A	Influenza
Norwalk and other parvovirus-like agents	Epidemic, acute nonbacterial gastroenteritis
Rotavirus	Nonbacterial, endemic, infantile gastroenteritis; epidemic vomiting and diarrhea
Protozoa	
Balantidium coli	Balantidiasis (balantidial dysentery)
Cryptosporidium	Cryptosporidiosis
Entamoeba histolytica	Amoebiasis (amoebic dysentery)
Giardia lamblia	Giardiasis — mild, acute, or chronic diarrhea
Helminths	
Ascaris	Ascariasis (roundworm infection)
Ancylostoma	Hookworm infection
Clonorchis	Clonorchiasis (Chinese liver fluke infection)
Diphyllobothrium	Diphyllobothriasis (broadfish tapeworm infection)
Dracunculus mediensis	Dracontiasis (Guinea worm infection)
Fasciola	Fascioliasis (sheep liver fluke infection)
Fasciolopsis	Fasciolopsiasis (giant intestinal fluke infection)
Paragonimus	Paragonimiasis (Oriental lung fluke infection)
Spirometra mansoni	Sparganosis (plerocercoid tapeworm larvae infection)
Taenia	Taeniasis (tapeworm infection)
Trichostrongylus	Trichostrongyliasis
Trichuris	Trichuriasis (whipworm infection)

Source: Harrison, R. M. (Ed.), *Pollution: Causes, Effects, and Control,* 2nd ed., Royal Society of Chemistry, Cambridge, 1990. With permission.

TABLE 3.5
Human Pathogenic Microorganisms Potentially Water-Contact Transmitted

Pathogen	Clinical Syndrome
Bacteria	
Aeromonas hydrophila	Wound and ear infections, septicemia, meningitis, endocarditis, corneal ulcers
A. sobria	Wound and ear infections
Chromobacterium violaceum	Septicemia
Clostridium perfringens	Wound infection — gas gangrene
Klebsiella pneumoniae	Pneumonia, bacteremia
Legionella spp.	Legionellosis (Legionnaires' disease)
Leptospira spp.	Leptospirosis (Weil's disease — jaundice, hemorrhages, aseptic meningitis)
Mycobacterium marinum	Skin infection ("swimming pool granuloma")
M. ulcerans	Skin infection (progressive subcutaneous ulceration)
Pseudomonas aeruginosa	Otitis externa and media; follicular dermatitis (pruritic pustular rash)
P. pseudomallei	Meliodosis (glanders-like infection)
Staphylococcus aureus	Wound and skin infections
Halophilic vibrios (inc.*Vibrio parahaemolyticus, Vibrio alginolyticus,* lactose-positive *Vibrio)*	Wound and ear infections, conjunctivitis, salpingitis, pneumonia, septicemia
Viruses	
Adenovirus	Pharyngoconjunctivitis (swimming pool conjunctivitis), respiratory infection
Adenosatellovirus	Associated with adenovirus type 3 conjunctivitis and respiratory infection in children but etiology not clearly established
Protozoa	
Naegleria fowleri	Primary amoebic meningoencephalitis (PAME)
Helminths	
Schistosoma spp.	Schistosomiasis (bilharzia)
Avian schistosomes *(Trichobilharzia, Austrobilharzia)*	Schistosome dermatitis (swimmer's itch)
Ancylostoma duodenale	Hookworm infection
Necator americanus	Hookworm infection

Source: Compiled from McNeil, A. R., Australia Water Resources Tech. Pap. No. 85, Australian Government Publishing Service, Canberra, 1985, 561.

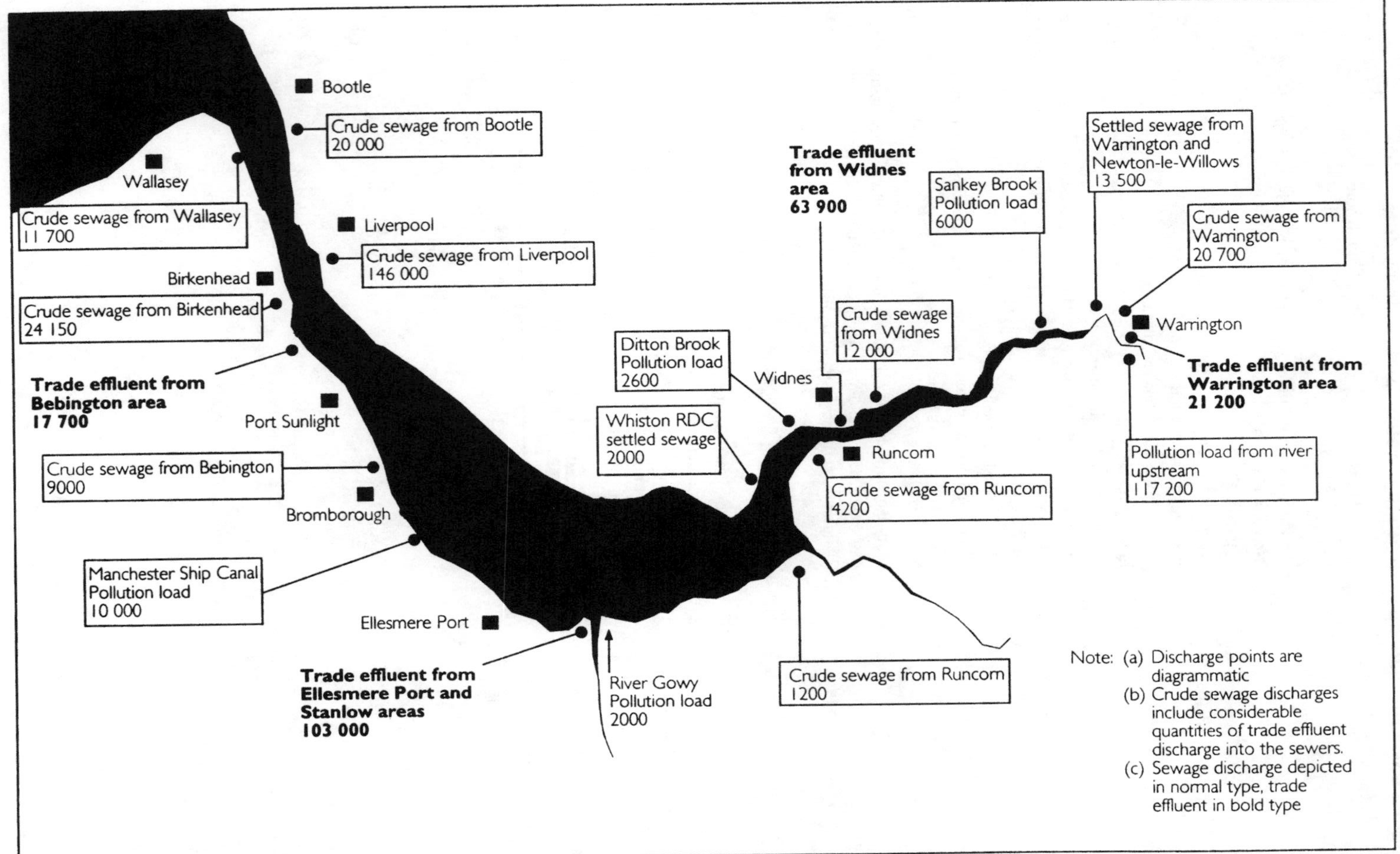

FIGURE 3.6. Sewage and industrial waste inputs (in lb BOD per day) to the Mersey estuary. (From Clark, R. B., *Marine Pollution*, 3rd ed., Clarendon Press, Oxford, 1992. With permission.)

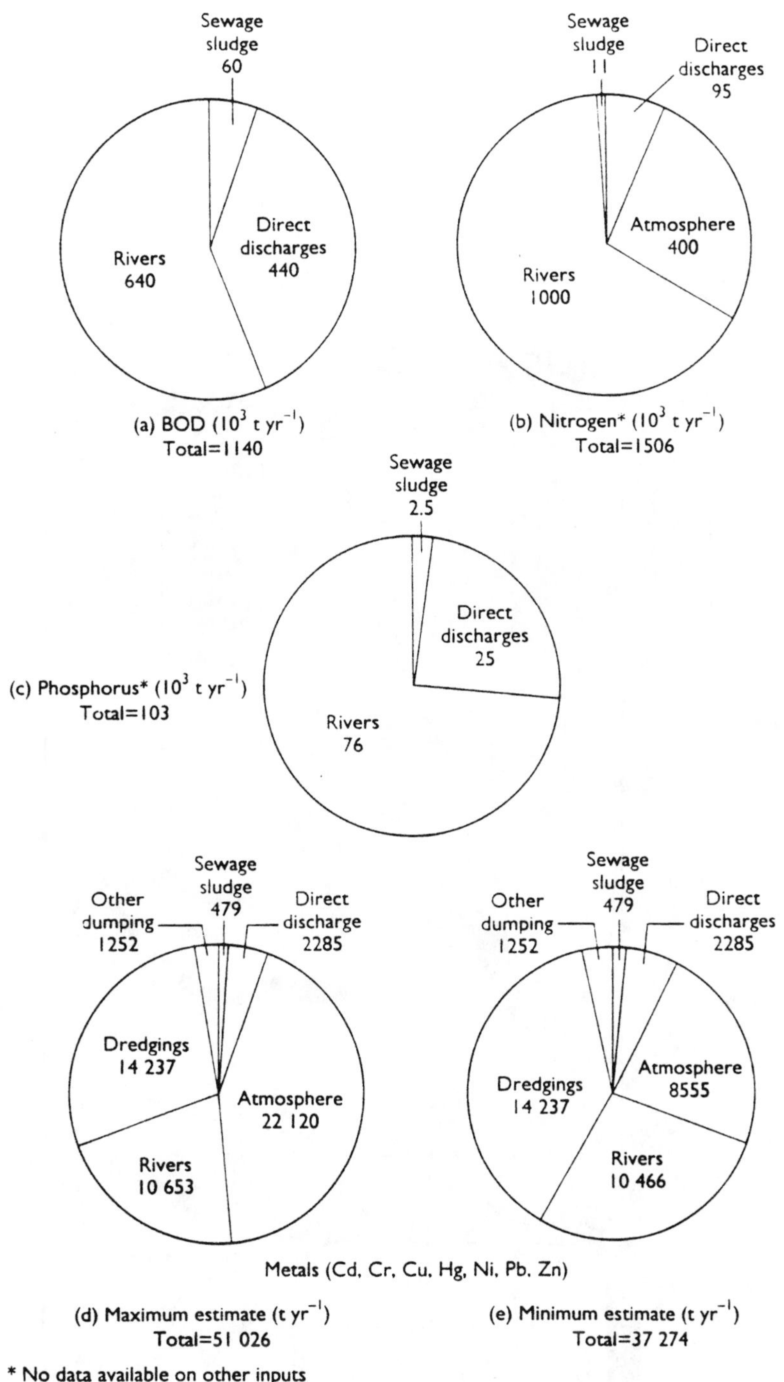

FIGURE 3.7. Estimated inputs of sewage sludge and other wastes to the North Sea by various pathways. (From Clark, R. B., *Marine Pollution,* 3rd ed., Clarendon Press, Oxford, 1992. With permission.)

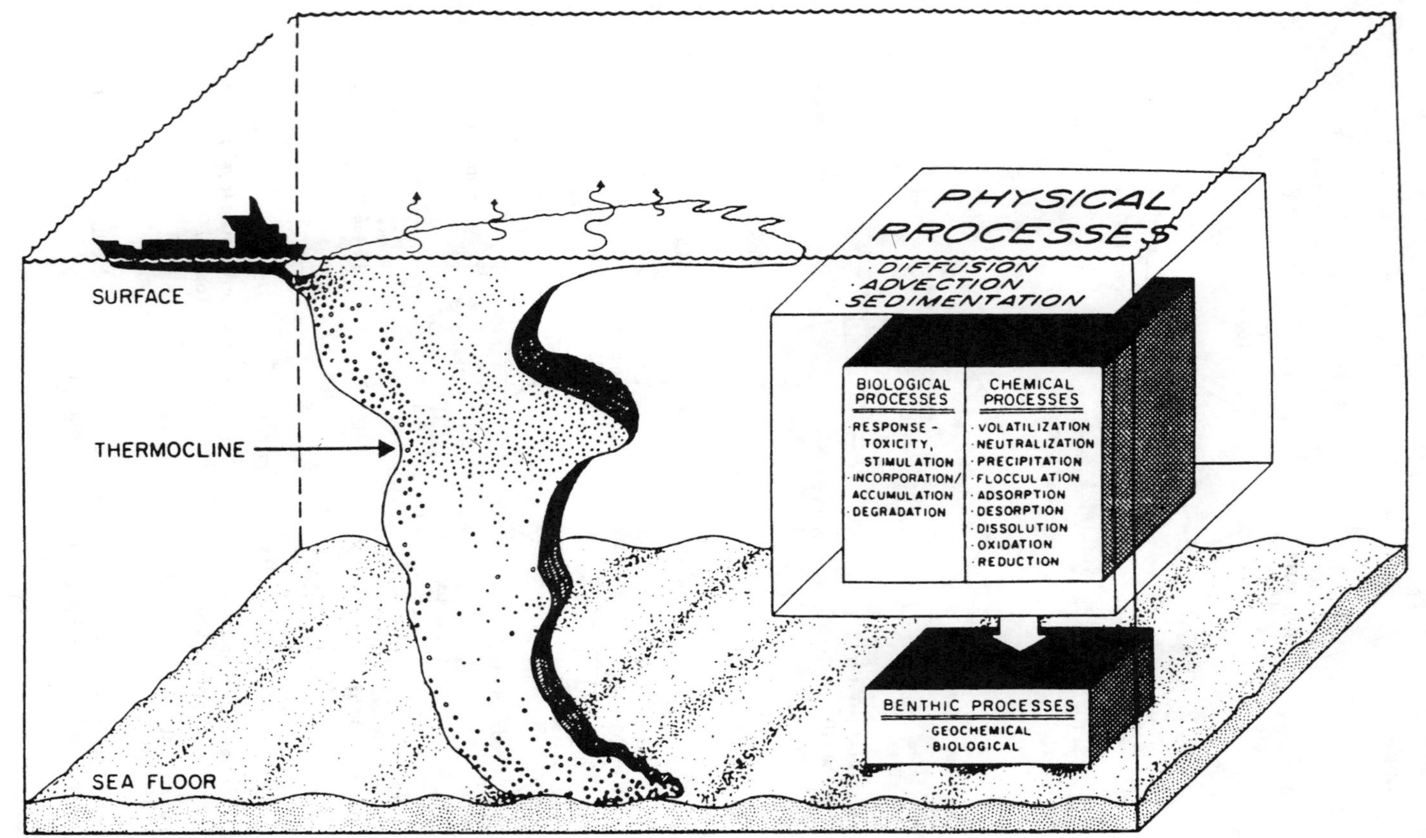

FIGURE 3.8. Schematic of the distribution of waste in the water column after a dumping event and some processes that will affect distribution and fate of the waste. (From Duedall, I. W., Ketchum, B. H., Park, P. K., and Kester, D. R., in *Wastes in the Ocean,* Vol. 1, *Industrial and Sewage Wastes in the Ocean,* Duedall, I. W., Ketchum, B. H., Park, P. K., and Kester, D. R. (Eds.), John Wiley & Sons, New York, 1983, 4. With permission.)

TABLE 3.6
Summary of Minimum Effective Toxic Concentrations (Mortality, Hatching, Feeding, Growth, Respiration, Filtration, Fertilization, Germination) and LC_{50} Averages of Sewage Sludge on Marine Organisms

Organisms	Life Stage	Duration of Test[a]	Effect	Effective Concentration Range (%)
Pisces				
Agonus cataphractus	Adult	96 h	LC_{50}	0.28–>1.0
Clupea harengus	Larvae	24 h	Mortality	1.0
	Larvae	96 h	Mortality	0.1–1.0
	Embryo	6 d	Hatching	0.1–1.0
Cyprinodon variegatus	Larvae	96 h	LC_{50}	5.0–7.5
Gadus morhua	Larvae	24 h	Mortality	0.1–1.0
	Larvae	4 d	Feeding	0.1–1.0
	Embryo	6 and 9 d	Hatching	1.0
	Juvenile	79 d	Growth	0.01
Limanda limanda	Juvenile	85 d	Growth	0.03
Menidia menidia	Adult	96 h	LC_{50}	0.72–2.87
	Adult	96 h	LC_{50}	(0.56–3.95)
	Juvenile	96 h	LC_{50}	1.0–2.6
	Larvae	96 h	LC_{50}	3.9–6.5
	Juvenile?	96 h	LC_{50}	0.11–16.5
Menidia beryllina	Larvae	96 h	LC_{50}	2.3–8.1
Crustacea				
Anostraca				
Artemia salina	Larvae	24 h	LC_{50}	8.5–11.75
Copepoda				
Acartia clausii	Adult	48 h	LC_{50}	0.1–2.0
Eurytemora herdmani	Adult	96 h	LC_{50}	<0.02–0.36
Pseudocalanus minutus	Adult	96 h	LC_{50}	0.11–0.19
Temora longicornis	Adult	24 h	LC_{50}	0.1
Decapoda (Natania, shrimps)				
Crangon crangon	Adult	96 h	LC_{50}	0.21–>1.0
		24 h	LC_{50}	0.16–>1.0
	Adult	96 h	LC_{50}	0.42–1.69
	Larvae	24 h	LC_{50}	0.0003–0.1
	Larvae	24 h	LC_{50}	0.36–>5.0
Palaemonetes pugio	Adult	96 h	LC_{50}	1.51–7.85
	Adult	96 h	LC_{50}	(1.87–8.40)
Mysidacea				
Mysidopsis bahia	Adult	96 h	LC_{50}	0.01–1.64
	Adult	96 h	LC_{50}	(0.36–2.80)
	Juvenile	96 h	LC_{50}	0.51–2.6
	Juvenile?	96 h	LC_{50}	0.005–4.2
Mollusca				
Cerastoderma edule	Adult	96 h	LC_{50}	0.54–>1.0
Littorina littorea	Adult	96 h	LC_{50}	>1.0
Mytilus edulis	Adult	96 h	LC_{50}	>1.0
Mytilus edulis	Adult	31 d	Respiration	0.021–0.42
			Clearance rate	0.021–0.42
			Condition	0.42
Echinodermata				
Arbacia punctatula	Sperm	1 h	Fertilization	0.15–1.63

TABLE 3.6 (continued)
Summary of Minimum Effective Toxic Concentrations (Mortality, Hatching, Feeding, Growth, Respiration, Filtration, Fertilization, Germination) and LC_{50} Averages of Sewage Sludge on Marine Organisms

Organisms	Life Stage	Duration of Test[a]	Effect	Effective Concentration Range (%)
Polychaeta				
Malacoceros fuliginosus	Adult	96 h	LC_{50}	>20.0
Scolelepis squamata	Adult	96 h	LC_{50}	19.8
Hydrozoa				
Laeomedea flexuosa	Adult	d?	Growth	1.0–2.0
Algae				
Isocrysis galbana	Adult	48 h	Loss from suspension	1.0
Skeletonema costatum	Adult	96 h	EC_{50}	0.77–42.0
Macrocystis pyrifera	Zoospore	48 h	Germination	0.56–1.0
Mesocosm				
Phytoplankton	Assemblage	4 mo	Depressed bloom	1.0–10.0

Note: Effective concentration is expressed as percentage sludge in the test sea water. Results of tests on the liquid fraction only are in parentheses.

[a] h = hours; d = days; mo = months.

Source: Costello, M. J. and Read, P., *Mar. Environ. Res.*, 37, 23, 1994. With permission.

TABLE 3.7
Distribution of Classified Estuarine Waters, 1985 to 1990

	Percent Classified							
	Approved		Prohibited		Conditional		Restricted	
Region	1985	1990	1985	1990	1985	1990	1985	1990
North Atlantic	87	69	10	29	1	1	2	1
Middle Atlantic	82	79	11	13	3	4	4	4
South Atlantic	75	71	22	21	3	4	<1	4
Gulf of Mexico	54	48	24	34	17	16	6	1
Pacific	42	53	40	31	18	11	1	5
Total	69	63	19	25	9	9	4	3

Source: NOAA, *The 1990 National Shellfish Register of Classified Estuarine Waters*, National Oceanic and Atmospheric Administration, Rockville, MD, 1991.

TABLE 3.8
Pollution Sources Affecting Harvest-Limited Acreage, 1990[a,b]

	North Atlantic		Middle Atlantic		South Atlantic		Gulf of Mexico		Pacific		Nationwide	
Sources	**Acres**	**%**	**Acres**	**%**	**Acres**	**%**	**Acres**	**%**	**Acres**	**%**	**Acres**	**%**
Point sources												
Sewage treat plants	238	67	641	57	374	44	973	27	75	25	2307	37
Combined sewers	21	6	224	20	0	0	211	6	0	0	457	7
Direct discharge	1	<1	84	7	5	1	920	25	6	2	1015	16
Industry	21	7	223	20	180	21	522	14	129	42	1077	17
Nonpoint sources												
Septic systems	91	26	123	11	288	34	1763	48	57	19	2322	37
Urban runoff	75	23	655	58	290	34	1276	35	110	36	2412	38
Agricultural runoff	5	3	130	12	233	28	301	8	41	13	718	11
Wildlife	19	7	112	10	306	36	1115	30	39	13	1597	25
Boats	55	17	353	31	146	17	507	14	47	15	1113	18
Upstream sources												
Sewage treat plants	2	1	104	9	9	1	1174	32	45	16	1334	21
Combined sewers	0	0	5	<1	0	0	134	4	0	0	0	2
Urban runoff	3	1	72	6	8	1	793	22	43	14	918	15
Agricultural runoff	0	0	1	<1	0	0	435	12	0	0	436	7
Wildlife	0	0	28	2	35	4	210	6	0	0	273	4

[a] Acres are times 1000; % is percent of all harvest-limited acreage in region.
[b] Since the same percentage of a shellfish area can be affected by more than one source, the percentages shown above cannot be added. They will not sum to 100.

Source: NOAA, *The 1990 National Shellfish Register of Classified Estuarine Waters,* National Oceanic and Atmospheric Administration, Rockville, MD, 1991.

TABLE 3.9
North Atlantic Pollution Sources Affecting Harvest-Limited Acreage, 1990[a,b]

	Maine		New Hampshire		Massachusetts	
Sources	**Acres**	**%**	**Acres**	**%**	**Acres**	**%**
Point sources						
Sewage treat plants	115	57	9	100	120	85
Combined sewers	0	0	1	11	21	15
Direct discharge	0	0	0	0	1	1
Industry	11	5	4	44	9	6
Nonpoint sources						
Septic systems	82	40	2	22	7	5
Urban runoff	24	12	6	67	50	36
Agricultural runoff	0	0	6	67	5	4
Wildlife	0	0	6	67	19	14
Boats	17	8	5	56	38	22
Upstream sources						
Sewage treat plants	0	0	0	0	2	1
Combined sewer	0	0	0	0	0	0
Urban runoff	0	0	0	0	3	2
Agricultural runoff	0	0	0	0	0	0
Wildlife	0	0	0	0	0	0

[a] Acres are times 1000; % is percent of all harvest-limited acreage in state.

[b] Since the same percentage of a shellfish area can be affected by more than one source, the percentages shown above cannot be added. They will not sum to 100.

Source: NOAA, *The 1990 National Shellfish Register of Classified Estuarine Waters*, National Oceanic and Atmospheric Administration, Rockville, MD, 1991.

TABLE 3.10
Pollution Sources of Estuaries in the Northeastern U.S. — North Atlantic (acres × 1000)

	Point				Nonpoint					Upstream				
Estuary	**STP**	**CSO**	**DD**	**IND**	**SEP**	**URO**	**ARO**	**WL**	**BTG**	**STP**	**CSO**	**URO**	**ARO**	**WL**
Passamaquoddy Bay	4	—	—	—	2	—	—	<1	—	—	—	—	—	—
Englishman Bay	3	—	—	—	—	—	—	—	—	—	—	—	—	—
Narraguagas Bay	—	—	—	—	1	—	—	—	—	—	—	—	—	—
Blue Hill Bay	—	—	—	—	<1	—	—	—	—	—	—	—	—	—
Penobscot Bay	39	—	<1	2	39	7	—	—	—	—	—	—	—	—
Muscongus Bay	4	—	—	<1	3	2	—	—	—	—	—	—	—	—
Sheepscot Bay	15	—	—	6	12	1	—	—	<1	—	—	—	—	—
Casco Bay	23	—	—	—	10	2	—	—	17	—	—	—	—	—
Saco Bay	8	—	—	1	<1	4	—	—	—	—	—	—	—	—
Great Bay	2	—	—	1	2	1	—	—	—	—	—	—	—	—
Merrimack River	2	—	—	<1	<1	<1	—	<1	<1	2	—	2	—	—
Massachusetts Bay	89	9	—	2	1	12	—	<1	4	—	—	1	—	—
Boston Bay	*2*	*11*	—	*6*	*<1*	*29*	—	*1*	*28*	—	—	—	—	—
Cape Cod Bay	<1	<1	<1	<1	2	3	1	6	2	—	—	—	—	—
Other	21	<1	—	3	19	14	4	12	4	—	—	—	—	—
North Atlantic total	238	20	—	21	91	75	5	19	55	2	—	3	—	—
% harvest-limited acreage	68	6	0	6	26	21	1	5	16	1	0	1	0	0
National total	2299	382	1011	1047	2325	2385	699	1552	1125	1337	142	1013	312	269
% harvest-limited acreage	36	6	16	16	36	37	11	24	18	21	2	16	5	4

Note: Subestuaries are in italics. STP, sewage treatment plant; CSO, combined sewer outfall; DD, direct discharge; IND, industry; SEP, septics; URO, urban runoff; ARO, agricultural runoff; WL, wildlife; BTG, boating; —, no acreage affected.

Source: NOAA, *The 1990 National Shellfish Register of Classified Estuarine Waters,* National Oceanic and Atmospheric Administration, Rockville, MD, 1991.

TABLE 3.11
Middle Atlantic Pollution Sources Affecting Harvest-Limited Acreage, 1990[a,b]

Sources	Massachusetts		Rhode Island		Connecticut		New York		New Jersey		Delaware		Maryland		Virginia	
	Acres	%	Acres	%	Acres	%	Acres	%	Acres	%	Acres	%	Acres	%	Acres	%
Point sources																
Sewage treat plants	10	11	23	55	78	68	212	79	109	67	14	23	16	13	179	68
Combined sewers	4	5	7	17	26	23	135	50	52	32	0	0	0	0	0	0
Direct discharge	0	0	9	21	7	6	68	25	0	0	0	0	0	0	0	0
Industry	0	0	6	14	8	7	1	<1	32	20	3	5	6	5	167	63
Nonpoint sources																
Septic systems	8	9	2	5	7	6	11	4	34	21	4	7	32	26	25	9
Urban runoff	11	13	7	17	61	54	250	93	121	74	5	8	38	31	162	61
Agricultural runoff	0	0	1	2	2	2	5	2	23	14	11	18	60	49	28	11
Wildlife	8	9	0	0	5	4	11	4	32	20	15	25	40	33	1	<1
Boats	7	8	16	38	48	42	32	12	62	38	0	0	15	12	173	66
Upstream sources																
Sewage treat plants	11	13	11	26	51	45	0	0	5	3	0	0	0	0	26	10
Combined sewer	0	0	0	0	3	<1	0	0	2	<1	0	0	0	0	<1	<1
Urban runoff	10	11	17	40	9	8	0	0	5	3	0	0	5	4	26	10
Agricultural runoff	0	0	0	0	0	0	0	0	1	1	0	0	0	0	0	0
Wildlife	10	11	0	0	2	2	0	0	0	0	0	0	0	0	16	6

[a] Acres are times 1000; % is percent of all harvest-limited acreage in state.
[b] Since the same percentage of a shellfish area can be affected by more than one source, the percentages shown above cannot be added. They will not sum to 100.

Source: NOAA, *The 1990 National Shellfish Register of Classified Estuarine Waters,* National Oceanic and Atmospheric Administration, Rockville, MD, 1991.

TABLE 3.12
South Atlantic Pollution Sources Affecting Harvest-Limited Acreage, 1990[a,b]

	North Carolina		South Carolina		Georgia		Florida	
Sources	**Acres**	**%**	**Acres**	**%**	**Acres**	**%**	**Acres**	**%**
Point sources								
Sewage treat plants	167	35	47	54	38	31	122	73
Combined sewers	0	0	0	0	0	0	0	0
Direct discharge	0	0	0	0	5	4	0	0
Industry	83	17	46	53	43	36	8	5
Nonpoint sources								
Septic systems	57	12	22	25	48	40	161	96
Urban runoff	77	16	39	45	34	28	140	84
Agricultural runoff	222	47	3	3	8	7	0	0
Wildlife	149	31	17	20	42	35	98	59
Boats	64	13	30	34	37	31	15	9
Upstream sources								
Sewage treat plants	0	0	7	8	2	2	0	0
Combined sewer	0	0	0	0	0	0	0	0
Urban runoff	0	0	6	7	2	2	0	0
Agricultural runoff	0	0	0	0	0	0	0	0
Wildlife	0	0	19	22	16	13	0	0

[a] Acres are times 1000; % is percent of all harvest-limited acreage in state.
[b] Since the same percentage of a shellfish area can be affected by more than one source, the percentages shown above cannot be added. They will not sum to 100.

Source: NOAA, *The 1990 National Shellfish Register of Classified Estuarine Waters*, National Oceanic and Atmospheric Administration, Rockville, MD, 1991.

TABLE 3.13
Gulf of Mexico Pollution Sources Affecting Harvest-Limited Acreage, 1990[a,b]

	Florida		Alabama		Mississippi		Louisiana		Texas	
Sources	**Acres**	**%**	**Acres**	**%**	**Acres**	**%**	**Acres**	**%**	**Acres**	**%**
Point sources										
Sewage treat plants	394	45	86	27	27	17	265	18	201	24
Combined sewers	7	1	0	0	0	0	204	14	0	0
Direct discharge	2	<1	5	2	0	0	912	60	1	<1
Industry	205	24	0	0	39	25	218	14	60	7
Nonpoint sources										
Septic systems	697	80	0	0	15	10	580	38	471	56
Urban runoff	466	54	0	0	32	20	643	43	135	16
Agricultural runoff	4	<1	18	6	0	0	59	4	220	26
Wildlife	528	61	41	13	8	5	415	28	123	15
Boats	64	7	1	<1	94	60	225	15	123	15
Upstream sources										
Sewage treat plants	131	15	2	1	3	2	1038	69	0	0
Combined sewer	7	<1	3	<1	0	0	13	<1	114	3
Urban runoff	7	<1	211	67	3	2	562	37	10	1
Agricultural runoff	0	0	211	67	0	0	3	<1	221	26
Wildlife	141	16	0	0	0	0	3	<1	66	8

[a] Acres are times 1000; % is percent of all harvest-limited acreage in state.

[b] Since the same percentage of a shellfish area can be affected by more than one source, the percentages shown above cannot be added. They will not sum to 100.

Source: NOAA, *The 1990 National Shellfish Register of Classified Estuarine Waters,* National Oceanic and Atmospheric Administration, Rockville, MD, 1991.

TABLE 3.14
Pacific Pollution Sources Affecting Harvest-Limited Acreage, 1990[a,b]

	California		Oregon		Washington		Alaska		Hawaii	
Sources	**Acres**	**%**	**Acres**	**%**	**Acres**	**%**	**Acres**	**%**	**Acres**	**%**
Point sources										
Sewage treat plants	16	13	5	18	53	40	0	0	1	6
Combined sewers	0	0	0	0	0	0	0	0	0	0
Direct discharge	0	0	6	21	0	0	0	0	0	0
Industry	86	68	0	0	37	28	0	0	6	33
Nonpoint sources										
Septic systems	11	9	9	32	37	28	0	0	0	0
Urban runoff	26	20	12	43	54	41	0	0	18	100
Agricultural runoff	18	14	8	29	15	11	0	0	0	0
Wildlife	18	14	0	0	4	3	0	0	17	94
Boats	25	20	6	21	10	8	0	0	6	33
Upstream sources										
Sewage treat plants	0	0	2	7	43	33	0	0	0	0
Combined sewer	0	0	0	0	0	0	0	0	0	0
Urban runoff	0	0	0	0	43	33	0	0	0	0
Agricultural runoff	0	0	0	0	0	0	0	0	0	0
Wildlife	0	0	0	0	0	0	0	0	0	0

[a] Acres are times 1000; % is percent of all harvest-limited acreage in state.

[b] Since the same percentage of a shellfish area can be affected by more than one source, the percentages shown above cannot be added. They will not sum to 100.

Source: NOAA, *The 1990 National Shellfish Register of Classified Estuarine Waters,* National Oceanic and Atmospheric Administration, Rockville, MD, 1991.

TABLE 3.15
Summary of Pollutant Concentration Susceptibility in Estuaries

Most Susceptible Systems	Least Susceptible Systems
General population	General population
Brazos River**	Willapa Bay**
Ten Thousand Islands	St. Catherines/Sapelo Sound**
San Pedro Bay**	Penobscot Bay**
North-South Santee Rivers**	Humboldt Bay**
Galveston Bay**	Broad River**
Suisun Bay**	Hood Canal**
Sabine Lake**	Coos Bay**
St. Johns River*	Casco Bay**
Apalachicola Bay	Grays Harbor
San Antonio Bay**	Chincoteague Bay*
Connecticut River*	Bogue Sound**
Great South Bay*	St. Andrew/St. Simons Sound*
Merrimack River	Sheepscot Bay*
Atchafalaya/Vermillion Bays*	Apalachee Bay
Matagorda Bay*	Rappahannock River
Heavy industry	St. Helena Sound
Brazos River**	Puget Sound*
North/South Santee Rivers**	Heavy industry
Galveston Bay**	St. Catherines/Sapelo Sound**
Sabine Lake**	Hood Canal**
San Pedro Bay**	Penobscot Bay**
Connecticut River*	Casco Bay**
Calcasieu Lake	Humboldt Bay**
Hudson River/Raritan Bay	Buzzards Bay
Charleston Harbor	Boston Bay
Perdido Bay	Coos Bay**
Potomac River	Broad River**
San Antonio Bay**	Willapa Bay**
Mobile Bay	Bogue Sound**
Suisun Bay**	Puget Sound*
Great South Bay*	Narragansett Bay
Baffin Bay	Santa Monica Bay
Chesapeake Bay	Saco Bay
Agricultural activities	St. Andrew/St. Simons Sound*
Brazos River**	Agricultural activities
Suisun Bay**	Humboldt Bay**
North/South Santee Rivers**	Hood Canal**
St. Johns River*	Penobscot Bay**
Matagorda Bay*	Coos Bay**
Atchafalaya/Vermillion Bays*	St. Catherines/Sapelo Sound**
San Pedro Bay**	Chincoteague Bay*
Sabine Lake**	Bogue Sound**
Corpus Christi Bay	Long Island Sound
Galveston Bay**	Casco Bay**
San Antonio Bay**	Willapa Bay**
Winyah Bay	Broad River**
Albemarle Sound	Sheepscot Bay*
Neuse River	Klamath River
Laguna Madre	

*,** Systems that are present in all three categories are marked with two asterisks; systems present in two categories are marked by one asterisk.

Source: Biggs, R. B., DeMoss, T. B., Carter, M. M., and Beasley, E. L., *Rev. Aquat. Sci.*, 1, 203, 1989. With permission.

FIGURE 3.9. Map showing the location of the sewage sludge dumpsite in the New York Bight Apex. Inset shows concentrations of coliforms in sediments. (From Studholme, A. L., Ingham, M. C., and Pacheco, A. (Eds.), *Response of the Habit and Biota of the Inner New York Bight to Abatement of Sewage Sludge Dumping*, Third Annual Progress Report, NOAA Tech. Mem. NMFS-F/NEC-82, Woods Hole, MA, 1991.)

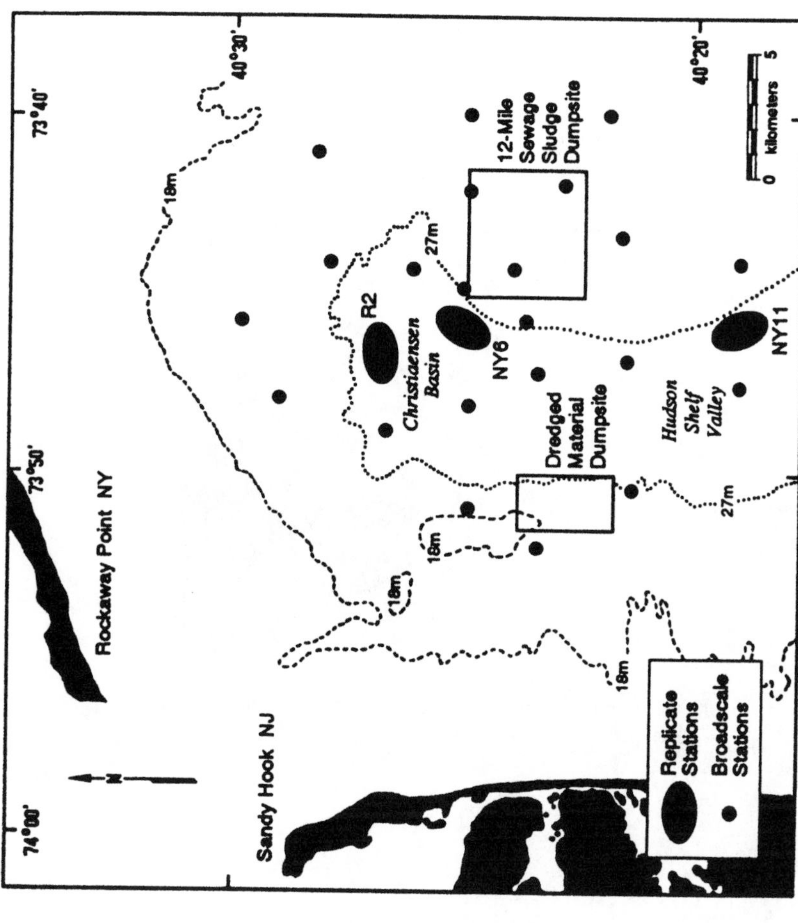

FIGURE 3.10. Major bathymetric features of the New York Bight Apex showing the location of the 12-mile sewage sludge dumpsite relative to the dredged material dumpsite. (From Draxler, A. F. J., in *Effects of the Cessation of Sewage Sludge Dumping at the 12-Mile Site*, Studholme, A. L., O'Reilly, J. E., and Ingham, M. C. (Eds.), NOAA Tech. Rep. NMFS 124, U.S. Department of Commerce, Seattle, WA, 1995, 133.)

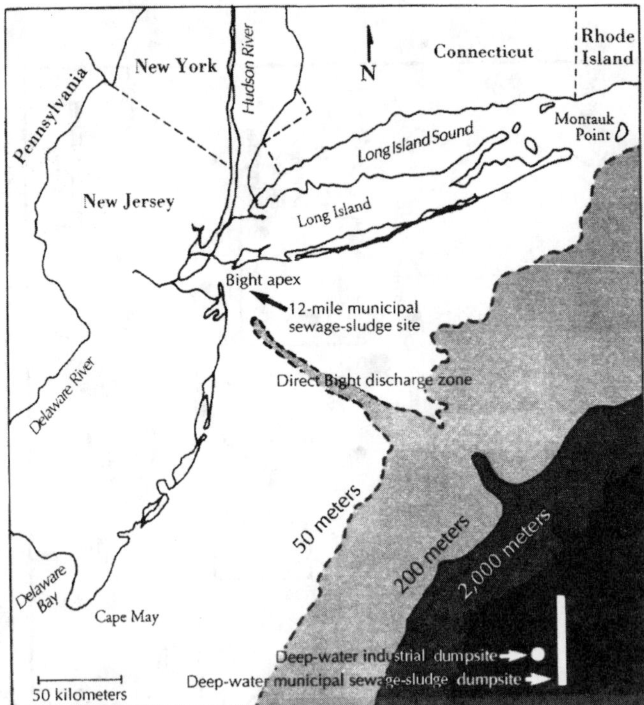

FIGURE 3.11. Map showing the location of the deep water dumpsites (lower right). (From Stegeman, J. J., *Oceanus,* 33, 54, 1990. With permission.)

APPENDIX 4. OXYGEN DEPLETION

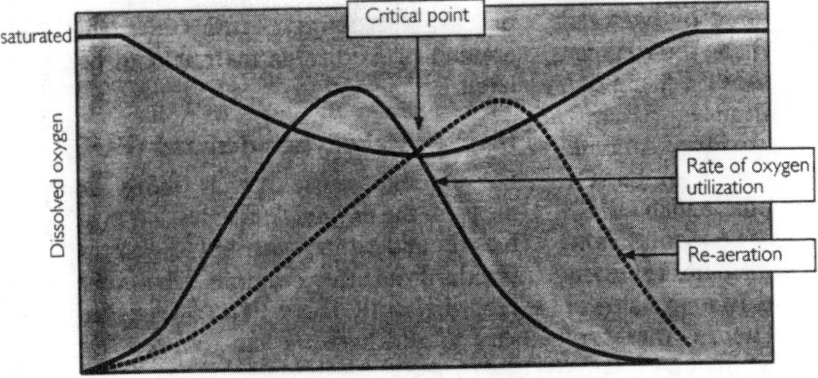

FIGURE 4.1. Oxygen budget for a river or body of water receiving organic wastes. The horizontal axis represents time for a single input of waste to a static body of water or distance downstream from a continuous input to a river. (From Clark, R. B., *Marine Pollution,* 3rd ed., Clarendon Press, Oxford, 1992. With permission.)

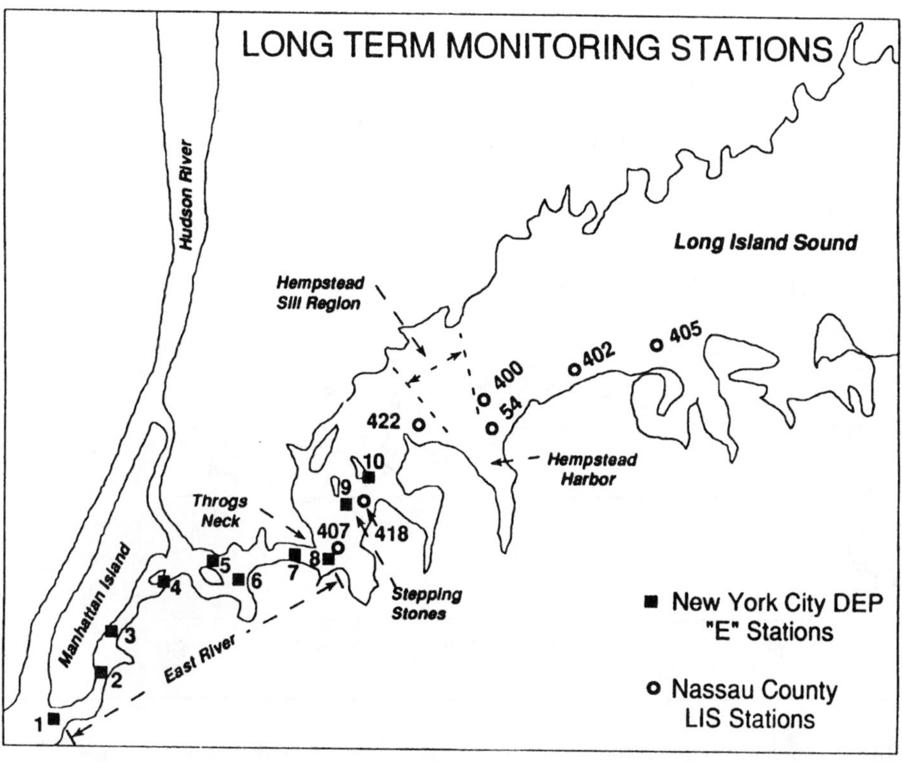

FIGURE 4.2. Long-term monitoring stations for dissolved oxygen in Long Island Sound. (From Parker, C. A. and O'Reilly, J. E., *Estuaries,* 14, 248, 1991. With permission.)

TABLE 4.1
Analysis for Temporal Trends in Bottom Dissolved Oxygen in the East River and Western Extremes of Long Island Sound Over the Period 1970–1990

		Summer Minimum				Summer Average			
Sta.	n	Slope	SE	Prob > F	R^2	Slope	SE	Prob > F	R^2
E1	21	0.084	0.019	0.0003	0.50	0.092	0.016	0.0001	0.63
E2	21	0.088	0.020	0.0004	0.50	0.091	0.018	0.0001	0.57
E3	21	0.083	0.019	0.0004	0.50	0.098	0.016	0.0001	0.67
E4	21	0.074	0.012	0.0001	0.68	0.083	0.015	0.0001	0.60
E5	21	0.054	0.017	0.0044	0.35	0.054	0.018	0.0081	0.32
E6	21	0.057	0.017	0.0034	0.37	0.084	0.015	0.0001	0.62
E7	21	0.019	0.025	0.4578	0.03	0.065	0.019	0.0029	0.38
E8	21	–0.027	0.033	0.4242	0.03	0.023	0.020	0.2527	0.07
E9	21	–0.065	0.039	0.1080	0.13	–0.019	0.026	0.4743	0.03
E10	21	–0.060	0.037	0.1166	0.12	–0.017	0.022	0.4324	0.03

Note: The slopes (mg l^{-1} yr^{-1}) and associated statistics were derived by linear regressions of DO against year (NYC-DEP data).

Source: Parker, C. A. and O'Reilly, J. E., *Estuaries,* 14, 248, 1991. With permission.

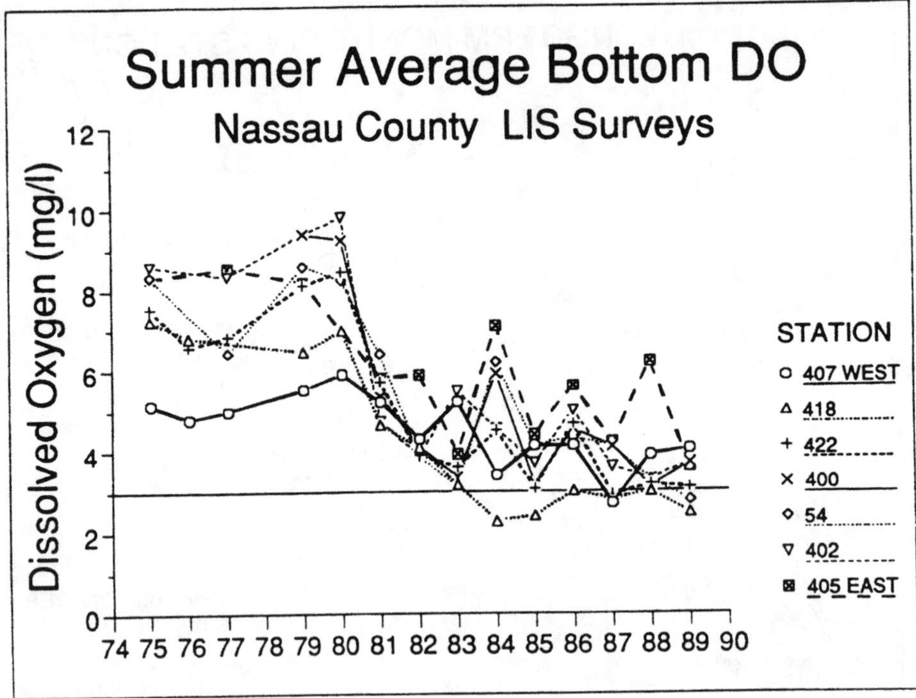

FIGURE 4.3. Summer average bottom dissolved oxygen values recorded at Nassau County Long Island Sound monitoring stations between 1975 and 1989. (From Parker, C. A. and O'Reilly, J. E., *Estuaries,* 14, 248, 1991. With permission.)

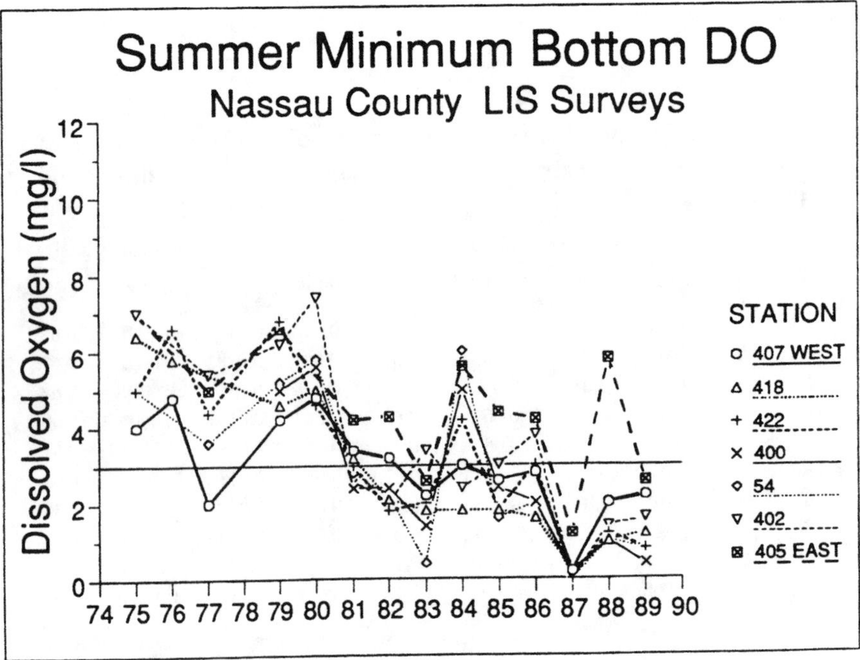

FIGURE 4.4. Summer minimum bottom dissolved oxygen values recorded at Nassau County Long Island Sound monitoring stations between 1975 and 1989. (From Parker, C. A. and O'Reilly, J. E., *Estuaries,* 14, 248, 1991. With permission.)

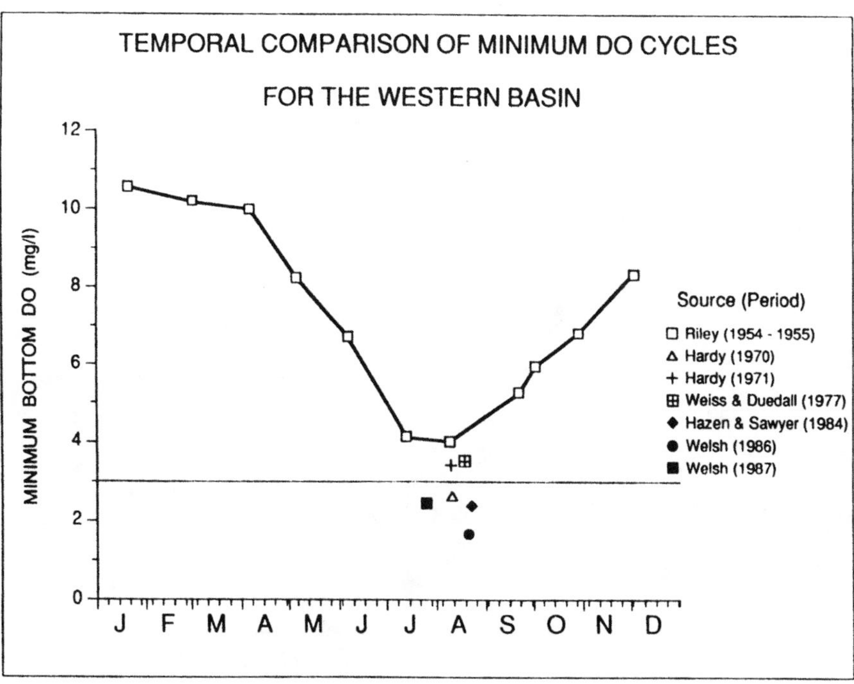

FIGURE 4.5. Temporal comparison of minimum bottom dissolved oxygen conditions for the western basin of Long Island Sound. (From Parker, C. A. and O'Reilly, J. E., *Estuaries,* 14, 248, 1991. With permission.)

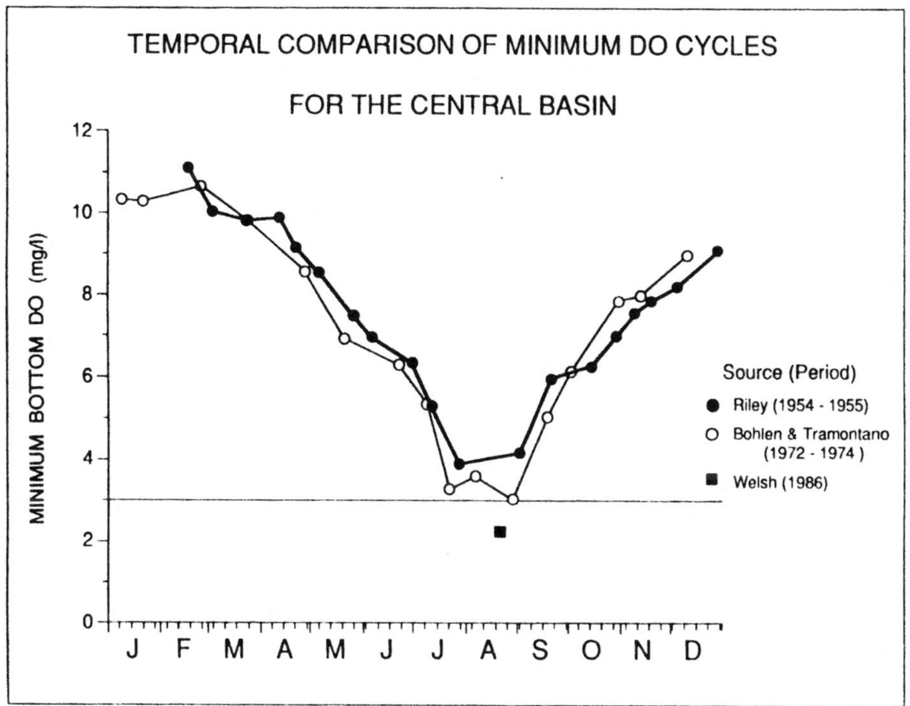

FIGURE 4.6. Temporal comparison of minimum bottom dissolved oxygen conditions for the central basin of Long Island Sound. (From Parker, C. A. and O'Reilly, J. E., *Estuaries,* 14, 248, 1991. With permission.)

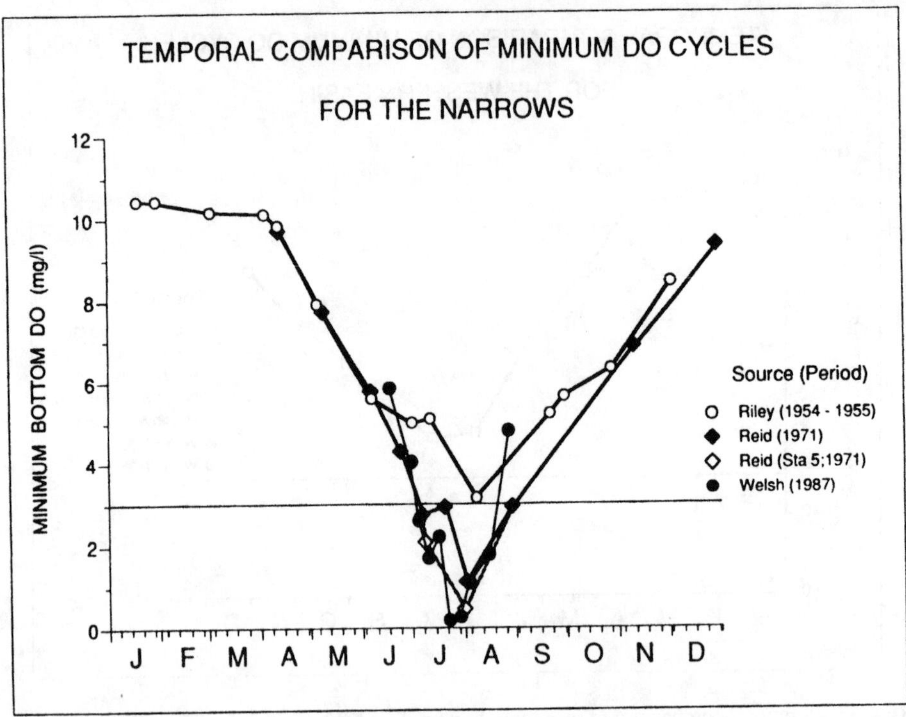

FIGURE 4.7. Temporal comparison of minimum bottom dissolved oxygen conditions for the Narrows of Long Island Sound. (From Parker, C. A. and O'Reilly, J. E., *Estuaries,* 14, 248, 1991. With permission.)

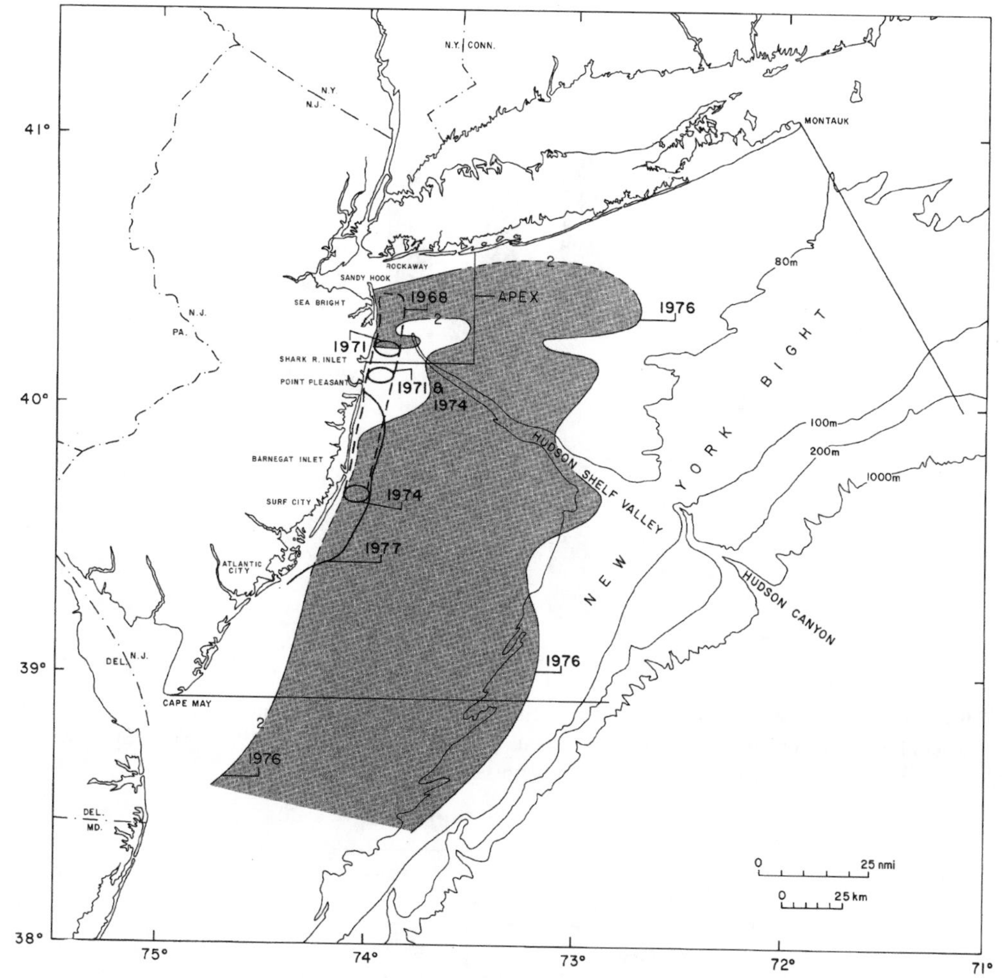

FIGURE 4.8. Approximate regions of oxygen-depleted (<2 ml/l) bottom waters in the New York Bight in 1968, 1971, 1974, 1976, and 1977. (From O'Connor, J. S., A perspective on natural and human factors, MESA New York Bight, State University of New York, Stony Brook, 1976. With permission.)

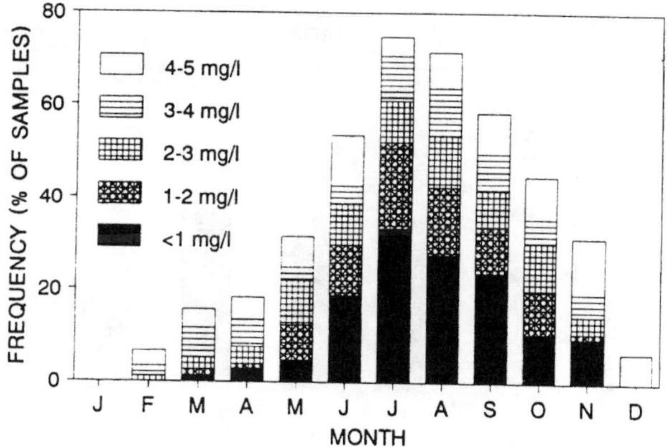

FIGURE 4.9. Frequency of five dissolved oxygen concentration ranges recorded at four monitoring stations in the Pamlico River estuary for the period 1975 to 1989. (From Stanley, D. W. and Nixon, S. W., *Estuaries,* 15, 270, 1992. With permission.)

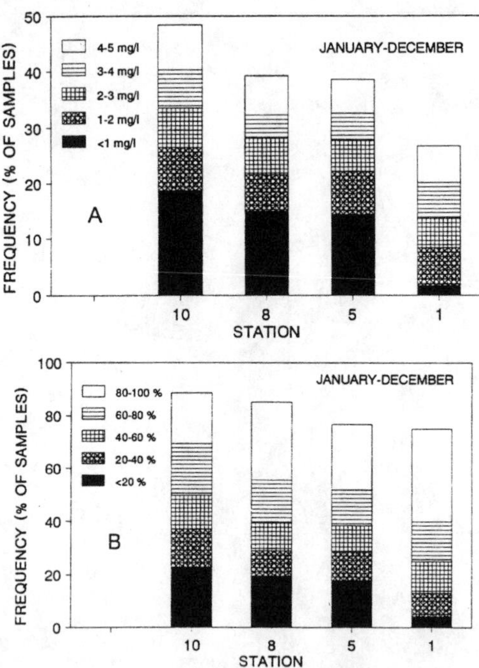

FIGURE 4.10. Frequency of five dissolved oxygen ranges (A) and percent saturation ranges (B) for each of four monitoring stations in the Pamlico River estuary for the period 1975 to 1989. (From Stanley, D. W. and Nixon, S. W., *Estuaries,* 15, 270, 1992. With permission.)

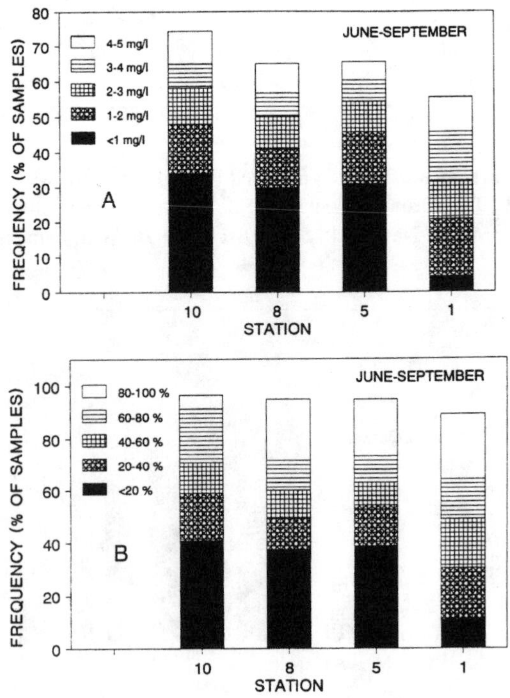

FIGURE 4.11. Frequency of five dissolved oxygen ranges (A) and percent saturation ranges (B) for each of four monitoring stations in the Pamlico River estuary for the period 1975 to 1989 (June to September data only). (From Stanley, D. W. and Nixon, S. W., *Estuaries,* 15, 270, 1992. With permission.)

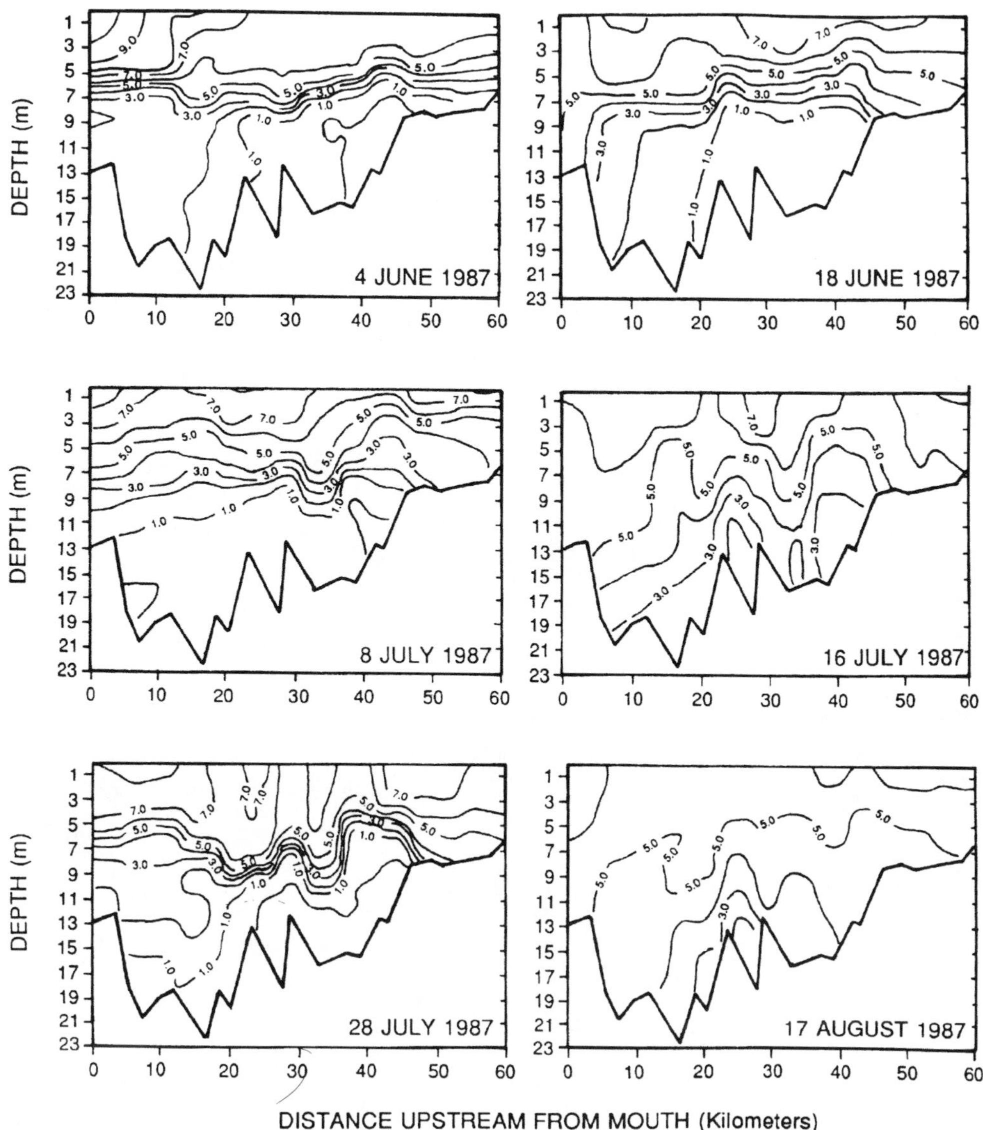

FIGURE 4.12. Dissolved oxygen concentrations (mg/l) during June, July, and August 1987 in the lower Rappahannock River. (From Kuo, A. Y., Park, K., and Moustafa, M. Z., *Estuaries,* 14, 113, 1991. With permission.)

3 Oil Pollution

INTRODUCTION

Acute and insidious environmental problems are associated with the presence of petroleum hydrocarbons in the marine hydrosphere. The sensitivity of estuarine and marine communities to oil contamination is well established.[1-3] Major changes in the structure of these communities due to the effects of polluting oil may persist from days to years. The oil can severely impact faunal and floral populations by physical (smothering, reduced light), habitat (altered pH, decreased dissolved oxygen, decreased food availability), and toxic actions.[4] In addition, chemical dispersants, solvents, and other agents used to clean up oil spills may be detrimental to estuarine and marine life.[5]

Oil pollution in the sea arises from multiple sources, notably spills from oil tankers, discharges during marine transportation, leakages from drilling operations, discharges from coastal refineries and marine terminals, municipal and industrial discharges, urban and river runoff, atmospheric deposition, and natural seeps.[6,7] Of the oil entering the sea from anthropogenic activities, most (65.2%) originates from discharges of municipal and industrial wastes, urban and river runoff, ocean dumping, and atmospheric fallout. An additional 26.2% of the oil derives from discharges related to transportation (e.g., tanker accidents, deballasting, dry docking). Only about 8.5% of the anthropogenic input is attributable to releases from fixed installations (e.g., coastal refineries, offshore production facilities, marine terminals). The total oceanic input of petroleum hydrocarbons from man's activities amounts to approximately 2.115 million tons (mt)/yr, which far exceeds that from natural sources (0.25 mt/yr) such as oil seeps.[6]

The world crude oil production is approximately 3 billion tons (bt)/yr, with roughly 50% of this production transported by sea.[6] The amount of crude oil lost in the sea has decreased significantly during the last 15 years owing to improved methods of deballasting and reductions in accidental oil spills. Hence, the amount of crude oil that entered the sea between 1978 and 1983 was approximately four times greater than that between 1984 and 1990.[8] Nearly 250 mt of oil were accidently spilled at sea in 1989.[9]

About 75% of the accidental oil spills in the U.S. occurs in coastal areas (primarily estuaries, enclosed bays, and wetlands), and most of the chronic oil pollution is associated with routine operations of oil refineries and installations, as well as discharges of municipal and industrial wastes.[3,10] An estimated 33% of the entire input of oil to the marine hydrosphere enters estuaries.[11] Based on the data compiled on petroleum intentionally or unintentionally released to estuarine and marine waters, oil inputs originating from the users of petroleum products far surpass those from the oil extraction and transport industries. Since petroleum hydrocarbon compounds are hydrophobic contaminants that readily sorb to fine particulate matter suspended in estuarine and inshore waters, they tend to accumulate in bottom sediments of these coastal ecosystems.[12,13]

COMPOSITION AND CHARACTERISTICS

Crude oils, consisting of a complex mixture of hydrocarbon and nonhydrocarbon compounds, vary widely in chemical composition and physical properties. Hundreds of hydrocarbon and nonhydrocarbon compounds are present in any particular grade of crude oil. While hydrocarbons comprise more than 75% by weight of most crude oils, nonhydrocarbons (compounds containing oxygen, nitrogen, sulfur, and metals such as copper, iron, nickel, and vanadium) can predominate in heavy crude oils.

Four major classes of hydrocarbons occur in crude oils depending on how the atoms bind together in the molecules: (1) straight-chain alkanes (*n*-alkanes or *n*-paraffins), (2) branched alkanes (isoparaffins), (3) cycloalkanes (cycloparaffins), and (4) aromatics. Although the alkenes (olefins) are common in many petroleum products as a result of the refining process, they are relatively rare in crude oils. Low-molecular-weight compounds of each class dominate in crude oils.[4] Besides hydrocarbons, sulfur compounds are important components. The total amount of sulfur in crude oils may amount to 5% by weight. The oxygen and nitrogen content may reach 2% and 1% by weight, respectively. Sulfur, oxygen, and nitrogen are found primarily in the heterocyclic compounds.[11]

Aside from their chemical composition, crude oils can be classified according to certain physical properties. For example, they are often classified as light, medium, or heavy based on specific gravity values, which range from 0.73 to 0.95. The physical properties and chemical composition of crude oil often differ dramatically from one oil field to another.

OIL TOXICITY

Crude oils contain a wide array of toxic substances such as benzene, toluene, xylene, and other low-molecular-weight aromatics, in addition to acids, phenols, sulfur compounds (e.g., sulfides, thiols, and thiophenes), and PAHs (e.g., 1,2-benzanthracene; 3,4-benz(a)pyrene; 1,2-benzphenanthrene; diphenylmethane; fluorene; and phenanthrene). Among the hydrocarbon constituents, the smaller molecules tend to be most toxic. When proceeding along the series paraffins, naphthenes and olefins to aromatics, the toxicity rises. The light aromatic fractions, although most toxic, also evaporate most rapidly. Thus, during the first 24 to 48 hours of an oil spill at sea the lower-molecular-weight volatile components are readily lost from the oil. This evaporative loss of the volatile hydrocarbons substantially lowers the overall toxicity of crude oils in seawater, thereby reducing their potential impacts on marine communities and habitats.

ENVIRONMENTAL FATE

The form and behavior of oil spilled at sea are contingent upon its composition, as well as various abiotic (e.g., water temperature, wind, tides, wave action, currents) and biotic (e.g., microbial activity) factors. The oil spreads quickly over the sea surface, often covering extensive areas as slicks varying from micrometers to a centimeter or more in thickness. Oil slicks travel downwind at 3 to 4% of the wind speed, spreading at a rate dependent on the water temperature and composition of the oil. Light oils spread faster than heavy oils.[6] Three distinct spreading phases are recognized: (1) an initial phase in which gravitational and inertial resistance forces control the spreading, (2) an intermediate phase in which gravitational and viscous drag forces predominate, and (3) a final phase in which surface tension and viscous drag forces play key roles.[14]

Several physical-chemical processes — evaporation, photochemical oxidation, emulsification, and dissolution — initiate changes in the composition of oil on the sea surface within hours of a spill. Low-molecular-weight volatile fractions evaporate, hydrocarbons undergo photooxidation, water soluble constituents dissolve in seawater, and immiscible components become emulsified. The lighter, more toxic and volatile components of the oil are lost relatively rapidly due to evaporation and dissolution. Evaporative effects are greatest for light crude oils and refined products (No. 2 fuel oil, gasoline), amounting to as much as 75% of the spilled oil for these fractions. In contrast, evaporation amounts to only about 10% of the spilled oil for heavy crudes and refined products (No. 6 fuel oil). Still smaller losses (~5%) are ascribable to dissolution in seawater, primarily the low-molecular-weight aromatics and polar nonhydrocarbons. Degradation of the oil is accelerated by photooxidation, which converts some of the hydrocarbons to polar oxidized compounds. The high-molecular-weight aromatic hydrocarbons are particularly susceptible to this type of alteration.[4]

Advective and turbulent diffusive processes coupled to wave and current activity break up large spills into smaller patches of oil which are then transported away from the primary spill site. Horizontal eddy diffusion effectively separates the oil patches, facilitating their movement. Oil-in-water emulsions or water-in-oil emulsions form as agitation provided by waves and currents mix the oil and seawater. A 50 to 80% water-in-oil emulsion may generate viscid pancake-like masses termed "chocolate mousse". Tar balls measuring 1 mm to 25 cm in diameter comprise the heaviest residues of the crude oil. The formation of water-in-oil emulsions can exacerbate the impact of polluting oil. Mousse formation, for example, enables oil to persist for months at sea, where it can be transported many kilometers from the spill site, endangering communities of organisms and sensitive habitat areas in remote regions. In addition, mousse and tar balls degrade extremely slowly in marine waters, which increases the probability of the beaching of these degradation products. The stability of water-in-oil emulsions is related to the chemical components in nonvolatile residues, namely the simultaneous presence of asphaltenes and paraffins.[15]

As the density of spilled oil approaches that of seawater, it tends to sink. Several processes increase oil density leading to its settlement to the sea floor. These include: (1) evaporation and dissolution of lower-molecular-weight compounds, (2) photooxidation and subsequent dissolution of oxygenated materials, (3) formation of viscous and higher density water-in-oil emulsions, (4) incorporation of particulates and the agglomeration of oil particulate mixtures, and (5) incorporation of higher density microorganisms and even macroorganisms (e.g., barnacles).[16] The density of the parent oil, which can vary appreciably, has much to do with its capacity to sink. Heavy crude oils and refined products (e.g., No. 6 fuel oil) have greater densities than other fractions and, therefore, tend to sink more rapidly, even without much degradation. In general, however, substantial alteration occurs prior to the settlement of oil products.

Sedimentation of oil is facilitated by the sorption of hydrocarbons to particulate matter suspended in the water column. Since large sediment loads typically occur in estuaries and shallow embayments, this process predominates in these coastal environments. Adsorption of oil onto suspended particulate matter results in a high specific gravity mixture more than twice that of seawater alone (1.025 g/cm^3). The high specific gravity of these particles causes deposition of the sorbed oil.

Microbes (bacteria, yeasts, filamentous fungi) play a pivotal role in the degradation of crude oil, often being the dominant factor controlling the fate of toxic hydrocarbons in aquatic environments.[17] Collectively, they can degrade as much as 40 to 80% of a crude oil spill.[4] Marine bacteria are the most important biological agents in the breakdown process since they are capable of degrading all component compounds in the oil, albeit at varying rates. Neilson[18] describes the pathways of microbial degradation of oil components. Light and structurally simple hydrocarbon and nonhydrocarbon compounds degrade most readily. The heavy and more complex constituents of the oil resist microbial attack and ultimately settle to the sea floor. Tar balls, consisting of high-molecular-weight compounds, degrade very slowly.

In addition to the composition of the oil, several other factors influence the biodegradation rates, most prominently water temperature, nutrient availability, oxygen levels, and salinity. Higher water temperatures accelerate the rate of biodegradation, as do greater concentrations of nutrients. Anaerobic conditions severely restrict microbial degradation.[19] The optimal salinity range of most marine bacteria is 25 to 35 ppt.

Microbial breakdown of oil is most rapid in shallow nearshore environments where an abundant supply of nutrients and organic substrates fosters the proliferation of dense microbial populations. While the residence time of oil in the water column is typically less than 6 months, it may exceed 10 years in the bottom sediments of sheltered estuaries and wetlands.[20,21] Polluting oil generally persists for much longer periods of time along shorelines in high latitude regions than along shorelines of temperate and tropical zones due to reduced temperatures. In the deep sea, microbial action on petroleum hydrocarbons declines substantially at depths below 1000 m.[22] Oil retention times in these extreme environments can be considerable.

BIOLOGICAL EFFECTS

Marine fauna and flora are highly susceptible to the toxic effects of oil and oil fractions. Most aquatic organisms readily absorb aliphatic and polycyclic aromatic hydrocarbon fractions of dissolved petroleum. These fractions are highly lipid soluble and tend to bioconcentrate in fish and shellfish, thereby posing an immediate threat to fisheries resources.[23] Since these compounds also accumulate in sediments, they pose a chronic threat especially to benthic organisms long after spill events.[24]

Estuarine and marine organisms assimilate hydrocarbons by active uptake of dissolved or dispersed substances; by ingestion of petroleum-sorbed particles, including live and dead organic matter; and by the drinking or gulping of water containing the compounds, as in the case of fish. Algae can assimilate and metabolize hydrocarbons.[4,25] Marine animals exhibit variable capacities to metabolize petroleum hydrocarbons. Many invertebrates (e.g., crustaceans, polychaetes, echinoderms), fish, mammals, and birds metabolize and excrete at least some of the hydrocarbons ingested during feeding, grooming, and respiration.[4] Once ingested, hydrocarbon compounds may follow one of three general pathways in these animals: the compounds may be metabolized, stored with possible elimination at a later time, or excreted unchanged. Fish and some crustaceans, for example, have well developed mixed function oxygenase (MFO) capability and, hence, do not accumulate hydrocarbons to any great extent, except perhaps in heavily polluted areas. In contrast, bivalve mollusks, which have poorly developed MFO systems, tend to accumulate hydrocarbon compounds quite readily and depurate them slowly.[25,26] Some zooplankton likewise metabolize oil very poorly or not at all. Mammals, reptiles, and birds have well developed MFO capability, but the database on their responsiveness to petroleum hydrocarbons is not substantial. In general, those organisms with high lipid contents and poor MFO systems sequester significant concentrations of hydrocarbon compounds.[26]

The severity of oil impacts on estuarine and marine organisms depends on many factors, the most notable being (1) the amount of the oil; (2) composition of the oil; (3) form of the oil (i.e., fresh, weathered, or emulsified); (4) occurrence of the oil (i.e., in solution, suspension, dispersion, or adsorbed onto particulate matter); (5) duration of exposure; (6) involvement of neuston, plankton, nekton, or benthos in the spill or release; (7) juvenile or adult forms involved; (8) previous history of pollutant exposure of the biota; (9) season of the year; (10) natural environmental stresses associated with fluctuations in temperature, salinity, and other variables; (11) type of habitat affected; and (12) cleanup operations (e.g., physical methods of oil recovery and the use of chemical dispersants).[1,5,6,10,11,14,27-29] Lethal and sublethal effects of oil contamination are manifested in both acute and chronic responses of biota. Organisms which are trapped, smothered, and suffocated by an oil spill, for example, suffer essentially immediate lethal effects. The oil typically interferes with cellular and subcellular processes in these individuals soon after the spill. Those individuals surviving the physical impact of the oil may lose normal physiological or behavorial function if coated, thus predisposing them to greater long-term risk of death. Sublethal effects, such as impairment of the ability of the organisms to obtain food or to escape from predators after being coated by the oil, likewise contribute to increased mortality within days or weeks of a spill. Sublethal doses of toxins to eggs and juveniles commonly contribute to even greater long-term damage to populations. Sublethality can be as devastating as lethal effects to a community by adversely affecting reproduction, growth, distribution, and behavior of organisms resulting in gradual shifts in species composition, abundance, and diversity that can persist for many years in an area.

Long-term adverse effects of polluting oil on estuarine and marine communities often arise indirectly via habitat degradation. In this regard, coastal shoreline and estuarine habitats are assigned the highest priority of biological protection, and offshore areas the lowest priority.[2] The coastal regions of the world are critically important not only for their fisheries and energy resources, but also for their value in recreation, transportation, and tourism.[30] Salt marshes, seagrasses, mangroves, mudflats, and other low-energy habitats tend to trap oil. Suspended clay and silt in these habitats sorb hydrocarbons and other oil components. Once the oil settles with these particulates to the sea floor and permeates through the bottom sediments, it commonly persists for years, creating a long-term problem in regard to the stability and health of benthic communities. Oil accumulating as a layer on the sediment surface has a much smaller effective surface area than oil dispersed in the water column, which consequently reduces the rate of microbial degradation.[11] Low oxygen concentrations in deeper sediment layers also hinder bacterial degradation of the oil, further delaying the overall recovery of the habitat. The contaminants may then be re-released years later, especially during storms and other major episodic events that roil bottom sediments. Hence, these habitats may remain inhospitable to many organisms for long periods of time, perhaps a decade or more.[6,31-35]

PLANTS

Phytoplankton, because of differences in species sensitivity to hydrocarbons, show a wide variation in community response to oil spills.[4] Laboratory studies have revealed that the effects of oil on natural phytoplankton assemblages are not only dependent on the species investigated, but also on temperature, light intensity, nutritional physiology of the algae, and the type of oil.[36] The lack of mobility and generally close proximity of some phytoplankton to oil floating on the sea surface can have hazardous consequences; however, the high turnover rates and patchy distribution of phytoplankton tend to mitigate these factors, and most impacts are very transient. As a result, phytoplankton populations typically recover quickly from a surface slick of oil and usually return rapidly to population levels that existed prior to a spill.

While high concentrations of petroleum hydrocarbons are toxic to nearly all phytoplankton, low concentrations may actually increase primary production.[37] Petroluem hydrocarbon concentrations below 50 ng/g purportedly enhance photosynthesis by algal populations in culture. Above 50 ng/g, however, photosynthesis by algae is progressively depressed.[6] Enhanced or reduced biomass and photosynthetic activity of natural phytoplankton communities exposed to polluting oil are reported elsewhere.[38,39] The concentration of oil necessary to cause death of phytoplankton ranges from about 1 to 10^{-4} ml/l.[40]

Salt marsh, seagrass, and mangrove biotopes are particularly vulnerable to oil pollution. Once oil penetrates deeply into the bottom sediments of these plant systems, extending below the level of the roots and rhizomes, significant damage usually ensues. Recovery of the plants is quite rapid when oil does not penetrate to the roots and rhizomes. Destruction of salt marsh, seagrass, and mangrove vegetation due to oil spills is well chronicled.[2,4,41-45] Most local plant populations require several weeks to about 5 years to recover from the effects of oil spills.[4] Repeated oilings are much more devastating than single oil spillages, with spill events during the winter causing more damage than those during the summer.

Annual salt marsh plants tend to be much less resilient to polluting oil than perennials. They typically require two or three years to recover from the toxic effects of petroleum hydrocarbons. Although large numbers of perennial plants can also be lost after a single oil spill, responses vary widely among species. The most susceptible forms are shallow-rooted plants with little or no food reserves (e.g., *Salicornia* spp. and *Sueda maritima*). Those plants with large food reserves have lower mortality, and they recover more readily from repeated oilings.

Seagrasses likewise exhibit a range of reactions to oil. In the subtropical Atlantic, for example, the seagrass *Syringodium filiforme* is the least tolerant form to high concentrations of oil; *Thalassia testudium*, the most tolerant form. *Halodule wrightii* appears to be slightly more tolerant to oil than *S. filiforme*. Based on comparative Atlantic-Pacific research, cogenetors of the Indo-Pacific are believed to behave similarly.[2]

Mangroves are especially susceptible to oil spills. Once petroleum hydrocarbons penetrate into sediments of the tropical intertidal zone, mangroves often suffer chronic exposure to these xenobiotics.[46,47] Lenticels on aerial roots or proproots of these plants, which play a critical role in oxygen uptake, may become clogged with oil, thereby greatly reducing gaseous transport to the submerged parts of the plants. The blockage of air exchange through surface pores has caused the mortality of many hectares of mangroves in tropical regions.[2,6] Oil spills have eliminated broad expanses of mangroves in Nigeria, Indonesia, Panama, Kenya, Puerto Rico, and elsewhere due to other lethal or sublethal effects as well, including the direct contact of oil with the plants, systematic uptake of hydrocarbon compounds by the plants, and chemical or physical alteration of seawater and sediments (e.g., depletion of oxygen and nitrogen, pH change, and decreased light penetration).[2,4] Extensive areas of mangroves throughout the world also have experienced reduced growth and impaired reproduction due to the effects of oil pollution. Mangrove forests may require from 10 to 15 years for complete recovery from the effects of infiltrating crude oil.

Impacts of oil on benthic algae have been documented both under spill conditions in nature and experimental conditions in the laboratory. Benthic microalgal populations (cyanophytes, diatoms), which commonly live on upper layers of sediments in salt marshes, mangroves, and tidal flats, are often completely destroyed by oil spreading across intertidal habitats. Such was the case of the *Amoco Cadiz* oil spill off the coast of northern Brittany (France) in March 1978.[33,34] The release of 223,000 mt of oil from this tanker damaged about 320 km of coastline over a 2-week period. The oil eliminated benthic microalgal populations in oil-affected sites which remained underpopulated years after the event.[48]

Benthic macroalgae have mixed responses to oil contamination. The green algae *Chaetomorpha aerea*, *Enteromorpha intestinalis*, and *Ulva angusta* were only slightly damaged in mid- and high intertidal zones subsequent to the Santa Barbara oil spill in 1969.[49] Much more extensive damage occurred to *Fucus spiralis* on the rocky shores of Chedabucto Bay, Nova Scotia, due to Bunker C oil spilled from the tanker *Arrow* in February 1970.[50] Peckol et al.,[51] investigating the effects of oiling from the grounding of the tanker *World Prodigy* on condition, growth rate, and pigment acclimation in the subtidal kelps *Laminaria saccharina* and *L. digitata*, found no evidence of detrimental effects. Laboratory studies have revealed that crude oil alone may be less toxic to benthic macroalgae than dispersant-oil mixtures.[52]

Invertebrates

Marine invertebrates have been the target of numerous studies of oil spill impacts.[53-60] Invertebrates in the water column (zooplankton), while obviously sensitive to dispersed oil, usually recover relatively rapidly and return to pre-spill conditions faster than benthic invertebrates. Even low concentrations of oil and dispersants have been shown to adversely affect the fertilization, embryonic development, and feeding of copepods.[53,61] Zooplankton may ingest significant amounts of petroleum hydrocarbons, as demonstrated by oil-ingesting copepods during the *Argo Merchant* spill on Nantucket Shoals in December 1969. Oil-laden fecal pellets produced by zooplankton which settle through the water column serve as a potentially important pathway for removal of hydrocarbon contaminants to the sea floor.

Because of the relatively limited mobility of benthic invertebrates, they are prone to contamination by oil accumulating in bottom sediments. The entire benthic community may be smothered when the volume of oil reaching the sea floor is high. The West Falmouth oil spill caused by the grounding of the barge *Florida* in September 1969 resulted in mass mortality of the benthos and detectable impacts on the community lasting for more than a decade. Twenty years after the spill, weathered and biodegraded fuel oil, aromatic hydrocarbons, and cycloalkanes were still detected in trace concentrations in some intertidal and subtidal sediment samples down to 15 cm depth.[32] Recovery of benthic populations (e.g., mussels, green abalones) on rocky substrates from the effects of the 1957 *Tampico Maru* spill was incomplete more than 15 years after the spill.[62] More recently, the *Exxon Valdez* oil spill in March 1989 contaminated many hectares of subtidal and rocky intertidal habitats along the south coast of Alaska. Owing to the slow decomposition of the oil and its persistence in bottom sediments, benthic invertebrates in Prince William Sound likely will be exposed to various products of the oil for years. In January 1996, a damaged barge that spilled more than 3 million liters of oil along the southern coast of Rhode Island after running aground off a wildlife refuge in a storm killed more than 10,000 lobsters and many other benthic organisms. This spill, the largest in Rhode Island state history, also resulted in the closure of a 27,200-ha area to all fishing. Adverse biotic effects of the spill are expected for years.

Shellfish often remain contaminated for extended periods due to oil spills. For example, shellfish beds in Falmouth, MA, remained closed to shellfishing for more than 8 years following the *Florida* oil spill. Oysters in the Aber-Benoit and in the Baie de Morlaix (Carantec) of France contained significant levels of residual aromatic hydrocarbons 7 years after the *Amoco Cadiz* oil spill.[63] The oil caused a number of sublethal effects, such as necrosis, inflammation, and atrophy of gonadal cells. Brown et al.[60] found increased incidence of lesions in the digestive gland, gills, mantle, heart, and kidney of *Mya arenaria* exposed to #2 fuel oil in a tidal mud flat of the Arthur Kill in Elizabeth, NJ.

Shellfish generally display variable levels of tolerance to oil pollution.[7] For example, mussels (*Mytilus edulis*) withstand higher concentrations of oil than scallops (*Pecten opercularis*) and cockles (*Cardium edule*). Hard clams (*Mercenaria mercenaria*) tolerate higher levels of oil contamination than oysters (*Crassostrea virginica*). The tolerance of shellfish to oil contamination may be reduced by dispersants. Stromgren,[64] for example, conveyed that oil treated with dispersants produces a mixture more toxic to mussels than oil alone.

Early life stages of benthic invertebrates are more sensitive to the toxic effects of oil than juveniles and adults. This is demonstrated by studies of early life stages of bivalves, crabs, shrimp, and lobsters.[7,65] Eggs and meroplankton of these organisms typically drift in the water column for weeks, occasionally in surface waters where they are at greater risk of contact with surface slicks of oil. Egg viability and meroplankton success appear to be reduced in areas affected by spills. However, the most detrimental effects of oil on these early life stages are derived from experimental investigations in the laboratory rather than observations in nature. Once oil covers the substrate, larval settlement can be compromised. For instance, Smith and Hackney[66] recorded significantly lower spat densities of the American oyster (*Crassostrea virginica*) on oil treatments vs. control and gas-treated shells in the high intertidal zone of a southeastern North Carolina estuary.

The recolonization of oil-disturbed habitats usually is initiated by opportunistic species. For instance, soon after the West Falmouth oil spill, the opportunistic polychaete worm *Capitella capitata* invaded the defaunated habitat and rapidly reached high densities. Other benthic fauna began to repopulate the impacted sites within about 1 year. Polychaetes, which often comprise the pioneering forms, rank among the most tolerant benthic invertebrates to oil contamination; however, repeated oilings can eliminate them from impacted habitats as well. The recolonization of oil-laden habitats by benthic populations ultimately leads to infaunal reworking and bioturbation of bottom sediments, contributing to potentially chronic releases of oil products at the benthic boundary layer.

Fish

Oil pollution in coastal waters represents a serious threat to recreational and commercial fisheries. Many coastal regions sustain valuable fisheries resources that may be at risk during large oil spills. Because of their mobility, juvenile and adult fish generally are able to avoid oil slicks in open seas. However, the rapid advection of large volumes of oil into estuaries and embayments can trap fish populations, culminating in substantial mortality.[28]

Eggs, larvae, and early juvenile stages of fish are most sensitive to oil. The contact of egg masses and ichthyoplankton with surface oil can be lethal. Apart from the lethal effects of oil on embryos and larvae, sublethal effects on these early life stages commonly include abnormal development, reduced growth, premature and delayed hatching of eggs, and cellular abnormalities.[67] Adult fish tolerate much higher concentrations of petroleum hydrocarbons than eggs and larvae. Sublethal effects of oil typically manifested in adult fish are changes in heart and respiratory rates, gill hyperplasia, enlarged liver, reduced growth, fin erosion, impaired endocrine system, behavioral modification, as well as alterations in feeding, migration, reproduction, swimming activity, schooling, and burrowing behavior. A variety of biochemical, blood, and cellular changes also may arise.[4,7]

It is difficult to assess impacts of oil on eggs and larvae of fish populations in nature because the variation of natural mortality of these early life stages often exceeds that of mortality observed in the largest of spills. This effect has been demonstrated for several species such as the Atlantic cod (*Gadus morhua*), haddock (*Melanogrammus aeglefinus*), and Atlantic herring (*Clupea harengus*). In addition, natural variation in recruitment can obfuscate the impacts of a major oil spill on a single year class.[4] Nevertheless, some changes in fish populations have been linked to oil spills in nature. For example, reduced recruitment and catches of fish were registered in affected bay waters off Brittany 1 year after the *Amoco Cadiz* oil spill.[28] Following the *Torrey Canyon* oil spill off the coast of England, Smith[68] discerned 50 to 90% mortality of pilchard (*Sardina pilcardus*) eggs in affected waters. In areas of Georges Bank and Nantucket shoals influenced by oil from the *Argo Merchant* spill, 25% of the cod embryos and 46% of the pollock embryos were dead, and sand lance larvae were reduced by 80%.[69]

Because fish rapidly convert aromatic hydrocarbons from petroleum to a variety of metabolic products, these compounds usually cannot be assessed by directly analyzing fish tissues and fluids.[70] However, the cytochrome P-450 monooxygenase ("mixed-function oxidase") induction response in fish is fairly specific of oil compounds and other organic contaminants (e.g., organochlorines) and reportedly has been extremely useful in detecting sublethal biological effects of pollutants in small sample sizes.[71] Cytochrome P-450 enzymes appear to play a major role in fish metabolism of petroleum hydrocarbons.[72] Several regional and national monitoring programs, including the National Benthic Surveillance Project of the Status and Trends Program of the National Oceanic and Atmospheric Administration (NOAA), have effectively analyzed hepatic cytochrome P-4501A and associated monooxygenase activities in fish from Prince William Sound following the *Exxon Valdez* oil spill.[73] The measurement of P-4501A is considered to be a valuable bioindicator of oil contaminant exposure in estuarine and marine waters.

Birds

The physical coating of coastal and marine birds by floating oil often results in direct mortality via drowning or hypothermia. Diving ducks, auks, penguins, and sea ducks appear to be particularly vulnerable, because they spend a considerable amount of time on the sea surface. While external oiling commonly kills many seabirds soon after large spills, an array of sublethal effects of the oil can lead to seabird death from other causes, such as starvation, disease, and predation. Feather preening, drinking, consumption of oil-contaminated food, and the inhalation of fumes from evaporating oil cause an array of maladies including gastrointestinal irritation, pneumonia, red blood cell damage, immune system depletion, hormonal imbalance, inhibited reproduction, retarded growth in young, and abnormal parental behavior.[4,74] As is true of most marine fauna, the early life stages (i.e., embryos and young juveniles) of seabirds are very sensitive to oil.[75,76] Peakall et al.[77] demonstrate that even small amounts of oil (10 to 20 μl) account for acute embryotoxicity in susceptible forms.

Seabird casualties can be substantially reduced by the application of dispersants to mitigate surface exposures to the oil. The rapid dispersing of oil into the water column can lower oiling by a factor of 25 to 5000.[77] However, oil-dispersant mixtures also can adversely affect the reproductive success of bird populations.[7,77]

Some oil spills cause substantial losses of birds. For example, an estimated 100,000 to 300,000 birds died after the grounding of the *Exxon Valdez* in Prince William Sound in 1989.[78,79] Significant reductions in at least seven local populations of birds were noted. In some cases, however, the effects of single or periodic catastrophic oil spills may be obscured by death from natural causes, weather, food availability, movement of birds, and human activites (e.g., commercial fishing).

Hence, it is generally difficult to accurately evaluate the long-term impact of oil spills on bird populations at the local and regional levels.

Conducting a statistical analysis of 45 oil spills, Burger[80] showed that a weak log-log correlation exists between oil spill volume and the number of seabirds killed. He stressed, however, that this relationship cannot be used to predict mortality, and thus each spill should be examined independently. Comparative investigations are needed on oiled seabirds and oiled habitat areas.[81]

MAMMALS

Cetaceans and other marine mammals appear to be at less risk to oil pollution than seabirds and other marine fauna. For example, cetaceans and seals have considerable capacity for hydrocarbon metabolism, and they generally exhibit rapid renal clearance subsequent to oil ingestion. However, the external coating of oil on some marine mammals, such as the Alaska fur seal (*Callorhinus ursinus*), hair seal pups (Phocidae), polar bear (*Ursus maritimus*), and sea otter (*Enhydra lutris*), commonly causes matting of fur, loss of water repellency, and the amelioration of thermal insulation, which can culminate in hypothermia and death. The *Exxon Valdez* oil spill, for instance, affected a number of mammal populations — particularly sea otters and harbor seals — over an extensive area.[82] In addition to the direct impacts of external oiling, the ingestion and inhalation of oil may generate a host of sublethal effects in marine mammals, notably gastrointestinal and blood disorders, respiratory problems, changes in enzymatic activity in the skin, renal deficiencies, interferences with swimming, eye irritation, and lesions.[5,83-86] Marine mammals suffering from these sublethal effects often cannot effectively pursue prey and are themselves prone to predation. Indirect effects of oil spills (e.g., an alteration in the composition of prey species and reduction in food supply) can also create unfavorable conditions for mammal populations. Despite many cleanup efforts, surprisingly little is known about the long-term or chronic impact of oiling on marine mammals.

OIL CLEANUP

There are four major methods of cleaning oil spills: (1) mechanical cleanup, (2) shoreline cleanup, (3) no spill control, and (4) dispersants.[2] A number of devices have been developed to physically remove oil from the water surface. The most useful types include floating booms and skimmers which have been successfully used in many harbors and other inshore areas. Floating booms are designed to contain oil or channel oil into a confined area where it can be pumped out or physically removed by other means. "Slick-lickers", fitted with continuously rotating belts of absorbent, extract the oil and pass it onto a ship or barge. Both floating booms and "slick-lickers" have been most effective in mopping up small oil spills in sheltered waters.[6] However, even in many successful operations, much oil can escape from these devices.

When oil is stranded on a beach or rocky shoreline, it may be pumped, scraped, or physically removed by other means. These methods of removal are costly, time consuming, frequently damaging to habitat, and commonly unsuccessful since a considerable amount of oil typically remains behind on the substrate. Removal of sand from the beach can accelerate erosion. The use of high pressure hoses, steam, or dispersants may be more hazardous to the fauna and flora than the oil itself.

No spill control may be the best strategy to adapt when an oil slick is moving out to sea and/or the spill remains at sea and does not threaten the shoreline.[2] Dispersants have been of value in managing oil spills at sea, especially in temperate waters; however, some dispersants are toxic to marine organisms. In addition, dispersants are of little value once the oil is stranded on a sandy beach because the oil sinks and saturates the sand grains. Dispersants are generally more effective for the cleanup of light rather than heavy oil spills.

REFERENCES

1. Boesch, D. F. and Rabalais, N. N. (Eds.), *Long-Term Environmental Effects of Offshore Oil and Gas Development*, Elsevier Science, New York, 1987.
2. Thorhaug, A., Oil spills in the tropics and subtropics, in *Pollution in Tropical Aquatic Systems*, Connell, D. W. and Hawker, D. W. (Eds.), CRC Press, Boca Raton, FL, 1992.
3. Kennish, M. J., Pollution in estuaries and coastal marine waters, *J. Coastal Res.*, Special Issue No. 12: Coastal Hazards, 1994, 27.
4. Albers, P. H., Petroleum and individual polycyclic aromatic hydrocarbons, in *Handbook of Ecotoxicology*, Hoffman, D. J., Rattner, B. A., Burton, G. A., Jr., and Cairns, J., Jr. (Eds.), Lewis Publishers, Boca Raton, FL, 1994, 330.
5. National Research Council, *Using Oil Spill Dispersants on the Sea*, National Academy Press, Washington, D.C., 1989.
6. Clark, R. B., *Marine Pollution*, 3rd ed., Clarendon Press, Oxford, 1992.
7. Kennish, M. J., *Ecology of Estuaries: Anthropogenic Effects*, CRC Press, Boca Raton, FL, 1992.
8. Bonin, P., Rambeloarisoa Ranaivoson, E., Raymond, N., Chalamet, A., and Bertrand, J. C., Evidence for denitrification in marine sediment highly contaminated by petroleum products, *Mar. Pollut. Bull.*, 28, 89, 1994.
9. Madany, I. M., Al-Haddad, A., Jaffar, A., and Al-Shirbini, E.-S., Spatial and temporal distributions of aromatic petroleum hydrocarbons in the coastal waters of Bahrain, *Arch. Environ. Contam. Toxicol.*, 26, 185, 1994.
10. National Academy of Sciences, *Petroleum in the Marine Environment*, National Academy Press, Washington, D.C., 1975.
11. Carlberg, S. R., Oil polluting of the marine environment — with an emphasis on estuarine studies, in *Chemistry and Biogeochemistry of Estuaries*, Olausson, E. and Cato, I. (Eds.), John Wiley & Sons, Chichester, 1980, 367.
12. Pelletier, E., Ouellet, S., and Paquet, M., Long-term chemical and cytochemical assessment of oil contamination in estuarine intertidal sediments, *Mar. Pollut. Bull.*, 22, 273, 1991.
13. Abu-Hilal, A. H. and Khordagui, H. K., Petroleum hydrocarbons in the nearshore marine sediments of the United Arab empirates, *Environ. Pollut.*, 85, 315, 1994.
14. Cormack, D., *Response to Oil and Chemical Marine Pollution*, Applied Science Publishers, London, 1983.
15. Payne, J. R. and Phillips, C. R., Petroleum spills in the marine environment, in *The Chemistry and Formation of Water-in-Oil Emulsions and Tar Balls*, Lewis Publishers, Chelsea, MI, 1985.
16. Jordan, R. E. and Payne, J. R., *Fate and Weathering of Petroleum Spills in the Marine Environment*, Ann Arbor Science Publishers, Ann Arbor, MI, 1980.
17. DeLaune, R. D., Gambrell, R. P., Pardue, J. H., and Patrick, W. H., Jr., Fate of petroleum hydrocarbons and toxic organics in Louisiana coastal environments, *Estuaries*, 13, 72, 1990.
18. Neilson, A. H., *Organic Chemicals in the Aquatic Environment: Distribution, Persistence, and Toxicity*, Lewis Publishers, Boca Raton, FL, 1994.
19. Gundlach, E. R., Boehm, P. D., Marchand, M., Atlas, R. M., Ward, D. M., and Wolfe, D. A., The fate of *Amoco Cadiz* oil, *Science*, 221, 122, 1983.
20. Page, D. S., Foster, J. C., Fickett, P. M., and Gilfillan, E. S., Long-term weathering of Amoco Cadiz oil in soft intertidal sediments, in *Proceedings of the 1989 Oil Spill Conference*, Publ. 4479, American Petroleum Institute, Washington, D.C., 1989, 401.
21. Corredor, J. E., Morell, J. M., and Del Castillo, C. E., Persistence of spilled crude oil in a tropical intertidal environment, *Mar. Pollut. Bull.*, 21, 385, 1990.
22. Bishop, J. M., *Applied Oceanography*, John Wiley & Sons, New York, 1984.
23. O'Connor, J. M. and Huggett, R. J., Aquatic pollution problems, North Atlantic coast, including Chesapeake Bay, *Aquat. Toxicol.*, 11, 163, 1988.
24. Gunster, D. G., Gillis, C. A., Bonnevie, N. L., Abel, T. B., Wenning, R. J., Petroleum and hazardous chemical spills in Newark Bay, New Jersey, U.S.A. from 1982 to 1991, *Environ. Pollut.*, 82, 245, 1993.
25. Eisler, R., Polycyclic Aromatic Hydrocarbon Hazards to Fish, Wildlife, and Invertebrates: A Synoptic Review, Biological Report 85(1.11), U.S. Fish and Wildlife Service, Washington, D.C., 1987.
26. McElroy, A. E., Farrington, J. W., and Teal, J. M., Bioavailability of polycyclic aromatic hydrocarbons in the aquatic environment, in *Metabolism of Polycyclic Aromatic Hydrocarbons in the Aquatic Environment*, Varanasi, U. (Ed.), CRC Press, Boca Raton, FL, 1989.

27. Evans, D. R. and Rice, S. D., Effects of oil on marine ecosystems: a review for administrators and policy makers, *Fish. Bull.*, 72, 625, 1974.
28. Teal, J. M. and Howarth, R. W., Oil spill studies: a review of ecological effects, *Environ. Manage.*, 8, 27, 1984.
29. Doerffer, J. W., *Oil Response in the Marine Environment*, Pergamon Press, Oxford, 1992.
30. May, R. F., Marine conservation reserves, petroleum exploration and development, and oil spills in coastal waters of western Australia, *Mar. Pollut. Bull.*, 25, 147, 1992.
31. Sanders, H. L., Grassle, J. F., Hampson, G. R., Morse, L. S., Garner-Price, S., and Jones, C. C., Anatomy of an oil spill: long-term effects from the grounding of the barge *Florida* off West Falmouth, Massachusetts, *J. Mar. Res.*, 38, 265, 1980.
32. Teal, J. M., Farrington, J. W., Burns, K. A., Stegeman, J. J., Tripp., B. W., Woodin, B., and Phinney, C., The West Falmouth oil spill after 20 years: fate of fuel oil compounds and effects on animals, *Mar. Pollut. Bull.*, 24, 607, 1992.
33. Page, D. S., Foster, J. C., Fickett, P. M., and Gilfillan, E. S., Identification of petroleum sources in an area impacted by the *Amoco Cadiz* oil spill, *Mar. Pollut. Bull.*, 19, 107, 1988.
34. Dauvin, J-C. and Gentil, F., Conditions of the peracarid populations of subtidal communities in northern Brittany ten years after the *Amoco Cadiz* oil spill, *Mar. Pollut. Bull.*, 21, 123, 1990.
35. Reish, D. J., Oshida, P. S., Mearns, A. J., and Ginn, T. C., Effects on saltwater organisms, *J. Water Pollut. Cont. Fed.*, 61, 1042, 1989.
36. Harrison, P. J., Cochlan, W. P., Acreman, J. C., Parsons, T. R., Thompson, P. A., and Dovey, H. M., The effects of crude oil and Corexit 9527 on marine phytoplankton in an experimental enclosure, *Mar. Environ. Res.*, 18, 93, 1986.
37. Alldredge, A. L., Elias, M., and Gotschalk, C. C., Effects of drilling muds and mud additives on the primary production of natural assemblages of marine phytoplankton, *Mar. Environ. Res.*, 19, 157, 1986.
38. Johansson, S., Larsson, U., and Boehm, P., The *Tsesis* oil spill. Impact on the pelagic ecosystem, *Mar. Pollut. Bull.*, 11, 284, 1980.
39. Shailaja, M. S., The influence of dissolved petroleum hydrocarbon residues on natural phytoplankton biomass, *Mar. Environ. Res.*, 25, 315, 1988.
40. IMCO/FAO/UNESCO/WMO/WHO/IAEA/UN Joint Group of Experts on the Scientific Aspects of Marine Pollution (GESAMP), Impact of Oil on the Marine Environment, Reports and Studies No. 6, Food and Agriculture Organization, Rome, 1977.
41. Lugo, A., Teas, A. H., and Lewis, R., Effect of oil on mangroves and restoration, in *Restoration and Management of Marine Ecosystems Impacted by Oil*, Cairns, J. and Buikema, A. (Eds.), Butterworths, New York, 1984.
42. Lane, P. A., Vandermeulen, J. H., Crowell, M. J., and Patriquin, D. G., Impact of experimentally dispersed crude oil on vegetation in a northwestern Atlantic salt marsh — preliminary observations, in *Proceedings of the 1987 Oil Spill Conference,* American Petroleum Institute, Washington, D.C., 1987, 509.
43. Teas, H. J., Duerr, E. O., and Wilcox, J. R., Effects of South Louisiana crude oil and dispersants on *Rhizophora* mangroves, *Mar. Pollut. Bull.*, 18, 122, 1987.
44. Thorhaug, A. and Marcus, J. H., Oil spill cleanup: the effect of three dispersants on three subtropical/tropical seagrasses, *Mar. Pollut. Bull.*, 18, 124, 1987.
45. Thorhaug, A., Marcus, J., and Booker, F., Oil and dispersed oil on subtropical and tropical seagrasses in laboratory studies, *Mar. Pollut. Bull.*, 17, 357, 1986.
46. Klekowski, E. J., Jr., Corredor, J. E., Morell, J. M., and Del Castillo, C. A., Petroleum pollution and mutation in mangroves, *Mar. Pollut. Bull.*, 28, 166, 1994.
47. Corredor, J. E., Morell, J. M., and Del Castillo, C., Persistence of spilled crude oil in a tropical intertidal environment, *Mar. Pollut. Bull.*, 21, 385, 1990.
48. Campion-Alsumard, T., Plante-Cunz, M.-R., and Vacelet, E., Evolution des hydrocarbures et des populations bacteriennes et microphytiques dans les sediments des marais maritimes de I'Lle Grande pollues par l'*Amoco Cadiz*: 2 — Evolution des peuplements microphytiques, *Mar. Environ. Res.*, 11, 275, 1984.
49. Foster, M., Neushul, M., and Zingmark, R., The Santa Barbara oil spill. II. Initial effects on intertidal and kelp bed organisms, *Environ. Pollut.*, 2, 115, 1971.
50. Thomas, M. L. H., Effects of Bunker C oil on intertidal and lagoonal biota in Chedabucto Bay, Nova Scotia, *J. Fish. Res. Bd. Can.*, 30, 83, 1973.

51. Peckol, P., Levings, S. C., and Garrity, S. D., Kelp response following the *World Prodigy* oil spill, *Mar. Pollut. Bull.*, 21, 473, 1990.
52. Thelin, I., Effects in culture of two crude oils and one oil dispersant on zygotes and germlings of *Fucus serratus, Linnaeus, Fucales, and Phaeophyceae, Bot. Mar.*, 24, 515, 1981.
53. Falk-Peterson, I. B., Lonning, S., and Jakobsen, R., Effects of oil and dispersants on plankton organisms, *Astarte,* 12, 45, 1983.
54. Boucher, G., Long-term monitoring of meiofauna densities after the *Amoco Cadiz* oil spill, *Mar. Pollut. Bull.,* 16, 328, 1985.
55. Frithsen, J. B., Elmgren, R., and Rudnick, D. T., Responses of benthic meiofauna to long-term, low-level additions of No. 2 fuel oil, *Mar. Ecol. Prog. Ser.,* 23, 1, 1985.
56. Blaylock, W. M. and Houghton, J. P., Infaunal recovery at Edez Hook following the *Arco Anchorage* oil spill, in *Proceedings of the 1989 Oil Spill Conference,* Publ. 4479, American Petroleum Institute, Washington, D.C., 1989, 421.
57. Garrity, S. D. and Levings, S. C., Effects of an oil spill on the gastropods of a tropical intertidal reef flat, *Mar. Environ. Res.,* 30, 119, 1990.
58. Burger, J., Brzorad, J., and Gochfeld, M., Immediate effects of an oil spill on behavior of fiddler crabs (*Uca pugnax*), *Arch. Environ. Contam. Toxicol.,* 20, 404, 1991.
59. Nance, J. M., Effects of oil/gas field produced water on the macrobenthic community in a small gradient estuary, Hydrobiologia, 220, 189, 1991.
60. Brown, R. P., Cristini, A., and Cooper, K. R., Histopathological alterations in *Mya arenaria* following a #2 fuel oil spill in the Arthur Kill, Elizabeth, New Jersey, *Mar. Environ. Res.,* 34, 65, 1992.
61. Spooner, M. F. and Corkett, C. J., Effects of Kuwait oils on feeding rates of copepods, *Mar. Pollut. Bull.,* 10, 197, 1979.
62. Hall, C. A. S., Howarth, R., Moore, III, B., and Vorosmarty, C. J., Environmental impacts of industrial energy systems in the coastal zone, *Ann. Rev. Energy,* 3, 395, 1978.
63. Berthou, F., Balouet, G., Bodennec, G., and Marchand, M., The occurrence of hydrocarbons and histopathological abnormalities in oysters for seven years following the wreck of the *Amoco Cadiz* in Brittany (France), *Mar. Environ. Res.,* 23, 103, 1987.
64. Stromgren, T., Effect of oil and dispersants on the growth of mussels, *Mar. Environ. Res.,* 21, 239, 1987.
65. Capuzzo, J. M., Lancaster, B. A., and Saski, G., The effects of petroleum hydrocarbons on lipid metabolism and energetics of larval development and metamorphosis in the American lobster (*Homarus americanus*), *Mar. Environ. Res.,* 14, 201, 1984.
66. Smith, C. M. and Hackney, C. T., The effects of hydrocarbons on the setting of the American oyster, *Crassostrea virginica*, in intertidal habitats in southeastern North Carolina, *Estuaries,* 12, 42, 1989.
67. Malins, D. C. and Hodgins, H. O., Petroleum and marine fishes: a review of uptake, disposition, and effects, *Environ. Sci. Technol.,* 15, 1272, 1981.
68. Smith, J. E., *Torrey Canyon, Pollution, and Marine Life*, Cambridge University Press, Cambridge, 1970.
69. Grose, P. L. and Mattson, J. S., The *Argo Merchant* Oil Spill: A Preliminary Scientific Report, U.S. Department of Commerce and National Oceanic and Atmospheric Administration, Boulder, CO, 1977.
70. Krahn, M. M., Kittle, L. J., Jr., and MacLeod, W. D., Jr., Evidence for exposure of fish to oil spilled into the Columbia River, *Mar. Environ. Res.,* 20, 291, 1986.
71. Goksoyr, A., Solberg, T. S., and Serigstad, B., Immunochemical detection of cytochrome P450IA1 induction in cod larvae and juveniles exposed to a water soluble fraction of North Sea crude oil, *Mar. Pollut. Bull.,* 22, 122, 1991.
72. Stegeman, J. J., Detecting the biological effects of deep-sea waste disposal, *Oceanus,* 33, 54, 1990.
73. Collier, T. K., Connor, S. D., Eberhart, B.-T. L., Anulacion, B. F., Goksoyr, A., and Varanasi, U., Using cytochrome P450 to monitor the aquatic environment: initial results from regional and national surveys, *Mar. Environ. Res.,* 34, 195, 1992.
74. Eppley, Z. A. and Rubega, M. A., Indirect effects of an oil spill, *Nature,* 340, 513, 1989.
75. Hoffman, D. J., Embryotoxicity and teratogenicity of environmental contaminants to bird eggs, *Rev. Environ. Contam. Toxicol.,* 115, 39, 1990.
76. McOrist, S. and Lenghaus, C., Mortalities of little penguins (*Eudyptula minor*) following exposure to crude oil, *Vet. Rec.,* 130, 161, 1992.
77. Peakall, D. B., Wells, P. G., and Mackay, D., A hazard assessment of chemically dispersed oil spills and seabirds, *Mar. Environ. Res.,* 22, 91, 1987.

78. Piatt, J. F. and Lensink, C. J., *Exxon Valdez* bird toll, *Nature,* 342, 865, 1989.
79. Piatt, J. F., Lensink, C. J., Butler, W., Kenziorek, M., and Nysewander, D. R., Immediate impact of the *Exxon Valdez* oil spill on marine birds, *Auk,* 197, 387, 1990.
80. Burger, A. E., Estimating the mortality of seabirds following oil spills: effects of spill volume, *Mar. Pollut. Bull.,* 26, 140, 1993.
81. Dahlmann, G., Timm, D., Averbeck, C., Camphuysen, C., Skov, H., and Durinck, J., Oiled seabirds — comparative investigations on oiled seabirds and oiled beaches in the Netherlands, Denmark, and Germany (1990–93), *Mar. Pollut. Bull.,* 28, 305, 1994.
82. Loughlin, T. R. (Ed.), *Marine Mammals and the Exxon Valdez,* Academic Press, San Diego, CA, 1994.
83. National Research Council, *Oil in the Sea: Inputs, Fates, and Effects*, National Academy Press, Washington, D.C. 1985.
84. Engelhardt, F. R., Effects of petroleum on marine mammals, in *Petroleum Effects in the Arctic Environment*, Englehardt, F. R. (Ed.), Elsevier Applied *Science,* New York, 1985, 217.
85. Waldichuk, M., Sea and oil pollution, *Mar. Pollut. Bull.,* 21, 10, 1990.
86. Ekker, M., Lorentsen, S.-H., and Rov, N., Chronic oil-fouling of grey seal pups at the froan breeding ground, Norway, *Mar. Pollut. Bull.,* 24, 92, 1992.

APPENDIX 1. SOURCES AND FATE OF OIL POLLUTION

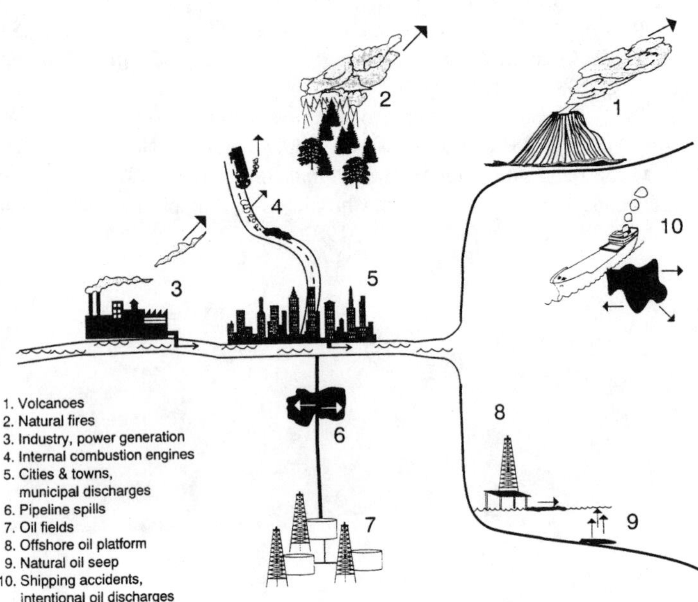

FIGURE 1.1. Sources of oil and polycyclic aromatic hydrocarbons in the marine environment. (From Albers, P. H., in *Handbook of Ecotoxicology,* Hoffman, D. J., Rattner, B. A., Burton, G. A., Jr., and Cairns, J., Jr. (Eds.), Lewis Publishers, Boca Raton, FL, 1994, 330. With permission.)

TABLE 1.1
Estimated World Input of Petroleum Hydrocarbons to the Sea (metric tons per year)

Source		Total
Transportation		
Tanker operations	0.158	
Tanker accidents	0.121	
Bilge and fuel oil	0.252	
Dry docking	0.004	
Nontanker accidents	0.020	
		0.555
Fixed installations		
Coastal refineries	0.10	
Offshore production	0.05	
Marine terminals	0.03	
		0.180
Other sources		
Municipal wastes	0.70	
Industrial wastes	0.20	
Urban runoff	0.12	
River runoff	0.04	
Atmospheric fallout	0.30	
Ocean dumping	0.02	
		1.380
Natural inputs		0.250
Total		2.365
Biosynthesis of hydrocarbons		
Production by marine phytoplankton		26,000
Atmospheric fallout		100–4000

Source: Clark, R. B., *Marine Pollution,* 3rd ed., Clarendon Press, Oxford, 1992. With permission.

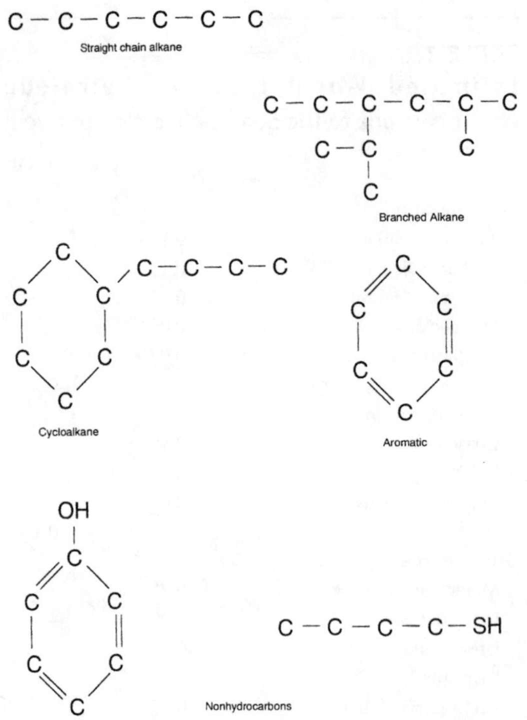

FIGURE 1.2. Types of molecular structures found in petroleum. Hydrogen atoms bonded to carbon atoms are omitted. (From Albers, P. H., in *Handbook of Ecotoxicology,* Hoffmann, D. J., Rattner, B. A., Burton, G. A., Jr., and Cairns, J., Jr. (Eds.), Lewis Publishers, Boca Raton, FL, 1994, 330. With permission.)

TABLE 1.2
Refinery "Cuts" of Crude Oil

	Boiling Range (°C)	Molecular Size (Number of Carbon Atoms)
Petroleum gases	30	3–4
Light gasoline, benzene	30–140	4–6
Naphtha	120–175	7–10
Kerosene	165–200	10–14
Gas oil (diesel)	175–365	15–20
Fuel oil and residues	350	20+

Source: Clark, R. B., *Marine Pollution,* 3rd ed., Clarendon Press, Oxford, 1992. With permission.

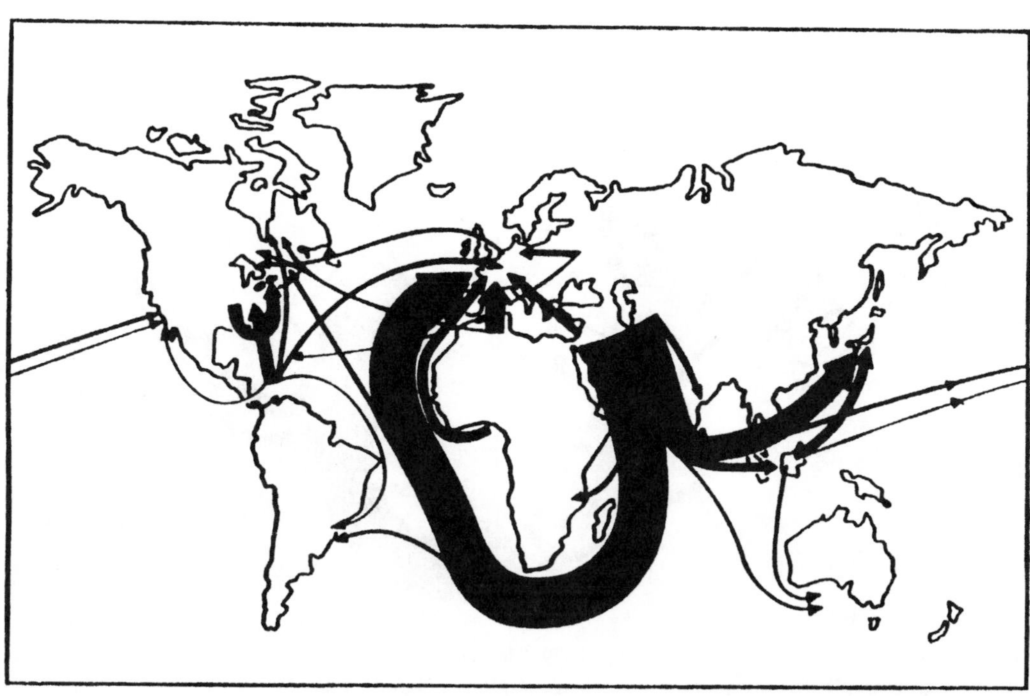

FIGURE 1.3. Major oil movements at sea in 1972. (From Cowell, E. B., in *Marine Pollution,* Johnson, R. (Ed.), Academic Press, London, 1976, 353. With permission.)

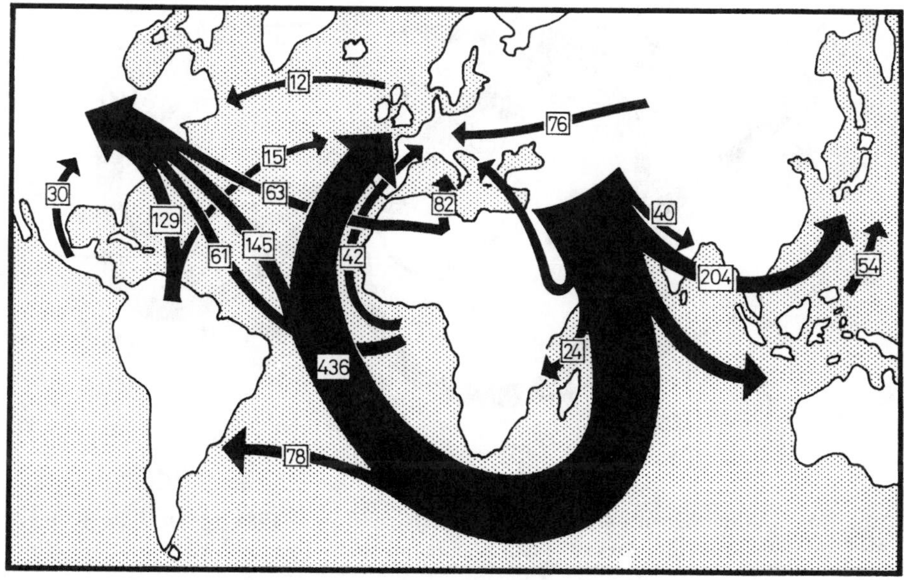

FIGURE 1.4. The most important sea lanes for petroleum transport in 1978/1979. Numbers indicate million tons per year. (From Gerlach, K., *Marine Pollution,* Springer-Verlag, Berlin, 1981. With permission.)

FIGURE 1.5. Major oil movements at sea in 1979. The width of the arrows represents the relative volume of oil. (From Clark, R. B., *Marine Pollution,* Clarendon Press, Oxford, 1986. With permission.)

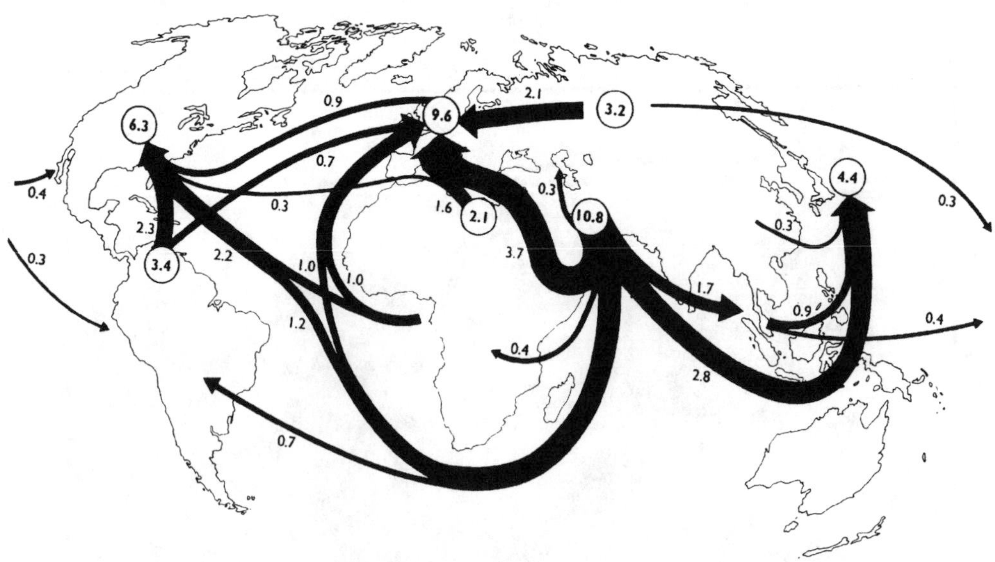

FIGURE 1.6. Major oil movements at sea (million barrels per day) in 1987. (From Clark, R. B., *Marine Pollution,* 3rd ed., Clarendon Press, Oxford, 1992. With permission.)

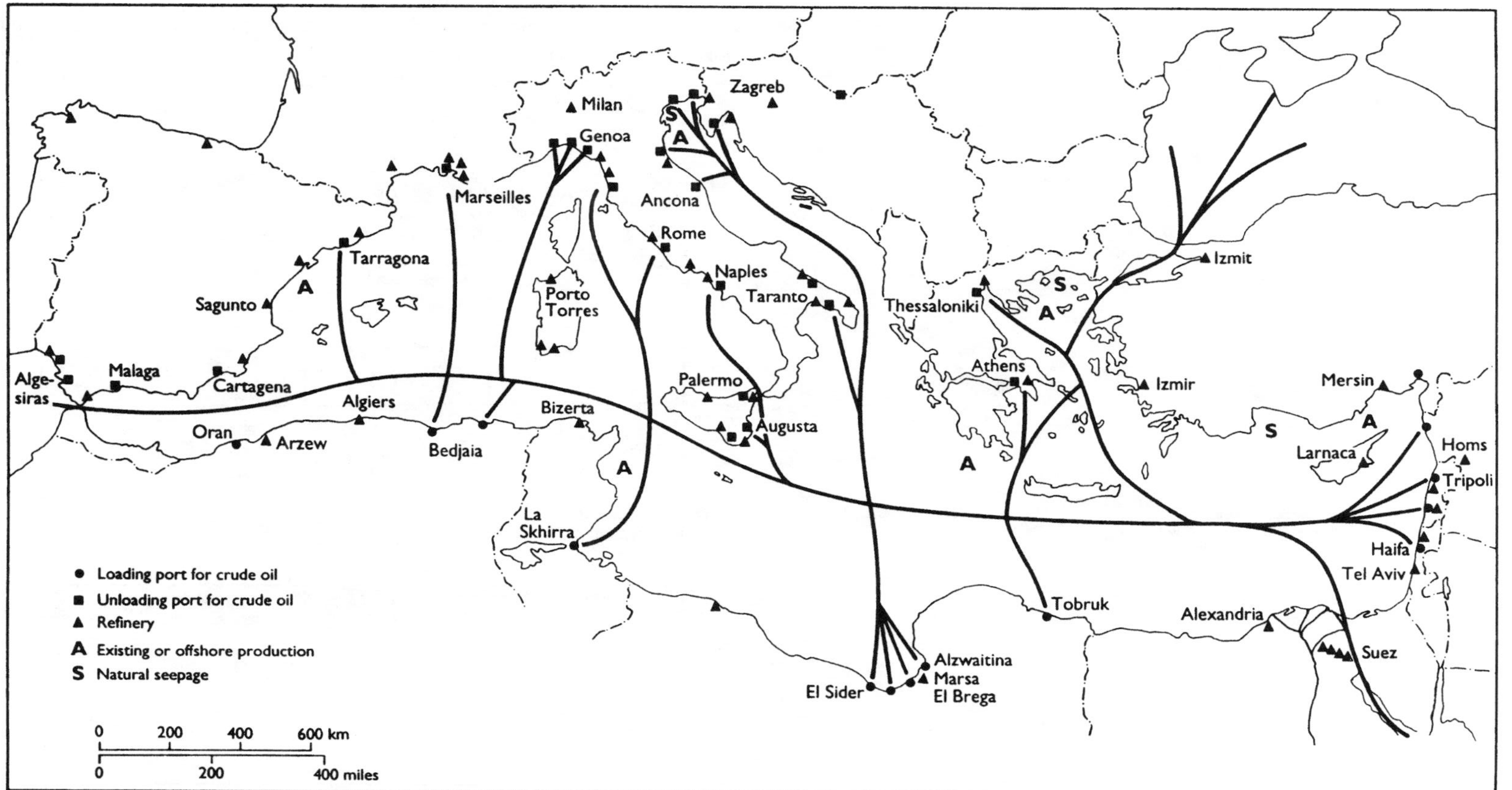

FIGURE 1.7. Activities of the oil industry in the Mediterranean Sea. (From Clark, R. B., *Marine Pollution,* 3rd ed., Clarendon Press, Oxford, 1992. With permission.)

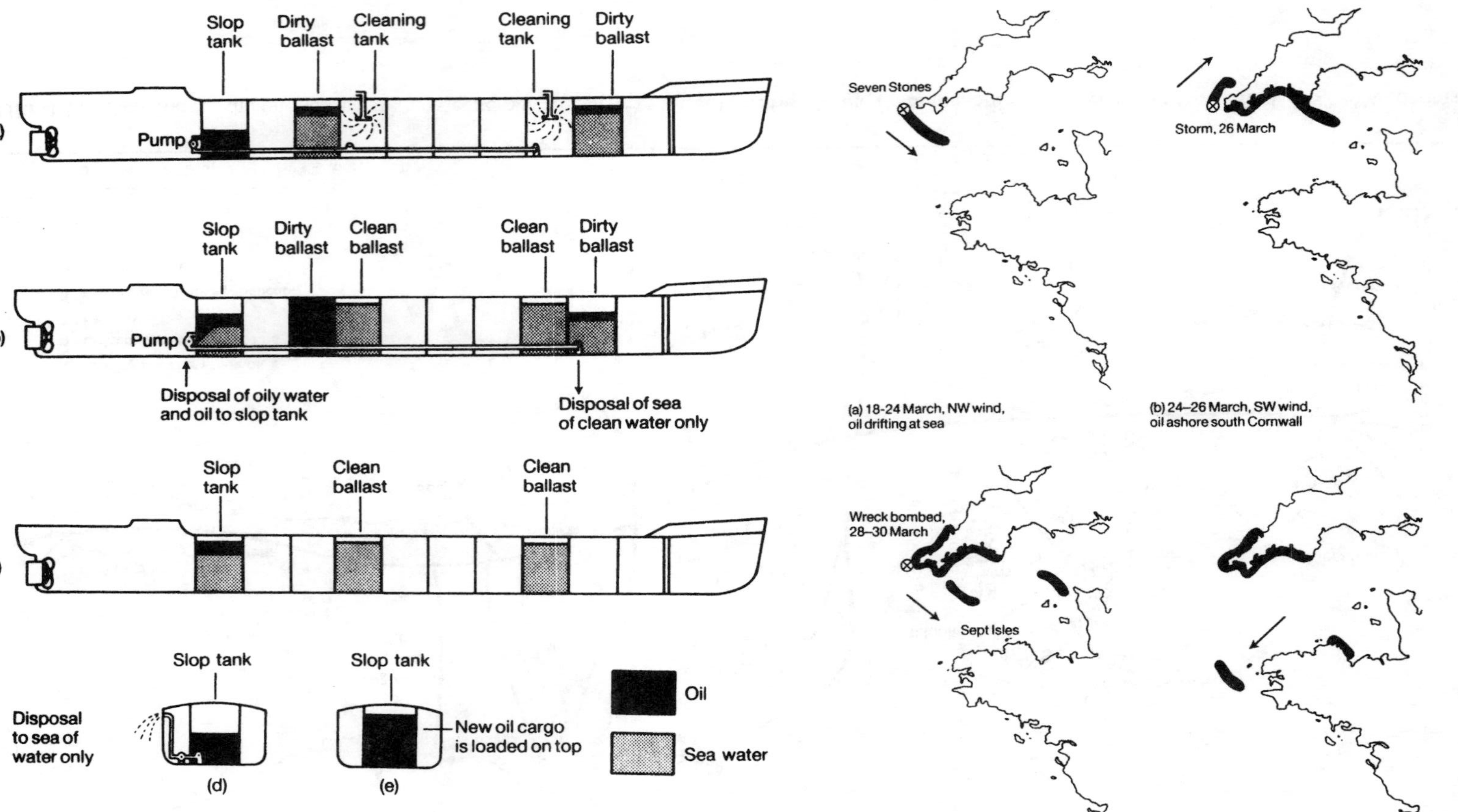

FIGURE 1.8. Load-on-top system of oil transport at sea. (a) Empty tanks are cleaned by water jet and the washings transferred to the slop tanks. (b) Oil floats to the top in tanks containing dirty ballast; the water is discharged to the sea and oil transferred to the slop tank. (c) Eventually the ship carries only clean ballast, and oil in the slop tanks floats to the surface. (d) Underlying water is pumped out. (e) New cargo is loaded on top of the oil remaining in the slop tank. (From Clark, R. B., *Marine Pollution,* 3rd ed., Clarendon Press, Oxford, 1992. With permission.)

FIGURE 1.9. Movement of oil spilled from the tanker *Torrey Canyon* off the southwest coast of England in 1967 in response to changing wind direction. (From Clark, R. B., *Marine Pollution,* 3rd ed., Clarendon Press, Oxford, 1992. With permission.)

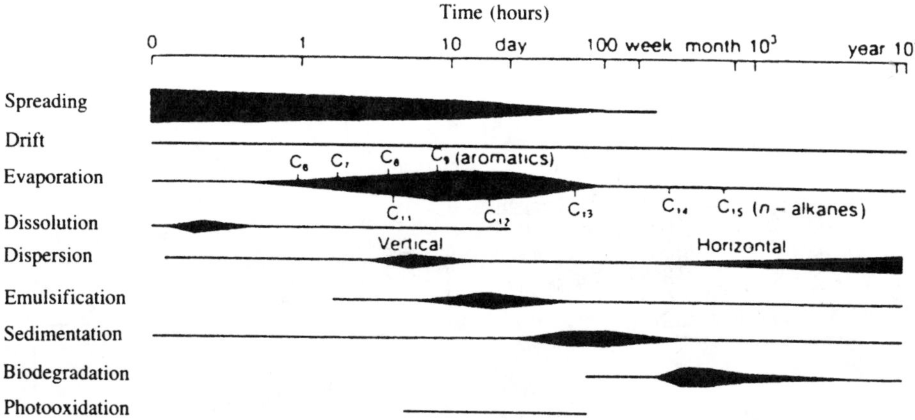

FIGURE 1.10. Time-course of factors affecting an oil spill at sea. (From Clark, R. B., *Marine Pollution,* 2nd ed., Clarendon Press, Oxford, 1989. With permission.)

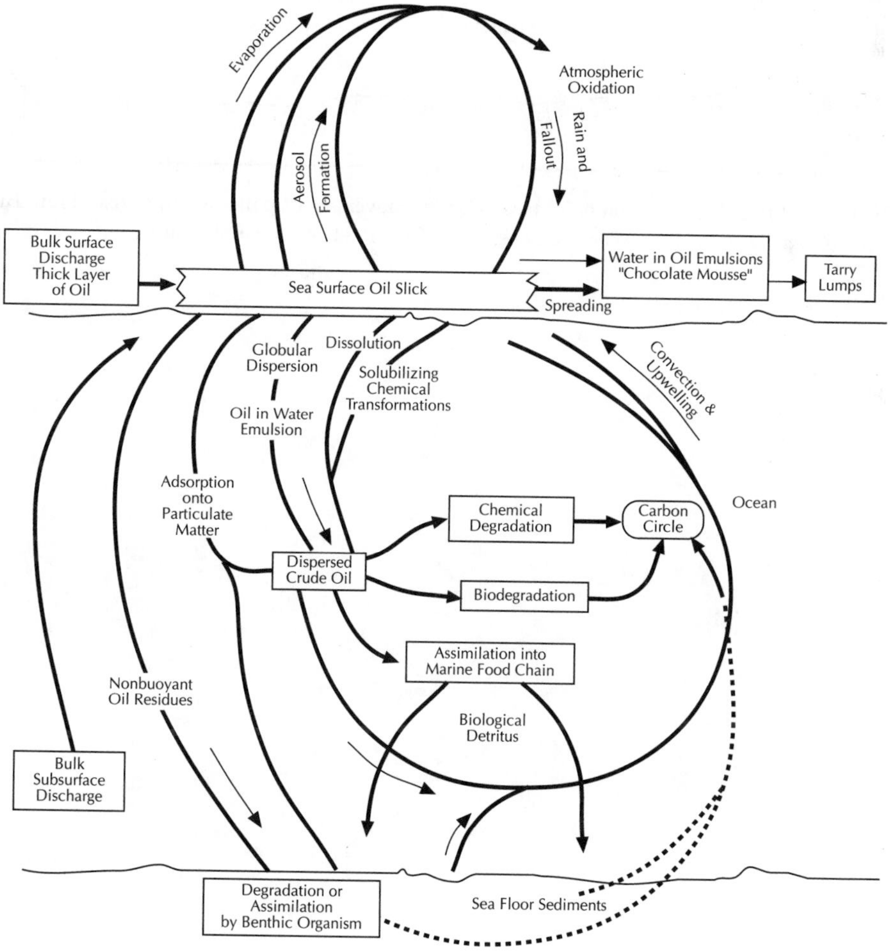

FIGURE 1.11. Fate and weathering of polluting oil in estuarine and marine waters showing various abiotic and biotic processes that act to alter the oil. (From Burwood, R. and Speers, G. C., *Estuarine Coastal Mar. Sci.,* 2, 117, 1974. With permission.)

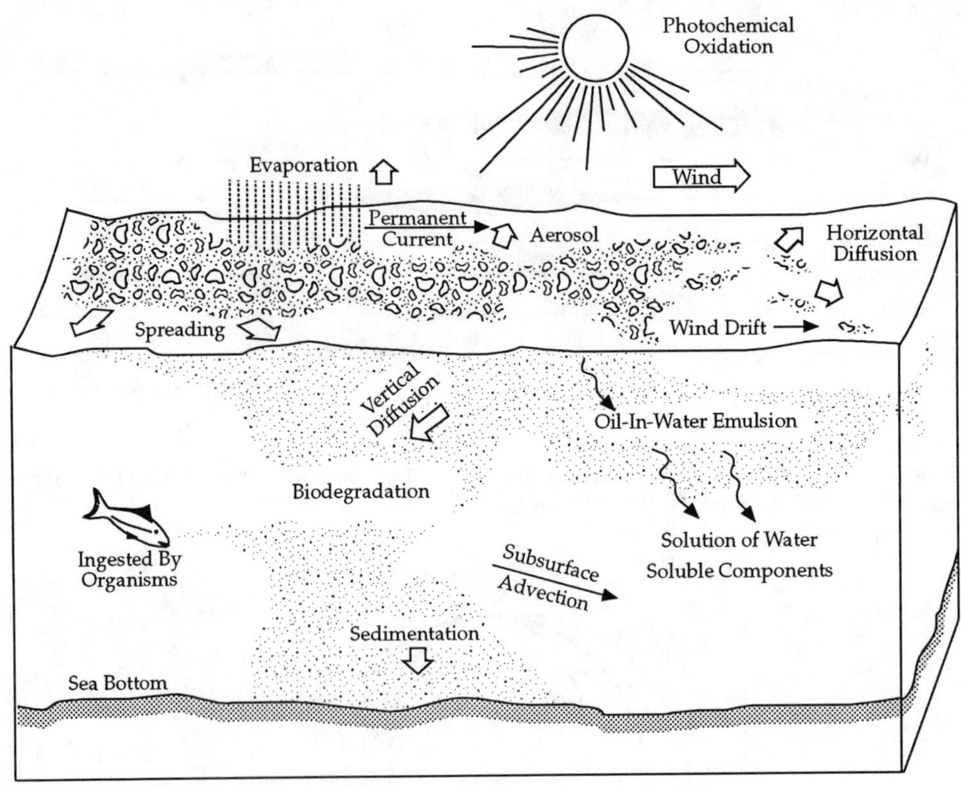

FIGURE 1.12. Effects of wind and other factors on the movement of polluting oil at sea. (From Bishop, J. M., *Applied Oceanography*, John Wiley & Sons, New York, 1984. With permission.)

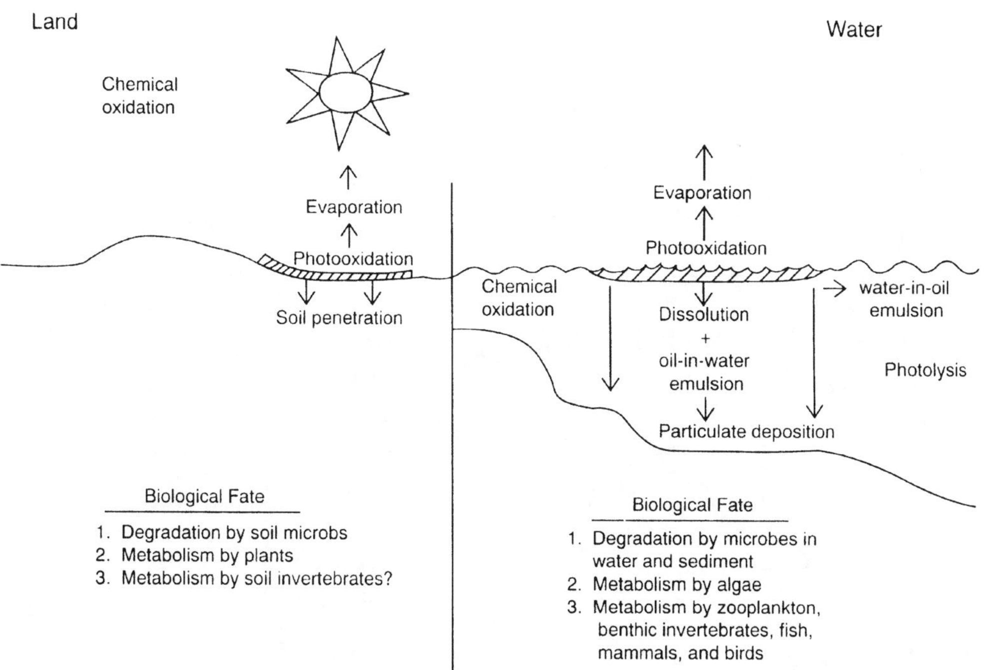

FIGURE 1.13. Chemical, physical, and biological fate of oil on land vs. in the sea. (From Albers, P. H., in *Handbook of Ecotoxicology,* Hoffman, D. J., Rattner, B. A., Burton, G. A., Jr., and Cairns, J., Jr. (Eds.), Lewis Publishers, Boca Raton, FL, 1994, 330. With permission.)

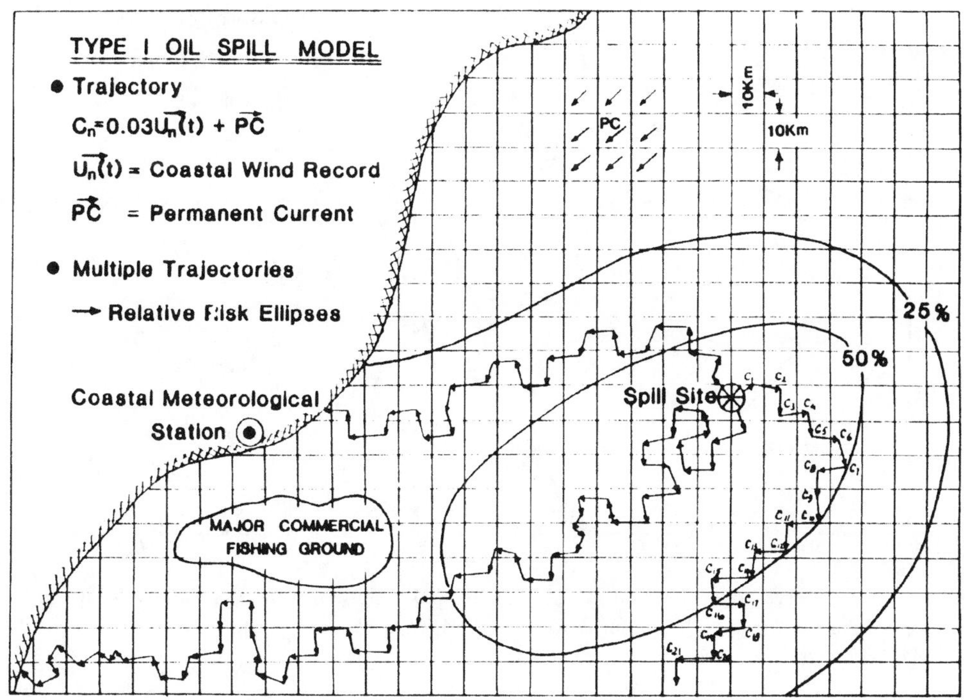

FIGURE 1.14. A type I climatological oil spill trajectory model calculation, showing relative risk ellipses. Numbers on relative risk ellipses are the result of a contour analysis of percentage "hits" in each grid area. (From Bishop, J. M., *Applied Oceanography,* John Wiley & Sons, New York, 1984. With permission.)

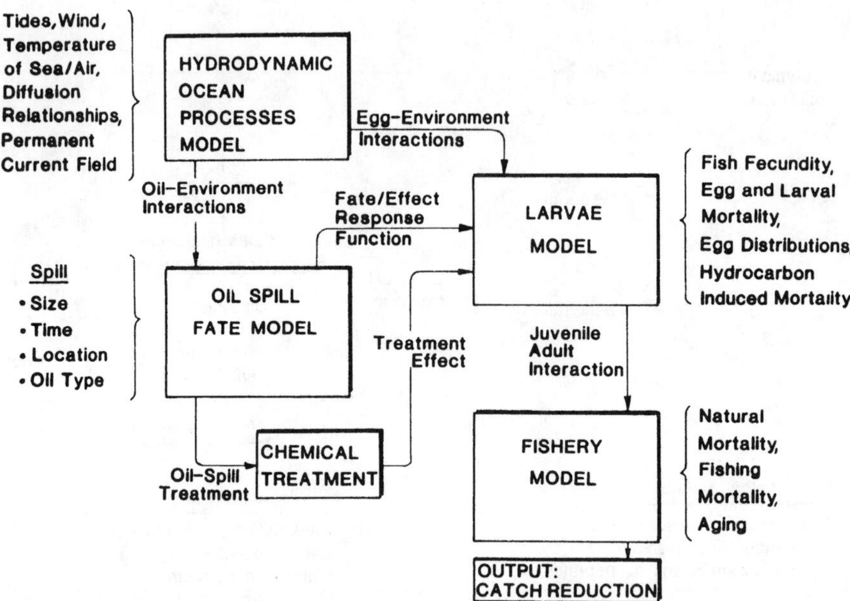

FIGURE 1.15. Components of a type II oil spill fate-and-effect model. (From Bishop, J. M., *Applied Oceanography,* John Wiley & Sons, New York, 1984. With permission.)

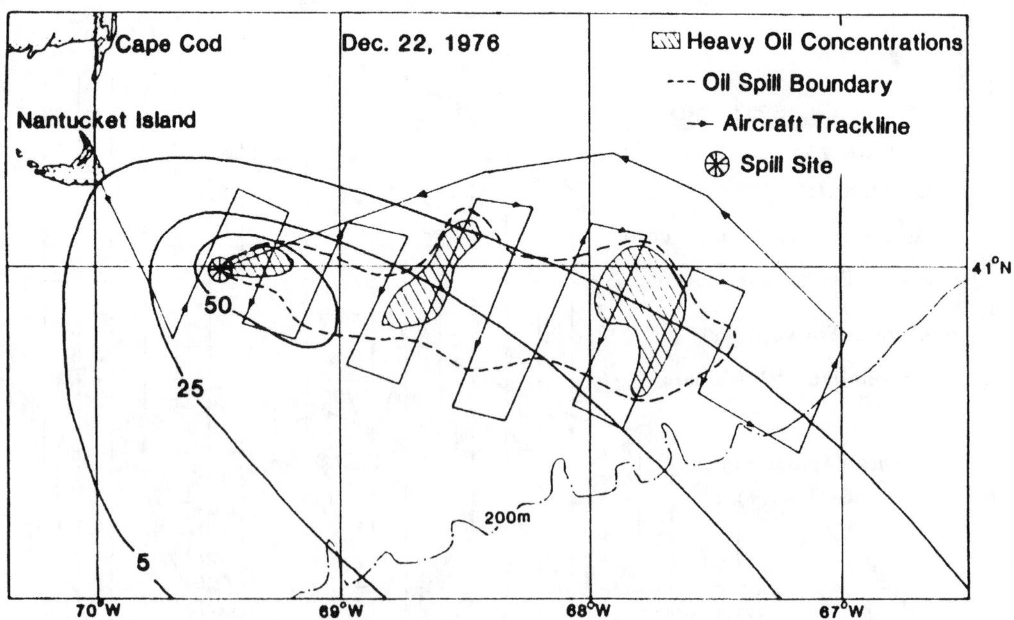

FIGURE 1.16. Observed oil concentrations and relative risk ellipses for the *Argo Merchant* oil spill off Cape Cod, MA, in December 1976. (From Bishop, J. M., *Applied Oceanography,* John Wiley & Sons, New York, 1984. With permission.)

TABLE 1.3
Polluting Incidents Reported In and Around U.S. Waters: 1970 to 1984

Item	Incidents	Gallons (1000)
1970	3711	15,253
1971	8736	8840
1972	9931	18,806
1973	13,328	24,315
1974	14,432	19,422
1975	12,781	22,243
1976	13,930	36,608
1977	15,330	11,248
1978	14,495	17,557
1979	13,134	13,661
1980	11,155	15,093
1981	10,564	19,773
1982	10,414	23,154
1983	11,346	30,076
1984	10,745	19,749
Breakdown of Incidents in 1984		
Vessel	2466	4915
Tankship	254	1830
Tank barge	545	2646
Other	1667	439
Nontransport facilities	2577	6060
Onshore	1108	5501
Offshore	198	499
Pipeline	554	1363
Marine facilities		
Onshore/offshore	521	342
Land vehicles	707	623
Land facilities	176	178
Other or unknown	3744	6266
Type of pollutant		
Crude oil	1983	1422
Diesel oil	2069	7106
Other oil	4859	5477
Other or unknown	1834	5744
Location		
Atlantic Coast	2212	6411
Gulf Coast	2306	436
Pacific Coast	1694	1877
Great Lakes	78	1251
Inland	4380	9717

Source: van der Leeden, F., Troise, F. L., and Todd, D. K., *The Water Encyclopedia,* 2nd ed., Lewis Publishers, Chelsea, MI, 1990, 515. With permission.

TABLE 1.4
Pollution Discharges in Navigable Waters of the U.S. in 1984[a]

Materials	Oil				Hazardous				Other				Total			
	No.	%	Quantity	%	No.	%	Quantity	%	No.	%	Quantity	%	No.	%	Quantity	%
Crude oil	3069	33.4	1,516,776	9.5	0	.0	0	.0	0	.0	0	.0	3069	29.6	1,516,776	8.7
Gasoline	642	7.0	593,768	3.7	0	.0	0	.0	0	.0	0	.0	642	6.2	593,768	3.4
Kerosene/fuel oil	157	1.7	579,169	3.6	0	.0	0	.0	0	.0	0	.0	157	1.5	579,169	3.3
Diesel oil	2142	23.3	7,220,236	45.2	0	.0	0	.0	0	.0	0	.0	2142	20.7	7,220,236	41.5
Fuel oil	428	4.7	3,131,237	19.6	0	.0	0	.0	0	.0	0	.0	428	4.1	3,131,237	18.0
Asphalt/tar/pitch	35	.4	57,151	.4	0	.0	0	.0	0	.0	0	.0	35	.3	57,151	.3
Other distillate	73	.8	21,793	.1	4	1.2	1031	.2	0	.0	0	.0	77	.7	22,824	.1
Solvents	6	.1	154	.0	9	2.7	343	.1	0	.0	0	.0	15	.1	497	.0
Animal/vegetable oil	196	2.1	2,277,334	14.3	0	.0	0	.0	0	.0	0	.0	196	1.9	2,277,334	13.1
Other oil	2436	26.5	566,884	3.5	0	.0	0	.0	0	.0	0	.0	2436	23.5	566,884	3.3
Chemical	0	.0	0	.0	315	96.0	672,191	99.8	0	.0	0	.0	315	3.0	672,191	3.9
Other substances	14	.2	6058	.0	0	.0	0	.0	829	100.0	748,845	100.0	843	8.1	754,903	4.3
Total	9198	100.0	15,970,560	100.0	328	100.0	673,565	100.0	829	100.0	748,845	100.0	10,355	100.0	17,392,970	100.0

[a] In gallons, by type of material discharged.

Source: U.S. Coast Guard, Polluting Incidents In and Around U.S. Waters, Calendar Year 1983 and 1984. COMDTINST M16450.2G; NTIS 87–186821.

TABLE 1.5
Polluting Incidents from Vessels in Canadian Waters, 1974 to 1983[a]

Year	Transfer Accident		Collision, Ground, Sinking		Other		Total	
	Events	Tons	Events	Tons	Events	Tons	Events	Tons
1974	60	371	21	4277	60	248	141	4896
1975	52	116	13	613	28	886	93	1615
1976	53	206	13	1613	19	160	85	1979
1977	47	249	11	931	38	294	96	1474
1978	51	154	15	1343	33	73	99	1570
1979	49	108	6	948	33	8186	88	9242
1980	68	145	12	121	68	213	148	479
1981	75	97	13	2296	33	931	121	3324
1982	58	199	16	2106	27	989	101	3294
1983	28	73	7	504	22	404	57	981
Total	541	1718	127	14,752	361	12,384	1029	28,854
Percent	53	6	12	51	35	43	100	100

[a] Tankers, bulk carriers, and other vessels; metric tons.

Source: Environment Canada, Summary of Spill Events in Canada, 1974–1983, EPS 5/SP/1.

TABLE 1.6
Severe Cases of Pollution by Crude Oil Residues in the Netherlands, Denmark, and Germany

Date	Country	Area	Type	Remarks
November 1990	NL	Whole coast	Bachaquero crude	Heavily polluted beaches
March/April 1990	DK	NW coast	Brega (crude) (Liberia)	Big pollution
February 1991	DK	NW coast	Nigerian crude	"Several tons" on the beaches
June 1991	D	Juist, Sylt	North Sea crudes (2 types)	Big pollution (in coincidence with *bis*-phenols)
July 1991	DK	NW coast	Ninian (?) crude (NS)	"Several hundred kilos"
August 1991	D	Amrum, Pellworm	Ekofisk crude (NS)	Big pollution
September 1991	DK	NW coast	South American	Big pollution
October 1991	DK	NW coast	Beatrice (?) crude	"16 tons"
November 1991	DK	NW coast	North Sea crude	"Hundreds of kilos"
March 1992	D	Amrum, Hörnum, St. Peter	Tia Juana light (Venezuela)	Heavily polluted beaches
March 1992	DK	N coast	Tia Juana light (Venezuela)	Big pollution
April 1992	DK	NW coast	Ekofisk crude (NS)	Big pollution

Note: NL = Netherlands; DK = Denmark; D = Germany; NS = North Sea.

Source: Dahlmann, G., Timm, D., Averbeck, C., Camphuysen, C., Skov, H., and Durinck, J., *Mar. Pollut. Bull.*, 28, 305, 1994. With permission.

TABLE 1.7
Severe Cases of Pollution by Bunker Oil Residues in the Netherlands, Denmark, and Germany

Date	Country	Area	Type	Remarks
February 1991	D	Helgoland	Bunker oil residues	Many birds oiled and killed
April 1991	DK	NW coast	Bunker oil residues (different types)	"500 kg"
June 1991	DK	NW coast	Bunker oil	"Some hundred kilos"
July 1991	NL		Bunker oil residues	Big pollution
December 1991	NL		Bunker oil residues	Big pollution
February 1992	NL	Texel	Bunker oil	Lot of oiled/killed birds and heavily polluted beaches
1991/92	D	Juist, Sylt	Different bunker oils	24 different bigger cases of oil pollution on German beaches

Note: D = Germany; DK = Denmark; NL = Netherlands.

Source: Dahlmann, G., Timm, D., Averbeck, C., Camphuysen, C., Skov, H., and Durinck, J., *Mar. Pollut. Bull.*, 28, 305, 1994. With permission.

TABLE 1.8
Special Cases of Pollution by Nontoxic Chemicals (Legal Discharges) in the Netherlands, Denmark, and Germany

Date	Country	Area	Type	Remarks
1990–1992	D/NL/DK		Paraffin wax	11 different cases of pollution on different sampling sites
August 1991	D	Pellworm, Langeoog, Juist, Amrum	Paraffin wax (petrolatum)	Big pollution
September 1991	D	Friedrichskoog	Paraffin wax	Big pollution
December 1991	D	Sylt	Paraffin wax	Big pollution
November 1992	D	St. Peter	Paraffin wax	About 8 t from a tank-washing
December 1992	D	Amrum	Paraffin wax	Big pollution
February/March 1993	NL	Northern part	Paraffin wax	"About 2000 birds killed", big beach pollution
February 1991	DK	Bornholm	Palm oil	Oiled birds
November 1991	NL		Palm oil	Big lumps, diameter up to 70 cm
April 1992	NL	Texel	Palm oil	Big pollution caused by a shipwreck
November 1992	NL		Palm oil	Bigger pollution
July 1991	D	Sylt	Palm oil	Bigger pollution
November 1991	D	Föhr	Palm oil	Pollution (8 kg)
May 1991	D	Helgoland	Palm oil	Big pollution
November 1992	D	Sylt	Palm oil	Pollution
November 1992	DK	W coast	C12 fatty acid	Pollution

Note: D = Germany; NL = Netherlands; DK = Denmark.

Source: Dahlmann, G., Timm, D., Averbeck, C., Camphuysen, C., Skov, H., and Durinck, J., *Mar. Pollut. Bull.*, 28, 305, 1994. With permission.

TABLE 1.9
Special Cases of Pollution by — More or Less — Toxic Chemicals in the Netherlands, Denmark, and Germany

Date	Country	Area	Type	Remarks
February 1990	NL		Dodecylphenols + lubricating oil	Polluted birds
March 1990	DK	Bornholm	Dodecylphenols + lubricating oil	Polluted birds
November 1990	NL		Dodecylphenols + lubricating oil	Polluted birds
	DK	Juist		
January 1991	D	Helgoland	Dodecylphenols + 1	Polluted birds
February/March 1991	NL		Thiophenes (?)	Polluted birds
March 1991	DK	Baltic	Phenols (insecticides)	Pollution
	VII			
April 1991	DK	NW coast	Alk. *bis*-phenols	"300 kg", beaches heavily polluted
June 1991	D	Juist, Sylt	Alk. *bis*-phenols	Several tons on the beaches
January/February 1992	D	Helgoland, Sylt	Alkylated aromatics (special fuel oil?)	Many polluted birds
	NL			
	N	Roga		
April 1992	DK	Bornholm	Plasticizer (phthalates)	Pollution
July 1991	DK	NW coast	Detergent (di-*t*-butyl-phosphin?)	Pollution

Note: NL = Netherlands; DK = Denmark; D = Germany.

Source: Dahlmann, G., Timm, D., Averbeck, C., Camphuysen, C., Skov, H., and Durinck, J., *Mar. Pollut. Bull.*, 28, 305, 1994. With permission.

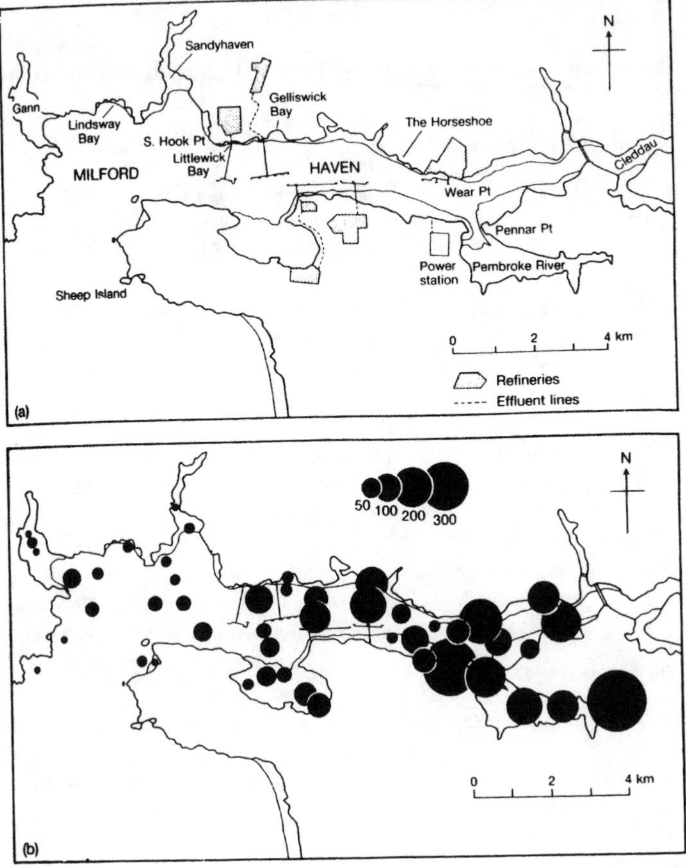

FIGURE 1.17. Location of oil terminals and refineries (a) as well as hydrocarbons in sediments (b) at Milford Haven, Pembrokeshire, England. (From Clark, R. B., *Marine Pollution,* Clarendon Press, Oxford, 1986. With permission.)

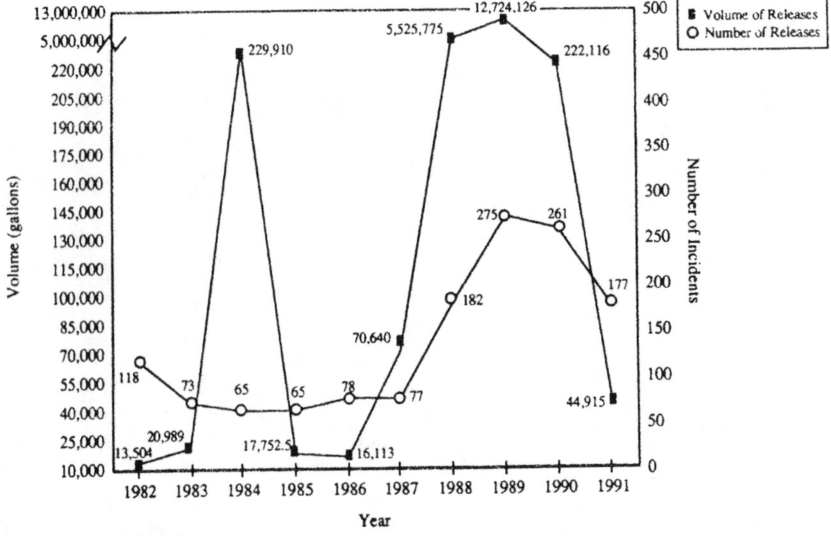

FIGURE 1.18. Total volume and frequency of petroleum products and hazardous chemicals annually released into the Newark Bay estuary between 1982 and 1991. (From Gunster, D. G., Gillis, C. A., Bonnevie, N. L., Abel, T. B., and Wenning, R. J., *Environ. Pollut.,* 82, 245, 1993. With permission.)

TABLE 1.10
Number of Petroleum and Chemical Spills in the Newark Bay Estuary Reported to the U.S. Coast Guard National Response Center Between 1982 and 1991[a]

Receiving Water	1982		1983		1984		1985		1986	
	No. of Incidents	Total Volume (U.S. gal)	No. of Incidents	Total Volume (U.S. gal)	No. of Incidents	Total Volume (U.S. gal)	No. of Incidents	Total Volume (U.S. gal)	No. of Incidents	Total Volume (U.S. gal)
Arthur Kill	61	2782	41	8324	26	217,306	27	2845.5	20	1178
Elizabeth River	1	95	3	1000	0	—	1	—	0	—
Hackensack River	2	63	3	5300	4	801	2	20	3	1500
Kill Van Kull	7	375	2	15	5	2186	5	243	4	—
Morses Creek	0	—	0	—	0	—	0	—	0	—
Newark Bay	3	90	6	348	4	5044	2	284	1	10
Passaic River	4	345	4	40	7	1020	7	473	13	8160
Rahway River	0	—	0	—	1	10	1	—	3	110
Woodbridge Creek	10	83	0	—	0	—	1	100	0	—
Unknown location[b]	22	2525	4	480	10	375	11	138	20	3600
Sewer[c]	8	7146	10	5482	9	3168	8	13,649	14	1555

TABLE 1.10 (continued)
Number of Petroleum and Chemical Spills in the Newark Bay Estuary Reported to the U.S. Coast Guard National Response Center Between 1982 and 1991[a]

	1987		1988		1989		1990		1991[d]	
Receiving Water	No. of Incidents	Total Volume (U.S. gal)	No. of Incidents	Total Volume (U.S. gal)	No. of Incidents	Total Volume (U.S. gal)	No. of Incidents	Total Volume (U.S. gal)	No. of Incidents	Total Volume (U.S. gal)
Arthur Kill	29	51,456	87	13,698	118	12,612,596	124	15,349	69	16,600
Elizabeth River	0	—	3	200	1	84	0	—	3	20
Hackensack River	3	100	9	5,504,500	16	59,450	13	660	4	1500
Kill Van Kull	10	168	18	1177	50	37,427	52	1982	37	3339
Morses Creek	0	—	2	3101	4	42	1	1	0	—
Newark Bay	4	NQ[e]	10	1097	18	1025	14	23	16	3506
Passaic River	10	15,210	20	357	11	NQ	8	10	11	310
Rahway River	1	50	5	NQ	2	NQ	3	50	0	—
Woodbridge Creek	2	30	2	12	2	NQ	2	126	1	NQ
Unknown location[b]	5	140	9	110	22	1912	7	22	11	392
Sewer[c]	13	3486	17	1523	31	11,590	37	203,893	25	19,248

[a] Total volumes reflect only those spills that were reported. Actual volumes are likely to be higher.
[b] The location of the spill incident was not reported in the USCG database.
[c] The USCG database identified municipal and industrial rivers as the ultimate receiving waterways for these spill incidents.
[d] Data for 1991 include all spills occurring on or before 24 November 1991.
[e] NQ indicates that the volume of the reported spills was not quantified.

Source: Gunster, D. G., Gillis, C. A., Bonnevie, N. L., Abel, T. B., and Wenning, R. J., *Environ. Pollut.*, 82, 245, 1993. With permission.

TABLE 1.11
Total Volumes of Petroleum Products and Hazardous Chemicals Accidentally Discharged to Newark Bay and Its Major Tributaries from 1982 to 1991 as Reported by the U.S. Coast Guard National Response Center[a]

Material	1982	1983	1984	1985	1986	1987	1988	1989	1990	1991[b]	Total Volume
Petroleum products											
No. 1 fuel oil		15	100								115
No. 2 fuel oil	3939	122	76	792	2645	156	2940	36,726	573	520	48,489
No. 4 fuel oil	5							40		2900	2945
No. 6 fuel oil	445	215	214,788	8	10	23	912	12,607,910	1083	3878	12,829,272
Asphalt	42			210			42				294
Crude oil						30					30
Diesel fuel	450	195	20	103	105	155	440	15,919	2673	1180	21,240
Gasoline	555	6052	97	1181	8235	1320	4128	5325	7554	14,369	48,816
Hydraulic oil							6	125	23		154
Jet fuel							66	40	4300	30	4436
Kerosene	22						10	41			73
Motor oil	0.1					3	6				9
Oil other	637	568	5	150	112	76	397	1463	203,059	366	206,833
Oil unknown	1230		3292	353	165	21	2850	463	544	70	7888
Transformer oil						15,000	109	40			15,149
Subtotal	6225	7167	218,378	2797	11,272	16,784	11,906	12,668,092	219,809	23,313	13,185,743
Hazardous chemicals											
Acids	1381	1000	1290	1150	20	22,700	5158		1310	375	35,384
Chemicals	201	5000	2627	50	3000	30,150	570	1417	210	3080	46,305
Polychlorinated biphenyls		40		10							50
Solvents	500	1500			1400				125	150	3675
Subtotal	2082	7540	3917	1210	4420	52,850	5728	1417	1636	3605	84,414
Other											
Other	5205	5776	7605	13,305	415	905	5,507,856	54,615	657	17,796	5,614,135
Unknown		500	10	442	5	100	240			201	1498
Subtotal	5205	6276	7615	13,747	420	1005	5,508,096	54,615	657	17,997	5,615,633
Total	13,512	20,983	229,910	17,754	16,112	70,639	5,525,730	12,724,124	222,111	44,915	18,885,790

[a] Total volumes reflect only those spills that were reported. Actual volumes are likely to be higher.
[b] Data for 1991 include all spills occurring on or before 24 November 1991.
[c] The composition of the material spilled was not reported in the USCG database.

Source: Gunster, D. G., Gillis, C. A., Bonnevie, N. L., Abel, T. B., and Wenning, R. J., *Environ. Pollut.*, 82, 245, 1993. With permission.

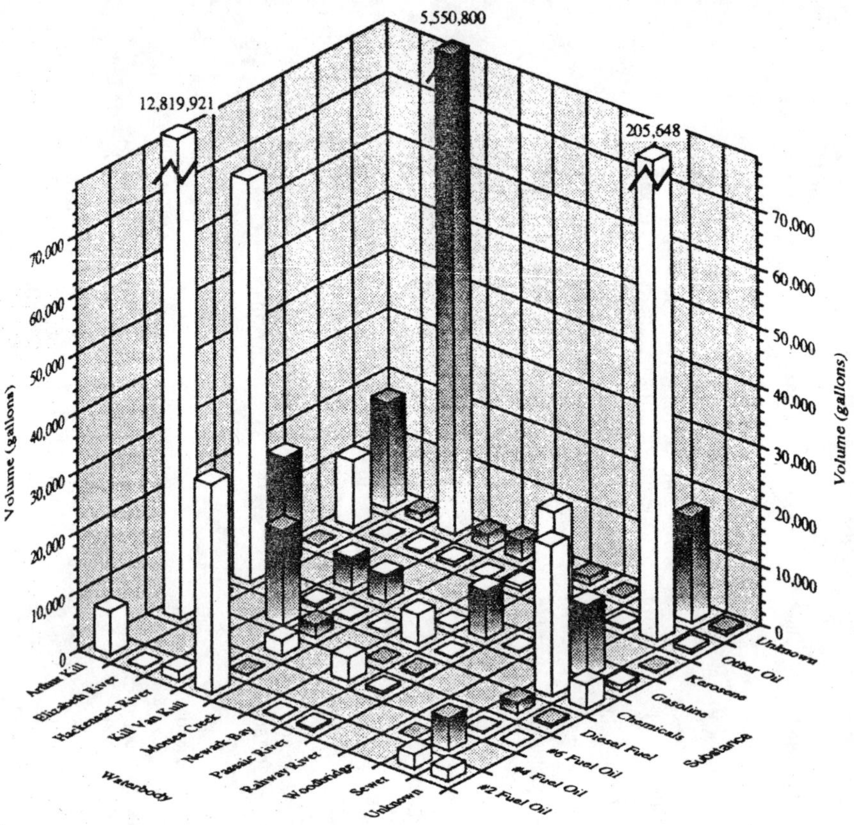

FIGURE 1.19. Total volume of petroleum products and hazardous chemicals released to selected waterways within the Newark Bay estuary between 1982 and 1991. (From Gunster, D. G., Gillis, C. A., Bonnevie, N. L., Abel, T. B., and Wenning, R. J., *Environ. Pollut.*, 82, 245, 1993. With permission.)

TABLE 1.12
Oil and Gas Production in the Gulf of Mexico

	U.S. Total	U.S. Offshore	Gulf of Mexico
No. of wells drilled			
1954–1984	—	—	23,414
July 1985–June 1985	—	—	721
Producing wells, 1984	—	11,000	8983
Production platforms, 1984	—	3896	3866
Crude oil (million m^3 d^{-1})			
1985	1.412	0.1696	0.156 (11.1%)[d]
1987	1.32, 1.339	—	—
Natural gas			
1985 (bcm)[a]	490.0	113.30	111.90 (22.8%)[e]
	(17,304 bcf)[c]	(4001.0 bcf)	(3951.8 bcf)
1987 (bcm)	453.1	—	—
	(16,000 bcf)	—	—
1985 (mmcm d^{-1})[b]	1343	310.41	306.6
	(47,410 mmcf d^{-1})[f]	(10,962 mmcf d^{-1})	(10,827 mmcf d^{-1})

[a] bcm: billion cubic meters.
[b] mmcm d^{-1}: million cubic meters per day.
[c] bcf: billion cubic feet.
[d] 11.1% of the U.S. total.
[e] 22.8% of the U.S. total.
[f] mmcf d^{-1}: million cubic feet per day.

Source: Fang, C. S., *Estuaries,* 13, 89, 1990. With permission.

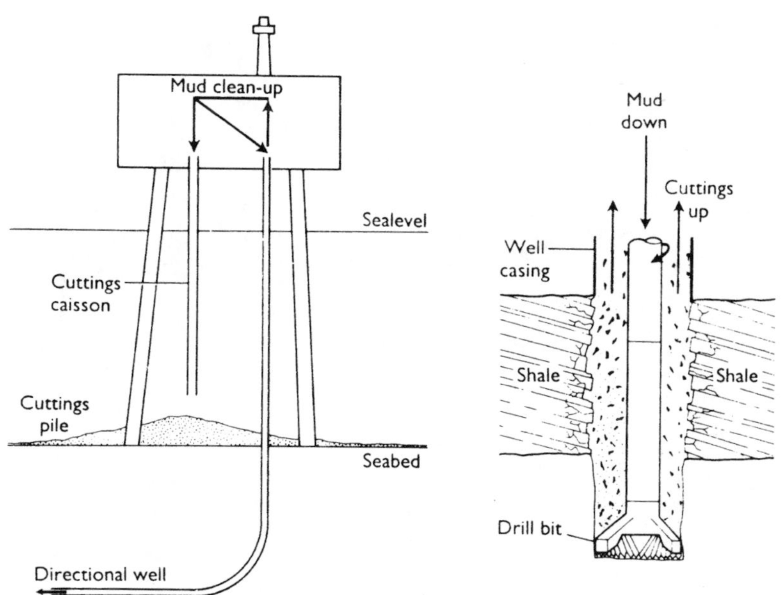

FIGURE 1.20. Diagrammatic representation of typical cutting discharges and drilling operations in the North Sea. (From Clark, R. B., *Marine Pollution,* 3rd ed., Clarendon Press, Oxford, 1992. With permission.)

TABLE 1.13
Characteristics of Drilling Fluid[a]

Total suspended solids ($g\ l^{-1}$)	450
Density ($g\ cm^{-3}$)	1.25
pH	11
Lignosulfonate (% of whole fluid)	0.5
Ba (%, dry weight)	9
$CaCO_3$ (%, dry weight)	6
Fe (%, dry weight)	2.7
Cr ($\mu g\ g^{-1}$)	500
Zn ($\mu g\ g^{-1}$)	370
Mn ($\mu g\ g^{-1}$)	290
Cu ($\mu g\ g^{-1}$)	43
Percentage of solids with a grain size	
>62 μm	2
2–62 μm	40
<2 μm	58

[a] Discharged from an 1800-m-deep bore hole in the outer continental shelf of the northwestern Gulf of Mexico.

Source: Trefry, J. H., Trocine, R. P., and Prioni, J. R., in *Wastes in the Ocean,* Vol. 4, *Energy Wastes in the Ocean,* Duedall, I. W., Kester, D. R., Park, P. K., and Ketchum, B. H. (Eds.), John Wiley & Sons, New York, 1985, 195. With permission.

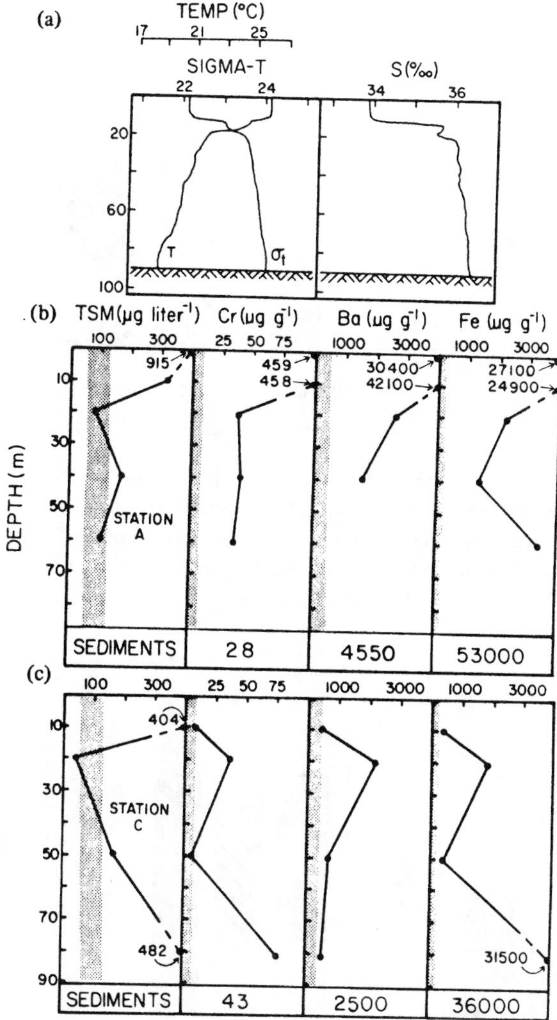

FIGURE 1.21. (a) Salinity, temperature, density (Sigma-t), and background suspended matter concentrations at the northwestern Gulf of Mexico drilling-fluid discharge site; (b) total suspended matter (TSM) concentrations and particulate Cr, Ba, and Fe concentrations at a discharge sampling station 0.2 km from the drilling rig, 2 h into the 3-h discharge; and (c) at a sampling station 0.8 km from the drilling rig, 8 h following the end of the discharge. Shaded areas represent background levels. (From Trefry, J. H., Trocine, R. P., and Prioni, J. R., in *Wastes in the Ocean,* Vol. 4, *Energy Wastes in the Ocean,* Duedall, I. W., Kester, D. R., Park, P. K., and Ketchum, B. H. (Eds.), John Wiley & Sons, New York, 1985, 195. With permission.)

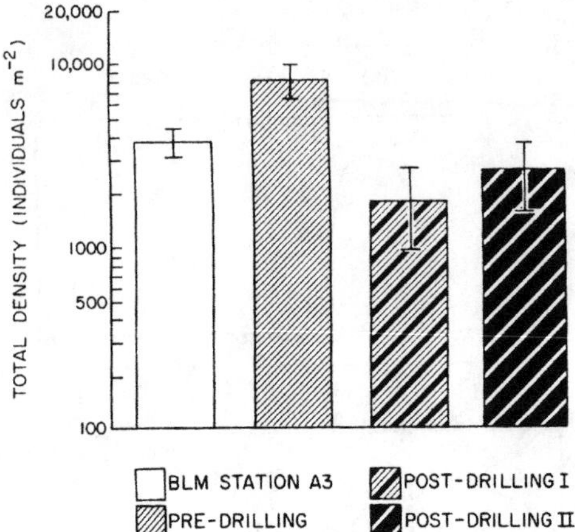

FIGURE 1.22. Density of macrobenthos observed during pre-drilling and two post-drilling surveys at a sampling station in the northwestern Gulf of Mexico. Bars indicate standard deviation. (From Gillmor, R. B., Menzie, C. A., Mariani, G. M., Levin, D. R., Ayres, R. C., Jr., and Sauer, T. C., Jr., in *Wastes in the Ocean,* Vol. 4, *Energy Wastes in the Ocean,* Duedall, I. W., Kester, D. R., Park, P. K., and Ketchum, B. H. (Eds.), John Wiley & Sons, New York, 1985, 243. With permission.)

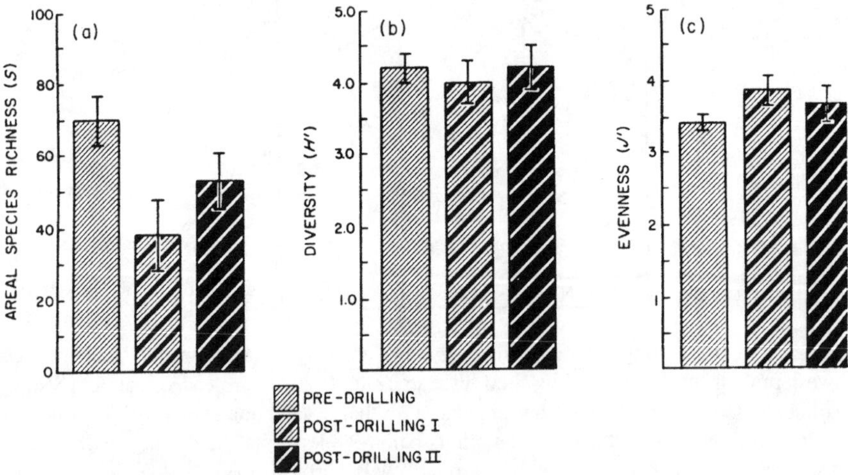

FIGURE 1.23. (a) Areal richness, (b) diversity, and (c) evenness of the macrofaunal assemblages observed during pre-drilling and two post-drilling surveys at a sampling station in the northwestern Gulf of Mexico. Bars indicate standard deviation. (From Gillmor, R. B., Menzie, C. A., Mariani, G. M., Levin, D. R., Ayres, R. C., Jr., and Sauer, T. C., Jr., in *Wastes in the Ocean,* Vol. 4, *Energy Wastes in the Ocean,* Duedall, I. W., Kester, D. R., Park, P. K., and Ketchum, B. H. (Eds.), John Wiley & Sons, New York, 1985, 243. With permission.)

TABLE 1.14
Concentration of Major and Minor Metals in Drilling Fluids

Drilling Rig Location	Total Solids ($g\ ml^{-1}$)	Concentration ($mg\ g^{-1}$, dry weight) Ba	Fe	Cr	Mn	Zn	Cu	Ni
Baltimore Canyon	0.30	310	22	1.1	0.44	0.5	0.03	0.08
Mobile Bay								
Sample 1 (15 May 1979)	0.28	57	46	2.5	0.47	0.8	0.04	0.06
Sample 2 (4 September 1979)	0.76	340	9	4.6	0.43	1.1	0.06	0.06
Galveston Island	0.76	290	25	0.8	0.80	0.4	0.11	0.10

Source: Gilbert, T. R., Penney, B. A., Liss, R. G., and Wayne, D. A., in *Wastes in the Ocean,* Vol. 4, *Energy Wastes in the Ocean,* Duedall, I. W., Kester, D. R., Park, P. K., and Ketchum, B. H. (Eds.), John Wiley & Sons, New York, 1985, 271. With permission.

TABLE 1.15
Wastes Produced from Present Fuel Cycles

Fuel	Drilling and Mining Wastes	Middle Wastes	Combustion and Dismantlement Wastes
Oil	Drilling fluid	Ballast and wash water	CO_2, nitrogen oxides, sulfur oxides; fine particulate matter
		Spills from accidents	
	Formation water (brine solutions)	H_2S and SO_2 from petroleum production	Heat; wastes from combustion of refined petroleum and a wide variety of other petroleum-derived products; discarded drilling and production platforms, pipelines
Natural gas	Drilling fluid, formation water	H_2S and SO_2 from petroleum processing	CO_2, nitrogen oxides, sulfur oxides; heat; discarded drilling and production platforms, pipelines
Coal	Acid mine-drainage waters	Seepage from coal-storage areas	CO_2, nitrogen oxides, sulfur oxides, fly ash, FGD[a] sludge, bottom ash, wastes from other coal-derived products; heat
	Coal dust	Residue from coal cleaning	
		Coal dust	
		Wastes from coal-processing systems	
Nuclear	Mine and mill tailings	Discarded equipment	Radioactive gases, chlorine, spent fuel rods, heat, discarded plants

[a] FGD = flue-gas desulfurization.

Source: Duedall, I. W., Kester, D. R., Park, P. K., and Ketchum, B. H., in *Wastes in the Ocean,* Vol 4, *Energy Wastes in the Ocean,* Duedall, I. W., Kester, D. R., Park, P. K., and Ketchum, B. H. (Eds.), John Wiley & Sons, New York, 1985, 3. With permission.

APPENDIX 2. EFFECTS OF OIL POLLUTION ON ORGANISMS

TABLE 2.1
Some of the Commonly Reported Effects of Petroleum and Individual PAHs on Living Organisms[a]

Effects	Plant	Invertebrate	Fish	Reptile and Amphibian	Bird	Mammal
Individual organisms						
Death	X	X	X	X	X	X
Impaired reproduction	X	X	X	X	X	
Reduced growth and development	X	X	X	X	X	
Impaired immune system						X
Altered endocrine function			X		X	
Altered rate of photosynthesis	X					
Malformations			X		X	
Tumors and lesions		X	X	X		X
Cancer			X	X		X
Altered behavior		X	X	X	X	X
Blood disorders		X	X	X	X	X
Liver and kidney disorders			X		X	X
Hypothermia					X	X
Inflammation of epithelial tissue					X	X
Altered respiration or heart rate		X	X	X		
Impaired salt gland function				X	X	
Gill hyperplasia			X			
Fin erosion			X			
Groups of organisms[a,b]						
Local population changes	X	X			X	
Altered community structure	X	X			X	
Biomass change	X	X				

[a] Some effects have been observed in the wild and in the laboratory, whereas others have only been induced in laboratory experiments.

[b] Populations of chlorophyllous (microalgae) and nonchlorophyllous plants (bacteria, filamentous fungi, yeast) can increase or decrease in the presence of petroleum, whereas animal populations decrease.

Source: Albers, P. H., in *Handbook of Ecotoxicology,* Hoffman, D. J., Rattner, B. A., Burton, G. A., Jr., and Cairns, J., Jr. (Eds.), Lewis Publishers, Boca Raton, FL, 1994, 330. With permission.

TABLE 2.2
Summary of the Effects of Oil Spills on Oceanic Communities

Community	Effect	Period of Impact[a]
Plankton		
Phytoplankton biomass and primary production	Increase due to diminished grazing; depression of chlorophyll *a*	Days to weeks, during occurrence of slicks
Zooplankton	Population reduction; contamination	
Fish eggs	Decreased hatching and survival	
Benthos		
Amphipods, isopods, ostracods	Initial mortality; population decrease	Weeks to years, depending on oil-retentive characteristics of habitat
Mollusks, especially bivalves	Initial mortality; contamination; histopathology	
Opportunistic polychaetes	Population increase	
Overall macrobenthic community	Decreased diversity	
Intertidal and littoral		
Meiofaunal crustaceans, crabs	Initial mortality; population decrease	Weeks to years, depending on oil-retentive characteristics of habitat
Mollusks	Initial mortality; contamination; histopathology	
Opportunistic polychaetes	Population increase	
Overall community	Decreased diversity	
Algae	Decreased biomass; species replacement	
Phanerogams	Initial die-back	
Fish		
Eggs and larvae	Decreased hatching and survival	Weeks to months
Adults	Initial mortality; contamination; histopathology	
Birds		
Adults	Mortality; population decrease	Years

[a] Period of impact depends on the scale and duration of the spill and on the oceanic characteristics of the specific system.

Source: Wolfe, D. A., in *Wastes in the Ocean,* Vol. 4, *Energy Wastes in the Ocean,* Duedall, I. W., Kester, D. R., Park, P. K., and Ketchum, B. H. (Eds.), John Wiley & Sons, New York, 1985, 45. With permission.

TABLE 2.3
Sublethal Effects of Petroleum Exposure on Marine Organisms in the Laboratory[a]

Effect	Species	Exposure Regime and Effective Concentration[b]
Reduced avoidance response	Spotted seatrout larvae (*Cynoscion nebulosus*)	No. 2 fuel WSF for 48 h at 0.01–1.0 μg g^{-1} (nominal)
50% reduction in defense response to predator	Sea urchin (*Strongylocentrotus droebachiensis*)	Prudhoe Bay Crude WSF for 15 min at 50 ng g^{-1} (measured)
Oil detection, indicated by antennular flicking rate	Blue crabs (*Callinectes sapidus*)	Prudhoe Bay Crude WSF for 2 min at 2 × 10^{-6} μg g^{-1} (median threshold detection limit)
		Naphthalene for 3 min at 10^{-7} μg g^{-1} (median threshold detection limit)
Oil detection, indicated by antennular flicking rate	Dungeness crabs (*Cancer magister*)	Prudhoe Bay Crude WSF for 1 min at 0.1 μg g^{-1} (median threshold detection limit)
		Naphthalene for 1 min at 0.01 μg g^{-1} (median threshold detection limit)
Decreased burying or increased surfacing in sediment	Intertidal clams (*Macoma balthica*)	Prudhoe Bay Crude OWD for 180 d at 0.3 μg g^{-1} (response threshold)
Resurfacing within 3 to 5 d	Intertidal clams (*Macoma balthica*)	Prudhoe Bay Crude WSF for 6 d at ~0.3 μg g^{-1} (median response)
Failure to burrow	Intertidal clams (*Macoma balthica*)	Prudhoe Bay Crude WSF for 6 d at ~2.5 μg g^{-1}
38% depression of feeding rate	Copepods (*Eurytemora affinis*)	Aromatic heating oil WSF for 24 h at 0.52 μg g^{-1}
Depression of feeding rate	Copepods (*Acartia clausi* and *Acartia tonsa*)	No. 2 fuel oil WAF for 18 h at 250 ng g^{-1}
Depression of feeding rate	Amphipods (*Boeckosimus affinis*)	Prudhoe Bay Crude WSF for 16 weeks at ≤0.2 μg g^{-1}
Food detection and finding slowed by one half	Mud snails (*Ilyanassa obsoleta*)	No. 2 fuel oil OWD for 48 h at 0.015 μg g^{-1}
	Mud snails (*Ilyanassa obsoleta*)	Kerosene WSF with a continuous exposure to 4 ng g^{-1}
Feeding rate and growth rate depressed	Rock-crab larvae (*Cancer irroratus*)	No. 2 fuel oil WAF for 27 d at 0.1 μg g^{-1}
Food detection (reduced antennular flicking in response to food extract)	Dungeness crab (*Cancer magister*)	Prudhoe Bay Crude WSF for 24 h at 0.25 μg g^{-1}
Inhibition of attraction to reproductive aggregation	Nudibranchs (*Onchidoris bilamellata*)	Prudhoe Bay Crude WSF for 24 h at 10 ng g^{-1}
		Toluene for 24 h at 32 ng g^{-1}
Reproduction (inhibited reproductive aggregation, delayed and reduced egg deposition)	Nudibranchs (*Onchidoris bilamellata*)	Prudhoe Bay Crude WSF for 14 d at 278 ng g^{-1}
Histopathology	Mummichogs (*Fundulus heteroclitus*)	
Lateral line necrosis		Naphthalene for 15 d at 0.02 μg g^{-1}
Ischemia in brain, liver		Naphthalene for 15 d at 0.2 μg g^{-1}
Inclusions in muscle columnar cells	Chinook salmon (*Oncorhynchus tshawytscha*)	Mixed aromatic compounds for 28 d at 5 μg g^{-1} in food
Delayed overall embryonic development	Pacific herring embryos (*Clupea harengus pallasi*)	Benzene for 24 h at 0.9 ng g^{-1}

[a] The effects documented here were selected based on the degree of realism of exposure conditions relative to chronic discharges and the potential significance of the measured effect to survival and reproduction of the organism. Many other sublethal effects have been documented at higher exposure concentrations.

[b] Exposure concentrations and compositional characteristics depend strongly on the techniques for preparing the oil-water mixtures. This table retains the designations used by the original authors for their preparations. WSF is water-soluble fraction, OWD is oil-water dispersion, and WAF is the water-accommodated fraction.

Source: Wolfe, D. A., in *Wastes in the Ocean,* Vol. 4, *Energy Wastes in the Ocean,* Duedall, I. W., Kester, D. R., Park, P. K., and Ketchum, B. H. (Eds.), John Wiley & Sons, New York, 1985, 45. With permission.

TABLE 2.4
Some Biotic Effects Associated with Low Concentrations of Oil

Organism	Exposure Period	Type of Hydrocarbon	Lowest Concentration Tested	Concentration (ha ng) Effect	Effects
Phytoplankton, various spp.	Hours	Oil and oil products	10 ppb	10–100 ppb	Growth inhibition
Phytoplankton, *Thalassiosira pseudonana*	Hours	#2 fuel oil	40 ppb	40 ppb	Growth inhibition
Phytoplankton, various spp.	19 days	#2 fuel oil	10 ppb	20 ppb	Microflagellates replaced diatoms
Plaice eggs	Days	Oil and oil products	10 ppb	10 ppb	40% mortality
Amphipod, *Gammarus oceanicus*	20–23 days	Venezuelan crude oil	1 ppm added, 0.3–0.4 ppm detected	0.3–0.4 ppm	Reduced reproduction
Gastropod, *Nassarius obsoletus*	Minutes	Kerosene extract	4 ppb	4 ppb	Interference with chemoreception
Fish, *Fundulus* and *Stenotomus*	Up to 1 month	#2 fuel oil	125–200 ppb	125–200 ppb	Altered metabolic systems
Oyster, *Crassostrea virginica*	50 days	#2 fuel oil	106 ppb	106 ppb	20-fold increase in hydrocarbon content
Copepod, *Calanus helgolandicus*	Hours	Naphthalene	0.1 ppb	0.1 ppb	Storage of naphthalene
Oyster, *Crassostrea virginica*	Months	Unreported oil constituents	10 ppb	10 ppb	Tainted flesh

Source: Hall, C. A. S., Howarth, R., Moore, III, B., and Vorosmarty, C. J., *Ann. Rev. Energy,* 3, 395, 1978. With permission.

TABLE 2.5
Influence of Crude Oil on Above-Ground Biomass, New Shoots, and Stem Density of *Spartina alterniflora* in a Louisiana Saltmarsh

Crude Oil ($l\ m^{-2}$)	Above-Ground Biomass ($g\ m^{-2}$)		New Shoots (number per m^2)	Stem Density (number per m^2)
	23 September 1976	16 September 1977	18 April 1977	16 September 1977
0	2000 ± 213	1001 ± 133	116 ± 25	205 ± 30
1	1908 ± 300	822 ± 72	102 ± 13	199 ± 19
2	2015 ± 153	1161 ± 277	111 ± 10	215 ± 21
4	1819 ± 265	935 ± 110	101 ± 10	219 ± 2
8	1832 ± 153	991 ± 162	98 ± 19	197 ± 20
Least significant difference ($\alpha = 0.05$)	316	236	22	28

Source: Delaune, R. D., Patrick, W. H., Jr., and Buresh, R. J., *Environ. Pollut.*, 20, 21, 1979. With permission.

TABLE 2.6
Sediment Hydrocarbon Content in Mangrove Habitat Along the Southwest Coast of Puerto Rico ($\mu g\ g^{-1}$)

Sample	n	Resolved Alkanes		UCM		Total Aliphatic		PAH		Total Hydrocarbons	
		Mean	SD	Mean	SD	Mean	SD	Mean	SD	Mean	SD
Guayanilla Thermal Cove	3	40.4	14.5	1792.4	1169.3	1832.8	1178.5	58.8	7.8	1891.6	1177.2
Bahía Sucia	6	16.7	30.1	92.7	93.9	109.4	117.0	41.7	40.1	151.1	155.5
Canal Parguera	3	90.2	79.4	188.1	119.0	278.3	129.6	6.0	1.3	284.3	130.9
Phosphorescent Bay	3	45.8	21.2	93.7	55.6	139.5	52.4	1.1	0.5	140.6	52.2
Bahía Montalva	3	37.3	17.2	33.9	20.9	71.2	34.1	0.9	0.6	72.1	33.8
Cabo Rojo West	2	9.5	5.0	25.6	5.1	35.1	0.1	0.6	0.1	35.7	0.2
Guayanilla West	3	34.0	48.8	79.8	70.3	113.8	117.8	0.5	0.1	114.4	117.9

Note: UCM = unresolved complex mixture; PAH = polycyclic aromatic hydrocarbons.

Source: Klekowski, E. J., Jr., Corredor, J. E., Morell, J. M., and Del Castillo, C. A., *Mar. Pollut. Bull.*, 28, 166, 1994. With permission.

TABLE 2.7
Tropical and Subtropical Seagrass Dispersant Oil and Oil Effects on Seagrasses

Location	Type	Dispersant Used (Dilution)	Type and Conc. of Dispersed Oil	Amount of Spill	Date	Resource Affected	Impact	Dispersant Effect
Miami, FL	Lab outdoors	Corexit 9527 (1:20)	50 ppm oil, 1:20, 24 h	50 ppm oil, lab	1984	*Thalassia testudinum*	LD_{50} 12- and 96-h bioassays, oil and dispersed oil	Oil with dispersant has lower toxicity than without dispersant
Miami, FL	Lab outdoors	Corexit 9527 (1:20)	LA crude, Murban	Lab	1983–1984	*Thalassia*	LD_{50} vs. time and conc. at 5 to 100 h	At medium conc.
						Halodule		High
						Syringudium		High
Miami, FL	Lab outdoors	Arco D-609 (1:10)	LA crude, Murban	Lab		*Thalassia*	LD_{50} 5 h 100 h	Low to medium
						Halodule		Low to medium
						Syringudium		Low to medium at 75 and 125 ml
Miami, FL	Lab outdoors	Conco K (K) (1:10)	LA crude, Murban	Lab		*Thalassia*	LD_{50} at 5 and 100 h	Medium to high
						Halodule		High
						Syringudium		High
Panama	Field	Corexit 9527	Prudhoe Bay crude, 50 ppm 24 h	Lab	1985	*T. testudinum*	None to *Thalassia*	No effect on *Thalassia*
Miami, FL	Lab outdoors	Corexit 9550 (1:20)	LA crude 125 and 75 ml oil 1:20 disp. in 100,000 cc SW	Lab	1986	*Thalassia*	LD_{50} at 100 h	Low
						Halodule		Medium
						Syringudium		Low to medium
Miami, FL	Lab outdoors	OFC-D-607 (1:10)	LA crude, 75 and 125 ml	Lab	1986	*Thalassia*	LD_{50} at 100 h	Low
						Halodule		Low
						Syringudium		Medium
Miami, FL	Lab outdoors	Cold Clean 500 (1:10)	LA crude, 75 and 125 ml in 100,000 cc SW	Lab	1986	*Thalassia*	LD_{50} 100 h	Low
						Halodule		Low to medium
						Syringudium		Low
Miami, FL	Lab	Finsol OSP-7 (1:10)	LA crude, 75 and 125 ml in 100,000 cc SW	Lab	1986	*Thalassia*	LD_{50} 100 h	Medium
						Hodule		Low
						Syringudium		Low to medium low

Note: LA = Louisiana; SW = sea water.

Source: Thorhaug et al., in *Pollution in Tropical Aquatic Systems,* Connell, D. W. and Hawker, D. W. (Eds.), CRC Press, Boca Raton, FL, 1992, 110. With permission.

TABLE 2.8
A Comparison of the Effect of Murban Oil vs. Louisiana Crude Plus Conco K(K) Dispersant on *Thalassia, Halodule,* and *Syringodium* at a Variety of Exposure Times and Volumes in 100 l Sea Water

				Thalassia testudinum				*Halodule wrightii*				*Springodium filiforme*			
Dosage (ppm)	Volume of Dispersant (ml)	Exposure (hr)	Dispersant/Oil	n	G 95% C.L.	CV (%)	Mortality (%)	n	G 95% C.L.	CV (%)	Mortality (%)	n	G 95% C.L.	CV (%)	Mortality (%)
Murban oil															
75.0	7.5	5	1/10	15	3.80 ± 0.21	42	13	15	3.34 ± 0.21	34	13	15	3.25 ± 0.61	31	13
75.0	7.5	100	1/10	15	2.78 ± 0.33	61	40	15	1.11 ± 0.25	29	87	15	1.35 ± 0.38	25	100
125.0	12.5	5	1/10	15	3.21 ± 0.46	38	20	15	3.04 ± 0.61	58	20	15	3.15 ± 0.43	61	27
125.0	12.5	100	1/10	15	1.21 ± 0.65	29	73	15	0.68 ± 0.21	30	100	15	0.93 ± 0.25	44	100
0	0	100	—	15	4.15 ± 0.21	18	0	15	4.51 ± 0.65	24	0	15	4.15 ± 0.31	25	0
Louisiana crude															
75.0	7.5	5	1/10	15	4.21 ± 0.11	28	0	15	3.74 ± 0.23	32	0	15	3.55 ± 0.28	24	7
75.0	7.5	100	1/10	15	3.10 ± 0.61	41	33	15	1.98 ± 0.23	41	73	15	1.65 ± 0.88	45	87
125.0	12.5	5	1/10	15	3.98 ± 0.25	13	7	15	3.25 ± 0.45	35	7	15	3.10 ± 0.36	31	13
125.0	12.5	100	1/10	15	2.51 ± 0.41	37	40	15	1.66 ± 0.41	48	100	15	1.05 ± 0.47	52	100
0	0	100	—	15	4.25 ± 0.21	21	0	15	4.10 ± 0.21	21	0	15	3.99 ± 0.24	21	7

Note: Mean specific growth rates (G % per day, 95% confidence limits (C.L.)) and coefficient of variation in growth rates (CV%) are given on number of blades measured (n) for each species of seagrass.

Source: Thorhaug, A. and Marcus, J., in *Proceedings of the 1987 Oil Spill Conference,* Publ. 4452, American Petroleum Institute, Washington, D.C., 1987, 223. With permission.

TABLE 2.9
Seagrasses in Jamaica vs. Dispersant Toxicity

Dispersant Product	Dispersability Ratio[a]	Cost[b]	Mortality: 125 ml; 6 h		Mortality: 75 ml; 6 h		Mortality: 12.5 ml; 6 h	
			Thal	*Hal*	*Thal*	*Hal*	*Thal*	*Hal*
Conco	0.580	0.59	100	82	48	70	17	35
OFC D609	0.007	0.08	70	63	25	70	20	25
Corexit 9527	0.009	0.11	89	93	42	72	10	22
Kemrarine	—	—	63	68	42	68	17	28
ADP 7	—	—	50	46	18	68	21	30
Corexit 9550	0.009	0.11	40	46	15	20	7	7
Jansolv	—	—	0	0	0	0	0	0
Elastosol	—	—	15	46	10	11	8	10
Cold Clean	—	—	0	10	0	8	0	5
Finasol	0.038	0.28	0	8	0	0	0	0
Oil only	0	0	30	28	10	16	10	12
Control	0	0	11	9	7	5	0	5

[a] Dispersability is the ratio of dispersant to oil required to disperse 90% of the oil.

[b] Cost is the relative effective cost of sufficient dispersant to disperse 90% of 1 gal of oil under the conditions of the Mackay apparatus.

[c] Concentrations in 100,000 cc of sea water were 10:1 of oil to dispersant product.

Source: Thorhaug, A., Carby, B., Reese, R., Rodriguez, M., McFarlane, J., Teas, H., Sidrak, G., Anderson, M., Aiken, R., McDonald, F., Miller, B., Gordon, V., and Gayle, P., in *Proceedings of the 1991 Oil Spill Conference,* American Petroleum Institute, Washington, D.C., 1991. With permission.

TABLE 2.10
Dispersed Oil and Oil Effects on Mangroves

Location	Type	Dispersant Used and Dilution	Type of Oil	Amount of Spill	Date	Resource Affected	Impact	Dispersant Effect
Panama	Field	Corexit 9527, 24 h, 1:20	50 ppm Prudhoe Bay crude	Exp.	1984	Mangroves	Defoliation, death	Dispersed oil before it reached mangroves
Coast on Caribbean side of Panama	Accidental	Corexit 9527, ca. 21,000 l, 1:20	Medium weight crude	55,000–60,000	April 27, 1986	*Rhizophora mangle*	Defoliation, death	
Coast on Caribbean side of Panama	Experimental	Corexit 9527, 1:20	Prudhoe Bay crude 50 ppm	Exp.	1985	*R. mangle*	28% trees defoliated	No defoliation at sites with dispersant
South Florida: Turkey Pt., Biscayne Bay	Field	Corexit 9527, 1:20	LA crude concentrated	Exp.	1982–1986	*R. mangle*		
Panama	Field spill	Corexit 9527, 1:20	Medium weight crude		Fall 1986	Mangroves *R. mangle*	Observed mangrove death	If dispersed before oil on mangroves, less mortality
Jamaica	Lab	11 dispersants	Venezuelan	Exp.	1988–1989	*Rhizophora* *Avicennia* *Laguncularia*	Defoliations, death of root	Various at 1250 ppm, not low

Note: Exp. = experimental; LA = Louisiana.

Source: Connell, D. W. and Hawker, D. W. (Eds.), *Pollution in Tropical Aquatic Systems,* CRC Press, Boca Raton, FL, 1992, 106. With permission.

TABLE 2.11
The Mortality Percent of Jamaican Mangroves Exposed for 10 h at 1250 ppm Dispersed Oil

Dispersants	*Rhizophora* (Red) (%)	*Avicennia* (Black) (%)	*Laguncularia* (White) (%)
Conco K	28.6	14.3	80
OFC D609	42.9	14.3	25
Corexit 9527	71.4	0	40
Corexit 9550	0	0	20
Wonder-O	42.9	14.3	50
ADP-7	28.6	28.6	50
Jansolv	14.3	0	60
Cold Clean	28.6	14.3	20
Finasol	0	14.3	40
V-25	71.4	14.3	40
LTX	nd	nd	nd
Oil only	14.3	28.6	40
Control	0	14.3	14.3

Source: Thorhaug, A., Carby, B., Reese, R., Rodriguez, M., McFarlane, J., Teas, H., Sidrak, G., Anderson, M., Aiken, R., McDonald, F., Miller, B., Gordon, V., and Gayle, P., in *Proceedings of the 1991 Oil Spill Conference,* American Petroleum Institute, Washington, D.C., 1991. With permission.

TABLE 2.12
Dispersed Oil and Oil Effects on Corals

Location	Type	Dispersant Used and Dilution	Conc. of Dispersant	Amount of Spill	Date	Resource Affected	Impact	Dispersant Effect
Bermuda	Field and lab	Corexit 9527 BP 1100 WD	1:20 1:10	Arabian light crude	1981–1986	Corals	6–24 h after, 1–50 ppm on *Diploria strigosa*	No effect to brief exposures; when oil dispersed 20 ppm polychaetes, bivalves, crustacea intolerant; unclear after 9 mo. whether dispersant had effect or not
Arabian Gulf	Field	Corexit 9527 (20:1)		Arabian light crude experiment	1980	Corals	No impact immediately, some death after 6 mo. during winter cold	
Panama	Field	Corexit 9527	50 ppm, 20:1	Prudhoe Bay crude experiment	1985	Corals, seagrasses, mangroves	No coral death at 24-h exposure	No death of corals with dispersant
Panama	Spill	Corexit 9527	20:1	50,000 med. weight crude	1986	Corals, seagrasses, mangroves	Coral death	Reports intertidal reefs, extensive mortality; subtidal to 2 m mortality
Jamaica	Lab	10 dispersants	1:10	Venezuela light	1988–1989	Corals	Various	3 nontoxic, 5 highly toxic

Source: Connell, D. W. and Hawker, D. W. (Eds.), *Pollution in Tropical Aquatic Systems,* CRC Press, Boca Raton, FL, 1992. With permission.

TABLE 2.13
Jamaican Coral Mortality and Seagrass Mortality, Dispersed Oil, and Oil at Various Concentrations and Times

	125 ml, 6 h			75 ml, 10 h	
	Porietes porietes	*Montastrea annularis*	*Acropora palmata*	*Porietes porietes*	*Montastrea annularis*
Conco	100	100	100	100	100
OFC D609	100	100	100	91	91
Corexit 9527	90	86	100	88	76
Kemarine	90	57	100	85	90
ADP 7	90	72	100	85	90
Corexit 9550	64	14	100	0	12
Jansolv	0	0	50	0	0
Elastosol	15	0	4	0	0
Cold Clean	15	0	15	8	0
Finasol	8	0	15	0	0
Oil only	52	52	100	12	0
Control	0	0	29	0	0

Source: Thorhaug, A., McDonald, F., Miller, B., McFarlane, J., Carby, B., Anderson, M., Gordon, V., and Gayle, P., in *Proceedings of the 1989 Oil Spill Conference,* American Petroleum Institute, Washington, D.C., 1989, 455. With permission.

TABLE 2.14
The Mean Mortality of Three Jamaican Fish Species, Three Jamaican Seagrasses, and Three Jamaican Coral Species to Dispersed Oil

	Fish[a]				Seagrasses[b]				Coral[c]			
	1	2	3	$\bar{X}$	1	2	3	$\bar{X}$	1	2	3	$\bar{X}$
Conco	100	100	100	100	100	82	100	94	100	100	100	100
OFC D609	40	100	100	80	70	63	nd	67	100	100	100	100
Corexit 9527	40	100	100	80	89	93	nd	91	72	100	100	91
V-25	100	100	100	100	nd	nd	nd	nd	100	100	100	100
Wonder-O	100	100	100	100	100	100	100	100	100	100	100	100
Kemarine	nd	nd	nd	nd	63	68	nd	66	100	100	100	100
ADP-7	100	100	100	100	50	46	nd	48	100	100	100	100
Jansolv	100	100	100	100	0	0	0	0	0	0	73	24
LTX	100	100	100	100	nd	nd	nd	nd	nd	nd	nd	nd
Corexit 9550	0	60	0	20	40	46	nd	43	43	0	100	47
Cold Clean	0	60	80	47	0	0	0	0	0	0	21	7
Finasol	0	0	20	7	0	0	0	0	0	0	11	4
Oil	0	0	0	0	30	28	30	29	52	52	58	54
Control	0	0	0	0	11	9	10	10	0	0	0	0

Note: At 6-h exposure of 125 ppm dispersed oil for the seagrasses and coral; at 3-h exposure of 125 ppm dispersed oil for the fish. $\bar{X}$ is mean of toxicity of three species. nd = no data.

[a] Fish 1, *Holocentrus rufus;* fish 2, *Acanthurus* sp.; fish 3, *Haemulon* sp.

[b] Seagrass 1, *Thalassia testudinum;* seagrass 2, *Halodule wrightii;* seagrass 3, *Syringodium filiforme.*

[c] Coral 1, *Porites porites;* coral 2, *Montastrea annularis;* coral 3, *Acropora palmata.*

Source: Thorhaug, A., Carby, B., Reese, R., Rodriguez, M., McFarlane, J., Teas, H., Sidrak, G., Anderson, M., Aiken, R., McDonald, F., Miller, B., Gordon, V., and Gayle, P., in *Proceedings of the 1991 Oil Spill Conference,* American Petroleum Institute, Washington, D.C., 1991, 142. With permission.

TABLE 2.15
Studies of the Effect of Prudhoe Bay Crude Oil, Corexit 9527, and Combinations on Avian Reproduction

Stage of Reproductive Cycle	Species	Protocol	Combination Studied	Finding
Hatchability of eggs	Mallard (*Anas platyrhynchos*)	Applied to egg surface with syringe	Oil; Corexit, 5:1, 30:1 mixture	Toxicity ranking Corexit = 5:1 mixture; oil, 30:1 mixture
		Oil slick on water sprayed with Corexit	Oil, Corexit 10:1 combination	Corexit alone similar to control; oil and combination showed similar decrease in hatchability
Weight gain and survival of nestlings	Leach's Petrel (*Oceanodroma leucorhoa*)	Emulsion or oil painted on plumage or given internally	Oil, Corexit 10:1 combination	Combination applied externally to adults caused greater decrease of survival and weight gain of chicks than did oil alone
	Mallard	Given in diet	Oil, Corexit 10:1 combination	Weight gain and survival were not affected
	Herring Gull (*Larus argentatus*)	Single internal dose	Oil, Corexit 10:1 combination	Corexit alone similar to control; oil and combination decreased weight gain to a similar extent
		Single internal dose; birds food stressed	Oil, Corexit 10:1 combination	Oil and combination birds both lost weight faster than control
		External, painting on feathers; not food stressed	Oil, Corexit 10:1 combination	Birds exposed to combination lost weight; oil and control birds maintained weight

Source: Peakall, D. B., Wells, P. G., and Mackay, D., *Mar. Environ. Res.*, 22, 91, 1987. With permission.

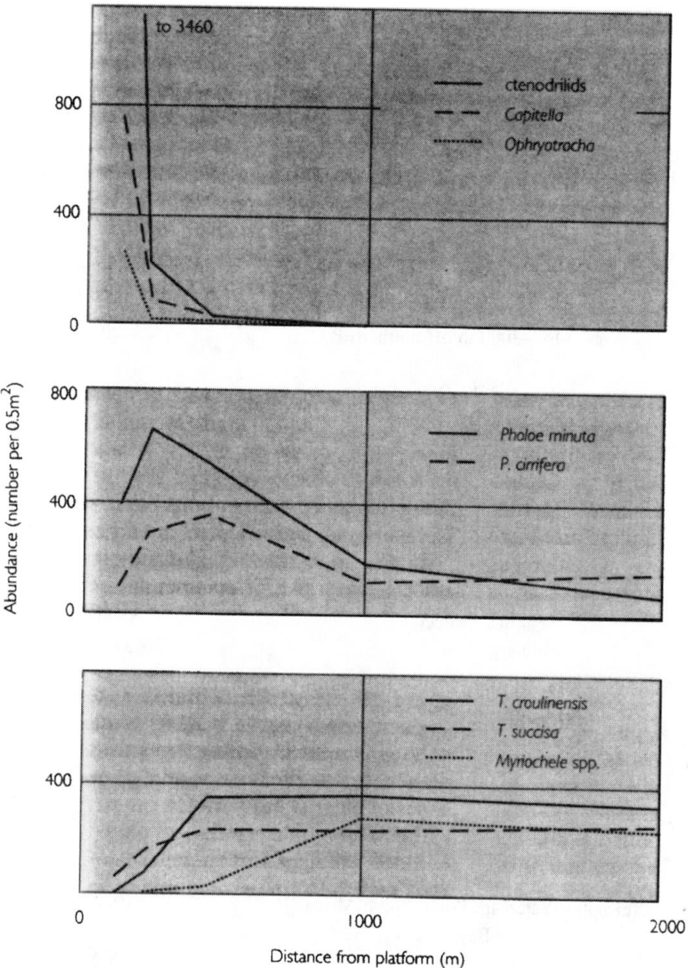

FIGURE 2.1. Three types of abundance variation of benthic annelid worms influenced by disturbance at North Sea oil platforms. (From Clark, R. B., *Marine Pollution,* 3rd ed., Clarendon Press, Oxford, 1992. With permission.)

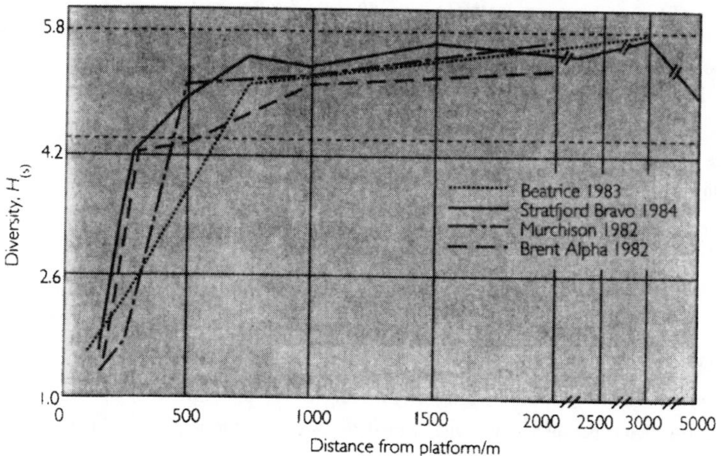

FIGURE 2.2. Diversity of the benthic fauna (measured by the Shannon-Wiener index) in relation to distance from North Sea oil production platforms. Broken lines indicate the maximum and minimum values obtained in studies conducted before the initiation of operations. (From Clark, R. B., *Marine Pollution,* 3rd ed., Clarendon Press, Oxford, 1992. With permission.)

TABLE 2.16
Mortality of Seabirds Resulting from Oil Spills

Year	Vessel Name or Source of the Oil	Site	Spill Volume (t)	Oiled Birds: No. Found	Oiled Birds: Estimated Mortality
1991	Tenyo Maru	Off Vancouver Island, British Columbia	330	4300	
1989[a]	Exxon Valdez	Prince William Sound, Alaska	36,400	31,000	350,000–390,000
1988[a]	Nestucca	Gray's Harbor, WA	770	12,535	56,000
1988	Barge MCN5	Anacortes, WA	240	None reported	
1987	Stuyvesant	150–300 km off northern B.C.	2000	Not known	
1986[a]	Apex Houston	S. California	87	4198	10,577
1985	Arco Anchorage	Port Angeles, WA	800	1917	4000
1984[a]	Puerto Rican	San Francisco Bay	4900	1300	4815
1984	Unknown	Whidbey Island Puget Sound, WA	17	>406	>1500
1984	Mobil Oil	Columbia River and WA coast	660	450	?
1983	Swedish tanker	Kattegat, Denmark	500		50,000
1981	Deivos	Helgoland, Norway	1000	>3000	14,000
1979	Kurdistan	Cape Breton, NB	7900	1697	
1979		Douarnenez Bay, France	30–60	ca. 100	
1979	Russian tanker	Ventpils, Sweden	5500	3053	
1978	Amoco Cadiz	Brittany, France	200,000	4572	20,000
1978	Pantelis a Lemos	Cape coast, South Africa	300	ca. 100	
1978	Outfall	Dounreay, UK	68	650	>1000
1976	Olympic Games	Delaware River, PA	450		
1976	Barge STC-101	Chesapeake Bay	833		20,000–50,000
1975	Olympic Alliance	Dover Spit, UK	2000	>199	
1974	Oriental Pioneer	Struisbaai, South Africa	200	"Thousands"	
1974	Metula	Magellan Strait	50,000	3000	
1972	Dewdale	Cromarty Firth, UK	30		1000
1972	Oswego Guardian, Texanita	Ystervark Point, South Africa	10,000	>400	
1971	Barge U17	Padilla Bay, WA	767	>374	Not known
1971	Collision	San Francisco	2700	7380	20,000
1971	Wafra	Cape Agulhas, South Africa	6000–10,000	>1216	
1970	Kazimah	Robben Island, South Africa	1000	>560	
1970	Arrow	Cape Breton, NB	10,000	567	7000
1970	Irving Whale	SE Newfoundland	<30	625	5000
1969[a]	Hamilton Trader	Irish Sea, UK	700	4400	5900–10,600
1969	Palva	Uto, Finland	150	1000	3000
1969		Loch Indaal, UK	115	449	
1969		Waddensee, Netherlands	150	14,564	35,000–41,000
1968	Esso Essen	Cape Peninsula, South Africa	4000	1250	14,000–19,000
1968	Tank Duchess	Tay Estuary, UK	87	1368	
1967	Torrey Canyon	English Channel	119,328	7815	30,000
1966	Seestern	Medway, UK	1700	2772	5000
1961	Collision	Poole, UK	300	487	
1959		Lower Weser, Germany	360	7032	14,132
1956	Seagate	Olympic Peninsula, WA	Not known		>3000
1955	Gerda Maersk	Elbe, Germany	8000		500,000
1952	Fort Mercer and Pendleton	Monomoy, Mass.	22,400		>3500
1937	Frank Buck	San Francisco	11,800		10,000

[a] Examples of studies in which experimental data and/or modeling were used to improve estimates of bird mortality.

Note: 50 barrels = 7 tonnes; 1 barrel = 0.14 tonnes; 1 tonne = approx. 300 gallons (U.S.); 1 tonne = approx. 1 ton; 1 tonne = approx. 1100 liters.

Source: Burger, A. E., *Mar. Pollut. Bull.*, 26, 140, 1993. With permission.

APPENDIX 3. OIL CLEANUP AND ECOSYSTEM RECOVERY

TABLE 3.1
Recovery of Various Ecosystems Subjected to Catastrophic Oil Spills

Ecosystem Type	Time Between Major Stresses (in Years)				
	3	**5**	**10**	**20**	**100**
River					
Headwaters	50–70% of species recovered	Recovery less than 95% of species	Recovered	Recovered	Recovered
Middle reach	50–75% of species recovered	State of constant recovery less than 95% of species	Recovered	Recovered	Recovered
Slow	50–75% of species recovered	Recovery less than 95% of species	Recovered	Recovered	Recovered
Lakes	Most species would not be recovered	Biological integrity not maintained	State of constant recovery	Final state of recovery	Recovered
Estuaries	Principally clams and mollusks are recovered	Clam and mollusk populations still reduced	Recovered	Recovered	Recovered
Marine					
Beaches	Beaches are in state of final repopulation	Repopulated and probably recovered	Recovered	Recovered	Recovered
Rock Shore	Colony communities not recovered	Colony communities generally recovered	Recovered	Recovered	Recovered
Tidal Flat	Principally bivalves not recovered	Bivalves still reduced	Recovered	Recovered	Recovered
Marshes	Annual plants and short life span	Long-lived plants not reestablished; most organisms recovered	Final stages of recovery	Recovered for very large systems	Recovery depends upon size of area affected
Open Water	Very small area repopulated	Long life-span organisms in recovery	Most species present	Recovered except for very large systems	Recovery depends upon size of area affected

Source: Cairns, J., Jr. (Ed.), *Rehabilitating Damaged Ecosystems,* Vol. 2, CRC Press, Boca Raton, FL, 1988. With permission.

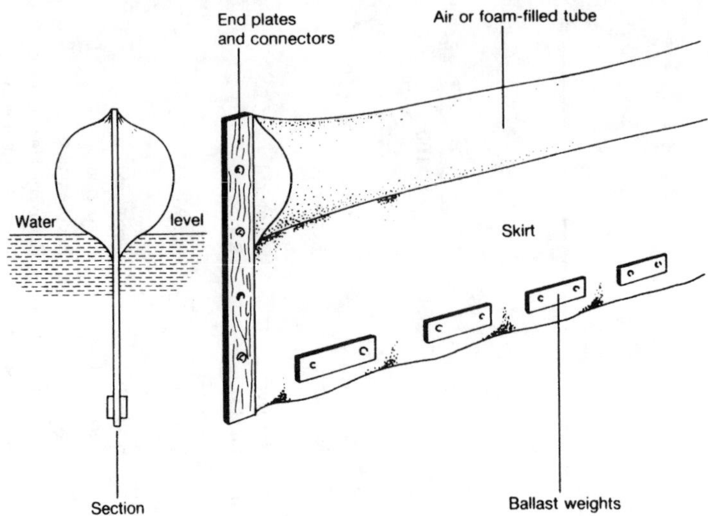

FIGURE 3.1. Design of one type of floating boom in oil cleanup operations. (From Clark, R. B., *Marine Pollution,* 3rd ed., Clarendon Press, Oxford, 1992. With permission.)

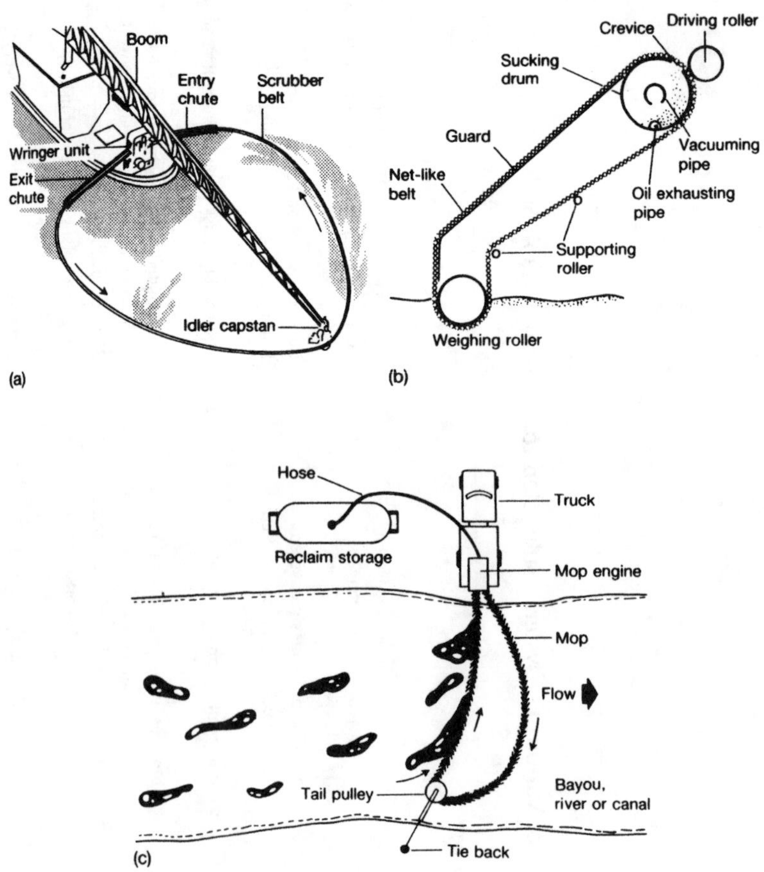

FIGURE 3.2. "Slick-lickers": devices used for removing floating oil from estuarine and marine waters. (From Clark, R. B., *Marine Pollution,* 3rd ed., Clarendon Press, Oxford, 1992. With permission.)

TABLE 3.2
Costs of Oil Spill Cleanup Techniques

Type	Cost per Barrel
Mechanical	$65–5000
Dispersant	$15–65
Shoreline	$650–7000
Gels	Above $7000

Source: Connell, D. W. and Hawker, D. W. (Eds.), *Pollution in Tropical Aquatic Systems,* CRC Press, Boca Raton, FL, 1992, 122. With permission.

TABLE 3.3
Short-Term Oil Spill Cleanup and Indirect Long-Term Costs

	Capitalization Costs	Direct Costs	Indirect Waste Removal	Indirect Environmental Costs	Socioeconomic Costs
Mechanical	High	$1.50–0.60/gal	$2/gal		None, if successful; high, if not successful
Beach cleanup	Low	$8–17/gal	$4/gal	Very high in productive ecosystems (mangroves, reefs)	High, if tourism is involved or if mangroves/coral die; low, if rock or sand not used
No action	0	0	0	About $100,000/acre if mangroves, seagrasses, corals are killed	Very high, if fisheries or tourism are affected or if mangroves or corals die
Dispersant	Medium to low	$0.06–0.30/gal	0	None, if nontoxic dispersants used; some, if toxic dispersants used	Low

Source: Connell, D. W. and Hawker, D. W. (Eds.), *Pollution in Tropical Aquatic Systems,* CRC Press, Boca Raton, FL, 1992, 122. With permission.

4 Polycyclic Aromatic Hydrocarbons

INTRODUCTION

Among the most ubiquitous organic xenobiotics in estuarine and marine environments are compounds termed polycyclic aromatic hydrocarbons (PAHs) which have become increasingly important because of their potential carcinogenicity, mutagenicity, and teratogenicity to aquatic organisms and man. Also known as polynuclear aromatic hydrocarbons or polycyclic organic matter, PAHs tend to be absorbed and accumulated by marine organisms from waters polluted by industrial and municipal wastes.[1] Elevated levels of PAHs are commonly found in estuarine and coastal marine waters near heavily populated areas.[2] While not all PAH compounds are potent carcinogens, mutagens, and teratogens, the environmental impact of the majority of PAHs remains uncertain, and only a rudimentary knowledge exists with regard to the physical, chemical, and biological processes controlling their behavior in aquatic environments.[3] Hence, they have been the focus of ongoing investigations of their fate in various environmental compartments and in marine biota.

Most PAH compounds released into the environment originate from anthropogenic activities, although some PAHs are synthesized by bacteria, plants, and fungi or derive from natural products and processes, such as coal and oil, grass and forest fires, and marine seep and volcanic emissions.[4] The formation of fossil fuels via low- to moderate-temperature diagenesis of sedimentary organic matter produces significant quantities of PAHs. Notable anthropogenic sources of these contaminants include sewage and industrial effluents, waste incineration, oil spills (crude oils contain 0.2 to 7% PAHs), asphalt production, creosote oil, and the combustion of fossil fuels. PAH compounds originating from the production, transport, and use of oil by man exceed those derived from natural seepage. A large fraction of the PAHs that enter estuarine and marine waters results from the pyrolysis of organic matter, especially fossil fuels. Although direct discharges, urban and agricultural runoff, and groundwater flow transport substantial quantities of PAHs to these environments, atmospheric deposition of PAH compounds generated by the pyrolysis of fossil fuels appears to be the primary delivery system. Many domestic and industrial activities contribute to the pyrosynthesis of PAHs (e.g., cigarette smoking, internal combustion engines, coal coking, petroleum refining). These contaminants can be translocated to aquatic environments in airborne particulates or in solid or liquid byproducts of the pyrolytic process.

In both atmospheric and aquatic environments, PAHs readily sorb to suspended particulate matter. They have strong adsorption affinity for particulate surfaces because of their hydrophobicity, low water solubilities, relatively low vapor pressures, and aromaticity.[5] Thus, as PAHs enter the marine hydrosphere, they typically partition out of the water column and onto suspended particulates and bottom sediments.[4,6] PAHs are less affected by photochemical or biological oxidation in bottom sediments than in the water column; consequently, they can persist for long periods of time and may accumulate to high concentrations within seafloor sediments.[3] Estuarine and marine sediments serve as a major repository of PAH compounds and a continual source of contamination for biotic communities.

Polycyclic aromatic hydrocarbon concentrations in aquatic organisms are highly variable. Reported values range from approximately 0.01 to greater than 5000 μg/kg dry weight for individual PAH contaminants. Elevated concentrations of PAHs in marine organisms often occur in areas receiving chronic hydrocarbon discharges.[4]

CHEMICAL STRUCTURE

Polycyclic aromatic hydrocarbons are a group of compounds consisting of hydrogen and carbon arranged in the form of two or more fused aromatic (benzene) rings in linear, angular, or cluster arrangements with unsubstituted groups possibly attached to one or more rings.[7] The compounds range from naphthalene ($C_{10}H_8$, two rings) to coronene ($C_{24}H_{12}$, seven rings).[8] Common PAH compounds include six two-ring compounds (biphenyl, naphthalene, 1-methylnaphthalene, 2-methylnaphthalene, 2,6-dimethylnaphthalene, and acenaphthene); three-ring compounds (flourene, phenanthrene, 1-methylphenanthrene, and anthracene); four-ring compounds (flouranthene, pyrene, and benz[a]anthracene); and five-ring compounds (chrysene, benzo[a]pyrene, benzo[e]pyrene, perylene, and dibenz[a,h]anthracene). Together, PAHs comprise a homologous series of fused-aromatic ring compounds of increasing environmental concern.[9] The low-molecular-weight PAH (LMWpah) compounds, containing two or three rings, are acutely toxic but noncarcinogenic to a broad spectrum of marine organisms. The high-molecular-weight PAH (HMWpah) compounds, containing four, five, and six rings, are less toxic but have greater carcinogenic potential.[7,10] Hence, the LMWpah compounds are sometimes classified separately from the HMWpah varieties. Examples of LMWpah compounds that tend to be toxic are anthracene, fluorene, naphthalene, and phenanthrene. HMWpah compounds that are carcinogenic include benzo(a)pyrene, benzo(c)phenanthrene, dibenzo(a,i)pyrene, and 3-methylcholanthrene.

TRANSFORMATION OF POLYCYCLIC AROMATIC HYDROCARBONS

Photooxidation, chemical oxidation, and biotic metabolism are important processes in the transformation or degradation of PAH compounds in estuarine and marine systems.[10] Photo-induced transformation of PAHs in seawater may take place by direct photolysis reactions, as well as by photooxygenated reactions involving singlet oxygen ozone, OH radicals, and other oxidizing agents. Photooxidation is a key factor in PAH removal from the water column. Aside from the photoreactive nature of PAHs in the upper part of the water column, chemical reactions may effectively reduce PAH levels. For example, ozone reacts with aqueous PAHs to yield aromatic aldehydes, carboxylic acids, and guinones. Chlorination also eliminates aqueous PAHs.

Biological transformation by bacteria, fungi, and aquatic fauna can be significant. Microbial metabolism of PAH compounds, especially by bacteria, typically far outweighs metabolism by other organisms, particularly in highly contaminated areas such as those impacted by oil spills.[11] The degradation of PAHs by microbes proceeds most rapidly under aerobic conditions.[12] Aquatic fauna exhibit variable capacities for metabolizing PAHs. The level of development of the cytochrome P-450-dependent mixed function oxidase system appears to be a major factor in the ability of animal populations to detoxify PAHs.

POLYCYCLIC AROMATIC HYDROCARBON PATHWAYS

The principal routes of entry of PAHs into estuarine and marine environments include atmospheric deposition, discharges of domestic and industrial wastes, runoff from land, and spillage and seepage of fossil fuels.[10] Endogenous sources of PAHs (i.e., biosynthesis) in these environments may be significant only in anoxic sediments, and reliable estimates of the global input of PAHs by this pathway generally are lacking.[13] Atmospheric deposition is a major route of PAH entry into marine waters, as is the discharge of domestic and industrial wastes. Nearly all PAHs in atmospheric fallout are associated with airborne particulate matter and aerosols. Rain, dry fallout, and vapor phase deposition represent the principal atmospheric processes responsible for the flux of PAHs on the world's oceans. Air masses passing over highly industrialized regions often accumulate high PAH loads (20 to 30 ng/m^3). Rates of chemical degradation and photooxidation, together with particulate settling rates, largely control the residence time of PAHs in the atmosphere.

Much of the PAH burden entering estuarine and coastal waters from land runoff derives from nonpoint sources, such as spilled crankcase oil and asphalt-covered roadways, making quantification difficult. Wastewater effluents from oil refineries, plastics facilities, and many other industries usually contain substantial amounts of PAHs. Raw sewage, storm sewer runoff, and other PAH-contaminated domestic wastewaters likewise have significant PAH loads which enter nearshore regions. Since most of the PAH burden transported to coastal habitats is sorbed to particulate matter, it remains relatively close to the source, thereby concentrating in bottom sediments of rivers, harbors, estuaries, and embayments. Waters near heavily industrialized metropolitan centers invariably receive the greatest concentration of PAHs. Farther offshore, other sources (i.e., oil spills, natural oil seep emissions, and atmospheric deposition) deliver the bulk of the PAH compounds.

POLYCYCLIC AROMATIC HYDROCARBONS IN WATER

As noted above, PAHs rapidly sorb to particulate matter in seawater and ultimately settle to the sea floor. Total PAH concentrations in river water flowing through heavily industrialized regions range from 1 to 5 μg/l. Unpolluted river water and seawater contain less than 0.1 μg/l total PAH.[13] Neff[10] reported PAH concentrations in surface waters of the Atlantic Ocean amounting to about 0.4 μg/l, with a range of 0.13 to 1.3 μg/l. PAH levels in marine waters decrease approximately logarithmically with distance from the source.

Polycyclic aromatic hydrocarbon concentrations are lower in the water column than in biota and sediments, owing in part to the low aqueous solubility of PAHs. Dissolved or colloidal organic matter (e.g., humic and fulvic acids) in seawater may act as PAH solubilizers. As the number of aromatic rings or molecular weight of the PAHs increases, the solubility decreases. For example, the solubility of naphthalene, a two-ring PAH, is about 30 ppm, whereas that of five-ring PAHs ranges from 0.5 to 5.0 ppb.[13]

POLYCYCLIC AROMATIC HYDROCARBONS IN SEDIMENTS

The partitioning behavior of PAHs onto sediments and suspended particulates results in an overwhelmingly greater concentration of these contaminants in seafloor sediments (usually by a factor of 1000 or more) than in the water column. Sediment PAH assemblages tend to be persistent, and they have been used as indices of the rate of PAH input to aquatic environments.[13] Sediment samples have an integrating effect on the temporal patterns of PAH input.[12] The aqueous fraction of PAHs typically consists of the most soluble low-molecular-weight components. Dissolved PAHs in the water column degrade rapidly by photooxidation, which is accelerated by higher temperatures, dissolved oxygen levels, and incidences of solar radiation.[6,14] Photochemical and biological oxidation of PAHs declines in seafloor sediments, most acutely in anoxic sediments.

Concentrations of total PAHs vary widely in estuarine and marine sediments. Areas receiving drainage from industrialized centers may have total PAH levels of 100 mg/kg (ppm) or more in bottom sediments. In regions removed from anthropogenic activity, total PAH values in bottom sediments often are in the low ppb range.[10,13] Many of these remote regions probably receive PAHs from long-range aerosols, forest fire particulates, or oil seeps.

Polycyclic aromatic hydrocarbon concentrations have been determined in bottom sediments of various U.S. estuarine and coastal marine systems. Johnson et al.[15] reported that the total PAHs in the sediments of Penobscot Bay, ME, ranged from 286 to 8794 μg/kg dry weight, with PAH concentrations at the head of the estuary commonly exceeding 5900 μg/kg dry weight and those at the seaward margin <1000 μg/kg dry weight. Windsor and Hites[16] observed PAH concentrations ranging from 200 to 870 μg/kg in bottom sediments of the Gulf of Maine and 18 to 160 μg/kg in deep ocean sediments directly offshore. In Boston Harbor, the concentration of total PAHs in bottom sediments amounts to 120 mg/kg dry weight, which decreases to 160 μg/kg dry weight at a distance of 64 km seaward.[17] The Hudson-Raritan estuary, a highly impacted system, exhibits a broad range of PAH levels in bottom sediments from 9900 ng/g in the lower bay to 16,400 ng/g in the Gowanus Canal and 182,000 ng/g in Newtown Creek. In the New York Bight Apex, PAH levels in sediments of the Christiaensen Basin are also elevated (6000 ng/g), being particularly high (20,400 ng/g) in sewage sludge. Farther offshore in sediments of the outer Bight, PAH readings in seafloor sediments are much lower (22 ng/g).[18] Sediment PAH measurements in the Elizabeth River, a subestuary of Chesapeake Bay, are as high as 390 μg/g dry weight at a heavily contaminated site.[19] The concentration of total combustion-derived PAHs in Puget Sound sediments ranges from 16 to 2400 ng/g.[20] The highest PAH levels occur in proximity to influent systems draining coal-bearing strata, urban areas in central Puget Sound, and industrial facilities in northern Puget Sound.

Similar PAH concentrations are found in estuarine and coastal marine sediments of Europe. In surface sediments of the Tamar estuary (England), for instance, PAH readings are between 100 and 1000 ng/g.[21] The levels of PAHs in sediments of the Adriatic Sea range from 30 to 527 ng/g dry weight. Those in superficial sediments of the Rhone delta and the open western Mediterranean Sea range from 179 to 2427 ng/g.[22] The organic solvent extracts from "hot spot" areas of both European and U.S. systems can be highly toxic.[23]

Benzo(a)pyrene (BaP) concentrations in impacted European systems are as follows: (1) Severn estuary, 470 μg/kg dry sediment;[24] (2) southern regions of the North Sea, traces to 122.5 μg/kg dry sediment;[25] (3) English Channel coast, France, 160 to 1320 μg/kg dry sediment;[26] and (4) Bay of Naples, Italy, 1.4 to 3000 μg/kg dry sediment.[27] The highest BaP levels are obtained in bottom sediments near regions of high population density or intense industrial activity. However, BaP is often detected in small quantities in seafloor sediments from remote regions of the world.[10]

Human activity in remote areas often leads to PAH contamination of seafloor sediments in otherwise pristine areas. Such is the case on Rowley Shelf off Western Australia, where PAH concentrations of <0.005 μg/g dry sediment may be linked to fishing boats, commercial shipping, and barges associated with oilfield construction activities.[28] Similar contamination in Arthur Harbor sediments along the Antarctic Peninsula can be traced to diesel fuel spills, ship and boating activities, and runoff.[29]

POLYCYCLIC AROMATIC HYDROCARBONS IN BIOTA

The body burdens of PAHs differ widely among estuarine and marine organisms due to three principal factors: (1) variable concentrations of PAHs in coastal environments from highly polluted systems to those that are pristine, (2) different degrees of bioavailability of the compounds, and (3) variable capacities of the organisms to metabolize them. Because PAHs tend to accumulate in bottom sediments, benthic organisms may be continuously exposed to the contaminants, especially in areas receiving large pollutant loads. However, sediment-sorbed PAHs have only limited bioavailability to marine organisms which greatly reduces their toxicity potential.[7,10] During the past few decades, there has been a large number of PAH measurements on estuarine and marine animals, especially bivalve mollusks, which often are used as sentinel organisms since they rapidly accumulate PAHs and have little capacity for metabolizing them. Of all PAH contaminants in aquatic organisms, the most extensive database exists on the five-ring PAH, benzo(a)pyrene (BaP).[2,6,10,30,31] Except near point sources of PAH contamination, BaP concentrations in marine animals are generally in the low ppb range. A high degree of correlation of BaP contamination in sentinel organisms (mainly mussels, *Mytilus* spp.) with industrial, urban, and recreational uses of coastal water has been demonstrated along the Pacific coast of Canada and the U.S.

Bacteria and fungi oxidize PAHs to dihydrodiols and catechols. Further oxidation of these substances yields carbon dioxide and water. Fungi metabolize PAHs by means of a cytochrome P-450-dependent mixed function oxidase (MFO) system similar to that found in mammals. Marine invertebrates display variable levels of MFO activity. For example, annelids and arthropods possess significant MFO activity, with most activity localized in the liver, gills, and kidney.[32,33] A number of endogenous factors influence MFO activity and PAH metabolism, including the age, sex, nutritional status, and period of the molt cycle (in arthropods) of estuarine and marine animals. Among exogenous factors of importance are temperature, season of the year, currents, and prior history of exposure to inducers or inhibitors of different components of the microsomal PAH-metabolizing system. Cytochrome P-450 has proven to be useful in monitoring contaminants such as PCBs in aquatic environments.[34]

Eisler,[7] compiling data on PAH contamination in tissues of marine finfish and shellfish, reported PAH levels ranging from barely detectable up to 1600 FW. Highest PAH concentrations were recorded in bivalves of the New York Bight region. Bivalves typically bioaccumulate PAHs with little alteration.[35] Since finfish rapidly metabolize PAH compounds, they generally retain lower concentrations of PAHs in their tissues than do shellfish. Perhaps more importantly, they do not appear to accumulate HMWpah compounds, which are more carcinogenic than the LMWpah varieties. This provides a significant advantage since, once assimilated, heavy PAH compounds are the most difficult types to excrete regardless of MFO capability.[7,36] In summary, those fauna with high lipid contents, poor MFO systems, and distributions that coincide with the location of the hydrocarbon source are most likely to accumulate PAHs.

The acute toxicity of PAHs to marine organisms increases as the molecular weight (MW) increases up to 202 (fluoranthene, pyrene). HMWpah compounds are not acutely toxic, presumably due to a concomitant decline in solubility which lowers the aqueous concentrations of the contaminants. However, even at very low concentrations, HMWpah compounds may result in sublethal effects such as inhibited growth, abnormal cellular development, prevalence of chronic diseases, reproductive impairment, and reduced lifespan. Acute responses of marine biota to PAHs are observed at concentrations of approximately 0.2 to 10 ppm, and sublethal responses at levels of about 5 to 100 ppb. The toxicity of PAHs arises from their interference with cellular membrane function and membrane-associated enzyme systems.[10]

Common PAH-induced diseases observed in fish include various types of liver maladies, such as neoplasia. For example, English sole in the Duwamish Waterway, a highly contaminated estuary in Seattle, WA, show a prevalence of hepatic neoplasms (hepatocellular carcinoma, hepatocellular adenoma, and cholangiocellular carcinoma).[37] This bottom-dwelling species is exposed to elevated PAH concentrations which lead to a prevalence of contaminant-associated hepatic lesions. Exposure to these contaminants may also be responsible for reproductive impairment of the species.

Bioaccumulation of PAHs in other benthic fish (e.g., *Pseudopleuronectes americanus*) purportedly increases hepatic mono-oxygenase activity.[38] The monitoring of hepatic mono-oxygenase activity is especially useful in evaluating the environmental effects of oil contamination.[39] Hepatic activities of xenobiotic metabolizing enzymes appear to be sensitive indicators of organic xenobiotic exposure.[40]

CASE STUDIES

Estuaries and nearshore coastal waters in proximity to urban and industrial centers are major repositories of PAHs. For example, coastal waters near Boston, Providence, Baltimore, San Francisco, and Seattle exhibit elevated PAH readings. Larsen[41] registered high PAH values in Casco, Penobscot, and Massachusetts bays, averaging 4300, 2600, and 1600 ppb, respectively. Boston Harbor and Salem Harbor, two bodies of water surrounded by high population densities, have even higher PAH levels. High concentrations of PAHs in the northern perimeter of Narragansett Bay, particularly near Providence, RI, are coupled to sewage discharges, fossil fuel combustion, and urban runoff.[42] The northern reaches of Chesapeake Bay similarly contain high PAH concentrations, most notably in bottom sediments near Baltimore Harbor. Concentrations gradually decrease down estuary. The elevated PAH measurements in Chesapeake Bay are ascribed primarily to combustion or high temperature pyrolysis of carbonaceous fuels.[43] Chronic PAH contamination derived from petrogenic, pyrogenic, and urban sources is present in San Francisco Bay. In addition, more than 150 municipal and industrial facilities contribute point sources of wastes which may exacerbate PAH contamination in the estuary.[44] Anthropogenic sources are also responsible for high PAH measurements in some areas and biota of Puget Sound.[45]

The National Status and Trends (NS&T) Program of the National Oceanic and Atmospheric Administration (NOAA) monitors concentrations of PAHs and other contaminants in estuarine and coastal marine waters nationwide utilizing mussels and oysters as sentinel organisms (see Chapter 7).[46] Sampling sites in the NS&T Mussel Watch Program are selected to characterize the overall concentration of contaminants in these coastal systems, away from known point sources of contamination.[47] They are approximately 20 km apart in estuaries and embayments and 100 km apart along open coastlines.[48]

Results of the NS&T Mussel Watch Program in 1990 — which sampled oysters and mussels at 214 sites nationwide — revealed a mean concentration of total PAHs amounting to 260 ng/g-dry.[46] In a review of the first 5 years of the Mussel Watch Program in the Gulf of Mexico (1986 to 1990), Jackson et al.[47] compiled two major groups of data. The first group consisted of sites with lower PAH concentrations probably due to background contamination (i.e., stormwater runoff, sewage effluents, atmospheric deposition). The second group included sites with higher concentrations of PAHs associated with local sources of PAH input (i.e., small oil spills, sewage wastes, etc.). Sources of local variation in PAH concentrations in this region can be significant.[49]

During the first 3 years of the NS&T Mussel Watch Program, temporal trends in PAH contaminant concentrations in mollusks were followed at several sites.[50] Locations where LMWpah concentrations in mollusks decreased between 1986 and 1988 included the Hudson-Raritan estuary (NY), New York Bight (NJ), Tampa Bay (FL), San Simeon Point (CA), and Point St. George (OR). Those locales where LMWpah measurements increased over this interval were Buzzards Bay (MA), Chesapeake Bay (MA), Mississippi Sound (MS), Matagorda Bay (TX), and Corpus Christi (TX). HMWpah values in mollusk tissues declined in the Hudson-Raritan estuary, Tampa Bay, St. Andrew Bay (FL), and Sinclair Inlet (WA). Higher HMWpah levels were obtained in samples from Long Island Sound (CN), Chesapeake Bay, and Choctawhat Bay (FL).

The HMWpah compounds originate principally from fossil fuel combustion; the LMWpah compounds generally derive from relatively fresh, unburned petroleum. Efforts to lower PAH contamination in U.S. estuarine and coastal marine waters in future years cannot rely simply on product bans, as in the case of certain chlorinated hydrocarbon compounds (e.g., DDT and PCBs). Both point- and nonpoint control measures must continue to be applied in order to minimize the input of PAH compounds to these environments.

REFERENCES

1. Cocchieri, R. A., Arnese, A., and Minicucci, A. M., Polycyclic aromatic hydrocarbons in marine organisms from Italian central Mediterranean coasts, *Mar. Pollut. Bull.*, 21, 15, 1990.
2. den Besten, P. J., O'Hara, S. C. M., and Livingstone, D. R., Further characterization of benzo(a)pyrene metabolism in the sea star, *Asterias rubens* L., *Mar. Environ. Res.*, 34, 309, 1992.
3. Guzzella, L. and De Paolis, A., Polycyclic aromatic hydrocarbons in sediments of the Adriatic Sea, *Mar. Pollut. Bull.*, 28, 159, 1994.
4. McElroy, A. E., Farrington, J. W., and Teal, J. M., Bioavailability of polycyclic aromatic hydrocarbons in the aquatic environment, in *Metabolism of Polycyclic Aromatic Hydrocarbons in the Aquatic Environment*, Varanasi, U. (Ed.), CRC Press, Boca Raton, FL, 1989, 1.
5. Onuska, F. I., Analysis of polycyclic aromatic hydrocarbons in environmental samples, in *Analysis of Trace Organics in the Aquatic Environment*, Afghan, B. K. and Chau, A. S. Y. (Eds.), CRC Press, Boca Raton, FL, 1989, 205.
6. Kennish, M. J., *Ecology of Estuaries: Anthropogenic Effects*, CRC Press, Boca Raton, FL, 1992.
7. Eisler, R., *Polycyclic Aromatic Hydrocarbon Hazards to Fish, Wildlife, and Invertebrates: A Synoptic Review,* Biological Report 85 (1.11), U.S. Fish and Wildlife Service, Washington, D.C., 1987.
8. Albers, P. H., Petroleum and individual polycyclic aromatic hydrocarbons, in *Handbook of Ecotoxicology*, Hoffman, D. J., Rattner, B. A., Burton, G. A., Jr., and Cairns, J., Jr. (Eds.), Lewis Publishers, Boca Raton, FL, 1994, 330.

9. Wijayaratne, R. D. and Means, J. C., Sorption of polycyclic aromatic hydrocarbons by natural estuarine colloids, *Mar. Environ. Res.*, 11, 77, 1984.
10. Neff, J. M., *Polycyclic Aromatic Hydrocarbons in the Aquatic Environment: Sources, Fates, and Biological Effects*, Applied *Science*, London, 1979.
11. Cerniglia, C. E. and Heitkamp, M. A., Microbial degradation of PAH in the aquatic environment, in *Metabolism of Polycyclic Aromatic Hydrocarbons in the Aquatic Environment*, Varanasi, U. (Ed.), CRC Press, Boca Raton, FL, 1989, 41.
12. Delaune, R. D., Hambrick, G. A., III, and Patrick, W. H., Jr., Degradation of hydrocarbons in oxidized and reduced sediments, *Mar. Pollut. Bull.*, 11, 103, 1980.
13. Neff, J. M., Polycyclic aromatic hydrocarbons, in *Fundamentals of Aquatic Toxicology*, Rand, G. M. and Petrocelli, S. R. (Eds.), Hemisphere, New York, 1985, 416.
14. Bauer, J. E. and Capone, D. G., Degradation and mineralization of the polycyclic aromatic hydrocarbons anthracene and naphthalene in intertidal marine sediments, Appl. Environ. Microbiol., 50, 81, 1985.
15. Johnson, A. C., Larsen, P. F., Gadbois, D. F., and Humason, A. W., The distribution of polycyclic aromatic hydrocarbons in the surficial sediments of Penobscot Bay (Maine, USA) in relation to possible sources and to other sites worldwide, *Mar. Environ. Res.*, 15, 1, 1985.
16. Windsor, J. G. and Hites, R. A., Polycyclic aromatic hydrocarbons in Gulf of Maine sediments and Nova Scotia soils, *Geochim. Cosmochim. Acta*, 43, 27, 1979.
17. McLeese, D. W., Ray, S., and Burridge, L. E., Accumulation of polynuclear aromatic hydrocarbons by the clam *Mya arenaria*, in *Wastes in the Ocean*, Vol. 6, *Nearshore Waste Disposal*, Ketchum, B. H., Capuzzo, J. M., Burt, W. V., Duedall, I. W., Park, P. K., and Kester, D. R. (Eds.), John Wiley & Sons, New York, 1985, 81.
18. O'Connor, J. M., Klotz, J. B., and Kneip, T. J., Sources, sinks, and distribution of organic contaminants in the New York Bight ecosystem, in *Ecological Stress and the New York Bight: Science and Management*, Mayer, G. F. (Ed.), Estuarine Research Federation, Columbia, SC, 1982, 631.
19. Bieri, R. H., Hein, C., Huggett, R. J., Shou, P., Slone, H., Smith, C., and Su, C.-W., Toxic Organic Compound in Surface Sediments from the Elizabeth and Patapsco Rivers and Estuaries, Tech. Rep., Virginia Institute of Marine *Science*, Gloucester Point, 1982.
20. Barrick, R. C. and Prahl, F. G., Hydrocarbon geochemistry of the Puget Sound region. III. Polycyclic aromatic hydrocarbons in sediments, *Est. Coastal Shelf Sci.*, 25, 175, 1987.
21. Readman, J. W., Mantoura, R. F. C., and Rhead, M. M., A record of polycyclic aromatic hydrocarbon (PAH) pollution obtained from accreting sediments of the Tamar estuary, U.K.: evidence for non-equilibrium behavior of PAH, *Sci. Tot. Environ.*, 66, 73, 1987.
22. Lipiatou, E. and Saliot, A., Hydrocarbon contamination of the Rhone Delta and western Mediterranean, *Mar. Pollut. Bull.*, 22, 297, 1991.
23. Demuth, S., Casillas, E., Wolfe, D. A., and McCain, B. B., Toxicity of saline and organic solvent extracts of sediments from Boston Harbor, Massachusetts, and the Hudson River-Raritan Bay estuary, New York, using the Microtox® Bioassay, *Arch. Environ. Contam. Toxicol.*, 25, 377, 1993.
24. Thompson, S. and Eglinton, G., Composition and sources of pollutant hydrocarbons in the Severn estuary, *Mar. Pollut. Bull.*, 9, 133, 1978.
25. Mallet, L., Perdriau, V., and Perdriau, S., Extent of pollution by polycyclic aromatic hydrocarbons of the benzo-3,4-pyrene type in the North Sea and the glacial Arctic Ocean, *Bull. Acad. Nat. Med. (Paris)*, 147, 320, 1963.
26. Mallet, L., Lima-Zanghi, C., and Brisou, J., Investigations of the possibilities of biosynthesis of polybenzene hydrocarbons of the benzo-3,4-pyrene type by a *Clostridium putride* in the presence of marine plankton lipids, *C. R. Acad. Sci. (Paris) Ser. D*, 264, 1534, 1967.
27. Boucart, J. and Mallet, L., Marine pollution of the shores of the central region of the Tyrrhenian Sea (Bay of Naples) by benzo-3,4-pyrene-type polycyclic aromatic hydrocarbons, *C. R. Acad. Sci. (Paris) Ser. D*, 260, 3729, 1965.
28. Pendoley, K., Hydrocarbons in Rowley Shelf (western Australia) oysters and sediments, *Mar. Pollut. Bull.*, 24, 210, 1992.
29. Kennicutt, M. C., McDonald, T. J., Denoux, G. J., and McDonald, S. J., Hydrocarbon contamination on the Antarctic Peninsula. I. Arthur Harbor — subtidal sediments, *Mar. Pollut. Bull.*, 24, 499, 1992.
30. McElroy, A. E. and Kleinow, K. M., In vitro metabolism of benzo(a)pyrene and benzo(a)pyrene-7,8-dihydrodiol by liver and intestinal mucosa homogenates from the winter flounder (*Pseudopleuronectes americanus*), *Mar. Environ. Res.*, 34, 279, 1992.

31. Michel, X., Salaun, J.-P., Galgani, F., and Narbonne, J.-F., Benzo(a)pyrene hydroxylase activity in the marine mussel *Mytilus galloprovincialis*: a potential marker of contamination by polycyclic aromatic hydrocarbon-type compounds, *Mar. Environ. Res.*, 38, 257, 1994.
32. Foster, G. D. and Wright, D. A., Unsubstituted polynuclear aromatic hydrocarbons in sediments, clams, and clam worms from Chesapeake Bay, *Mar. Pollut. Bull.*, 19, 459, 1988.
33. Stegeman, J. J., Nomenclature for hydrocarbon-inducible cytochrome P450 in fish, *Mar. Environ. Res.*, 34, 133, 1992.
34. Collier, T. K., Connor, S. D., Eberhart, B.-T.L., Anulacion, B. F., Goksoyr, A., and Varanasi, U., Using cytochrome P450 to monitor the aquatic environment: initial results from regional and national surveys, *Mar. Environ. Res.*, 34, 195, 1992.
35. Wade, T. L., Kennicutt, M. C., II, and Brooks, J. M., Gulf of Mexico hydrocarbon seep communities. Part III. Aromatic hydrocarbon concentrations in organisms, sediments, and water, *Mar. Environ. Res.*, 27, 19, 1989.
36. Varanasi, U., Stein, J. S., and Nishimoto, M., Biotransformation and disposition of PAH in fish, in *Metabolism of Polycyclic Aromatic Hydrocarbons in the Aquatic Environment*, Varanasi, U. (Ed.), CRC Press, Boca Raton, FL, 1989, 93.
37. Collier, T. K. and Varanasi, U., Hepatic activities of xenobiotic metabolizing enzymes and biliary levels of xenobiotics in English sole (*Parophrys vetulus*) exposed to environmental contaminants, *Arch. Environ. Contam. Toxicol.*, 20, 462, 1991.
38. Addison, R. F., Willis, D. E., and Zinck, M. E., Liver microsomal mono-oxygenase induction in winter flounder (*Pseudopleuronectes americanus*) from a gradient of sediment PAH concentrations at Sydney Harbor, Nova Scotia, *Mar. Environ. Res.*, 37, 283, 1994.
39. Hellou, J., Payne, J. F., Upshall, C., Fancey, L. L., and Hamilton, C., Bioaccumulation of aromatic hydrocarbons from sediments: a dose-response study with flounder (*Pseudopleuronectes americanus*), *Arch. Environ. Contam. Toxicol.*, 27, 477, 1994.
40. Payne, J. F., Fancey, L. L., Rahimtula, A. D., Porter, E. L., Review and perspective on the use of mixed-function oxygenase enzymes in biological monitoring, *Comp. Biochem. Physiol.*, 86C, 233, 1987.
41. Larsen, P. F., Marine environmental quality in the Gulf of Maine, *Rev. Aquat. Sci.*, 6, 67, 1992.
42. Lake, J. L., Norwood, C., Dimock, C., and Bowen, R., Origins of polycyclic aromatic hydrocarbons in estuarine sediments, *Geochim. Cosmochim. Acta*, 43, 1847, 1979.
43. Huggett, R. J., De Fur, P. O., and Bieri, R. H., Organic compounds in Chesapeake Bay, *Mar. Pollut. Bull.*, 19, 454, 1988.
44. Pereira, W. E., Hostettler, F. D., and Rapp, J. B., Bioaccumulation of hydrocarbons derived from terrestrial and anthropogenic sources in the Asian clam, *Potamocorbula amurensis*, in San Francisco Bay estuary, *Mar. Pollut. Bull.*, 24, 103, 1992.
45. Klauda, R. J. and Bender, M. E., Contaminant effects on Chesapeake Bay finfishes, in *Contaminant Problems and Management of Living Chesapeake Bay Resources*, Majumdar, S. K., Hall, L. W., Jr., and Austin, H. M. (Eds.), Pennsylvania Academy of *Science*, Easton, PA, 1987, 321.
46. O'Connor, T. P., *Recent Trends in Coastal Environmental Quality: Results from the First Five Years of the NOAA Mussel Watch Project*, Spec. Publ., NOAA/NOS/ORCA, Rockville, MD, 1990.
47. Jackson, T. J., Wade, T. L., McDonald, T. J., Wilkinson, D. L., and Brooks, J. M., Polynuclear aromatic hydrocarbon contaminants in oysters from the Gulf of Mexico (1986-1990), *Environ. Pollut.*, 83, 291, 1994.
48. O'Connor, T. P., Price, J. E., and Parker, C. A., Results from NOAA's National Status and Trends Program on distributions, effects, and trends of chemical contamination in the coastal and estuarine United States, *Oceans 89 Conf. Proc.*, 2, 569, 1989.
49. Ellis, M. S., Choi, K. S., Wade, T. L., Powell, E. N., Jackson, T. J., and Lewis, D. H., Sources of local variation in polynuclear aromatic hydrocarbon and pesticide body burden in oysters (*Crassostrea virginica*) from Galveston Bay, Texas, *Comp. Biochem. Physiol.*, 106C, 689, 1993.
50. NOAA, *A Summary of Data on Tissue Contamination from the First Three Years (1986–1988) of the Mussel Watch Project*, NOAA Tech. Mem. NOS OMA49, National Oceanic and Atmospheric Administration, Rockville, MD, 1989.

APPENDIX 1. STRUCTURE AND PROPERTIES OF POLYCYCLIC AROMATIC HYDROCARBON COMPOUNDS

TABLE 1.1
Structure of PAH and Other Aromatic Hydrocarbon Compounds

Structure	1957 I.U.P.A.C. Name	Other Names	Mol. Weight	Relative Carcino-genicity[a]	Common Abbreviation (if any)
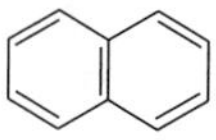	Naphthalene	–	128	–	–
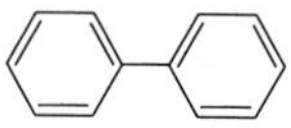	Biphenyl	–	154	–	–
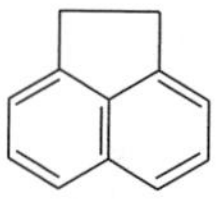	Acenaphthene	–	154	–	–
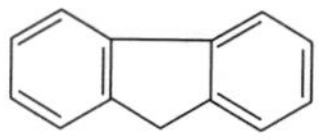	Fluorene	–	166	–	–
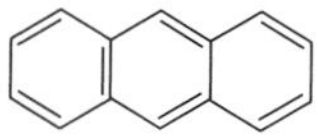	Anthracene	–	178	–	–
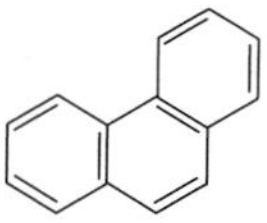	Phenanthrene	–	178	–	–
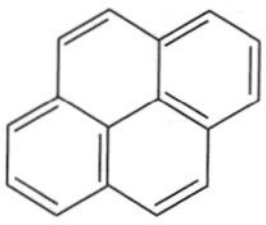	Pyrene	–	202	–	–
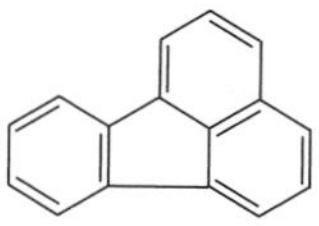	Fluoranthene	–	202	–	–
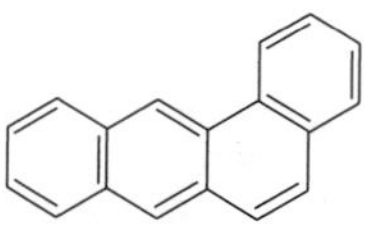	Benzo(a)anthracene	1,2 Benzanthracene	228	< +	B(a)A

TABLE 1.1 (continued)
Structure of PAH and Other Aromatic Hydrocarbon Compounds

Structure	1957 I.U.P.A.C. Name	Other Names	Mol. Weight	Relative Carcino-genicity[a]	Common Abbreviation (if any)
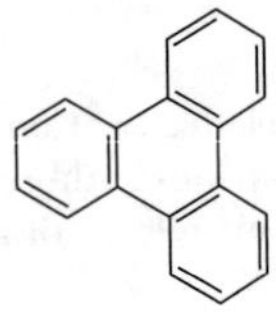	Triphenylene	–	228	–	–
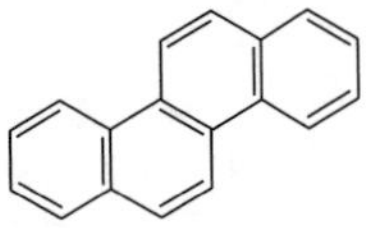	Chrysene	–	228	< +	–
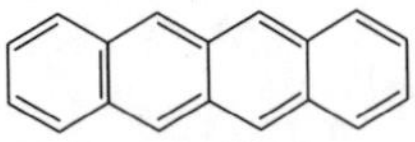	Naphthacene	Tetracene	228	–	–
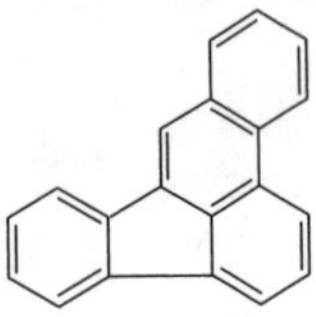	Benzo(b)fluoranthene	3,4 Benzfluoranthene	252	+ +	B(b)F
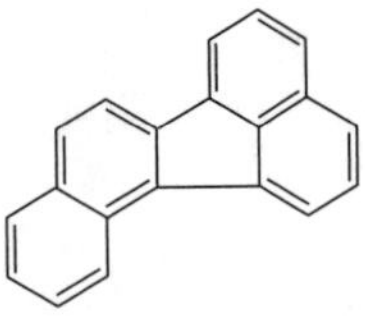	Benzo(j)fluoranthene	10,11 Benzfluoranthene	252	+ +	B(j)F
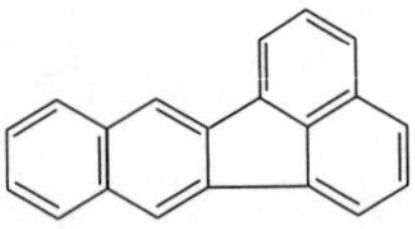	Benzo(k)fluoranthene	11,12 Benzfluoranthene	252	–	B(k)F
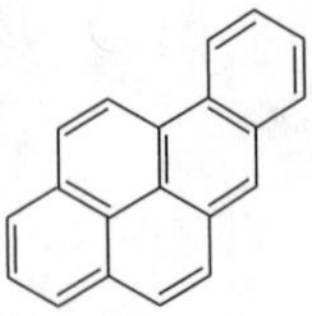	Benzo(a)pyrene	3,4 Benzopyrene	252	+ + + +	B(a)P
	Benzo(e)pyrene	1,2 Benzopyrene	252	< +	B(e)P
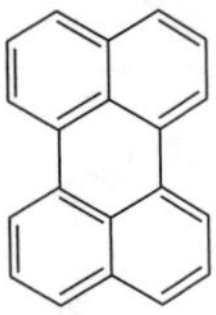	Perylene	–	252	–	–

TABLE 1.1 (continued)
Structure of PAH and Other Aromatic Hydrocarbon Compounds

Structure	1957 I.U.P.A.C. Name	Other Names	Mol. Weight	Relative Carcino-genicity[a]	Common Abbreviation (if any)
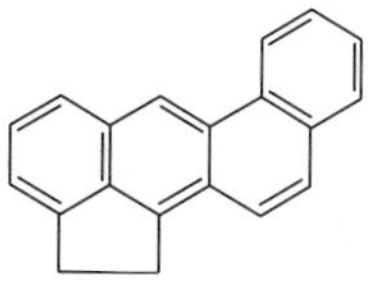	Cholanthrene	–	254	–	–
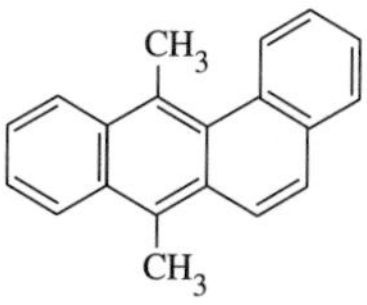 CH_3 / CH_3	7,12 Dimethylbenz-(a)anthracene	7,12 Dimethyl-1,2-benzan-thracene	256	+ + + + +	–
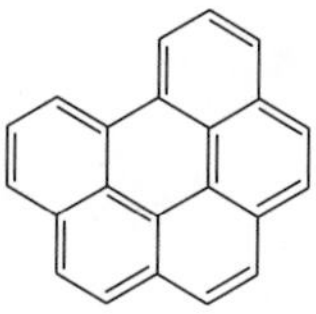	Benzo(ghi)perylene	1,12 Benzperylene	276	–	–
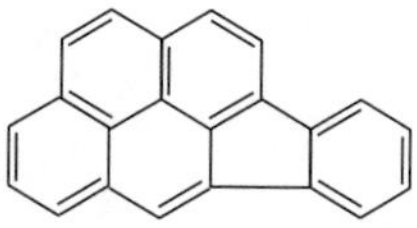	Indeno(1,2,3-cd)pyrene	o-Phenylenepyrene	276	+	IP
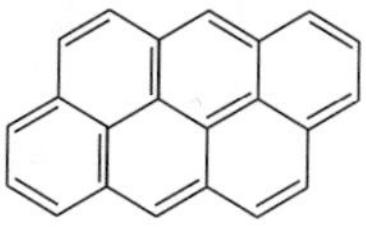	Anthanthrene	–	276	< +	–
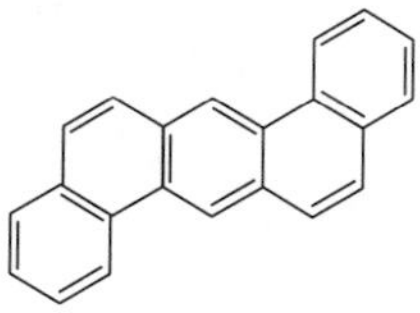	Dibenz(a,h)anthracene	1,2,5,6 Dibenzanthracene	278	+ + +	–
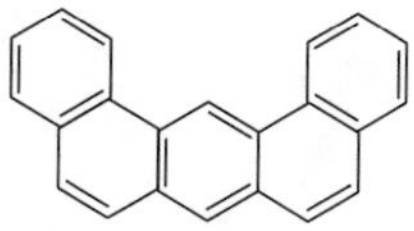	Dibenz(a,j)anthracene	1,2,7,8 Dibenzanthracene	278	–	–
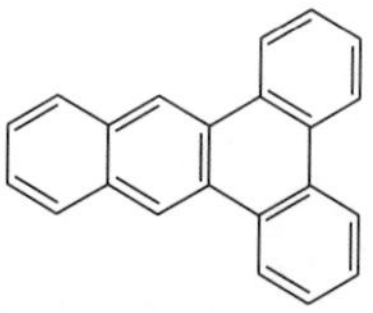	Dibenz(a,c)anthracene	1,2,3,4 Dibenzanthracene	278	–	–
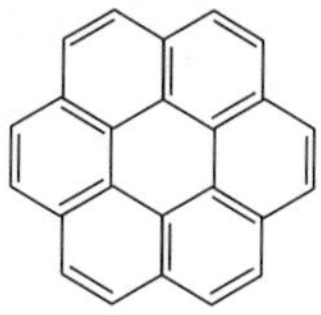	Coronene	–	300	–	–

[a] +++++ = extremely active; ++++ = very active; +++ = active; ++ = moderately active; + = weakly active; < = less than; — = inactive or unknown.

Source: Futoma, D. J., Smith, S. R., Smith, T. E., and Tanaka, J., *Polycyclic Aromatic Hydrocarbons in Water Systems,* CRC Press, Boca Raton, FL, 1981, 2. With permission.

TABLE 1.2
Major Sources of PAH in Atmospheric and Aquatic Environments

Ecosystem and Sources	Annual Input (metric tons)
Atmosphere	
Total PAH	
Forest and prairie fires	19,513
Agricultural burning	13,009
Refuse burning	4769
Enclosed incineration	3902
Heating and power	2168
Benzo(a)pyrene	
Heating and power	
Worldwide	2604
U.S. only	475
Industrial processes (mostly coke production)	
Worldwide	1045
U.S. only	198
Refuse and open burning	
Worldwide	1350
U.S. only	588
Motor vehicles	
Worldwide	45
U.S. only	22
Aquatic environments	
Total PAH	
Petroleum spillage	170,000
Atmospheric deposition	50,000
Wastewaters	4400
Surface land runoff	2940
Biosynthesis	2700
Total benzo(a)pyrene	700

Source: Eisler, R., *Polycyclic Aromatic Hydrocarbon Hazards to Fish, Wildlife, and Invertebrates: A Synoptic Review.* Biological Report 85 (1.11), U.S. Fish and Wildlife Service, Washington, D.C., 1987.

TABLE 1.3
Physical Properties of Polycyclic Aromatic Hydrocarbons

Compound	Density	Vapor Pressure at 25°C	Sat. Conc. (ng/m^3)	Vapor Equilibrium –10°C	Vapor 30°C
Fluorene	1.203	—	—	—	—
Anthracene	1.25	$2.6 \cdot 10^{-5}$	$1.9 \cdot 10^{7}$	—	—
Phenanthrene	1.79	$9.1 \cdot 10^{-5}$	$6.5 \cdot 10^{7}$	—	—
Fluoranthene	1.252	—	—	—	—
Pyrene	1.271	$9.1 \cdot 10^{-8}$	$7.4 \cdot 10^{4}$	$58 \cdot 10^{2}$	$1.4 \cdot 10^{5}$
Benzo(a)anthracene	—	$1.5 \cdot 10^{-8}$	$1.3 \cdot 10^{3}$	3.4	$2.8 \cdot 10^{3}$
Chrysene	1.274	—	—	—	—
Benzo(k)fluoranthene	—	$1.3 \cdot 10^{-11}$	$1.3 \cdot 10^{1}$	$1.3 \cdot 10^{-2}$	$3.0 \cdot 10^{1}$
Benzo(a)pyrene	1.351	$7.3 \cdot 10^{-10}$	$7.5 \cdot 10^{1}$	$1.5 \cdot 10^{-1}$	$1.6 \cdot 10^{2}$
Benzo(e)pyrene	—	$7.4 \cdot 10^{-10}$	$7.5 \cdot 10^{1}$	$1.5 \cdot 10^{-1}$	$1.6 \cdot 10^{2}$
Perylene	1.35	—	—	—	—
Benzo(ghi)perylene	—	$1.3 \cdot 10^{-11}$	1.5	$1.8 \cdot 10^{-3}$	3.4
Dibenz(ghi,pqr)chrysene	1.377	$2.0 \cdot 10^{-13}$	$2.0 \cdot 10^{-2}$	$1.8 \cdot 10^{-6}$	$5.8 \cdot 10^{-2}$

Source: Afghan, B. K. and Chau, A. S. Y. (Eds.), *Analysis of Trace Organics in the Aquatic Environment,* CRC Press, Boca Raton, FL, 1989, 220. With permission.

TABLE 1.4
Solubility and Octanol-Water Partition Coefficients of PAH at 25°C

Compound	Solubility (μg/l)	K_{ow}	Compound	Solubility (μg/l)	K_{ow}
Fluorene	800	—	7,12-Dimethylbenz(a)anthracene	1.5	6.36
Anthracene	59	4.5	Benzo(b)fluoranthene	2.4	6.21
Phenanthrene	435	4.46	Benzo(j)fluoranthene	2.4	6.21
2-Methylanthracene	21.3	4.77	Cholanthene	2.0	6.28
9-Methylphenanthrene	261.0	4.77	Benzo(a)pyrene	3.8	6.04
1-Methylphenanthrene	269.0	4.77	Benzo(e)pyrene	2.4	6.21
Fluoranthene	260.0	5.03	Perylene	2.4	6.21
Pyrene	133.0	4.98	Dibenzo(a,h)fluorene	0.8	6.57
9,10-Dimethylanthracene	56.0	5.13	Dibenzo(a,g)fluorene	0.8	6.57
Benzo(a)fluorene	45.0	5.34	Dibenzo(a,c)fluorene	0.8	6.57
Benzo(b)fluorene	29.6	5.34	3-Methylcholanthene	0.7	6.64
Benzo(a)anthracene	11.0	5.63	Dibenz(a,j)anthracene	0.4	6.86
Naphthacene	1.0	5.65	Benzo(ghi)fluoranthene	0.5	6.78
Chrysene	1.9	5.63	Benzo(ghi)perylene	0.3	6.78
Triphenylene	43.0	5.63	Coronene	0.14	7.36

Note: Salinity 32% at 22°C.

Source: Afghan, B. K. and Chau, A. S. Y. (Eds.), *Analysis of Trace Organics in the Aquatic Environment,* CRC Press, Boca Raton, FL, 1989, 220. With permission.

TABLE 1.5
The Aqueous Solubilities of Some Aromatic Hydrocarbons as Determined by Several Investigators

		Solubilities (mg/kg)		Solubilities (mg/kg)			
		May et al.		Davis et al.	MacKay and Show	Schwarz	Wauchope and Getnom
Compound	**Mol. Weight**	**25°C**	**29°C**	**29°C**	**25°C**	**25°C**	**25°C**
Benzene	78.1	1791 ± 10					
Naphthalene	128.2	31.69 ± 0.23			31.7 ± 0.2	30.3 ± 0.3	31.2
Fluorene	166.2	1.685 ± 0.005			1.98 ± 0.04		1.90
Anthracene	178.2	0.0446 ± 0.0002	0.0570 ± 0.003	0.075 ± 0.005	0.073 ± 0.005	0.041 ± 0.0003	0.075
Phenanthrene	178.2	1.002 ± 0.011	1.220 ± 0.013	1.600 ± 0.050	1.290 ± 0.070	1.151 ± 0.015	1.180
2-Methylanthracene	192.3	0.0213 ± 0.003					
1-Methylphenanthrene	192.3	0.269 ± 0.003					
Fluoranthene	202.3	0.206 ± 0.002	0.264 ± 0.002	0.240 ± 0.020	0.260 ± 0.020		0.265
Pyrene	202.3	0.132 ± 0.001	0.162 ± 0.001	0.165 ± 0.007	0.135 ± 0.005	0.129 ± 0.002	0.148
Benzanthracene	228.3	0.0094 ± 0.0001	0.0122 ± 0.0001	0.011 ± 0.001	0.014 ± 0.0002		
Chrysene	228.3	0.0018 ± 0.00002	0.0022 ± 0.00003	0.0015 ± 0.0004	0.002 ± 0.0002		
Triphenylene	228.3	0.0066 ± 0.0001		0.038 ± 0.005	0.043 ± 0.001		

Source: May, W. E., Wasik, S. P., and Freeman, D. H., *Anal. Chem.*, 50, 997, 1978. With permission.

TABLE 1.6
Calculated and Measured Rate Constants for Photolysis of Aromatics in Sunlight at 40° N Latitude[a]

PAH	Measured	Winter	Spring	Summer	Fall	S/W[b]
Benz(a)anthracene	6.0×10^{-3} (Spring)	1.4×10^{-4}	2.2×10^{-4}	3.8×10^{-4}	2.2×10^{-4}	2.7
Benzo(a)pyrene	1.8×10^{-4} (Winter)	1.8×10^{-4}	2.8×10^{-4}	3.9×10^{-4}	2.3×10^{-4}	2.2
Quinoline	4.0×10^{-7} (Summer)	5.0×10^{-8}	2.3×10^{-7}	3.6×10^{-7}	1.3×10^{-7}	7.2
Benzo(f)quinoline	3.7×10^{-4} (Summer)	1.5×10^{-4}	2.4×10^{-4}	4.8×10^{-4}	2.0×10^{-4}	3.2
9H-carbazole	6.6×10^{-5} (Winter)	6.5×10^{-5}	1.0×10^{-4}	2.0×10^{-4}	9.0×10^{-5}	3.1
7H-dibenzo(o,g)carbazole	5.2×10^{-4} (Winter)	2.3×10^{-4}	3.9×10^{-4}	5.0×10^{-4}	3.2×10^{-4}	2.2
Benzo(b)thiophene	6.9×10^{-7} (Summer)	2.3×10^{-8}	2.7×10^{-7}	5.7×10^{-7}	1.2×10^{-7}	25
Dibenzothiophene	1.0×10^{-8} (Spring)	2.9×10^{-7}	1.1×10^{-6}	1.5×10^{-6}	9.1×10^{-7}	5.2

[a] K_{pE} in s^{-1}.
[b] S/W = summer K_{pE}/winter K_{pE}.

Source: Payne, J. F. and Phillips, C. R., *Environ. Sci. Technol.*, 19, 569, 1985. With permission.

APPENDIX 2. POLYCYCLIC AROMATIC HYDROCARBON CONCENTRATIONS IN SEDIMENTS AND COASTAL SYSTEMS

TABLE 2.1
Comparison of Total PAH Concentrations in Marine and Freshwater (FW) Surficial Sediments

Location	Total PAH, (ppb, wet weight)	No. of Stations	Depth (m)
North America			
Penobscot Bay	286–8974[a]	49	9.2–126.3
Casco Bay	215–14,425	32	2–43
Gulf of Maine	543[a]	1	—
Murray Basin	540	1	282
Jordan Basin	500	1	265
Wilkinson Basin	540–870	1	215
Franklin Basin	200	1	225
North Atlantic			
Continental rise	160	1	4150
Continental slope	120	1	1830
Abyssal plain	18–97	2	5250, 5465
Abyssal plain	55[a]	1	—
Charles River, MA	87,000[a]	1	—
	120,000	1	—
Massachusetts Bay	160–3400	3	90, 130, 155
Boston Harbor, MA	8500	1	6
Buzzards Bay, MA	800	1	17
	4000–5000	3	—
	803[a]	1	17
Falmouth Marsh, MA	800	1	Intertidal
New Bedford Harbor, MA	63,000	1	—
Pettaquamscutt River, RI	10,000	1	—
New York Bight	5830[a]	1	28
Pennsylvania Creek (FW)	100	4	0.3
Lake Erie (FW)	530–3750	7	—
Adirondack Lakes (FW)	4070–12,807	2	—
Alaska	5–113[a]	2	Intertidal
Mono Lake, California (FW)	157–399[a]	2	5–10
Europe			
Tamar Estuary (FW)	4900	8	—
Southampton Estuary	91,000–1,791,000	19	—
Severn Estuary drainage system	1600–25,700[a]	9	Intertidal
Mediterranean	198–372	2	6
Côte Bleue	1232–232,000	3	3–10
Les Embiez	13,000–15,000	2	3–10
Monaco	5200–12,100	2	3–10
Baltic Sea	258[a]	1	164
South Baltic Sea	50–2550[a]	7	—
Gulf of Finland	437[a]	1	60
Western Norway	284–99,452[a]	6	—
Neckar, Rhine, and Danube Rivers (FW)	600–44,560[a]	73	—

TABLE 2.1 (continued)
Comparison of Total PAH Concentrations in Marine and Freshwater (FW) Surficial Sediments

Location	Total PAH, (ppb, wet weight)	No. of Stations	Depth (m)
Other			
Walvis Bay, Africa	68[a]	1	—
Cariaco Trench	1756[a]	1	—
Amazon River system (FW)	ND–544	4	—
South Georgia Island	100	1	18

[a] ppb, dry weight.

Source: Johnson, A. C., Larsen, P. F., Gadbois, D. F., and Humason, A. W., *Mar. Environ. Res.*, 15, 1, 1985. With permission.

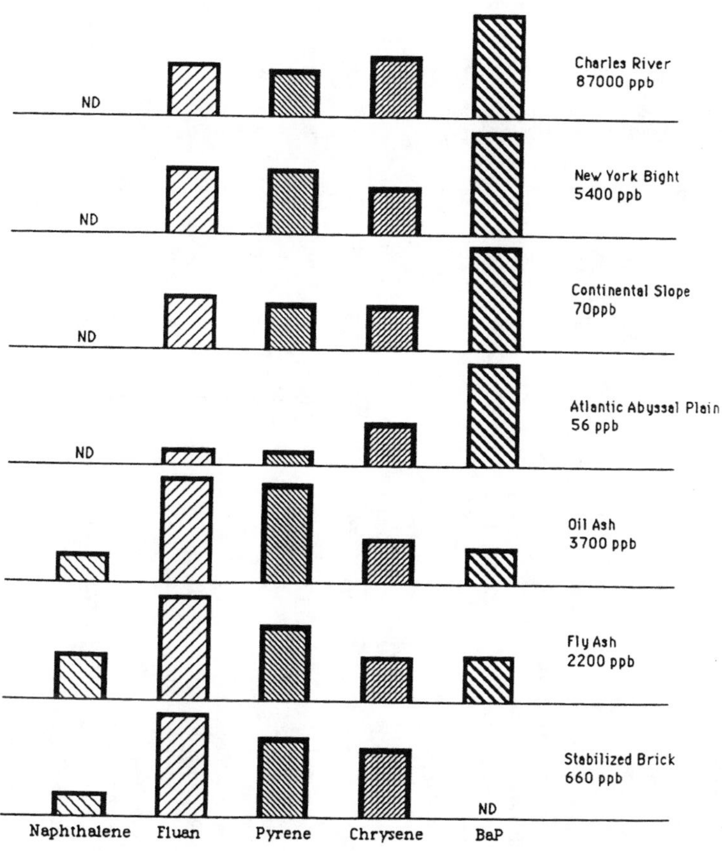

FIGURE 2.1. Relative distribution of PAH in typical world sediment samples vs. the distribution in oil ash, fly ash, and stabilized bricks. (From Frease, R. A. and Windsor, J. G., Jr., *Mar. Pollut. Bull.*, 28, 15, 1994. With permission.)

TABLE 2.2
Concentrations of Selected PAH Compounds in Sediments from Chesapeake Bay[a]

	Station and Year														
	LE5.5[b]					WE4.2					LE3.6				
Compound	**79S[c]**	**79F[d]**	**84**	**85[e]**	**86**	**79S**	**79F**	**84**	**85**	**86**	**79S**	**79F**	**84**	**85**	**86**
Phenanthrene	11	47	22	100	25	5	8	26	32	28	10	24	28	29	27
Fluoranthene	29	52	51	410	54	26	16	54	58	60	16	59	63	56	51
Pyrene	34	46	40	380	52	21	18	49	67	57	12	58	64	55	48
Benzo(a)fluorene	13	25	16	130	23	7	4	13	13	21	3	13	24	15	18
Benzo(a)anthracene	18	30	21	140	19	12	9	28	17	23	5	30	29	16	20
Chrysene/triphenylene	37	47	35	170	31	18	16	39	34	37	7	39	44	29	34
Benzo(e)pyrene	2	1	23	93	16	2	11	25	17	24	5	2	27	17	23
Benzo(a)pyrene	22	18	23	130	19	18	12	26	19	33	4	35	33	19	31
Pyrene	26	8	42	36	9	21	22	44	34	38	11	39	46	21	42
Benzo(g,h,i)perylene	15	6	18	46	9	15	8	31	23	20	3	17	28	12	26

	LE2.3				CB5.1				CB4.3C				CB3.3C[f]			
	79S	**79F**	**84**	**86**	**79S**	**79F**	**84**	**86**	**79S**	**79F**	**84**	**86**	**79S**	**79F**	**84**	**86**
Phenanthrene	19	42	54	64	17	49	47	50	44	68	26	11	280	220	300	240
Fluoranthene	34	70	89	88	35	81	85	74	60	82	42	14	370	220	370	300
Pyrene	29	57	72	87	33	73	79	66	43	70	38	4	360	220	370	290
Benzo(a)fluorene	10	16	24	34	10	27	24	23	11	38	13	2	120	98	150	77
Benzo(a)anthracene	10	23	28	25	10	32	26	20	8	25	16	>1	100	92	120	94
Chrysene/triphenylene	14	35	51	48	15	46	51	39	12	41	96	13	150	140	210	150
Benzo(e)pyrene	9	13	31	27	8	3	35	24	5	2	19	3	3	89	150	99
Benzo(a)pyrene	8	13	39	37	7	33	36	27	1	31	20	3	64	100	150	110
Pyrene	9	14	43	46	5	37	59	51	9	65	260	150	110	220	220	140
Benzo(g,h,i)perylene	7	5	27	21	5	18	35	19	4	21	21	10	38	56	96	79

[a] Concentrations in μg/kg.
[b] Southernmost sampling site.
[c] S = Summer.
[d] F = Fall.
[e] Northern Chesapeake Bay not sampled in 1985.
[f] Northernmost sampling site.

Source: Huggett, R. J., de Fur, P. O., and Bieri, R. H., *Mar. Pollut. Bull.*, 19, 454, 1988. With permission.

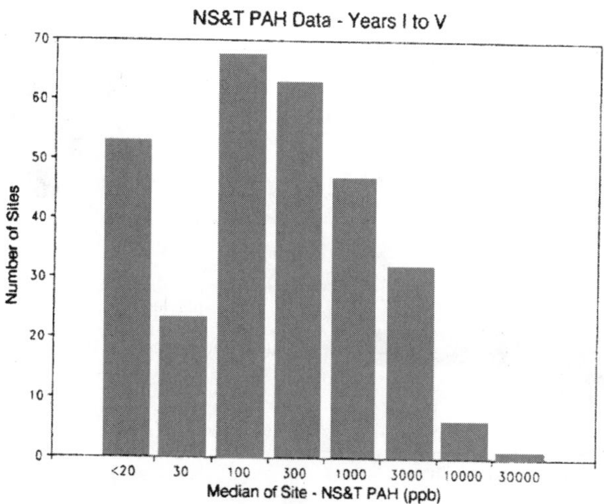

FIGURE 2.2. Frequency distribution of the median total National Status and Trends tPAH concentration in the Gulf of Mexico during the first 5 years of the program. (From Jackson, T. J., Wade, T. L., McDonald, T. J., Wilkinson, D. L., and Brooks, J. M., *Environ. Pollut.*, 83, 291, 1994. With permission.)

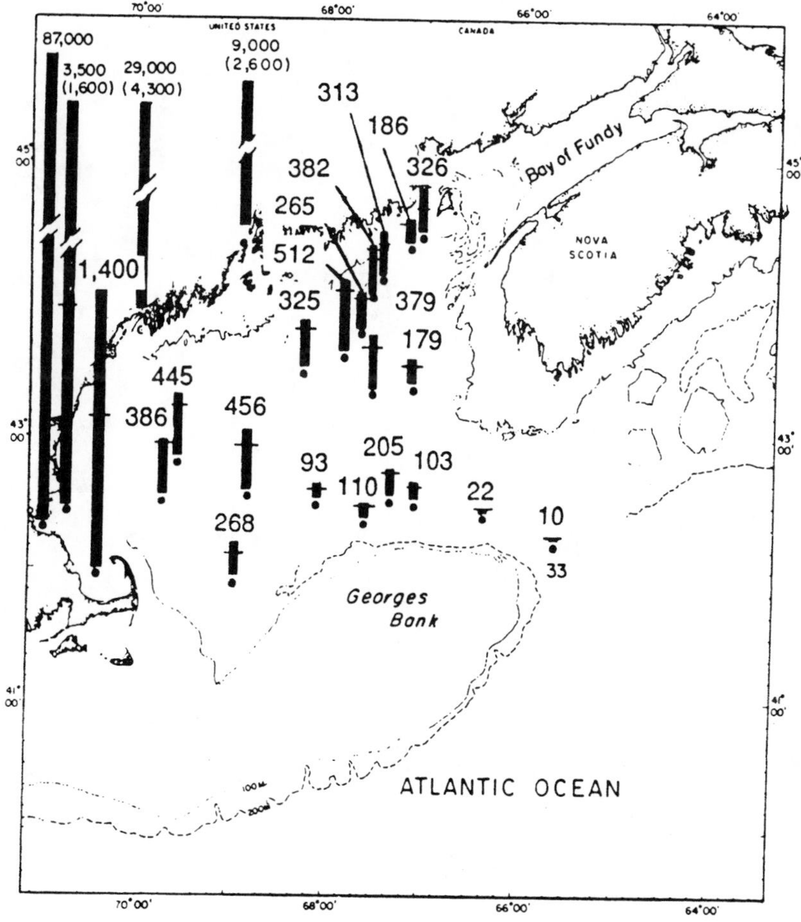

FIGURE 2.3. Composite results of several PAH surveys in the Gulf of Maine. The values on top of the bars are maximum values. Means are indicated in parentheses or as cross bars. Values are parts per billion dry weight. (From Larsen, P. F., *Rev. Aquat. Sci.*, 6, 67, 1992. With permission.)

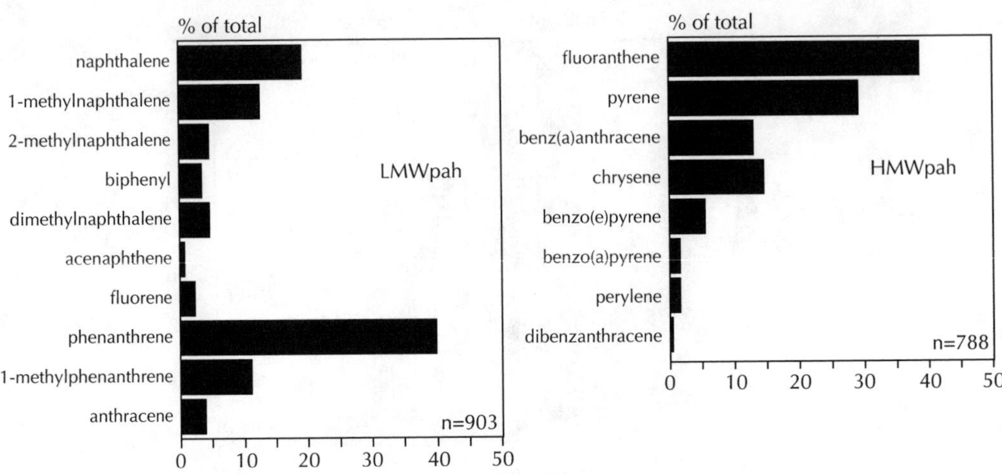

FIGURE 2.4. Mean percentage contributions of individual PAH compounds to low-molecular-weight (LMWpah) and high-molecular-weight (HMWpah) contaminant classes. (From NOAA, *A Summary of Data on Tissue Contamination from the First Three Years (1986–1988) of the Mussel Watch Project,* NOAA Tech. Mem. NOS OMA 49, National Oceanic and Atmospheric Administration, Rockville, MD, 1989.)

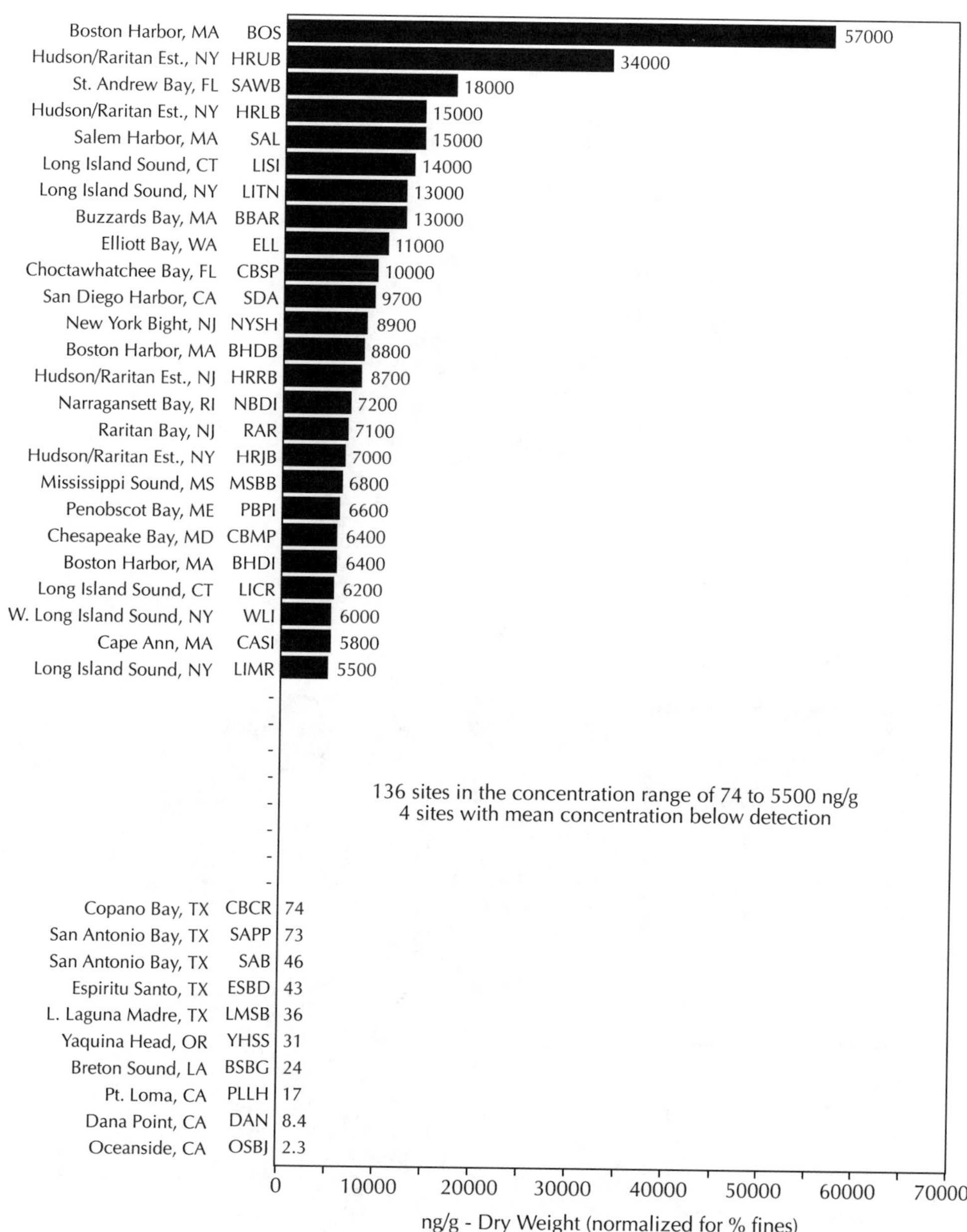

FIGURE 2.5. Total PAH concentrations in sediments of various estuarine and coastal systems of the U.S. (From NOAA, *A Summary of Data on Individual Organic Contaminants in Sediments Collected During 1984, 1985, 1986, and 1987*, Tech. Mem. NOS OMA 47, National Oceanic and Atmospheric Administration, Rockville, MD, 1989.)

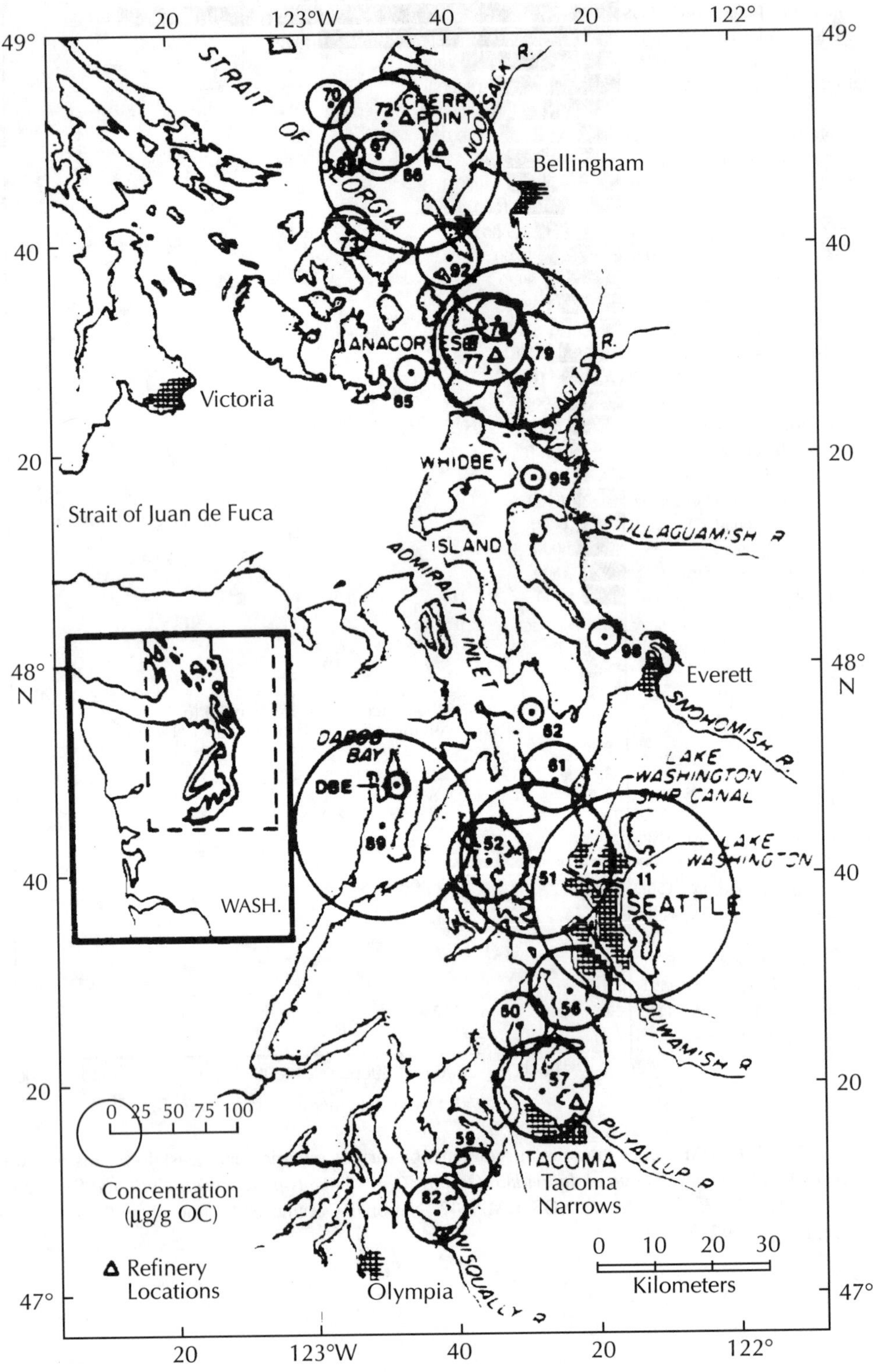

FIGURE 2.6. Total combustion-derived PAH at sampling stations (numbers) in Puget Sound. The radius of the circle about each station is proportional to the total concentration (µg/g OC) of the PAHs. (From Barrick, R. C. and Prahl, F. G., *Est. Coastal Shelf Sci.*, 25, 175, 1987. With permission.)

TABLE 2.3
PAH Content and Percent Organic Matter in Sediments of the Adriatic Sea[a]

Sampling Station	Total PAH (ng g^{-1} dry weight)	Organic Matter (%)
Po-Guarda Ven.	172	1.0
1	97	0.9
2	150	0.6
3	34	0.8
4	56	0.4
6	44	1.0
7	45	0.7
8	30	0.3
10	78	nd
11	53	0.6
13	527	1.1
15	355	1.2
17	163	1.3
18	164	0.6
19	41	0.0
20	59	0.6
21	65	1.0
23	33	0.8
24	102	1.0
25	102	1.1
26	128	1.6
27	116	0.9
27A	240	1.4
27B	214	1.3
27C	130	1.5
27D	346	1.1
28	126	1.2
29	225	1.4
30	37	0.7
31	27	0.4
32	31	0.4
33	92	0.8

[a] nd = not determined.

Source: Guzzella, L. and DePaolis, A., *Mar. Pollut. Bull.*, 28, 159, 1994. With permission.

TABLE 2.4
Annual Inputs of PAH via Urban Runoff of the Upper Narragansett Bay Watershed (kg/year)

Compound	Residential	Commercial	Industrial	Highway	Total
Naphthalene	0.18	1.7	4.8	2.6	9.3
2-Methylnaphthalene	7.5	1.0	12.1	4.9	25.5
1-Methylnaphthalene	0.35	0.79	14.3	1.9	17.3
Biphenyl	0.28	0.55	3.9	0.76	5.5
2-Ethylnaphthalene	0.02	0.66	12.6	2.2	15.4
Fluorene	0.04	0.44	9.7	7.9	18.1
Dibenzothiophene	0.18	0.30	20.4	3.9	24.8
Phenanthrene	1.7	2.1	32.4	32.2	68.4
Fluoranthene	20.8	9.4	40.7	101.2	172.1
Pyrene	12.0	6.7	32.4	35.1	86.2
Benzo(a)anthracene	2.9	1.7	16.3	58.7	79.2
Chrysene	5.2	7.7	8.8	86.0	107.7
Benzo(e)pyrene	5.7	3.3	4.8	10.1	23.9
Benzo(a)pyrene	4.8	1.1	14.4	12.4	32.7
Sum PAH	55.1	37.4	22.8	360.0	681.0
f_2	2200	3069	10,500	13,200	38,000
Total hydrocarbons	36,250	31,880	444,700	152,700	665,300
Sum PAH total hydrocarbons (%)	0.15	0.12	0.05	0.23	0.10
f_2 hydrocarbons/total hydrocarbons (%)	6.1	9.6	4.3	8.6	5.7

Source: Hoffman, E. J., Mills, G. L., Latimer, J. S., and Quinn, J. G., *Environ. Sci. Technol.*, 18, 580, 1984. With permission.

TABLE 2.5
Selected PAH and Total PAH Concentrations in Sediments from the Hudson-Raritan Estuary and the New York Bight and in Sewage Sludge[a]

Material and Location	Naphthalene	Phenanthrene	Anthracene	Benzo(a)anthracene	Total PAH
Sediment					
Hudson-Raritan estuary					
15 km north of the Battery	60	120	60	330	2000
Pierhead Channel	200	300	200	500	3200
Gowanus Canal	100	1000	500	3000	16,400
Newtown Creek	120,000	14,600	9600	5600	182,000
Lower Bay	100	600	300	2000	9900
New York Bight Region					
Christiaensen Basin	800	500	300	1000	6000
Sewage sludge dumpsite	80	70	40	200	1100
Outer Bight	0.6	3	N.D.[b]	3	22
Sewage sludge	2200	4400	1100	1000	20,400

[a] Concentrations in ng/g, dry weight.
[b] Not detected.

Source: O'Connor, J. M., Klotz, J. B., and Kneip, T. J., in *Ecological Stress and the New York Bight: Science and Management,* Mayer, G. F. (Ed.), Estuarine Research Federation, Columbia, SC, 1982, 631. Reprinted with permission of Estuarine Research Federation, copyright 1982.

APPENDIX 3. POLYCYCLIC AROMATIC HYDROCARBON COMPOUNDS IN ORGANISMS

TABLE 3.1
PAH Concentrations (μg/kg) in Selected Shellfish and Finfish from Estuarine and Coastal Marine Waters

Taxonomic Group, Compound, and Other Variables	Concentration
Shellfish	
Rock crab, *Cancer irroratus*,	
edible portions, 1980	
New York Bight	
Total PAH	1600 FW
BaP	1 FW
Long Island Sound	
Total PAH	1290 FW
BaP	ND
American oyster, *Crassostrea virginica*, soft parts,	
South Carolina, 1983, residential resorts	
Total PAH	
Spring months	
Palmetto Bay	520 FW
Outdoor resorts	247 FW
Fripp Island	55 FW
Summer months	
Palmetto Bay	269 FW
Outdoor resorts	134 FW
Fripp Island	21 FW
American lobster, *Homarus americanus*, edible portions, 1980	
New York Bight	
Total PAH	367 FW
BaP	15 FW
Long Island Sound	
Total PAH	328 FW
BaP	15 FW
Softshell clam, *Mya arenaria*	
Coos Bay, Oregon, soft parts 1978–1979	
Contaminated site	
Total PAH	555 FW
Phenanthrene = PHEN	155 FW
FL	111 FW
Pyrene = PYR	62 FW
BaP	55 FW
Benz(a)anthracene = BaA	42 FW
Chrysene = CHRY	27 FW
Benzo(b)fluoranthene = BbFL	12 FW
Others	<10 FW
Uncontaminated site	
Total PAH	76 FW
PHEN	12 FW
FL	10 FW
Others	<10 FW

TABLE 3.1 (continued)
PAH Concentrations (μg/kg) in Selected Shellfish and Finfish from Estuarine and Coastal Marine Waters

Taxonomic Group, Compound, and Other Variables	Concentration
Bay mussel, *Mytilus edulis,* Oregon, 1979–1980	
Soft parts, total PAH	
Near industrialized area	106–986 FW
Remote site	27–274 FW
Sea scallop, *Placopecten magellanicus,*	
Baltimore Canyon, east coast U.S.	
Muscle	
BaA	1 FW
BaP	<1 FW
PYR	4 FW
New York Bight, 1980	
Edible portions	
Total PAH	127 FW
BaP	3 FW
Clam, *Tridacna maxima*	
Australia, 1980–1982, Great Barrier Reef	
Soft parts, total PAH	
Pristine areas	<0.07 FW
Powerboat areas	Up to 5 FW
Mussel, *Mytilus* sp.	
Greenland	
Shell	60 FW
Soft parts	18 FW
Italy	
Shell	11 FW
Soft parts	130–540 FW
Bivalve mollusks, 5 spp., edible portions	6 (max. 36) FW
Decapod crustaceans, 4 spp., edible portions	2 (max. 8) FW
Softshell clam, *Mya arenaria,* soft parts	
Coos Bay, Oregon	
1976–1978	
Near industrialized areas	6–20 FW
Remote areas	1–2 FW
1978–1979	
Near industrialized areas	9 FW
Remote areas	4 FW
Finfish	
Fish, muscle	
Baltimore Canyon,	
East coast, U.S., 5 spp.	
BaA	Max. 0.3 FW
BaP	Max. <5 FW
PYR	Max. <5 FW
Smoked	
FL	3 FW
PYR	2 FW
Nonsmoked	
FL	Max. 1.8 FW
PYR	Max. 1.4 FW

TABLE 3.1 (continued)
PAH Concentrations (μg/kg) in Selected Shellfish and Finfish from Estuarine and Coastal Marine Waters

Taxonomic Group, Compound, and Other Variables	Concentration
Winter flounder, *Pseudopleuronectes americanus*, edible portions, 1980	
New York Bight	
Total PAH	315 FW
BaP	21 FW
Long Island Sound	
Total PAH	103 FW
BaP	ND
Windowpane, *Scopthalmus aquosus*, edible portions, 1980	
New York Bight	
Total PAH	536 FW
BaP	4 FW
Long Island Sound	
Total PAH	86 FW
BaP	ND
Red hake, *Urophycus chuss*, edible portions, 1980	
New York Bight	
Total PAH	412 FW
BaP	22 FW
Long Island Sound	
Total PAH	124 FW
BaP	5 FW
Fish	
Marine, edible portions	Max. 3 FW
9 spp.	15 FW
Greenland	65 FW
Italy	5–8 DW
Steak, charcoal broiled	11 DW

Note: ND, not detected; FW, fresh weight; DW, dry weight. See original source for specific references for these findings.

Source: Modified from Eisler, R., *Polycyclic Aromatic Hazards to Fish, Wildlife, and Invertebrates: A Synoptic Review,* Biological Report 85 (1.11), U.S. Fish and Wildlife Service, Washington, D.C., 1987.

TABLE 3.2
Selected PAH and Total PAH Concentrations in Finfish and Shellfish from the Hudson River and New York Bight[a]

Species (Location	Naphthalene	Phenanthrene	Anthracene	Biphenyl	Total PAH
Atlantic mackerel (*Scomber scombrus*) (New York Bight Apex)	ND	10	ND	ND	10
Winter flounder (*Pseudopleuronectes americanus*) (Christiaensen Basin)	2	ND	ND	6	8
Winter flounder (*P. americanus*) (Raritan Bay)	2	1	ND	ND	5
Striped bass (*Morone saxatilis*) (Montauk Point)	7	ND	ND	ND	19
Striped bass (*M. saxatilis*) (Hudson River)	4	ND	ND	4	8
Lobster (*Homarus americanus*) (New York Bight)	7	ND	ND	ND	7
Lobster (*H. americanus*) (Raritan Bay)	5	5	ND	ND	25
Lobster (*H. americanus*) (Raritan Bay)	7	ND	ND	ND	77
Blue mussel (*Mytilus edulis*) (Sandy Hook)	6	6	ND	4	250
Blue mussel (*M. edulis*)	20	10	1	40	120

[a] Concentrations in ng/g, dry weight; ND, not detected.

Source: O'Connor, J. M., Klotz, J. B., and Kneip, T. J., in *Ecological Stress and the New York Bight: Science and Management*, Mayer, G. F. (Ed.), Estuarine Research Federation, Columbia, SC, 1982, 631. Reprinted with permission of Estuarine Research Federation, copyright 1982.

TABLE 3.3
National Status and Trends: Oyster Polynuclear Aromatic Hydrocarbon Analytes

Aromatic Hydrocarbons	
Low Molecular Weight	**High Molecular Weight**
Biphenyl	Fluoranthene
Naphthalene	Pyrene
1-Methylnaphthalene	Benz(a)anthracene
2-Methylnaphthalene	Chrysene
2,6-Dimethylnaphthalene	Indeo(1,2,3-cd)pyrene[a]
1,6,7-Trimethylnaphthalene[a]	Benzo(a)pyrene
Acenaphthene	Benzo(e)pyrene
Acenaphthylene[a]	Perylene
Fluorene	Dibenz(a,h)anthracene
Phenanthrene	Benzo(g,h,i)perylene[a]
Anthracene	
1-methylphenanthrene	

[a] Analytes not used in tPAH summation.

Source: Jackson, T. J., Wade, T. L., McDonald, T. J., Wilkinson, D. L., and Brooks, J. M., *Environ. Pollut.,* 83, 291, 1994. With permission.

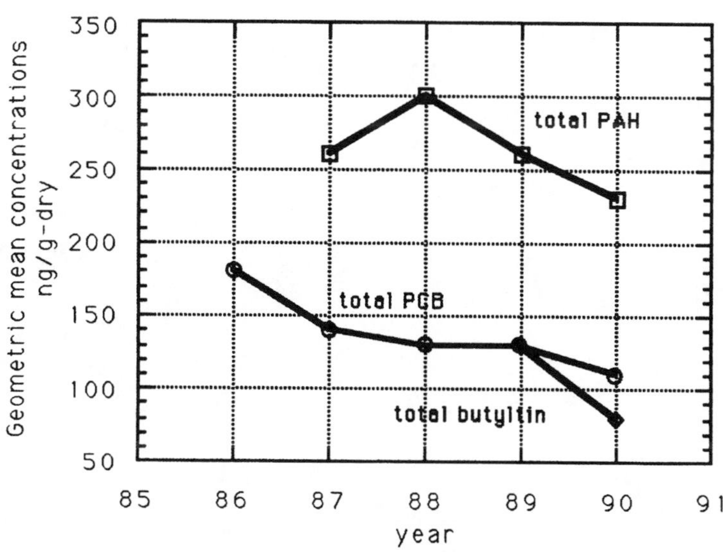

FIGURE 3.1. Mean concentrations of tPAH compounds and other selected chemical contaminants in mollusks as determined by NOAA's Mussel Watch Project. (From O'Connor, T. P., *Recent Trends in Coastal Environmental Quality: Results from the First Five Years of the NOAA Mussel Watch Project,* Spec. Publ., NOAA/NOS/ORCA, Rockville, MD, 1990.)

TABLE 3.4
Concentration of PAHs and Other Hydrocarbons in Mussels from Port Phillip Bay, Australia[a]

		Clean		Urb./Riv.		Oil Ref.		
Site		**SC**	**WS**	**HB**	**GB**	**CBN1**	**CBN2**	**CBN3**
Total HC	LIP	492	699	1060	873	3068	4443	3850
	DRY	67	136	146	152	380	760	674
Aromatic HC	LIP	260	341	593	383	2080	2181	1600
	DRY	35	66	82	67	258	373	280
Biogenic HC	LIP	49	78	45	20	<5	626	846
	DRY	7	15	6	3	<1	107	148
Total LABs	LIP	NM	NM	NM	NM	307	226	132
	DRY	NM	NM	NM	NM	38	39	23
B(b)F	LIP	16	53	53	63	226	437	276
	DRY	2.20	10	7.3	11	28	75	48
B(k)F	LIP	6.8	15	25	23	53	86	58
	DRY	0.93	2.9	3.5	4.1	6.6	15	10
B(a)P	LIP	3.2	4.9	9.6	13	52	172	116
	DRY	0.43	0.96	1.3	2.3	6.5	29	20

Note: NM = not measured; B(b)F = benzo(b)fluoranthene; B(k)F = benzo(k)fluoranthene; B(a)P = benzo(a)pyrene.

[a] Data expressed as mg/kg lipid weight (LIP) and mg/kg dry weight (DRY) of tissue. Values for PAHs are in μg/kg.

Source: Murray, A. P., Richardson, B. J., and Gibbs, C. F., *Mar. Pollut. Bull.*, 22, 595, 1991. With permission.

TABLE 3.5
Concentration of PAHs Obtained from Muscle Extract of Various Mammals[a]

Animals		Venezuelan Equivalents		Chrysene Equivalents	
		dry weight	wet weight	dry weight	wet weight
Beluga Whale	22	0.26	0.02	0.34	0.02
Hooded Seals	5	0.28	0.08	0.10	0.03
	4	0.35	0.09	0.12	0.03
	7	0.45	0.16	0.14	0.05
	8	0.45	0.17	0.16	0.06
	6	0.54	0.17	0.18	0.06
$\bar{x}$ (SD)		0.41 (0.10)	0.13 (0.05)	0.14 (0.03)	0.05 (0.02)
Sperm Whale	23	0.61	0.20	0.26	0.09
Minke Whale	30	0.69	0.27	0.21	0.08
Ringed Seal	11	0.73	0.27	0.18	0.07
	17	0.87	0.31	0.30	0.10
	19	1.13	0.30	0.38	0.11
	18	1.51	0.43	0.43	0.12
$\bar{x}$ (SD)		1.06 (0.34)	0.33 (0.07)	0.32 (0.11)	0.10 (0.02)
Common Dolphin	21	1.14	0.36	0.52	0.17
Harbour Seal	13	1.15	0.45	—	—
	1	2.22	0.66	0.30	0.10
	2	3.34	0.98	0.80	0.23
$\bar{x}$ (SD)		2.23 (1.09)	0.70 (0.27)	0.55 (0.35)	0.17 (0.09)
Harp Seal	26	1.00	0.32	0.29	0.09
	27	3.13	1.10	0.61	0.22
	25	3.22	1.12	0.72	0.25
$\bar{x}$ (SD)		2.45 (1.26)	0.85 (0.46)	0.54 (0.23)	0.19 (0.09)
White Sided Dolphin	10	4.47	1.73	1.07	0.41
Harbour Porpoise	12	5.50	2.04	1.21	0.45

Note: 1. Gas chromatographic (FID, MSD) results:

- Fluoranthene-pyrene (MW = 202): major PAH, detected in all samples
- Phenanthrene-anthracene (MW = 178): second major PAH, detected in all samples
- Acenaphthene (MW = 154), fluorene (MW = 166), benzanthracene-chrysene (MW = 228): minor PAH, detected in a few samples

2. The concentration of total lipids in the muscle tissues of harp seals was determined to be ~5% (dry weight). Lipid concentrations in the muscle of other marine mammals have been reported at ~2.5% for porpoise, ~1–13% for pothead whales, ~1% for beluga, ~0.05–0.85% for harbour seals, and ~1–13% for harp seals.

[a] Data in µb/g or ppm. NA = not available.

Source: Hellou, J., Stenson, G., Ni, I-H., and Payne, J. F., *Mar. Pollut. Bull.*, 21, 469, 1990. With permission.

TABLE 3.6
Toxicities of Selected PAH to Aquatic Organisms

PAH Compound, Organism, and Other Variables	Concentration in Medium[a]	Effect
Benzo(a)pyrene		
Sandworm, *Neanthes arenceodentata*	>1000	LC_{50} (96 h)
Chrysene		
Sandworm	>1000	LC_{50} (96 h)
Dibenz(a,h)anthracene		
Sandworm	>1000	LC_{50} (96 h)
Fluoranthene		
Sandworm	500	LC_{50} (96 h)
Fluorene		
Grass shrimp, *Palaemonetes pugio*	320	LC_{50} (96 h)
Amphipod, *Gammarus pseudoliminaeus*	600	LC_{50} (96 h)
Sandworm	1000	LC_{50} (96 h)
Sheepshead minnow, *Cyprinodon variegatus*	1680	LC_{50} (96 h)
Naphthalene		
Copepod, *Eurytemora affinis*	50	LC_{30} (10 d)
Pink salmon, *Oncorhynchus gorbuscha*, fry	920	LC_{50} (24 h)
Dungeness crab, *Cancer magister*	2000	LC_{50} (96 h)
Grass shrimp	2400	LC_{50} (96 h)
Sheepshead minnow	2400	LC_{50} (24 h)
Brown shrimp, *Penaeus aztecus*	2500	LC_{50} (24 h)
Amphipod, *Elasmopus pectenicrus*	2680	LC_{50} (96 h)
Coho salmon, *Oncorhyncus kisutch*, fry	3200	LC_{50} (96 h)
Sandworm	3800	LC_{50} (96 h)
Mosquitofish, *Gambusia affinis*	150,000	LC_{50} (96 h)
1-Methylnaphthalene		
Dungeness crab, *Cancer magister*	1900	LC_{50} (96 h)
Sheepshead minnow	3400	LC_{50} (24 h)
2-Methylnaphthalene		
Grass shrimp	1100	LC_{50} (96 h)
Dungeness crab	1300	LC_{50} (96 h)
Sheepshead minnow	2000	LC_{50} (24 h)
Trimethylnaphthalenes		
Copepod, *Eurytemora affinis*	320	LC_{50} (24 h)
Sandworm	2000	LC_{50} (96 h)
Phenanthrene		
Grass shrimp	370	LC_{50} (24 h)
Sandworm	600	LC_{50} (96 h)
1-Methylphenanthrene		
Sandworm	300	LC_{50} (96 h)

[a] Concentrations in μg/l.

Source: Modified from Eisler, R., *Polycyclic Aromatic Hazards to Fish, Wildlife, and Invertebrates: A Synoptic Review,* Biological Report 85 (1.11), U.S. Fish and Wildlife Service, Washington, D.C., 1987.

TABLE 3.7
Relative Carcinogenicity Index of Some PAH

Compound	Carcinogenicity Index
Benzo(a)anthracene	+
7,12-Dimethylbenz(a)anthracene	++++
Dibenz(a,j)anthracene	+
Dibenz(a,h)anthracene	+++
Dibenz(a,c)anthracene	+
Benzo(c)phenanthrene	+++
Dibenzo(a,h)fluorene	+
Dibenzo(a,h)fluorene	UC
Dibenzo(a,c)fluorene	+
Benzo(b)fluoranthene	++
Benzo(j)fluoranthene	++
Benzo(j)aceanthrylene	++
3-Methylcholanthrene	++++
Benzo(a)pyrene	+++
Dibenzo(a,l)pyrene	UC
Dibenzo(a,h)pyrene	+++
Dibenzo(a,i)pyrene	+++
Indeno(1,2,3-cd)pyrene	+
Chrysene	UC
Dibenzo(b,def)chrysene	++
Dibenzo(def.p)chrysene	+

Note: UC, unknown; + = low carcinogenicity; ++ = low to moderate carcinogenicity; +++ = moderate carcinogenicity; ++++ = high carcinogenicity.

Source: Afghan, B. K. and Chau, A. S. Y. (Eds.), *Analysis of Trace Organics in the Aquatic Environment,* CRC Press, Boca Raton, FL, 1989, 221. With permission.

TABLE 3.8
Ratio of Animal/Sediment Ratios for CAH/PAH and Mixed Function Oxidase (MFO) Activity for Various Phyla in Puget Sound (PS) and the New York Bight (NYB)

Animal (no. of samples)	CAH/PAH Ratio		MFO Activity (pmol m/ml m^{-1} mg^{-1})
	Median	Range	
Fish			
Winter flounder			213 ± 15
NYB winter flounder liver (1)	3400		
NYB winter flounder flesh (4)	800	120–2000	
PS sole liver (6)	2200	790–3300	
Polychaetes			
Sand worm			89 ± 24
NYB polychaetes (1)	210		
PS "worms" (4)	25	6–45	
Crustaceans			
Blue crab hepatopancreas			42 ± 15
PS crab hepatopancreas (5)	1300	450–2900	
NYB lobster digestive gland (1)	590		
NYB lobster flesh (3)	100	95–120	
NYB grass shrimp (1)	28		
PS shrimp (6)	19		
Mollusks			
Oyster			8 ± 2
Mussel digestive gland			3 ± 1
NYB sea scallops (3)	48	10–76	
NYB mussels (2)	50	32–65	
PS clams (7)	12	1–49	

Note: The geometric mean of the animal/sediment ratios for each PAH and CAH compound was used to calculate an overall CAH/PAH ratio.

Source: Connor, M. S., *Environ. Sci. Technol.*, 18, 31, 1984. With permission.

TABLE 3.9
Commonly Reported Effects of PAHs on Living Organisms

Effects	Plant	Invertebrate	Fish	Reptile and Amphibian	Bird	Mammal
Individual Organisms						
Death	X	X	X	X	X	X
Impaired reproduction	X	X	X	X	X	
Reduced growth and development	X	X	X	X	X	
Impaired immune system						X
Altered endocrine function			X		X	
Altered rate of photosynthesis	X					
Malformations			X		X	
Tumors and lesions		X	X	X		X
Cancer			X	X		X
Altered behavior		X	X	X	X	X
Blood disorders		X	X	X	X	X
Liver and kidney disorders			X		X	X
Hypothermia					X	X
Inflammation of epithelial tissue					X	X
Altered respiration or heart rate		X	X	X		
Impaired salt gland function				X	X	
Gill hyperplasia			X			
Fin erosion			X			
Groups of Organisms[a,b]						
Local population changes	X	X			X	
Altered community structure	X	X			X	
Biomass change	X	X				

[a] Some effects have been observed in the wild and in the laboratory, whereas others have been induced only in laboratory experiments.

[b] Populations of chlorophyllous (microalgae) and nonchlorophyllous plants (bacteria, filamentous fungi, yeast) can increase or decrease in the presence of petroleum, whereas animal populations decrease.

Source: Albers, P. H., in *Handbook of Ecotoxicology,* Hoffman, D. J., Rattner, B. A., Burton, G. A., Jr., and Cairns, J., Jr. (Eds.), Lewis Publishers, Boca Raton, FL, 1994, 330. With permission.

5 Halogenated Hydrocarbons

INTRODUCTION

Among the most persistent, ubiquitous, and toxic pollutants in estuarine and marine ecosystems are a number of hydrocarbon compounds containing chlorine, bromine, fluorine, or iodine (the halogens). The halogenated hydrocarbons include an array of substances ranging from low-molecular-weight volatile compounds, which have become common components of the atmosphere and surface waters of the ocean, to higher-molecular-weight compounds, which are widely distributed in estuarine and coastal marine waters worldwide. They often occur in significant quantities at sites near urban and industrial centers. The low-molecular-weight halogenated hydrocarbons generally are not perceived to be as serious a threat to estuarine and marine ecosystems as the higher-molecular-weight halogens, in part because they do not appear to accumulate in biota. Many of the higher-molecular-weight halocarbons (e.g., organochlorine compounds) do accumulate in the lipid-rich tissues of animals, and their toxic effects to nontarget organisms have been well chronicled. This chapter focuses on those halogenated hydrocarbons — particularly the higher-molecular-weight compounds — that have been shown to impact estuarine and marine environments.

LOW-MOLECULAR-WEIGHT COMPOUNDS

Various low-molecular-weight, volatile hydrocarbons manufactured as aerosol propellants, coolants, dry cleaning fluids, and industrial solvents have received considerable attention as potentially damaging atmospheric chemicals. The residence times of these halocarbons, on the order of years in the atmosphere, promote their global dispersal. Because they have been found throughout the surface waters of the world's oceans, there is growing concern of possible impacts of halocarbons on marine systems.

Several synthetic low-molecular-weight halogenated hydrocarbons have been manufactured in significant quantities over the years, and much of this production has been lost to the environment. One class of such compounds, the chlorofluorocarbons, has been implicated in the reduction of the ozone layer in the stratosphere. Trichlorofluoromethane (CCl_3F, Freon 11) and dichlorodifluoromethane (CCl_2F_2, Freon 12), the two principal members of this set of compounds, are used mainly as coolants, aerosol propellants, and constituents in foamed plastics. Carbon tetrachloride (CCl_4) and perchlorethylene ($Cl_2C{=}CCl_2$) constitute important dry-cleaning fluids. Other industrial solvents of significance are dichlorethane (CH_3CHCl), trichlorethane (CH_2Cl_3), trichlorethylene ($ClHC{=}CCl_2$), and vinyl chloride ($H_2C{=}CHCl$). The half-life values of these chemicals in the hydrosphere are in the range of months.

Some low-molecular-weight halocarbons have been detected in air samples even in polar regions. While these compounds continue to be perceived as significant atmospheric pollutants, their role in estuarine and marine ecosystems is less certain. Marine organisms appear to take up only small amounts of these contaminants, and there is as yet no indication of an acute threat of the contaminants to entire communities. However, relatively few detailed investigations of low-molecular-weight, volatile halogenated hydrocarbons have been conducted in estuarine and marine environments to date.

HIGH-MOLECULAR-WEIGHT COMPOUNDS

More ecotoxological attention has been given to the higher-molecular-weight chlorinated hydrocarbon compounds which are common contaminants in estuarine and marine environments. These compounds are mainly derived from pesticides and industrial chemicals (e.g., chlorinated aromatics, PCBs; chlorinated paraffins). While pesticides typically enter aquatic environments through agricultural and domestic applications, industrial chemicals are often introduced into these environments at manufacturing sites or through various disposal routes. The chlorinated hydrocarbons are broad-spectrum poisons that affect many different organisms in an area (e.g., plankton, benthic invertebrates, fish, mammals, birds), thereby threatening entire communities. The physiological responses to these compounds are numerous; adverse effects have been documented on the immune system, adrenal function, thyroid secretion, biogenic amines, and migratory condition of marine organisms. Their rapid distribution in marine environments has been facilitated by atmospheric dispersal and deposition, current transport, and migration of animals contaminated with the halocarbons.

It is the unique properties of chlorinated hydrocarbon contaminants — chemically-stable nature, great mobility, hydrophobicity, resistence to degradation, persistence in the environment, affinity for living systems, bioaccumulative capacity, and general toxicity — that have engendered apprehension and fostered numerous monitoring programs during the past two decades. Of particular concern are the potentially harmful effects of these contaminants on recreationally and commercially important finfish and shellfish, as well as on humans who consume them. Due to their lipophilicity, organochlorine compounds tend to concentrate in lipid-rich tissues of animals, and they biomagnificate through food webs. Hence, some marine fauna, such as mammals (e.g., whales, dolphins, porpoises, seals) situated at the uppermost trophic levels, carry very high contaminant residues.[1,2] The elevated concentrations result from the lipid solubility of the contaminants, large lipid pool (blubber) in these species, the lactational transfer of the compounds over generations, and the low capacity for xenobiotic degradation.[3]

The half-lives of chlorinated hydrocarbon compounds typically range from months to years in estuarine and marine environments. The residues of the more persistent compounds, however, may be present for decades or possibly centuries, depending on temperature, light, pH, rate of microbial degradation, and other conditions. The half-life of a contaminant within an organism is in part dependent on body composition and biological functions, such as the concentration of lipids and the reproductive activity of a species.

Chlorinated hydrocarbon contamination of estuarine and marine waters is inextricably linked to anthropogenic activities. Heaviest pollution loads occur in proximity to sites of urban and industrial development. Estuaries nearby metropolitan centers are highly susceptible. They serve as sinks or repositories for riverborne industrial chemicals, agricultural pesticides, organochlorine-containing sewage effluent, and other xenobiotics.

Despite their peak occurrence in localized "hot spots" in the coastal zone, chlorinated hydrocarbons tend to be found in a wide range of environmental compartments worldwide. Significant levels of some organochlorine compounds have been detected in air, water, plants, and wildlife from equatorial to polar regions.[4] The global distribution of persistent chlorinated hydrocarbons has been facilitated by atmospheric transport involving the cyclic processes of wet/dry deposition and sublimation or evaporation, combined with the net atmospheric flux of heat from equatorial regions. The tropical countries serve as the major emission sources of the persistent organochlorines contributing to global contamination through long-range atmospheric transport. Presently, the organochlorine residues in the estuarine and coastal marine atmosphere and waters of Asia exhibit higher levels in tropical than temperate and polar latitudes owing to continued use of chlorinated hydrocarbon pesticides by developing countries.[5] DDT usage, for example, appears to be greatest in developing countries in the tropics. While poleward transport of DDT and other organochlorine compounds can be rapid, their relative inertness and resistence to biodegradation and photolysis favor their long-distance transport and ultimate deposition in regions remote from the point of application.[6] Hence, although DDT has been banned for decades in the U.S., organisms still are being affected by the contaminant through airborne exposure from remote sources.

The Joint Group of Experts on the Scientific Aspects of Marine Pollution[7] has subdivided organochlorine compounds into five groups. These include:

- Three lower-molecular-weight groups (up to three carbons)
- Aliphatic and aromatic herbicides (up to six carbons)
- Long-chain chlorinated paraffins
- Chlorinated insecticides (e.g., Mirex and Camphenes)
- Chlorinated aromatics (e.g., PCBs)

PESTICIDES

Any substance used to control, repel, prevent, or destroy pests may be defined as a pesticide. Manno[8] classified pesticides by their use, identifying six classes:

1. Insecticides (e.g., chlorinated hydrocarbons, organophosphorus compounds, carbamates, pyrethroids, phenols)
2. Herbicides (chlorophenoxy compounds, bipyridylium compounds, triazines, thiocarbamates)
3. Fungicides (inorganic compounds, organometallic compounds, antibiotics, chloroalkylthio compounds, quinones, dithiocarbamates)
4. Rodenticides (fluoroacetate derivatives, thioureas, anti-vitamin K compounds)
5. Fumigants/nematocides (hydrocyanic acid, carbon disulfide)
6. Synergists (piperonyl butoxide, sulfoxide)

Unlike the first four classes, the fumigants are a rather loosely defined group of formulations, and the synergists are substances that enhance the activity of insecticides.

Ware[9] proposed a more extensive list of pesticide classes based on their function. These classes are the acaricides (kill mites), algicides (kill algae), avicides (kill or repel birds), bactericides (kill bacteria), fungicides (kill fungi), herbicides (kill weeds), insecticides (kill insects), larvicides (kill larvae, usually mosquitos), miticides (kill mites), molluscicides (kill snails, slugs, and other mollusks), nematicides (kill nematodes), ovicides (destroy eggs), pediculicides (kill lice), piscicides (kill fish), predicides (kill predators, notably coyotes), rodenticides (kill rodents), silvicides (kill trees and brush), slimicides (kill slimes), and termiticides (kill termites).

Another category of pesticide classes that do not actually kill pests but fit practically as well as legally under this umbrella term is also included in Ware's classification scheme. The chemical classes listed under this category are the attractants (attract insects), chemosterilants (sterilize pests), defoliants (remove leaves), dessicants (speed drying of plants), disinfectants (destroy or inactivate harmful microorganisms), growth regulators (stimulate or retard growth of plants or insects), pheromones (attract insects or vertebrates), and repellents (repel insects, mites, ticks or pest vertebrates).

The pesticides of principal interest here are the insecticides. Emphasis will be placed on the organochlorine, organophosphate, and carbamate insecticides, although the botanical insecticides will also be covered. In addition, the herbicides will be briefly addressed.

CHLORINATED HYDROCARBON INSECTICIDES

Chlorinated hydrocarbons, together with organophosphates, comprise the majority of all synthetic insecticides. The chlorinated hydrocarbon pesticides, also known as organochlorines or chlorinated insecticides, are extremely effective biocide agents, acting as nerve poisons to control the population sizes of target organisms. However, their presence in the marine hydrosphere has been linked to neurological and reproductive failure in many nontarget organisms as well, and neurological diseases also have been observed in some humans exposed to these chemicals.[10-12] The chlorinated hydrocarbon pesticides degrade much more slowly in the environment than the organophosphates (5 to 15 years compared to 1 week to several months).[13] As a result, most of the organochloride pesticides have been replaced by the organophosphate and carbamate pesticides during the past two decades.

The chlorinated hydrocarbon insecticides include a number of well-known synthetic compounds, such as DDT, aldrin, chlordane, dieldrin, endosulfan, endrin, lindane, heptochlor, chlordecone, mirex, perthane, and toxaphene. Smith[14] and Blus[15] subdivided the organochlorine insecticides into several major groups based on their chemical structure. These groups are DDT and its analogues, hexachlorocyclohexane (HCH), cyclodienes and similar compounds, toxaphene and related chemicals, and the caged structures mirex and chlordecone. Many of these pesticides were banned in the U.S. during the 1970s. For example, all interstate sale and transport of DDT in the U.S. were banned in 1973, except for emergency situations. Aldrin and dieldrin were banned between 1972 and 1974; chlordane and heptachlor were banned in 1975.[13] All registered uses of mirex were cancelled by the U.S. Environmental Protection Agency in 1977. Despite the banning of these chemicals in the U.S., they still are used by other countries, especially those in tropical regions which are plagued by infectious diseases transmitted by insects and other pests.

DDT

One of the most well known and ubiquitous chlorinated hydrocarbon contaminants is DDT (dichloro-diphenyl-trichloroethane), which belongs to the chemical class of diphenyl aliphatics consisting of an aliphatic or straight carbon chain, with two (di)phenyl rings attached. Introduced as an insecticide in 1939, DDT rapidly gained favor as the pesticide of choice in controlling a wide variety of insect pests. By 1961, more than 1200 formulations were available, being employed primarily for pest control in forestry and agriculture, but also for domestic applications. During its peak period of production in the U.S. during the late 1950s to early 1960s, DDT was registered for use on 334 crops against 240 species of agricultural pests.[16] Peak usage of DDT in the U.S. occurred in 1959. The dramatic early successes of DDT in agriculture and forestry were attributed to several factors: (1) its extreme toxicity to insects, (2) its great persistence, and (3) its low cost.

Ten DDT analogues have been used as commercial pesticides, including DDD (TDE), Bulan, chlorfenethol (DMC), dicofol, chlorobenzilate, chloropropylate, DFDT, etylan, methoxychlor, and Prolan. These compounds possess markedly different properties; however, because they were applied much more sparingly than DDT itself and were the focus of far fewer investigations, less ecotoxicological data are available on them.

Environmental problems with DDT began to surface during the 1950s and 1960s. Both acute and insidious effects of DDT and its metabolite DDE (dichloro-diphenyl-ethane) were noted as the contaminants accumulated in food chains and proved to be detrimental to many populations of terrestrial and aquatic organisms. By the end of the decade, DDT had been severely restricted or banned in much of North America. In estuarine and marine ecosystems, the biotic impacts of DDT were manifested most conspicuously in upper-trophic-level organisms, especially predatory fish and birds. Although not applied directly to estuarine and marine waters, DDT entered these systems via nonpoint source runoff from land, riverine discharges, anthropogenic disposal, precipitation, and airborne particulate deposition.

Solar radiation and the metabolic activities of animals degrade DDT to DDE and DDD. The combination of DDT and its degradation products DDE and DDD is referred to as total DDT (tDDT) or DDT residues. DDE is extremely important in estuarine and marine systems since it accounts for most of the tDDT in the sea and 80% of that in marine organisms. DDD is less toxic to marine organisms than DDT or DDE, and it rarely accumulates in them. Because it is less toxic to fish than DDT, it has been used occasionally in some countries in place of DDT to minimize impacts on fish populations. DDE is not employed as an insecticide because it has low toxicity to insects.[9]

DDT adversely affects the central nervous system of insects and other animals, precluding the normal conduction of nerve impulses by destroying the delicate balance of sodium and potassium within the neuron. It appears to inhibit a specific ATPase important in ion transport in nerves, while increasing the activity of microsomal enzymes. These effects culminate in hyperactivity, convulsions, paralysis, and death. DDT also affects estrogenic activity by inducing a breakdown of sex hormones that regulate the mobilization of calcium. DDT specifically inhibits carbonic anhydrase, an enzyme important in eggshell production of birds.[12] Hence, some bird populations exposed to DDT (e.g., brown pelicans, *Pelecanus occidentalis*; double-crested cormorants, *Phalacrocorax auritus*) have experienced declining population sizes, owing to their inability to properly metabolize calcium. Fish-eating birds, such as the brown pelican and herring gull (*Larus argentatus*), may have levels of DDT about 30- to 100-fold greater than those of their prey due to the effects of biomagnification.[15] Exposure of these birds to DDT has produced thin-shelled eggs which easily crack in the nest, resulting in prenatal death. At some locations, the birds have experienced total breeding failure due to eggshell collapse.

While not phytotoxic, DDT has been shown to inhibit oxidative phosphorylation in mitochondria and the Hill reaction in chloroplasts.[16] Estuarine and marine phytoplankton rapidly take up DDT from surrounding waters, and their growth may be adversely affected by its presence.[10] Plants containing even traces of DDT serve as a source of the contaminant for consumers at higher trophic levels. Since this organochlorine compound readily accumulates in adipose tissues of estuarine and marine animals and is poorly metabolized, it undergoes biomagnification in food webs.[9,10]

DDT is one of the most water-insoluble compounds ever synthesized. Its solubility amounts to approximately 6 ppb of water.[9] In contrast, DDT is highly fat soluble. Thus, it tends to partition out of the hydrosphere into biotic compartments where considerable biological damage may ensue. In aquatic environments, DDT is not readily degraded by heat, ultraviolet light, enzymes, or microorganisms.[16] It also tends to partition out of the hydrosphere into sedimentary compartments, sorbing to sediments and other particulates which subsequently may be ingested by pelagic and benthic organisms.[17]

Fowler[18] reported generally similar tDDT concentrations (approximately 0.005 to 0.06 ng/l) in open ocean waters of the northern and southern hemispheres, except for areas near the Asian continent (approximately 0.05 to 0.12 ng/l) and off Central America. He noted that in coastal waters DDT residues are generally less than 5 ng/l. Total DDT concentrations in the surface sediments from coastal and nearshore areas vary widely from <.01 to >1000 ng/g dry weight. The highest values are found in sediments of Palos Verdes, CA (1600 to 100,000 ng/g dry weight), a site affected by major sewage discharges from a highly populated and industrialized area.

Mussels and, in some cases, oysters have been used as bioindicators of DDT contamination in various coastal regions of the world. For example, tDDT residues in mussels from coastal waters of the northeast Pacific and northwest Atlantic (U.S.) amount to 5.4 to 1077 μg/kg dry weight and 2.8 to 1109 μg/kg dry weight, respectively.[18] The concentrations of tDDT in green mussels (*Perna viridis*) from the Gulf of Thailand range from 0.74 to 5.38 ng/g wet weight,[19] and those from the coastal waters of India, from 3 to 39 ng/g wet weight.[20] Elsewhere, higher-trophic-level organisms (dolphins, porpoises, whales,[1] seals,[3] and fish[21,22]) have substantially greater concentrations owing to bioaccumulation effects.

DDT has been detected in biotic, sediment, and water samples from every coastal state and from nearly every estuary in the U.S., as well as from many offshore and deep-sea locations. Geographic and long-term trends of DDT contamination in estuarine and marine organisms of the U.S. reveal elevated levels during the 1960s followed by dramatic declines since the mid-1970s.[10,23] DDT, along with other chlorinated hydrocarbons, reached maximum levels in estuarine and coastal marine sediments during the years 1950 to 1970.[24] Data collected on finfish and shellfish samples during the National Pesticide Monitoring Program, the U.S. Environmental Protection Agency Mussel Watch Project, and the NOAA Mussel Watch Program indicate a marked nationwide drop in tDDT contamination during the early to mid-1970s. In a historical assessment report in which data were reviewed from over 300 projects, monitoring programs, and fish surveys conducted between 1940 and 1985, Mearns et al.[23] inferred that tDDT concentrations decreased 80- to 100-fold in coastal and estuarine biota on a nationwide basis between 1965–1972 and 1984–1986.

During peak contamination levels in the late 1960s and early 1970s, the median tDDT concentration was 0.024 ppm wet weight in shellfish and 0.7 to 1.1 ppm wet weight in whole fish. The 1986 NOAA Mussel Watch Project obtained a median tDDT concentration of 0.003 ppm wet weight for shellfish at 145 sites.[25] The tDDT levels in finfish paralleled those in shellfish. The median value of tDDT in fish muscle from samples collected at 31 areas between 1972 and 1975 amounted to 0.11 ppm wet weight. This value is about tenfold greater than the median tDDT concentration of 0.013 ppm wet weight measured in fish muscle from samples collected at 19 areas in 1976 and 1977. The tDDT concentration in fish liver during the 1976 and 1977 sampling period amounted to 0.22 ppm wet weight, which is nearly fourfold greater than the median value of 0.06 ppm wet weight recorded in fish liver during the 1984 National Status and Trends Program.[23] These data reflect a significant drop of DDT concentrations in the mid-1970s, perhaps by a factor of 4 or 5. The concentrations continued to decline into the 1980s, albeit not as precipitously. Persistent "hot spots" in the 1980s were San Francisco Bay, San Pedro Harbor, and Palos Verdes (CA); Mobile Bay (AL); and Hudson-Raritan Bay (NY-NJ). A broad low-level gradient of DDT contamination also occurred along the East Coast of the U.S. radiating north and south of Delaware Bay and possibly secondarily from Long Island, NY.

On a national scale, data from the NOAA Mussel Watch Project demonstrate that the highest concentrations of DDT exist near urban areas on the east coast (e.g., Hudson-Raritan Bay) and west coast (e.g., San Pedro Harbor, Palos Verdes).[26] Elevated DDT measurements are relatively rare in coastal waters of the southeastern states and along the Gulf of Mexico.[27,28] Sericano et al.[29] detailed historical decreases in the mean tDDT concentration in oysters from the Gulf of Mexico.

Aside from data of the Mussel Watch Project, the National Status and Trends Program of NOAA has analyzed contaminant concentrations in surface sediments at nearly 300 estuarine and coastal marine sites throughout the U.S.[30] Similar to the results of the Mussel Watch Project, the highest concentrations for any particular contaminant in sediment samples were obtained at sites near urbanized areas, such as those in proximity to Boston, New York, San Diego, Los Angeles, and Seattle. Sites with high concentrations of tDDT (>37 ng/g fine grained sediment) included: the Hudson-Raritan estuary (NY-NJ), Delaware Bay (DE), Tampa Bay (FL), Panama City (FL), Choctawhatchee Bay (FL), Galveston Bay (TX), Point Loma (CA), Oceanside (CA), Newport Beach (CA), Anaheim Bay (CA), Long Beach (CA), San Pedro Bay (CA), San Pedro Harbor (CA), Palos Verdes (CA), West Monica Bay (CA), Marina Del Ray (CA), Point Dume (CA), Point Santa Barbara (CA), Monterey Bay (CA), and San Pablo Bay (CA).

Cyclodiene Compounds

Several chlorinated cyclic hydrocarbons belong to a class of compounds termed the cyclodiene pesticides. These compounds include the most toxic organochlorine insecticides, especially in terms of acute poisoning. The cyclodienes consist of chemicals developed after World War II (e.g., chlordane, 1945; aldrin and dieldrin, 1948; heptachlor, 1949; endrin, 1951; endosulfan, 1956). Chlordane and dieldrin have been produced in greatest quantities over the years because of their effectiveness, but less well-known cyclodiene compounds (e.g., alodan, bromodan, isobenzam, isodrin, and telodrin) also have been highly viable insecticides.

The cyclodienes share the common characteristics of low water solubility and extreme persistence. They are particularly stable in soil, and thus have been employed most commonly as soil insecticides. Some of the compounds (e.g., aldrin and heptachlor) are rapidly metabolized by organisms, but their metabolites (dieldrin and heptachlor epoxide) are as toxic and persistent as the parent compounds.[15] Like DDT, these insecticides are neurotoxins. Due to their general toxicity to insects as well as nontarget organisms (e.g., birds, fish, mammals), severe restrictions have been placed on their use in Canada and the U.S. Cyclodiene compounds tend to be particularly toxic to fish. The U.S. Environmental Protection Agency cancelled most agricultural uses of the cyclodienes between 1975 and 1980 in response to various environmental hazards that they pose.

Chlordane

Composed mainly of a mixture of polychloromethanoindenes, chlordane is a broad-spectrum poison that adversely affects many aquatic organisms. It was the first of the persistent, chlorinated cyclodiene insecticides with lipophilic properties, being introduced as an agricultural pesticide in the U.S. in 1945 and later utilized in other applications (e.g., as chemical protection against woodboring insects). Concerns over the potential environmental and human health impacts of chlordane led to restrictive regulations on its use in 1974. By 1983, the only remaining application of chlordane was as a termiticide, and by 1988 all uses of the chemical had been effectively eliminated in the U.S.[31] However, chlordane is probably still used by developing nations.

The National Pesticide Monitoring Program did not record chlordane above the detection limit of 0.01 ppm wet weight in more than 8000 shellfish samples collected from 1965 to 1972 and in 1977.[32,33] Between 1972 and 1976, however, the National Pesticide Monitoring Program detected chlordane in 39 samples of whole juvenile fish from five states. In 1984, the National Status and Trends Program documented an average chlordane (*alpha*-chlordane and *trans*-nonachlor) concentration of 0.038 ppm wet weight in fish livers from specimens taken from 48 sites. The 1986–1987 Bioaccumulation Study of the U.S. Environmental Protection Agency found variable levels of chlordane contamination in fish and shellfish sampled at about 60 estuarine and coastal marine sites in the U.S. The chlordane (*cis*-, *trans*-, and oxychlordane; *cis*- and *trans*-nonachlor; and heptachlor) concentrations in whole body samples of bottom-feeding and predatory fish ranged from 6.91 to 409 ppm wet weight and 7.50 to 42.5 ppm wet weight, respectively. Those in shellfish ranged from 7.50 to 11.9 ppm wet weight. The chlordane (*alpha*-chlordane, *trans*-nonachlor, and heptachlor) residues in estuarine and coastal marine sediment samples collected from 1986 to 1987 as part of the National Status and Trends Mussel Watch Project generally were low, averaging 1.51 ppm dry weight ($N = 393$) in 1986 and 2.07 ppm dry weight ($N = 366$) in 1987.

Results of the National Status and Trends Program between 1984 and 1988 suggest relatively minor chlordane contamination of estuarine and coastal marine environments in the U.S. During this period, urbanized areas of the northeastern Atlantic coast, industrialized parts of the Gulf of Mexico coast, and the Southern California Bight region exhibited the highest chlordane concentrations in biotic and sediment samples.[23,31,34] However, some of the highest levels of chlordane contamination were registered in Choctawhatchee Bay along the west coast of Florida.[31] Chlordane persists in aquatic environments because, like DDT and PCBs, it resists degradation. It has been difficult to determine nationwide trends in chlordane contamination from national surveys, however, because concentrations of the insecticide are often below detection levels or because chlordane is not always the target of the surveys.[23]

Polychloroterpene Insecticides

Strobane and toxaphene are the only two polychloroterpene compounds. Formed by the chlorination of bicyclic terpenes, both strobane and toxaphene behave similarly to the cyclodiene insecticides, acting as neurotoxins which cause an imbalance in sodium and potassium ions and an interference with the transmission of impulses. The polychloroterpenes exhibit less lipophilicity than DDT and are readily metabolized by mammals and birds. These compounds have low toxicity to insects, mammals, and birds, but fish appear to be highly susceptible to them.

Toxaphene

The most heavily used insecticide in U.S. history is toxaphene, a mixture of more than 250 polychlorinated compounds, polychloroboranes, and polydihydrocamphenes. Toxaphene was commonly used as a foliar material on cotton; as an agent for control of grasshoppers, crickets, and other insects in field crops; and as a chemical for rapid elimination of trash fish in lakes during the 1950s, 1960s, and 1970s. Manufactured by the chlorination of camphene, a pine tree derivative, toxaphene was introduced commercially in 1948 and quickly became an important contact insecticide in agriculture, often in combination with DDT.[9] However, due to its toxicity to nontarget organisms and bioaccumulative capacity (e.g., in fish), toxaphene was discontinued as a piscicide in the 1970s. Its production declined substantially by 1980, and the U.S. Environmental Protection Agency cancelled the registration of toxaphene for most uses in 1982. Despite its restrictive utilization in the 1980s, toxaphene remained a contaminant of estuarine and coastal marine sites throughout the 1980s owing to its persistence in sediments and water, with a documented half-life of several years.

Toxaphene is extremely toxic to fish, with the 96-h LC_{50} to estuarine fish being <1 μg/l.[35] The concentration of toxaphene deemed to be safe for protecting marine life is 0.07 μg/l.[36] Because it is reported to be toxic to wildlife, toxaphene also may be a hazard to warm-blooded animals utilizing estuarine and coastal marine habitats.[37]

Mearns et al.[23] analyzed toxaphene measurements in more than 12,000 biotic samples nationwide. Their data showed that toxaphene was consistently present above the detection limit of 0.25 ppm wet weight in only a few regions, such as southern Georgia and southern Laguna Madre, TX. Samples with secondary occurrences of the contaminant were observed in the San Francisco Bay, Delta area, East Bay, Los Angeles, CA, and Oso Bay, TX. Findings of the National Pesticide Monitoring Program surveys of 1965–1976 indicated contaminant concentrations below detectable levels in nearly all biotic samples. Consequently, toxaphene has not been a target chemical of nationwide estuarine or coastal surveys since that time. Nevertheless, toxaphene remained an important estuarine contaminant through the 1980s in "hot spots", notably in Georgia, California, and Texas. For example, the California Mussel Watch Program recorded concentrations of toxaphene in mussels from San Francisco Bay in 1985 approaching 0.2 ppm wet weight, which signaled a need to continue monitoring this contaminant at impacted sites.[38] Most of the major biotic problems with toxaphene were associated with its use as a piscicide.

Hexachlorocyclohexane (HCH)

Lindane

This pesticide consists of a mixture of eight steric isomers, including the well-known gamma isomer lindane, which is the main insecticidal component. Lindane, which acts much faster than DDT and the cyclodienes, is still widely used in the U.S. Though its mode of action is quite different from that of DDT, it is one of a group of neurotoxicants similar to DDT.[9] Upon ingestion by animals, it tends to be rapidly metabolized to water-soluble chlorophenols and chlorobenzenes. Lindane accumulates in food chains. Nevertheless, it has a much lower half-life in the tissues and eggs of birds than other organochlorine pesticides.[15]

The concentrations of lindane in open Atlantic surface waters (0.5 to 1.4 ng/l) are lower than those reported in the Pacific (1 to 10 ng/l).[18] As in the case of DDT, the higher concentrations of lindane in the Pacific Ocean may reflect enhanced inputs from pesticide usage in Asia. However, more sampling is required to confirm this observation.

Lindane degrades rapidly in the environment. Of the nearly 12,000 estuarine and marine fish and shellfish samples analyzed nationwide for lindane during the National Pollutant Monitoring Program, only a few samples contained the contaminant in concentrations above the detection limit (0.01 ppm wet weight). It was found in 44% of the 64 Mussel Watch Project samples collected in California during surveys from 1980–1981 and 1985–1986. Lindane does not appear to pose a contamination problem for estuarine and marine environments in the U.S. based on the findings of the National Status and Trends Benthic Surveillance Project.[23]

Hexachlorocyclopentadiene Pesticides

Mirex

Used primarily as a pesticide to control the fire ant in the southeastern states and secondarily as a fire retardent additive to polymers, mirex (dodecachlorooctahydro-1,3,4-methano-1H-cyclobuta[cd]pentalene) has been detected in migratory birds, mammals, and other biota throughout the U.S. and Canada.[39] It purportedly has been implicated in elevated mortalities of fish and wildlife. Some shellfish also appear to be impacted by this pesticide. For example, mirex concentrations of 0.01 to 10 ppb have been shown to adversely affect the development and survival of marine crab larvae.[40] Mirex residues may bioaccumulate at high rates in marine organisms and, because they are only partly metabolized and eliminated slowly, may create chronic toxic conditions.[15] Owing to the potential environmental dangers of mirex contamination, the U.S. Environmental Protection Agency banned its use in the U.S. in 1978.

While mirex was once considered to be a potentially serious contaminant of estuarine organisms in the southeast U.S., nationwide surveys conducted during the 1970s and 1980s reveal that this view most probably was not the case.[23] Of the more than 10,000 shellfish and finfish samples analyzed for mirex during surveys of the National Pesticide Monitoring Program and the Cooperative Estuarine Monitoring Program, no shellfish or adult fish contained the contaminant in concentrations above the detection limit of 0.01 ppm wet weight. In addition, no juvenile estuarine fish had mirex concentrations above the detection limit of 0.05 ppm wet weight. The National Status and Trends Benthic Surveillance Project of 1984 documented mirex concentrations in fish livers at 44 sites ranging from <0.001 to 0.003 ppm wet weight.[25] Results of these surveys indicate that mirex does not appear to be an important estuarine contaminant in the U.S.

Chlordecone

A degradation product of mirex is chlordecone (also commonly known as kepone) which has been linked to various biotic impacts. Chlordecone was responsible for significant mortality of organisms in the James River estuary during the mid-1970s due to illicit chemical plant discharges. Initially discovered in the estuary in 1973, chlordecone rapidly attained high concentrations in fish and shellfish, particularly in the lower James River, but also in adjacent areas of lower Chesapeake Bay. For instance, by 1975 chlordecone concentrations greater than 1.0 ppm wet weight were often observed in finfish samples, and in some cases, they exceeded 7.0 ppm wet weight.[10] These findings resulted in the banning of all forms of fishing in the estuary in December 1975. By the mid-1980s, the concentrations of chlordecone in crabs and fish in the estuary generally ranged from 0.2 to 0.8 ppm wet weight and those in oysters, usually below 0.1 ppm wet weight.[23] Chlordecone remains a concern primarily in the lower James River.

Chlorinated Benzenes and Phenols

Hexachlorobenzene

One of the most important fungicides is hexachlorobenzene (HCB), a chemical once widely used as a fumigant in grain storage against fungal attacks, as a soil fumigant, and as a component in wood preservatives. It occurs as a byproduct in the manufacture of carbon tetrachloride, vinyl chloride monomer, and pentachlorophenol.[41] It is a highly persistent contaminant, largely found in estuarine and marine environments sorbed to sedimentary particles.

Measurable concentrations of HCB have been detected in finfish and shellfish samples from the New York Bight, Upper Chesapeake Bay, Galveston Bay, Santa Monica Bay, Commencement

Bay, and Elliott Bay.[23] The National Status and Trends Benthic Surveillance Project commencing in 1984 documented HCB concentrations in fish liver samples ranging from 0.001 to 0.037 ppm wet weight, with a median value of 0.0013. Highest concentrations were found in the state of Washington, specifically Nisqually Reach (0.01 ppm wet weight), Commencement Bay (0.037 ppm wet weight), and Elliott Bay (0.007 ppm wet weight). Based on these data, HCB does not appear to be an important contaminant of U.S. estuarine and coastal marine biota.

Pentachlorophenol

A pesticide having broad applications, wide effectiveness, and multiple modes of action is the chlorinated phenol pentachlorophenol (PCP). A metabolite of HCB, PCP has been used as a fungicide for wood and textile preservation (fungicide), a slimicide in pulp mills, an insecticide for protection against termites and other insects, and a nonselective herbicide and preharvest defoliant. It is highly phytotoxic and, consequently, effectively employed in control of weeds.[16] PCP has periodically created widespread soil and water contamination problems at manufacturing and wood preservation sites.[42]

PCP concentrations generally have ranged from 0.008 to 0.02 ppm wet weight in estuarine and marine biota.[10] However, PCP has never been the target of any nationwide contaminant survey. When monitored during site-specific studies, PCP levels in estuarine and marine samples have been low.

Organophosphorus Insecticides

This group of pesticides, commonly known as the organophosphates, is characterized by compounds containing phosphorus (e.g., parathion, methyl parathion, diazinon, malathion, fenitrothion, TEPP, dicapthon, naled, ethion, fenthion, fonofos, crufomate, phorate, ronnel, and a number of others). They are all anticholinesterase insecticides, meaning that their toxic action derives from inactivation of acetylcholine, a nerve transmitter substance, by the inhibition of the enzyme cholinesterase.[9] In addition to cholinesterase, organophosphates may inhibit other enzymes. The primary symptom of acute organophosphate poisoning in insects is extreme hyperactivity of the nervous system manifested by severe twitching of appendages.

Phorate is an example of an organophosphorus compound that is highly toxic to marine organisms. Blus[15] reported that the 96-h LC_{50} for three species of shrimp exposed to this compound ranged from 0.1 to 1.9 μg/l. In the case of larvae of the hard clam (*Mercenaria mercenaria*) and the American oyster (*Crassostrea virginica*), the 96-h LC_{50} values amounted to 17 and 900 μg/l, respectively.

Organophosphates represent the largest and most diverse group of insecticides. Although they have negligible chronic toxicity, the organophosphates may display moderate to high acute toxicity.[35] They appear to be quite toxic to vertebrate animals. However, the organophosphates as a group tend to be less acutely toxic to aquatic biota than the organochlorine insecticides. They are rapidly metabolized or excreted by most animals and do not accumulate in food chains.[15] Because they are broad-spectrum poisons typified by chemical instability and nonpersistence in the environment, the organophosphate insecticides along with the carbamates have been long viewed as effective replacements for the more problematic organochlorines.

Carbamate Insecticides

The carbamates, a more recent group of insecticides, consist of compounds derived from carbamic acid. The first successful carbamate insecticide, carbaryl (Sevin), was introduced in 1956 and gained wide acceptance as a lawn and garden pesticide. It was followed by the development of a series of other compounds (i.e., carbofuran, aminocarb, methiocarb, bufencarb, metalkamate, methomyl, promecarb, propoxur, aldicarb, and a number of others). However, carbaryl has been the preferred insecticide, being used in greater quantities worldwide than all other carbamates combined.

Carbofuran is a carbamate compound that is highly toxic to fish, but considerably less toxic to marine shellfish. The 96-h LC_{50} value for sheepshead minnows (*Cyprinodon variegatus*) exposed to this compound is 386 μg/l. Four species of marine bivalves have 96-h LC_{50} values for carbofuran ranging from 3.75 (cockle, *Clinocardium nuttali*) to 125 mg/l (clam, *Rangia cuneata*).[15]

The carbamates are generally broad-spectrum poisons, as exemplified by carbamate and carbofuran. Like the organophosphorus insecticides, the carbamates act rapidly through interference with cholinergic nerve transmission, thereby incapacitating and killing target organisms. Carbamates are usually toxic to invertebrates; the 48- and 96-hr EC_{50} values for many invertebrates are less than 10 μg/l. They are also highly toxic to birds, and therefore, may pose a potential threat to waterfowl. However, they appear to be only moderately toxic to fish.[35] All are metabolized by plants and animals and do not bioaccumulate. In addition, they do not persist in aquatic environments.[14] Hence, the carbamates are probably much less of a hazard to estuarine and marine environments than the chlorinated hydrocarbon pesticides.

BOTANICAL INSECTICIDES

These pesticides are naturally occurring chemicals synthesized by plants, although in more recent years synthetic analogues have been developed. The botanical insecticides were more important during the pre-1945 era than they are today. As a group, the botanical insecticides have been in use longer than all other pesticides. Some of the most notable members of this group are pyrethrum, rotenone, sabadilla, ryania, and nicotine. Of these natural insecticides, pyrethrum is the only one of significance currently in use.[9] Utilization of botanical insecticides peaked in 1966 and has declined since that time. These compounds do not appear to pose a serious threat to estuarine or marine environments.

Being expensive to extract from plant tissues and often less effective than other types of insecticides, the botanical pesticides are seldom employed for agricultural purposes. The synthetic pyrethroids are more effective than many of the natural insecticides. Thus, they have been widely used during the last decade. While not persistent in the environment, botanical insecticides are highly toxic to fish. The 96-h LC_{50} values of the synthetic pyrethroid SBP-1382 range from about 1 to 5 μg/l for several salmonids.[35]

HERBICIDES

Herbicides are classified by several criteria. They may be broadly grouped into selective and nonselective chemicals. Selective herbicides kill weeds without harming particular species of vegetation; nonselective herbicides kill all vegetation. A second classification scheme subdivides herbicides into contact vs. translocated varieties. Contact herbicides kill those plant parts in contact with the chemical, whereas translocated herbicides affect plants after being absorbed either by the roots or the aboveground parts of the plant. A third method of classification is based on the timing of herbicide application relative to the stage of weed or crop development (i.e., categorizing them into preplanting, preemergence, and postemergence types). On the basis of their persistence, herbicides may be classified as nonpersistent (e.g., phenoxyderivatives), moderately persistent (e.g., dinitroanilines, phenylureas, triazines), and persistent (e.g., cationic herbicides) types.[43] Finally, herbicides are often grouped by their chemical composition (e.g., organic and inorganic herbicides).[9]

Manno[8] examined several of the more toxicologically important herbicides. These include the chlorophenoxy, bipyridium, and triazine compounds. The chlorophenoxy compounds, such as 2,4-dichlorophenoxyacetic acid (2,4-D), 2,4,5-trichlorophenoxyacetic acid (2,4,5-T), and 2-methyl-4-chlorophenoxyacetic acid (MCPA), are selective herbicides that have been most effective in destroying broadleaf and grass species. The bipyridiums (e.g., diquat, paraquat) are a group of contact herbicides capable of rapid plant tissue damage mediated by cell membrane destruction. Affected plants wilt and dessicate within hours of herbicide contact. The triazine herbicides (e.g., atrazine, cyanazine, propazine, simazine), a class of selective herbicides, have proven to be of commercial importance in agriculture, especially in weed control of crops (e.g., corn).

Compared to insecticides, herbicides have much less of an impact on estuarine and marine environments. In general, herbicides exhibit little toxicity to wildlife, livestock, and humans.[39] However, herbicides are acutely toxic to fish. There are few documented records of significant herbicide impacts on estuarine or marine habitats. The greatest threat, however, is to aquatic vegetation. For example, atrazine present at concentrations of 60 to 1040 μg/l has been shown to inhibit the growth of submerged aquatic vegetation.[44]

POLYCHLORINATED BIPHENYLS (PCBs)

Perhaps the most notable organochlorine contaminants in the marine biosphere are polychlorinated biphenyls (PCBs), a group of synthetic halogenated aromatic hydrocarbons, consisting of a complex mixture of chlorinated biphenyls that contain a varying number of substituted chlorine atoms on aromatic rings. By the substitution of one or several chlorine atoms at one or more of the ten carbons of the biphenyl molecule, up to 209 congeners are theoretically possible. The complex mixture of chlorobiphenyls contains 18 to 79% chlorine. Commercial PCB preparations are produced to physical, not chemical, specifications and may contain 100 or more individual compounds.[41] In the U.S., the resultant products are marketed as mixtures of chlorobiphenyls called Aroclors®. The type of Aroclor® is identified by the percentage of chlorine in the mixtures. Hence, Aroclors® 1221, 1232, 1242, 1248, 1254, 1260, 1262, and 1268 are manufactured by chlorinating biphenyl to a final chlorine content of 21 to 68%. PCBs have been marketed as Phenoclor and Pyralene in France, Apirolio and Fenclor in Italy, Clophen in Germany, Delor in Czechoslovakia, Soval in Russia, and Kanechlor and Santotherm in Japan.

An estimated 0.54 billion kg of PCBs were commercially produced in the U.S. between the time of initial (1929) and final production (1977). Unique physical and chemical properties of PCBs — thermal and chemical stability, miscibility with organic compounds, high dielectric constant, nonflammability, and low cost — enabled them to be widely used wherever such properties were desirable, such as in dielectric fluids of transformers and capacitors, heat exchange and hydraulic fluids, lubricants, fire retardents, plastics, and other materials. From a practical viewpoint, the most important properties of PCBs were found to be their general inertness and thermal stability.[45]

PCBs have been contaminants of the marine environment for more than 50 years. Over this time, they have become universally distributed in estuarine and marine environments and occur in nearly all marine plant and animal species. PCBs enter these environments by four primary pathways: (1) leaching from dumps; (2) volatilization by vaporization from plastics and inefficient burning in dumps and incinerators, followed by adsorption on particulates, transport, and eventual fallout; (3) adsorption on sediments and subsequent riverine transport to the sea; and (4) sedimentation in the sea. Atmospheric transport of PCBs to the sea is important. The concentration of chlorobiphenyls in air samples worldwide averages about 1 ng/m^3.[45]

PCBs are deleterious to marine life, especially upper-trophic-level organisms that tend to accumulate the compounds in their tissues. While the precise toxicological effects of PCBs are often unclear, they have been implicated in reproductive abnormalities in marine mammals (e.g., porpoises, seals, sea lions, whales).[1,46,47] In addition to being linked to a variety of chronic diseases in humans (e.g., skin lesions, reproductive disorders, liver damage), PCBs are suspected of being carcinogenic.

Many investigations of PCB contamination in marine biota have been undertaken since the 1960s due to major human health concerns. These investigations have revealed that PCBs are a potential hazard to marine life because of their great stability, persistence, and lipophilicity, and because they are poorly metabolized by biological systems.[48,49] Hence, they bioaccumulate in food chains and at elevated levels may cause chronic health effects in man. As a result, estuarine and marine monitoring programs conducted during the past three decades have focused on the uptake of PCBs by finfish and shellfish suitable for human consumption.

PCB concentrations in oceanic surface waters of the southern hemisphere (0.035 to 0.072 ng/l) are lower than those in the northern hemisphere where industrial production and use of the compounds have been greater.[18] The most recent reliable values of PCBs in surface waters of the Pacific, Indian, and North Atlantic oceans range from 0.04 to 0.5, 0.06 to 0.25, and 0.0066 to 0.021 ng/l, respectively.[50,51] PCB concentrations decrease with depth in the ocean. For example, Schulz et al.[51] reported substantially lower PCB levels at depth in the North Atlantic, where concentrations decrease from 6.6 to 21.0 pg/l in surface layers (10 to 250 m) to 1.5 to 2.0 pg/l in deep (3500 to 4000 m) waters. Similar to other chlorinated hydrocarbons, PCBs occur in highest concentrations in both open ocean and coastal areas within the surface microlayer, which is naturally enriched in lipid compounds.[18]

The concentrations of PCBs in coastal surface waters typically range from 1 to 10 ng/l.[52] A trend of decreasing PCB concentrations has been noted in these waters since the mid-1970s, as corroborated by independent studies from several countries. The most acute reduction in PCB concentrations, amounting to approximately a threefold decline, was registered between the mid-1970s and the early 1980s.

PCBs, having a high affinity for particulates, readily sorb to fine-grained sediments and organic matter and subsequently settle to the sea floor.[10,53,54] As a consequence, the highest concentrations of the contaminants in shallow water environments typically exist in bottom sediments, often derived from waste discharges of manufacturing facilities, industrial installations, and waste dumpsites. In the U.S., the highest concentrations of PCBs in coastal and nearshore bottom sediments have been recorded in New Bedford Bay, MA (8400 ng/g dry weight), Hudson-Raritan Bay, NY (286 to 1950 ng/g dry weight), the New York Bight off New Jersey (0.5 to 2200 ng/g dry weight), Escambia Bay, FL (<30 to 480,000 ng/g dry weight), and Palos Verdes, CA (80 to 7420 ng/g dry weight). Areas removed from point sources of PCBs display much lower residues of PCBs. This may explain why coastal marine sediments in England account for only 2.1% of the estimated contemporary environmental burden of tPCBs compared to 93.1% for British soils and 3.5% for sea water.[47] Sampling programs also have detected PCBs in deep-sea sediments, even at depths greater than 5000 m.[18] However, because of the paucity of PCB measurements in deep-sea sediments, spatial and temporal trends of PCB contamination cannot be deduced in these deeper habitats.

The concentrations of PCBs in estuarine and marine organisms depend on a number of factors, including but not limited to the type of species, its lipid content, organismal size, metabolism, feeding type, and diet. Primarily because of bioaccumulation, PCB concentrations increase by a factor of 10 to 100 times when proceeding upward on major trophic levels in a food chain (e.g., plankton to fish to birds). Laboratory studies indicate that diatoms are adversely affected by 0.1 µg/l of Aroclor® 1254.[49] Phytoplankton communities exposed to 10 µg PCB/l experience reduced biomass. Carbon fixation of phytoplankton populations tends to be inhibited at these contaminant levels due to reduced chlorophyll-*a* in the cells.[55] In an extensive review, Wassermann et al.[56] documented that PCB concentrations in marine zooplankton range from <.003 to 1.0 ppm. Substantially higher amounts of PCBs occur in fish and seals, the top consumers, with values ranging from 0.03 to 212 ppm. The PCB data of Wassermann et al.[56] are still generally valid today except for the highest values. The contaminant level producing lethality in fish, based on laboratory experiments, is in the 10 to 300 ppm range.[49] These findings suggest that survival of finfish populations may decrease at PCB concentrations that now exist in many estuarine and coastal marine systems.[45]

A striking feature of recent observations on marine biota is the persistence of PCB contamination long after many nations placed bans or restrictions on PCB use. Persistence of PCB residues is perhaps most clearly manifested among lipid-rich species of finfish and mammals. For instance, the concentrations of PCBs in cod-liver oil samples collected from the Baltic Sea decreased at a very slow rate from a mean value of 14 µg/g lipid weight during the 1971–1980 period to 10 µg/g lipid weight during the 1981–1989 period.[21] PCBs persist in striped bass (*Morone saxatilis*) of the Hudson River three decades after the contaminants were first discharged by an electric capacitor manufacturer. More than 2.5×10^5 kg of PCBs, primarily Aroclor® 1016, 1242, 1254, and 1260, contaminated the entire Hudson River estuary in the 1960s and 1970s, leading to the 1976 closure of the commercial striped bass gill net fishery and the issuance of annual fish consumption health advisories to recreational fishermen in New York. Similar consumption and health advisories on striped bass were issued by the New Jersey Department of Environmental Protection during the 1980s.[57] High levels of PCBs continue to be found in the edible flesh of striped bass (mean = 15 mg/kg) collected from the Hudson River estuary.[58]

In the Baltic Sea, diminishing populations of the common seal (*Phoca vitulina*), gray seal (*Halichoerus grypus*), otter (*Lutra lutra*), and common porpoise (*Phocoena phocoena*) during the last three decades have been ascribed to PCB contamination.[59] PCBs appear to have adversely affected reproductive processes and development of these mammals. The contaminants have persisted to this day in some invertebrates, fish, and mammals in regions far removed from PCB sources, such as waters off Ross Island, Antarctica,[60] and various sub-Antarctic areas.[61] PCBs have been commonly detected in Antarctic air samples.[60,62]

Birds are more resistant to either chronic or acute PCB toxicity than are mammals.[63] The LD_{50} for several avian species tested for PCB toxicity in the laboratory has ranged from approximately 60 to more than 6000 mg/kg of diet;[63] however, the primarily fish-consuming avian species (e.g., herring gulls, great blue herons, brown pelicans) generally accumulate higher levels of PCBs.[10,49]

Several organisms have been employed as bioindicators in PCB monitoring programs. Chief among these indicator (or sentinel) organisms in the U.S. are mussels (e.g., blue mussel, *Mytilus edulis*) and oysters (e.g., American oyster, *Crassostrea virginica*), which have been selected for national surveys because of their sessile habit and ability to filter and concentrate contaminants from surrounding waters. Both mussels and oysters have been of great practical value in quantifying PCB contamination of estuarine and marine environments.[23,26-28]

Four national surveys have analyzed PCBs in bivalves since the 1970s: (1) National Pollution Monitoring Program, 1970–1972; (2) National Pollution Monitoring Program, 1977; (3) U.S. Environmental Protection Agency Mussel Watch Program, 1976–1977; and (4) NOAA National Status and Trends Mussel Watch Program, 1984–1996. Results of these surveys, which have been reviewed by Mearns et al.,[23] indicate that the most contaminated sites of PCBs occur near metropolitan centers along the northeast coast (e.g., Boston Harbor, Long Island Sound, New York Bight) and in harbors of southern California. The U.S. Environmental Protection Agency Mussel Watch Survey of 86 nationwide sites in 1976 documented a grand national median for the PCB mixture Aroclor® 1254 of 0.009 ppm wet weight and a range of 0.0008 to 2.09 ppm wet weight. The grand national median value and range of tPCBs in the 1986 NOAA National Status and Trends Mussel Watch Survey of 144 sites was 0.015 and 0.0009 to .068 ppm wet weight, respectively. Accurate comparisons of these two data sets without applying a correction or adjustment is not possible because of analytical, site, and species differences in the two surveys.

Full nationwide sampling using bivalve monitoring was not implemented until 1976 with the onset of the U.S. Environmental Protection Agency Mussel Watch Program, since bivalve monitoring of PCBs in earlier surveys between 1970 and 1972 targeted only scattered localities. Nevertheless, several sites were identified with very high PCB concentrations in the early 1970s (e.g., Elizabeth River, VA; Palos Verdes, CA). From the mid-1970s to the early 1980s, sampling of urban embayments of the northeast and southwest uncovered relatively high PCB concentrations in bivalves (e.g., Buzzards Bay, MA; San Diego Bay, CA).[18] It is clear that PCB concentrations decreased in bivalves at specific localities between the 1970s and 1980s. For example, at the base of the White Point outfalls on the Palos Verdes Peninsula, PCB concentrations in mussels dropped 20-fold from approximately 0.4 in 1971 to <0.02 ppm wet weight in 1982. In Narragansett Bay, RI, Aroclor® 1254 concentrations in mussels ranged from 0.4 to 0.5 ppm dry weight between 1976 and 1979, but tPCB levels in these bivalves amounted to only 0.2 ppm dry weight by the early 1980s.[23] However, the National Status and Trends Mussel Watch Program from 1984 to 1986 continued to delineate locations of persistent PCB contamination, such as Buzzards Bay, MA, Hudson-Raritan Bay, NY, and the New York Bight.[64] Available data on PCB contamination in bivalves do not show dramatic reductions during the past decade.[10,65]

PCB residues in finfish, like shellfish, are highest in urban embayments, most notably in the northeast, northwest, and southern California regions. The concentrations of PCBs in finfish are generally greater than in shellfish. While the mean national value of PCBs in coastal fishes is suggested to be <0.10 ppm wet weight, tPCB concentrations >6.0 ppm wet weight have been recorded in some whole estuarine or coastal fish.[23] The Juvenile Estuarine Fish Monitoring Project of the National Pesticide Monitoring Program, conducting a national study of PCBs in whole fish from 1972 to 1976, detected PCBs in 331 of 1524 composite samples taken at 144 stations. The median concentration of PCBs amounted to <0.10 ppm wet weight, and samples collected at 58% of the stations had PCB levels below the detection limit of 0.05 ppm wet weight.

The Juvenile Estuarine Fish Monitoring Project was the only nationwide survey to systematically measure PCBs in whole fish. Since 1976, more PCB determinations have been made on fish muscle and liver than on whole fish. The NOAA/EPA Cooperative Estuarine Monitoring Program of 1976–1977 found that about 90% of the fish muscle samples analyzed for PCBs had concentrations below the rather high detection limit of 0.4 ppm wet weight (sum of four arochlor detection limits).[33] Nearly all PCB levels in flatfish muscle samples analyzed during the NOAA Southern California Coastal Research Project of 1976–1977 exceeded the detection limit of approximately 0.004 ppm wet weight. Total PCB levels in muscle of coastal flatfish sampled in 1980 from Atlantic and Pacific coastal regions were greater (>0.1 ppm wet weight) than those from the Gulf of Mexico (<0.1 ppm wet weight).[23] The highest concentrations (>2.0 ppm) were in samples from New Bedford Harbor, MA. In an analysis of PCB residues in pelagic and predatory fishes collected from 15 estuarine and coastal marine areas in 1979–1980, Gadbois and Maney[66] documented highest mean concentrations of PCBs in samples from the New York Bight Apex (1.1 ppm wet weight).

Higher PCB residues are observed in fish livers than whole fish or muscle. The 1984 National Status and Trends Benthic Surveillance Project reported a median PCB concentration in fish livers of samples collected at 42 Atlantic and Pacific coastal sites amounting to 0.58 ppm wet weight. Based on results of the national survey, the highest mean PCB values in fish livers were found in heavily contaminated systems — Boston Harbor, MA (2.62 ppm wet weight); Commencement Bay, WA (2.30 ppm wet weight), and Elliott Bay, WA (4.23 ppm wet weight).

Comparisons of nationwide surveys of fish liver samples collected in the 1970s (i.e., NOAA/EPA Cooperative Estuarine Monitoring Program, NOAA Southern California Coastal Water Research Project) and in the 1980s (i.e., NOAA Benthic Surveillance Project) clearly indicate no substantial decline in PCB contamination over this time period. Since the mid-1980s, there has been no clear evidence of any large-scale nationwide decrease of PCB contamination in estuarine and marine environments except in proximity to known industrial sources and other "hot-spot" areas.[10,65] Therefore, it is deemed necessary to continue to monitor PCB residues in biota, sediment, and water samples collected from many different habitats — even remote from U.S. coastal areas — to properly assess the long-term effects of these organochlorine contaminants.

CHLORINATED DIBENZO-*p*-DIOXINS AND DIBENZOFURANS

Two related classes of aromatic heterocyclic compounds, the chlorinated dibenzo-*p*-dioxins (CDDs) and chlorinated dibenzofurans (CDFs), cause considerable biological and toxic impacts on aquatic organisms. There are 75 possible CDD and 135 CDF compounds having from 1 to 8 chlorine substituents. The degree of toxicity varies in both groups, with some compounds (e.g., 2,3,7,8-TCDD; 2,3,7,8-TCDF) being extremely toxic to organisms.[49] 2,3,7,8-TCDD alone has been shown to be embryotoxic, teratogenic, carcinogenic, and cocarcinogenic to animals.[67] In general, the CDDs and CDFs are hydrophobic and lipophilic compounds that resist degradation. They tend to sorb to sediments and ultimately accumulate on the sea floor, which acts as a sink for the compounds.[68,69] Sediments, however, probably are not a significant direct source of CDDs and

CDFs for biota at higher trophic levels in estuarine and marine environments, since the food chain appears to be the most important pathway for bioaccumulation of these compounds.[49]

CDDs and CDFs occur as trace contaminants in industrial chemicals (e.g., phenoxy herbicides, chlorinated phenols, PCBs). Other primary sources of CDDs and CDFs are combustion products from municipal incinerators, wood burning, chemical waste, automotive emissions, pathological waste, and PCB-filled transformer fires.[70] Common routes of entry into estuarine and marine environments are sewer overflows, storm drains, and atmospheric deposition of particulates. Investigations during the past decade indicate that CDDs and CDFs are widely distributed in soil, air, sediments, and natural waters at low part-per-trillion concentrations, with highest concentrations nearby industrialized and heavily populated regions.[68,70]

Biological and toxicological responses of estuarine and marine organisms exposed to CDDs and CDFs are variable and highly species dependent. The most common responses include a wasting syndrome manifested by progressive weight loss, reproductive impairment, and induction of numerous enzymes (e.g., cytochrome P450 in fish). Other apparent impacts are immunosuppression, impaired liver function, cardiovascular changes, developmental abnormalities, histopathological alterations, and delayed onset of mortality.[49,70] More toxicity data on CDDs and CDFs are available on fish and mammals than on invertebrates.[71]

Cooper[71] conveyed that similar toxic effects are observed in most fish exposed to dioxins and dibenzofurans. In general, the smaller the fish, the lower the apparent amount of contaminant needed to induce toxic impacts, suggesting that toxicity is a function of total body mass. In eggs, juveniles, or adult fish, there is always a delay in the appearance of toxic effects. Toxicity in fish, as well as invertebrates, is often manifested by decreased reproductive success.

One of the first signs of toxicity in fish is the cessation of food intake (hypophagia) followed by decreased growth. Other manifestations of toxicity are petechial hemorrhages of the skin and fins, edema, skeletal abnormalities, and a high incidence of fungal infection. These responses may be initially discerned when seawater concentration of CDDs and CDFs are in the low nanograms-per-liter range.[71] Because of insufficient toxicological testing of CDD and CDF compounds on marine organisms, more investigations are required to assess the factors that alter the bioconcentration, toxicokinetics, and metabolism of these compounds.

In summary, CDDs and CDFs are globally-distributed, highly lipophilic, and environmentally persistent compounds produced by a number of industrial processes. They tend to be extremely toxic to various aquatic fauna and bioaccumulate in food webs.[72,73] Bioaccumulation appears to result primarily from food-chain passage rather than by direct uptake from water, suspended matter, or bottom sediments. However, direct accumulation of CDDs and CDFs has been documented in lower-food-chain organisms.[49] A major pathway of transport of the contaminants to the higher latitude regions is via the atmosphere from mid-latitude sources, with residues being deposited by precipitation or by particulate and aerosol fallout.[74,75] The effects of CDDs and CDFs on estuarine and marine organisms are far from well understood and need to be the focus of more intense research.

REFERENCES

1. Borrell, A., PCB and DDTs in blubber of cetaceans from the northeastern North Atlantic, *Mar. Pollut. Bull.*, 26, 146, 1993.
2. Storr-Hansen, E. and Spliid, H., Distribution patterns of polychlorinated biphenyl congeners in harbor seal (*Phoca vitulina*) tissues: statistical analysis, *Arch. Environ. Contam. Toxicol.*, 25, 328, 1993.
3. Tanabe, S., Sung, J.-K., Choi, D.-Y., Baba, N., Kiyota, M., Yoshida, K., and Tatsukawa, R., Persistent organochlorine residues in northern fur seal from the Pacific Coast of Japan since 1971, *Environ. Pollut.*, 85, 305, 1994.
4. Tanabe, S., Fate of persistent organochlorines in the marine environment, in *Contaminants in the Environment: A Multidisciplinary Assessment of Risks to Man and Other Organisms*, Renzoni, A., Mattei, N., Lari, L., and Fossi, M.C. (Eds.), Lewis Publishers, Boca Raton, FL, 1994, 19.

5. Iwata, H., Tanabe, S. Sakai, N., Nishimura, A., and Tatsukawa, R., Geographical distribution of persistent organochlorines in air, water, and sediments from Asia and Oceania, and their implications for global redistribution and lower latitudes, *Environ. Pollut.,* 85, 15, 1994.
6. Hargrave, B. T., Harding, G. C., Vass, W. P., Erickson, P. E., Fowler, B. R., and Scott, V., Organochlorine pesticides and polychlorinated biphenyls in the Arctic Ocean, *Arch. Environ. Contam. Toxicol.,* 22, 41, 1992.
7. GESAMP, The state of the marine environment, Group of Experts on the Scientific Aspects of Marine Pollution Reports and Studies, No. 39, 1990.
8. Manno, M., Toxicology and risk assessment of pesticides, in *Chemistry, Agriculture, and the Environment*, Richardson, M. L. (Ed.), The Royal Society of Chemistry, Cambridge, 1991, 466.
9. Ware, G. W., *Pesticides: Theory and Application*, W. H. Freeman and Company, New York, 1983.
10. Kennish, M. J., *Ecology of Estuaries: Anthropogenic Effects*, CRC Press, Boca Raton, FL, 1992.
11. Hellou, J., Warren, W. G., and Payne, J. F., Organochlorines including polychlorinated biphenyls in muscle, liver, and ovaries of cod, *Gadus morhua*, *Arch. Environ. Contam. Toxicol.,* 25, 497, 1993.
12. Pereira, W. E., Hostettler, F. D., Cashman, J. R., and Nishioka, R. S., Occurrence and distribution of organochlorine compounds in sediment and livers of striped bass (*Morone saxatilis*) from the San Francisco Bay-delta estuary, *Mar. Pollut. Bull.,* 28, 434, 1994.
13. Turk, J., *Introduction to Environmental Studies*, 2nd ed., Saunders College Publishing, Philadelphia, 1985.
14. Smith, A. G., Chlorinated hydrocarbon insecticides, in *Handbook of Pesticide Toxicology*, Vol. 3, Hayes, W. J., Jr. and Laws, E. R., Jr. (Eds.), Academic Press, San Diego, 1991.
15. Blus, L. J., Organochlorine pesticides, in *Handbook of Ecotoxicology*, Hoffman, D. J., Rattner, B. A., Burton, G. A., Jr., and Cairns, J., Jr. (Eds.), Lewis Publishers, Boca Raton, FL, 1995, 275.
16. McEwen, F. L. and Stephenson, G. R., *The Use and Significance of Pesticide in the Environment*, John Wiley & Sons, New York, 1979.
17. Domagalski, J. L. and Kuivila, K. M., Distribution of pesticides and organic contaminants between water and suspended sediment, San Francisco Bay, CA, *Estuaries,* 16, 416, 1993.
18. Fowler, S. W., Critical review of selected heavy metal and chlorinated hydrocarbon concentrations in the marine environment, *Mar. Environ. Res.,* 29, 1, 1990.
19. Ruangwises, S., Ruangwises, N., and Tabucanon, M. S., Persistent organochlorine pesticide residues in green mussels (*Perna viridis*) from the Gulf of Thailand, *Mar. Pollut. Bull.,* 28, 351, 1994.
20. Ramesh, A., Tanabe, S., Subramanian, A. N., Mohan, D., Venugopalan, V. K., and Tatsukawa, R., Persistent organochlorine residues in green mussels from coastal waters of South India, *Mar. Pollut. Bull.,* 21, 587, 1990.
21. Kannan, K., Falandysz, J., Yamashita, N., Tanabe, S., and Tatasukawa, R., Temporal trends of organochlorine concentrations in cod-liver oil from the southern Baltic proper, 1971–1989, *Mar. Pollut. Bull.,* 24, 358, 1992.
22. Miskiewicz, A. G. and Gibbs, P. J., Organochlorine pesticides and hexachlorobenzene in tissues of fish and invertebrates caught near a sewage outfall, *Environ. Pollut.,* 84, 269, 1994.
23. Mearns, A. J., Matta, M. B., Simececk-Beatty, D., Buchman, M. F., Shigenaka, G., and Wert, W. A., *PCB and Chlorinated Pesticide Contamination in U.S. Fish and Shellfish: A Historical Assessment Report,* NOAA Tech. Mem. NOS OMA 39, National Oceanic and Atmospheric Administration, Seattle, WA, 1988.
24. Vallette-Silver, N. J., The use of sediment cores to reconstruct historical trends in contamination of estuarine and coastal sediments, *Estuaries,* 16, 577, 1993.
25. NOAA, *National Status and Trends Program for Marine Environmental Quality: Progress Report and Preliminary Assessments of Findings of the Benthos Surveillance Project— 1984,* National Oceanic and Atmospheric Administration Office of Ocean Resources Conservation and Assessment, Rockville, MD, 1987.
26. O'Connor, T. P., *Recent Trends in Coastal Environmental Quality: Results from the First Five Years of the NOAA Mussel Watch Project,* National Oceanic and Atmospheric Administration Office of Ocean Resources Conservation and Assessment, Rockville, MD, 1992.
27. O'Connor, T. P., *Coastal Environmental Quality in the United States, 1990. Chemical Contamination in Sediments and Tissues,* Special NOAA 20th anniversary report, National Oceanic and Atmospheric Administration Office of Ocean Resources Conservation and Assessment, Rockville, MD, 1990.

28. O'Connor, T. P. and Ehler, C. N., Results from NOAA National Status and Trends Program on distributions and effects of chemical contamination in the coastal and estuarine United States, *Environ. Monit. Assess.*, 17, 33, 1991.
29. Sericano, J. L., Wade, T. L., Atlas, E. A., and Brooks, J. M., Historical perspective on the environmental bioavailabilty of DDT and its derivatives to Gulf of Mexico oysters, *Environ. Sci. Technol.*, 24, 1541, 1990.
30. NOAA, *National Status and Trends Program for Environmental Quality: Second Summary of Data on Chemical Contaminants in Sediments from the National Status and Trends Program,* NOAA Tech. Mem. NOS OMA 59, National Oceanic and Atmospheric Administration Office of Oceanography and Marine Assessment, Rockville, MD, 1991.
31. Shigenaka, G., *Chlordane in the Marine Environment of the United States: Review and Results from the National Status and Trends Program,* NOAA Tech. Mem. NOS OMA 55, National Oceanic and Atmospheric Administration, Seattle, WA, 1990.
32. Butler, P. A., Kennedy, C. D., and Schutzmann, R. L., Pesticide residues in estuarine mollusks, 1977 versus 1972 — national pesticide monitoring program, *Pestic. Monit. J.*, 12, 99, 1978.
33. Butler, P. A., EPA-NOAA Cooperative Estuarine Monitoring Program, Final Report, U.S. Environmental Protection Agency, Gulf Breeze, FL, 1978.
34. Sericano, J. L., Wade, T. L., Brooks, J. M., Atlas, E. L., Fay, R. R., and Wilkinson, D. L., National Status and Trends Mussel Watch Program: chlordane-related compounds in Gulf of Mexico oysters, 1986-90, *Environ. Pollut.*, 82, 23, 1993.
35. Murty, A. S., *Toxicity of Pesticides to Fish*, Vol. 2, CRC Press, Boca Raton, FL, 1986.
36. Eisler, R. and Jacknow, J., Toxaphene Hazards to Fish, Wildlife, and Invertebrates: A Synoptic Review, Biological Report 85 (1.4), U.S. Fish and Wildlife Service, Washington, D.C., 1985.
37. Sergeant, D. B. and Onuska, F. I., Analysis of toxaphene in environmental samples, in *Analysis of Trace Organics in the Aquatic Environment*, Afghan, B. K and Chau, A. S. Y. (Eds.), CRC Press, Boca Raton, FL, 1989, 69.
38. Stephenson, M., Smith, D., Ichikawa, G., Goetzl, J., and Marten, M., State Mussel Watch Program Preliminary Data Report 1985-1986, Report to the State Water Resources Control Board, California Department of Fish and Game, Monterey, CA, 1986.
39. Eisler, R., Mirex Hazards to Fish, Wildlife, and Invertebrates: A Synoptic Review, U.S. Fish and Wildlife Service, Biological Report 85, Washington, D.C., 1990, 42.
40. Goldberg, E. D., *The Health of the Oceans*, Unesco Press, Paris, 1976.
41. Clark, R. B., *Marine Pollution*, 3rd ed., Clarendon Press, Oxford, 1992.
42. Nebeker, A. V., Griffis, W. L., and Schuytema, G. S., Toxicity and estimated water quality criteria values in mallard ducklings exposed to pentachlorophenol, *Arch. Environ. Contam. Toxicol.*, 26, 33, 1994.
43. Galli, E., The role of microorganisms in the environment decontamination, in *Contaminants in the Environment: A Multidisciplinary Assessment of Risks to Man and Other Organisms*, Renzoni, A., Mattei, N., Lari, L., and Fossi, M. C. (Eds.), Lewis Publishers, Boca Raton, FL, 1994, 235.
44. Kemp, W. M., Means, J. C., Jones, T. W., and Stevenson, J. C., Herbicides in Chesapeake Bay and their effect on submerged aquatic vegetation in Chesapeake Bay and their effect on submerged aquatic vegetation, in *Chesapeake Bay Program Technical Studies: A Synthesis*, Part 4, U.S. Environmental Protection Agency, Washington, D.C., 1982, 503.
45. Oliver, B. G., Baxter, R. M., and Lee, H.-B., Polychlorinated biphenyls, in *Analysis of Trace Organics in the Aquatic Environment*, Afghan, B. K. and Chau, A. S. Y. (Eds.), CRC Press, Boca Raton, FL, 1989, 31.
46. Law, R. J., Allchin, C. R., and Dixon, A. G., Polychlorinated biphenyls in sediments downstream of contaminated industrial site in North Wales, *Mar. Pollut. Bull.*, 22, 492, 1991.
47. Harrad, S. J., Sewart, A. P., Alcock, R., Boumphrey, R., Burnett, V., Duarte-Davidson, R., Halsall, C., Sanders, G., Waterhouse, K., Wild, S. R., Jones, K. C., Polychlorinated biphenyls (PCBs) in the British environment: sinks, sources, and temporal trends, *Environ. Pollut.*, 85, 131, 1994.
48. Safe, S., Polyhalogenated aromatics: uptake, disposition, and metabolism, in *Halogenated Biphenyls, Terphenyls, Naphthalenes, Dibenzodioxins and Related Products*, 2nd ed., Kimbrough, R. D. and Jensen, A. A. (Eds.), Elsevier, Amsterdam, 1989, 131.
49. Rice, C. P. and O'Keefe, P., Sources, pathways, and effects of PCBs, dioxins, and dibenzofurans, in *Handbook of Ecotoxicology*, Hoffman, D. J., Rattner, B. A., Burton, G. A., Jr., and Cairns, J., Jr. (Eds.), Lewis Publishers, Boca Raton, FL, 1995, 424.

50. Tanabe, S. and Tatsukawa, R., Distribution, behavior, and load of PCBs in the oceans, in *PCBs and the Environment*, Vol. 1, Waid, J. S. (Ed.), CRC Press, Boca Raton, FL, 1986, 143.
51. Schulz, D. E., Petrick, G., and Duinker, J. C., Chlorinated biphenyls in North Atlantic surface and deep water, *Mar. Pollut. Bull.*, 19, 526, 1988.
52. Harding, G. C., Organochlorine dynamics between zooplankton and their environment, a reassessment, *Mar. Ecol. Prog. Ser.*, 33, 167, 1986.
53. Camacho-Ibar, V. F. and McEvoy, J., Total PCBs in Liverpool Bay sediments, *Mar. Environ. Res.*, 1996 (in press).
54. Klamer, H. J. C. and Fomsgaard, L., Geographical distribution of chlorinated biphenyls (CBs) and polycyclic aromatic hydrocarbons (PAHs) in surface sediments from the Humber Plume, North Sea, *Mar. Pollut. Bull.*, 26, 201, 1993.
55. Reutergardh, L., Chlorinated hydrocarbons in estuaries, in *Chemistry and Biogeochemistry of Estuaries,* Olausson, E. and Cato, I. (Eds.), John Wiley & Sons, Chichester, 1980, 349.
56. Wassermann, M., Wassermann, D., Cucos, S., and Miller, J. H., World PCBs map: storage and effects in man and his biologic environment in the 1970s, *Annu. N.Y. Acad. Sci.*, 320, 69, 1979.
57. Kennish, M. J., Belton, T. J., Hauge, P., Lockwood, K., and Ruppel, B., Polychlorinated biphenyls in estuarine and coastal marine waters of New Jersey: a review of existing pollution problems, *Rev. Aquat. Sci.*, 6, 275, 1992.
58. Bush, B., Streeter, R. W., and Sloan, R. J., Polychlorobiphenyl (PCB) congeners in striped bass (*Morone saxatilis*) from marine and estuarine waters of New York State determined by capillary gas chromatography, *Arch. Environ. Contam. Toxicol.*, 19, 49, 1989.
59. Olsson, M., PCBs in the Baltic environment, in *PCBs in the Environment*, Vol. 3, Waid, J. S. (Ed.), CRC Press, Boca Raton, FL, 1986, 181.
60. Larsson, P., Jarnmark, C., and Sodergren, A., PCBs and chlorinated pesticides in the atmosphere and aquatic organisms of Ross Island, Antarctica, *Mar. Pollut. Bull.*, 25, 9, 1992.
61. De Boer, J. and Wester, P., Chlorobiphenyls and organochlorine pesticides in various sub-Antarctic organisms, *Mar. Pollut. Bull.*, 22, 441, 1991.
62. Bidleman, T. F., Walla, M. D., Roura, R., Carr, E., and Schmidt, S., Organochlorine pesticides in the atmosphere of the Southern Ocean and Antarctica, January–March, 1990, *Mar. Pollut. Bull.*, 26, 258, 1993.
63. Eisler, R., Polychlorinated Biphenyl Hazards to Fish, Wildlife, and Invertebrates: A Synoptic Review, Biological Report 85 (1.7), U.S. Fish and Wildlife Service, Washington, D.C., 1986.
64. NOAA, *National Status and Trends Program for Marine Environmental Quality: Progress Report, a Summary of Selected Data on Chemical Contaminants in Tissues Collected during 1984, 1985, and 1986,* NOAA Tech. Mem. NOS OMA 38, National Oceanic and Atmospheric Administration, Rockville, MD, 1987.
65. Kennish, M. J., *Practical Handbook of Marine Science,* 2nd ed., CRC Press, Boca Raton, FL, 1994.
66. Gadbois, D. F. and Maney, R. D., Survey of polychlorinated biphenyls in selected finfish species from United States coastal waters, *Fish. Bull. U.S.*, 81, 389, 1983.
67. Belton, T. J., Hazen, R., Ruppel, B. E., Lockwood, K., Mueller, R., Stevenson, E. M., and Post, J. J., A Study of Dioxin (2, 3, 7, 8-Tetrachlorodibenzo-*p*-dioxin) Contamination in Select Finfish, Crustaceans, and Sediments of New Jersey, Technical Report, Office of Science and Research, N.J Department of Environmental Protection, Trenton, NJ, 1985.
68. Wenning, R. J., Paustenbach, D. J., Harris, M. A., and Bedbury, H., Principal components analysis of potential sources of polychlorinated dibenzo-*p*-dioxin and dibenzofuran residues in surficial sediments from Newark Bay, New Jersey, *Arch. Environ. Contam. Toxicol.*, 24, 271, 1993.
69. Ehrlich, R., Wenning, R. J., Johnson, G. W., Su, S. H., and Paustenbach, A., Mixing model for polychlorinated dibenzo-*p*-dioxins and dibenzofurans in surface sediments from Newark Bay, New Jersey using polytopic vector analysis, *Arch. Environ. Contam. Toxicol.*, 27, 486, 1994.
70. Clement, R. E. and Tosine, H. M., Analysis of chlorinated dibenzo-*p*-dioxins and dibenzofurans in the aquatic environment, in *Analysis of Trace Organics in the Aquatic Environment*, Afghan, B. K. and Chau, A. S. Y. (Eds.), CRC Press, Boca Raton, FL, 1989, 151.
71. Cooper, K. R., Effects of polychlorinated dibenzo-*p*-dioxins and polychlorinated dibenzofurans on aquatic organisms, *Rev. Aquat. Sci.*, 1, 227, 1989.
72. Cooper, K. R., Schell, J., Umbreit, T., and Gallo, M., Fish-embryo toxicity associated with exposure to soils and sediments contaminated with varying concentrations of dioxins and furans, *Mar. Environ. Res.*, 35, 177, 1993.

73. Hellou, J. and Payne, J. F., Polychlorinated dibenzo-*p*-dioxins and dibenzofurans in cod (*Gadus morhua*) from the Northwest Atlantic, *Mar. Environ. Res.*, 36, 117, 1993.
74. Norstrom, R. J. and Simon, M., Polychlorinated dibenzo-*p*-dioxins and dibenzofurans in marine mammals in the Canadian North, *Environ. Pollut.*, 1, 79, 1990.
75. Patton, G. W., Hinckley, D. A., Walla, M. D., and Bidleman, T. F., Airborne organochlorines in the Canadian High Arctic, *Tellus,* 41B, 243, 1989.

APPENDIX 1. ORGANOCHLORINE COMPOUNDS IN ESTUARINE AND MARINE WATERS

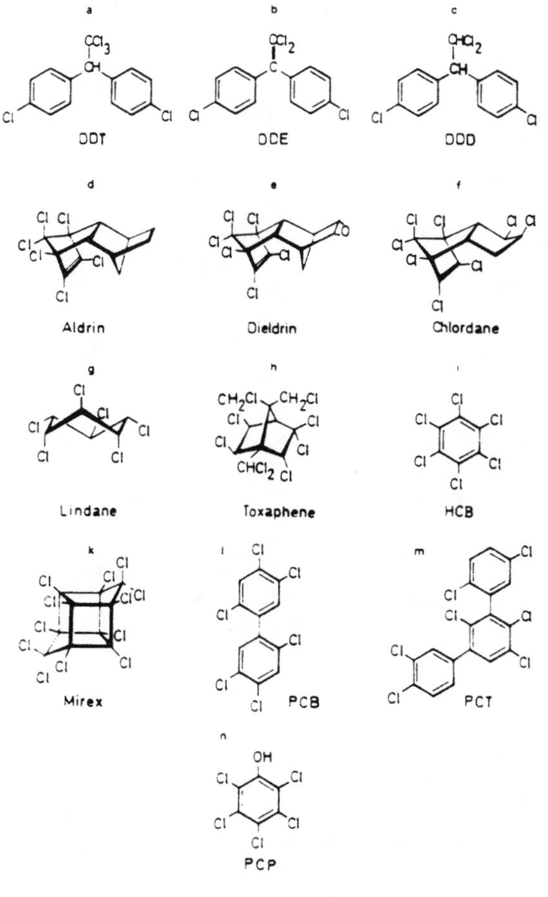

FIGURE 1.1. Chemical structure of common organochlorine compounds encountered in estuarine and marine environments. (From Reutergardh, L., in *Chemistry and Biogeochemistry of Estuaries,* Olausson, E. and Cato, I. (Eds.), John Wiley & Sons, Chichester, 1980, 349. With permission.)

TABLE 1.1
Classes of Pesticides by Use

Class	Example
Insecticides	Chlorinated hydrocarbons
	Organophosphorus compounds
	Carbamates
	Pyrethroids
	Phenols
Herbicides	Chlorophenoxy compounds
	Bipyridylium compounds
	Triazines
	Thiocarbamates
Fungicides	Inorganic compounds
	Organometallic compounds
	Antibiotics
	Chloroalkylthio compounds
	Quinones
	Dithiocarbamates
Rodenticides	Fluoroacetate derivatives
	Thioureas
	Antivitamin K compounds
Fumigants/nematocides	Hydrocyanic acid
	Carbon disulfide
Synergists	Piperonyl butoxide
	Sulfoxide

Source: Richardson, M. L. (Ed.), *Chemistry, Agriculture, and the Environment,* The Royal Society of Chemistry, Cambridge, 1991. With permission.

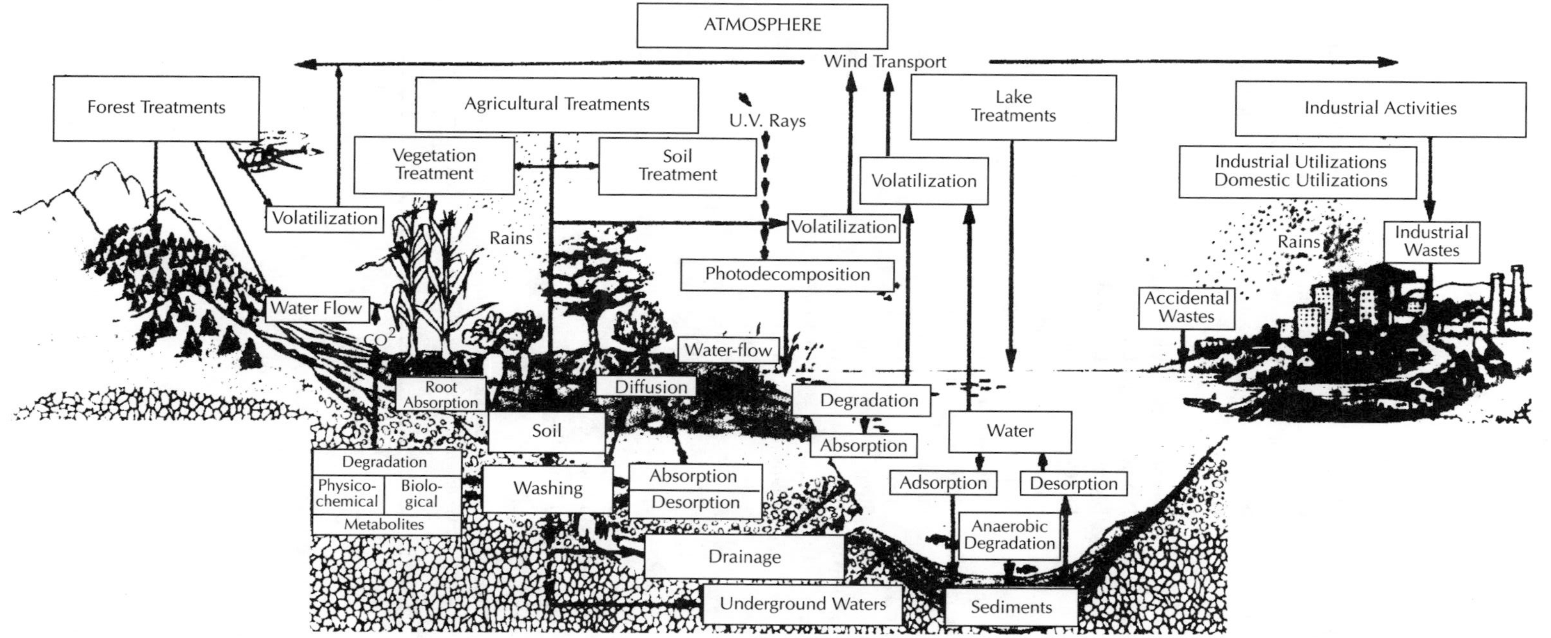

FIGURE 1.2. Pesticide dispersion in the environment. (From Boudou, A. and Ribeyre, F. (Eds.), *Aquatic Ecotoxicology: Fundamental Concepts and Methodologies,* CRC Press, Boca Raton, FL, 1989. With permission.)

TABLE 1.2
Relative Persistence of Some Pesticides in Natural Waters

Non-Persistent[a]	Slightly Persistent[b]	Moderately Persistent[c]	Persistent[d]
Azinphosmethyl	Aldrin	Aldicarb	Benomyl
Captan	Amitrole	Atrazine	Dieldrin
Carbaryl	CDAA	Ametryne	Endrin
Chlorpyrifos	CDEC	Bromacil	Hexachlorobenzene
Demeton	Chloramben	Carbofuran	Heptachlor
Dichlorvos	Chlorpropham	Carboxin	Isodrin
Dicrotophos	CIPC	Chlordane	Monocrotophos
Diquat	Dalapon	Chlorfenvinphos	
DNOC	Diazinon	Chloroxuron	
Endosulfan	Dicamba	Dichlorbenil	
Endothal	Disulfoton	Dimethoate	
Fenitrothion	DNBP	Diphenamid	
IPC	EPTC	Diuron	
Malathion	Fenuron	Ethion	
Methiocarb	MCPA	Fensulfothion	
Methoprene	Methoxychlor	Fonofos	
Methyl parathion	Monuron	Lindane	
Mevinphos	Phorate	Linuron	
Parathion	Propham	Prometone	
Naled	Swep	Propazine	
Phosphamidon	TCA	Quintozene	
Propoxur	Thionazin	Simazine	
Pyrethrum	Vernolate	TBA	
Rotenone		Terbacil	
Temephos		Toxaphene	
TFM		Trifluralin	
2,4-D			

[a] Half-life less than 2 weeks.
[b] Half-life 2 weeks to 6 weeks.
[c] Half-life 6 weeks to 6 months.
[d] Half-life more than 6 months.

Source: McEwen, F. L. and Stephenson, G. R., *The Use and Significance of Pesticide in the Environment,* John Wiley & Sons, New York, 1979. With permission.

TABLE 1.3
Comparison of Atmospheric and Riverine Input Rates of Organochlorine Compounds to the World Oceans

Compound	Atmospheric	Estimated Riverine	% Atmospheric
ΣHCH	4754	40–80	99
HCB	77.1	4	95
Dieldrin	42.9	4	91
ΣDDT	165	4	98
Chlordane	22.1	4	85
ΣPCB	239	40–80	80

Note: Input rates are $\times 10^6$ g a^{-1}.

Source: GESAMP (IMCO/FAO/UNESCO/WMO/WHO/IAEA/UN/UNEP Joint Group of Experts on Scientific Aspects of Marine Pollution). Reports and Studies, Inter-Governmental Maritime Consultative Organization, London, 1989.

TABLE 1.4
Summary of the Total Deposition of Some Organochlorine Compounds to the Oceans

	Atlantic		Pacific			
Compound	North	South	North	South	Indian	Mean Flux
ΣHCH	851	97	2640	471	698	4754
HCB	16.8	10	19.9	18.9	11.4	77.1
Dieldrin	16.6	2.0	8.9	9.5	6.0	42.9
ΣDDT	15.6	14.0	66.4	25.7	43.3	165
Chlordane	8.7	1.0	8.3	1.9	2.4	22.1
ΣPCB	99.7	13.8	35.5	29.1	52.1	239

Note: Deposition $\times 10^6$ g a^{-1}.

Source: GESAMP (IMCO/FAO/UNESCO/WMO/WHO/IAEA/UN/UNEP Joint Group of Experts on Scientific Aspects of Marine Pollution), Reports and Studies, Inter-Governmental Maritime Consultative Organization, London, 1989.

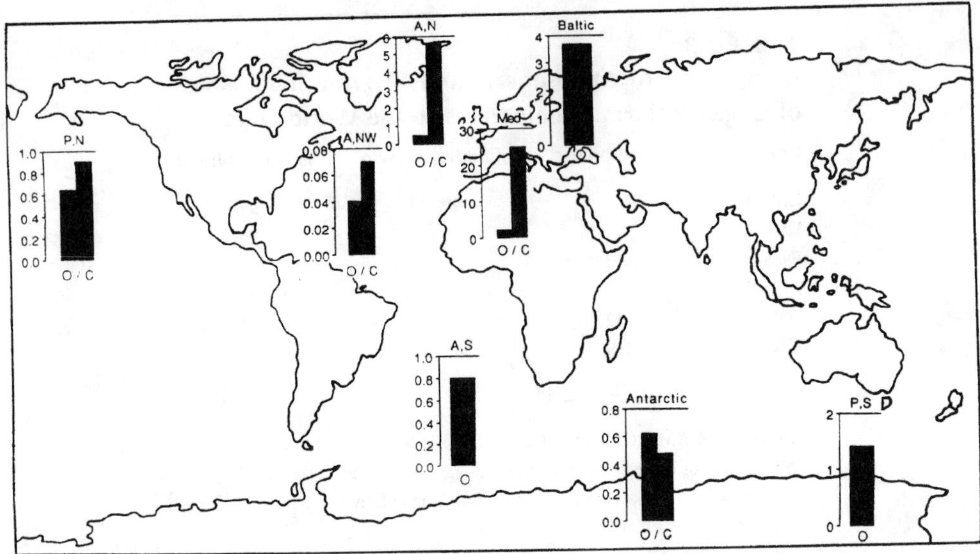

FIGURE 1.3. Mean concentrations of chlorinated hydrocarbons (ng/l) in open ocean (O) and coastal (C) surface waters. (From Davis, W. J., *Mar. Pollut. Bull.*, 26, 128, 1993. With permission.)

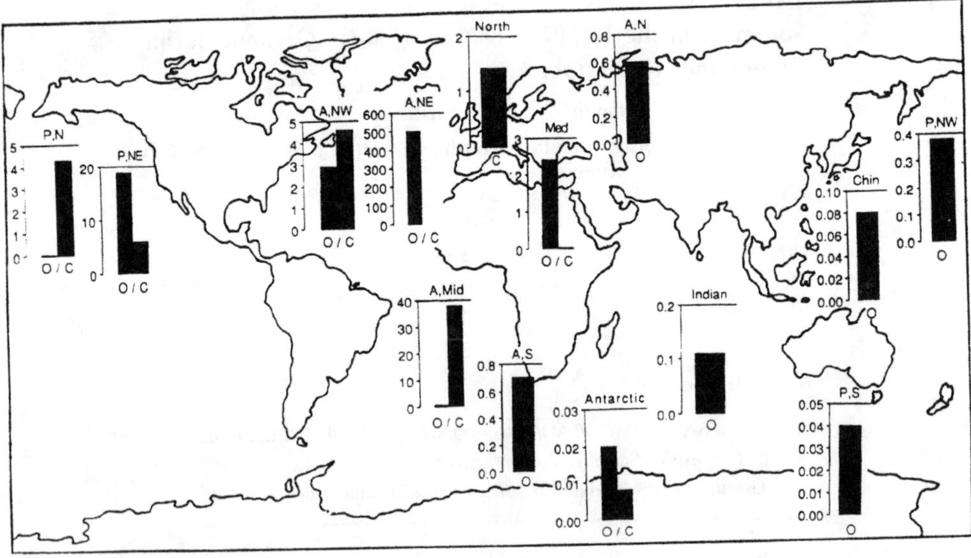

FIGURE 1.4. Mean concentrations of DDT and its metabolites (ng/l) in open ocean (O) and coastal (C) surface waters. Histograms are positioned in the approximate regions where measurements were made. Abbreviations: N, north; Chin., China Sea; NW, northwest; S, south. (From Davis, W. J., *Mar. Pollut. Bull.*, 26, 128, 1993. With permission.)

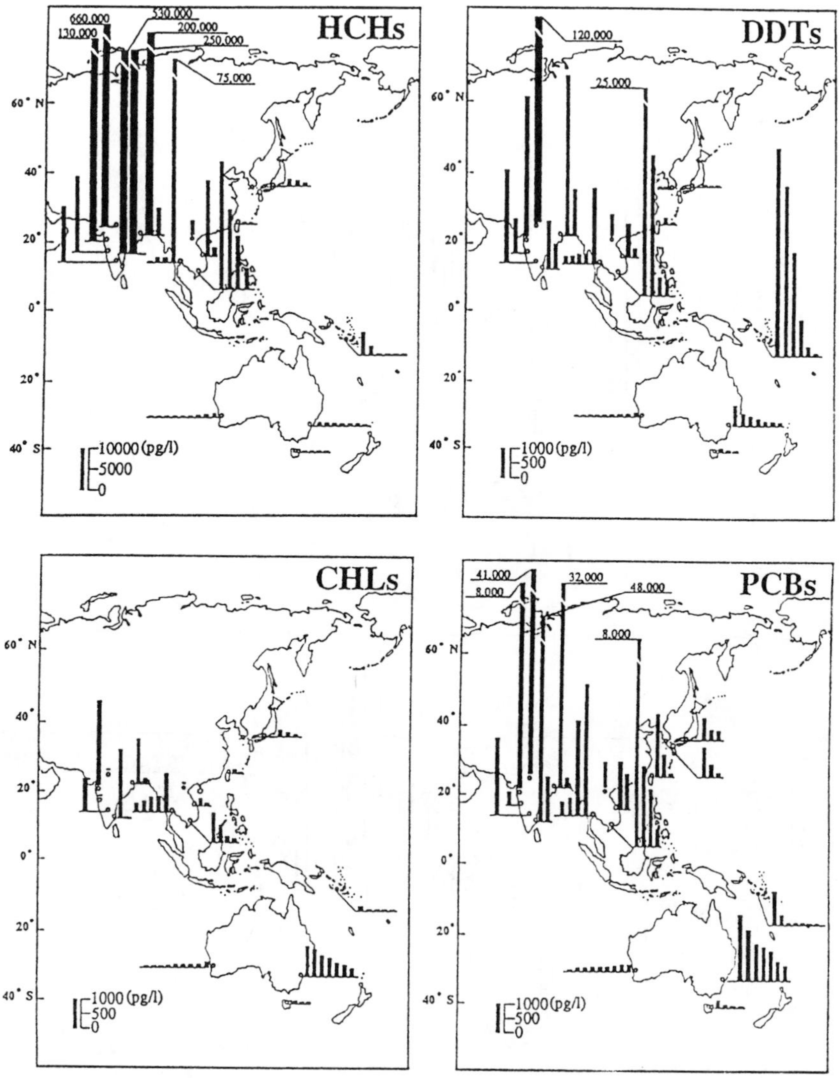

FIGURE 1.5. Distribution of persistent organochlorine residues in riverine and estuarine waters from urban areas of Asia and Oceania (1989 to 1990). (From Iwata, H., Tanabe, S., Sakai, N., Nishimura, A., and Tatsukawa, R., *Environ. Pollut.*, 85, 15, 1994. With permission.)

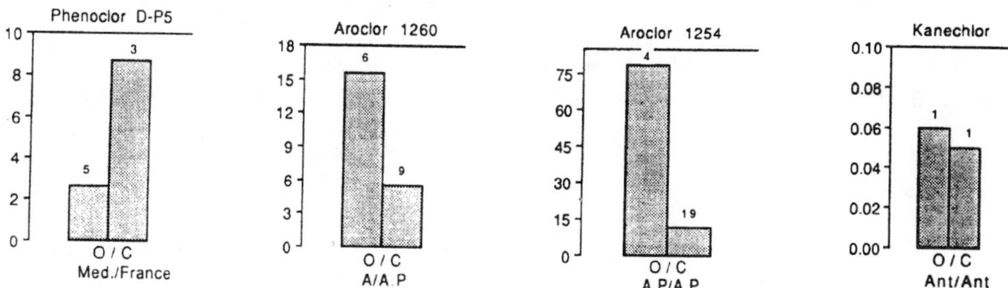

FIGURE 1.6. Mean concentrations of PCBs in ocean (O) and coastal (C) surface waters (ng/l). Abbreviations: Med., Mediterranean Sea; A., Atlantic; P., Pacific; Ant., Antarctic. Number above each bar represents the sample size. (From Davis, W. J., *Mar. Pollut. Bull.*, 26, 128, 1993. With permission.)

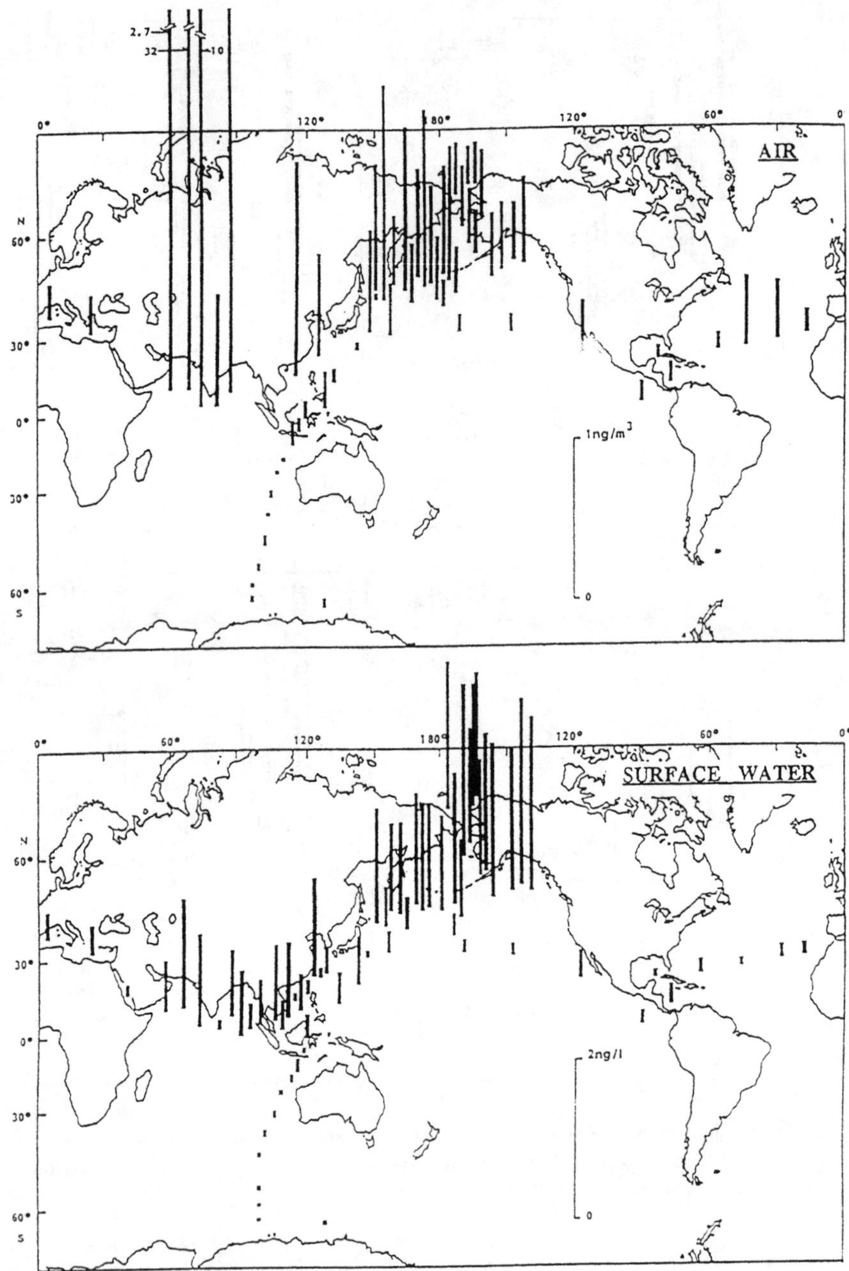

FIGURE 1.7. Distribution of HCH in air and surface sea water in the open ocean (1989 to 1990). (From Iwata, H., Tanabe, S., Sakai, N., and Tatsukawa, R., *Environ. Sci. Technol.*, 27, 1080, 1993. With permission.)

TABLE 1.5
PCB Concentrations in the Open Ocean Atmosphere

Location	Year	N	PCB Conc. (ng/m³) Range	Mean
North Atlantic				
Bermuda	1973	4	0.15–0.50	0.30
Bermuda	1973	8	0.21–0.65	0.51
Bermuda, U.S.	1973	4	0.72–1.6	0.99
Grand Banks (45°N, 52°W)	1973	5	0.05–0.16	0.086
Newfoundland	1977	6	0.042–0.15	0.12
Gulf of Mexico	1977	10	0.17–0.79	0.35
Barbados	1977–1978	17	<0.005–0.37	0.057
North Pacific				
Enewetak Atoll (12°N, 162°E)	1979	14	0.35–1.0	0.54
Western Pacific (3–35°N, 105–151°E)	1980–1981	7	0.089–0.74	0.25
Western Pacific (43–53°N, 154–172°E)	1981	2	0.041–0.061	0.051
Western Pacific (41–46°N, 144–174°E)	1982	5	0.022–0.095	0.043
Bering Sea	1981	3	0.026–0.059	0.041
South Pacific				
Western Pacific (1–46°S, 151–157°E)	1981	5	0.083–0.50[a]	0.27
Indian				
Eastern Indian (1–44°S, 104–125°E)	1980	5	0.066–0.33[a]	0.15
Western Indian (20–54°S, 48–57°E)	1982	4	0.060–0.24	0.16
Antarctic				
53–65°S, 125–161°E	1980–1981	5	0.056–0.18	0.091
54–68°S, 38–58°E	1982	4	0.076–0.11	0.091
Syowa Station (69°00′S, 39°35°E)	1981–1982	11	0.017–0.17	0.061

[a] Excluding the coastal regions.

Source: Tanabe, S. and Tatsukawa, R., in *PCBs and the Environment,* Vol. 1, Waid, J. S. (Ed.), CRC Press, Boca Raton, FL, 1986, 145. With permission.

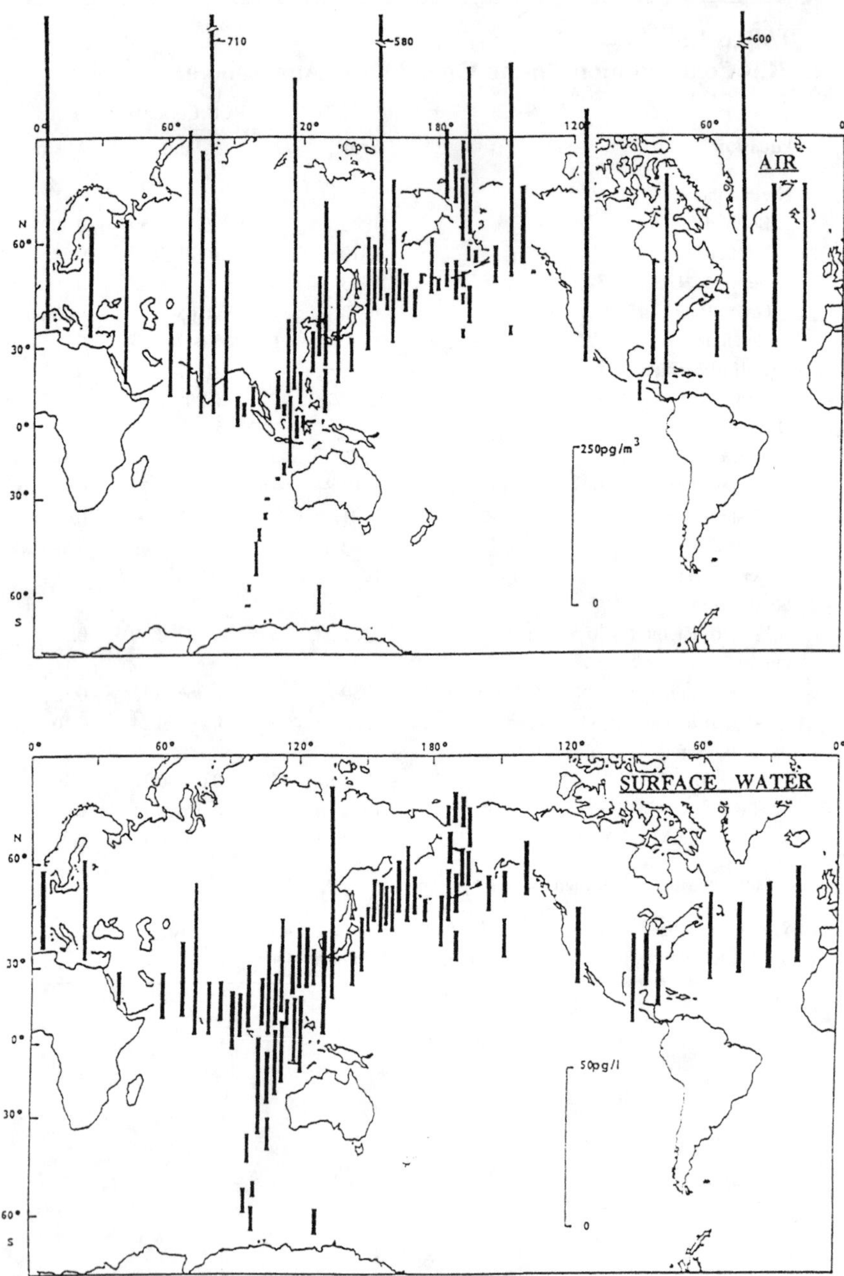

FIGURE 1.8. PCB residues in air and surface seawater in the open ocean (1989 to 1990). (From Iwata, H., Tanabe, S., Sakai, N., and Tatsukawa, R., *Environ. Sci. Technol.*, 27, 1080, 1993. With permission.)

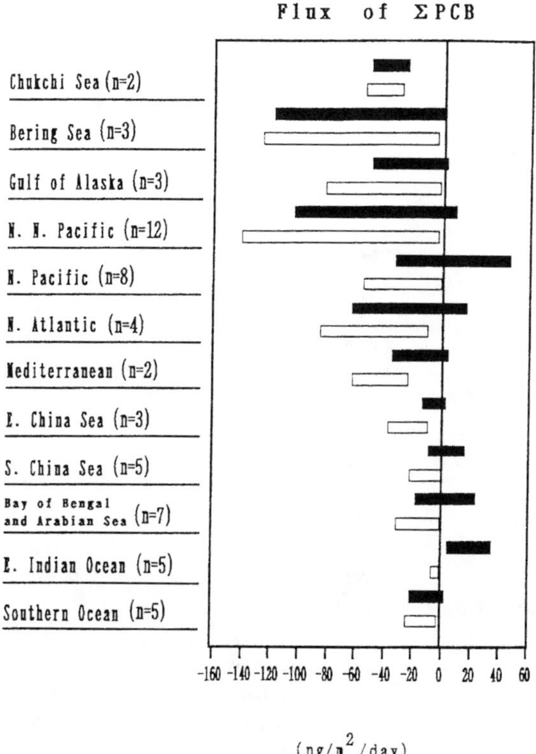

FIGURE 1.9. Fluxes by gas exchange of PCBs across the air-water interface in various seas and oceans. (From Tanabe, S., in *Contaminants in the Environment: A Multidisciplinary Assessment of Risks to Man and Other Organisms,* Renzoni, A., Mattei, N., Lari, L., and Fossi, M. C. (Eds.), Lewis Publishers, Boca Raton, FL, 1994, 19. With permission.)

TABLE 1.6
Ranges in PCB Concentrations (ng/l) Reported for Open Oceans, Coastal Waters, and Estuaries or Rivers

Area	Location	Range in PCB (ng/l)
Open oceans		
	North Atlantic	<1–150
		0.4–41
	Sargasso Sea	0.9–3.6
	North-South Atlantic	0.3–8.0
	Mediterranean Sea	0.2–8.6
Coastal waters		
	Southern California	2.3–36
	Northwest Mediterranean	1.5–38
	Atlantic coast, U.S.	10–700
	Baltic coasts	0.3–139
		0.1–28
	Dutch coast	0.7–8
Estuaries/rivers		
	Wisconsin rivers, U.S.	<10–380
	Rhine-Meuse system, Holland	10–200
	Tiber estuary, Italy	9–1000[a]
	Brisbane River, Australia	ND–50
	Hudson River, U.S.	$<100–2.8 \times 10^6$

Note: ND, not detectable (no limits quoted). Refer to *PCBs in the Environment* for particular studies.

[a] As decachlorinated biphenyl equivalents.

Source: Phillips, D. J. H., in *PCBs in the Environment,* Vol. 1, Waid, J. S. (Ed.), CRC Press, Boca Raton, FL, 1986, 130. With permission.

TABLE 1.7
Concentrations (ng/l) of PCBs in Coastal and Nearshore Surface Waters

Region	Year	N	ng/l	Quantified as
Baltic Sea				
Western Baltic Sea	1974	21	2.9 ± 1.2[a] (1.1–5.9)	Clophen A 30
Western Baltic Sea	1975	18	1.1 ± 0.8 (nd–3.9)	Clophen A 50
Western Baltic Sea	1976	14	7.2 ± 4.1 (1.1–15.6)	Clophen A 60
Western Baltic Sea	1978	12	5.7 ± 2.5 (3.5–11.9)	Clophen A 60
Hanö Bight	1975	8	0.9 ± 0.9 (0.3–3.0)	Clophen A 50
Mediterranean				
Marseille	1971		(100–210)	
French coastline	1975	11	13.1 ± 12.3 (1.5–38.0)	Phenoclor DP-5
French coastline	1984	14	<2–11	
Monaco	1981–1982	4	(0.2–1.2)	Aroclor® 1254
Northeast Atlantic				
British Isles (Liverpool Bay)	1974	31	0.4 ± 0.3 (<0.15–1.5)	Aroclor® 1254 and 1260
German Bight	1974	22	3.1 ± 0.9	Clophen A 30
German Bight	1975	17	2.1 ± 0.9 (0.8–3.6)	Clophen A 50
Holland	1976		0.7–8.1	
Oslofjord and Friefjord, Norway	1976–1977	3	5 ± 1 (4–<10)	Aroclor® 1254 and Clophen A 60
Brest	1977–1978	96	4.3 ± 2.8	
Gironde Estuary	1985		10 ± 6(3–25)	
Seine Bay (Channel)	1985		(3–6)	
Seine Estuary	1985		29 ± 12(40–370)	
Northwest Atlantic				
Narragansett Bay (U.S.)	1971		150 ± 40	Aroclor® 1254
New England	1974	6	0.8	Aroclor® 1260
Texas (Corpus Christi Bay)	1980	8	4.8 ± 10.7 (0.1–31.0)	Aroclor® 1260
Northwest Pacific				
Japan	1972		(0.3–13.9)	
Japan	1973		(0–6.4)	
Northeast Pacific				
California	1973		(5.4–16.3)	
California	1974	7	12.7 ± 10.6 (3.0–35.6)	Aroclor® 1254
San Francisco Bay	1978	1	0.66	
Farallon Islands	1978	2	(0.018–0.028)	
Golden Gate Bridge	1978	2	(0.145–0.160)	
Southern California Bight	1973	7	0.4 ± 0.10 (0.3–0.5)	Aroclor® 1254
Southern California Bight	1975	20	(0.04–2.0)[b]	Aroclor® 1254
Antarctic				
Syowa Station	1981–1982	6	0.05 ± 0.01 (0.03–0.07)	Kanechlor 300, 400, or 500

Note: See original source for particular studies.

[a] Mean ± SE (range).

[b] Pentachlorobiphenyls.

Source: Harding, G. C., *Mar. Ecol. Prog. Ser.*, 33, 167, 1986. With permission.

TABLE 1.8
PCBs in Open Ocean Surface Waters

Region	Year	N	ng/l	Quantified as
Mediterranean				
West and east basins	1975	37	2.9 ± 3.6[a] (0.2–19.0)	Phenochlor DP-5
West and east basins	1977–79		0.7(0.1–2.5)	
Northwestern Mediterranean	1982		1.5–5.1	
North Sea	mid-1980s		0.01–0.14	Σ of 18 congeners
North Atlantic				
Iceland to Nova Scotia	1971	8	25 ± ?	Aroclor® 1260
Azores to Barbados	1973	8	2 ± ?	Aroclor® 1260
Sargasso to New York Bight	1973	9	0.8 ± ?	Aroclor® 1260
Northeast Atlantic				
34°N–63°N	1972	19	39 ± 36 (1–150)	Aroclor® 1260
47°N 20°W	1986–87		0.0066–0.021	Σ of 18 congeners
Northwest Atlantic				
Gulf Stream, Sargasso	1972	15	27 ± 24 (1–88)	Aroclor® 1260
Sargasso Sea	1973	9	1.4 ± 0.9 (<0.09–3.6)	Aroclor® 1254
North Pacific (NE Pacific Gyre)	1972	2	2.6 ± 1.4 (1.6–3.6)	Aroclor® 1254
Northeast Pacific (off Mexico)	1975	23	(<10–<300)[b]	Aroclor® 1254
Northwest Pacific	1975	13	0.4 ± 0.10 (0.25–0.56)	Kanechlor 300, 400, and 500
Northwest Pacific	1976	8	0.54 ± 0.29 (0.29–1.11)	Kanechlor 300, 400, and 500
Northwest Pacific	1978	6	0.35 ± 0.13 (0.32–0.59)	Kanechlor 300, 400, and 500
Northwest Pacific	1979	5	0.33 ± 0.05 (0.22–0.38)	Kanechlor 300, 400, and 500
Indo Pacific	1980–81	18	0.12 ± 0.06 (0.04–0.25)	Kanechlor 300, 400, and 500
Antarctic (below Australia)	1980–81	9	0.06 ± 0.01 (0.04–0.08)	Kanechlor 300, 400, and 500

Note: See original source for particular studies.

[a] Mean ± SD (range).

[b] Pentachlorobiphenyls.

Source: Harding, G. C., *Mar. Ecol. Prog. Ser.*, 33, 167, 1986. With permission.

TABLE 1.9
Estimated Concentration and Load of PCBs in the Open Ocean Environment

Compartment Mass PCB Conc., Load	North Pacific	South Pacific	North Atlantic	South Atlantic	Indian	Total Antarctic	Load
Compartment mass[a]							
Air (×10[16] m[l])[b]	70	76	38	35	58	46	
Water (×10[19] l)	30	33	15	14	23	19	
Sediment (dry, ×10[15] g)[c]	70	76	38	35	58	46	
Plankton (wet, ×10[15] g)[d]			48 (whole ocean)				
Fish (fresh, ×10[14] g)[d]			26 (whole ocean)				
Mammals (fresh, ×10[12] g)[d]			85 (whole ocean)				
PCB concentration							
Air (ng/m³)	0.3	0.2	0.5	0.2	0.2	0.1	
Water (ng/l)	0.2	0.1	0.6	0.1	0.1	0.05	
Sediment (dry, ng/g)	0.4	0.2	1.0	0.2	0.2	0.1	
Plankton (wet, ng/g)	2.0	1.0	5.0	1.0	1.0	0.5	
Fish (fresh whole, ng/g)	10	2.0	30	2.0	2.0	0.2	
Mammals (fresh whole, μg/g)[e]	2.0	0.5	5.0	0.5	0.5	0.05	
PCB load (t)							
Air	210	150	190	70	120	50	790
Water[f]	60,000	33,000	90,000	14,000	23,000	10,000	230,000
Sediment	28	15	38	7	12	5	105
Plankton[g]			20–240 (whole ocean)				130[h]
Fish[g]			1–80 (whole ocean)				40[h]
Mammals[g]			2–200 (whole ocean)				100[h]
Total							231,165

[a] Data on the surface area and mean depth in respective oceans employed for the calculation of air mass, water mass, and sediment mass were adapted from Sugimura, personal communications, 1983 (see original source).
[b] Calculated in troposphere (10 km height).
[c] Calculated in upper 1-mm sediment layer. Sediment was assumed to contain about 50% water and to have a mass density of 2.
[d] Compiled in consideration of the following moisture contents: plankton 95%, fish 75%, and mammals 65%.
[e] Values show the PCB concentrations regarding male specimens. PCB concentrations on whole-body basis were calculated following the relationship of PCB concentrations between blubber and whole body obtained from the striped dolphin.
[f] PCB loads in water were estimated on the assumption of their having vertically uniform concentrations in the water column as shown by their vertical profiles in original source of table.
[g] PCB loads in these organisms were estimated from the following probable concentrations: plankton 0.5–5 ng/g, fish 0.2–30 ng/g, and mammals (male) 0.05–5, μg/g. The PCB load in mammals took into account that female specimens generally have lower concentrations of PCBs in their bodies than males because of parturitional and lactational losses.
[h] Median values.

Source: Tanabe, S. and Tatsukawa, R., in *PCBs and the Environment,* Vol. 1, Waid, J. S. (Ed.) CRC Press, Boca Raton, FL, 1986, 156. With permission.

TABLE 1.10
Estimated Residence Time of PCBs in the Open Ocean Mixing Layer (Upper 100 m of Water Column)

Ocean	Carbon Productivity ($g/m^2/year$)	Residence Time (day)
Oligotrophic		
Pacific Ocean (22°05′N, 145°02′E)	50–100	130–280
Eutrophic		
Southern Ocean (64°42′S, 124°15′E)	150–250	26–44

Source: Tanabe, S. and Tatsukawa, R., in *PCBs and the Environment,* Vol. 1, Waid, J. S. (Ed.), CRC Press, Boca Raton, FL, 1986, 157. With permission.

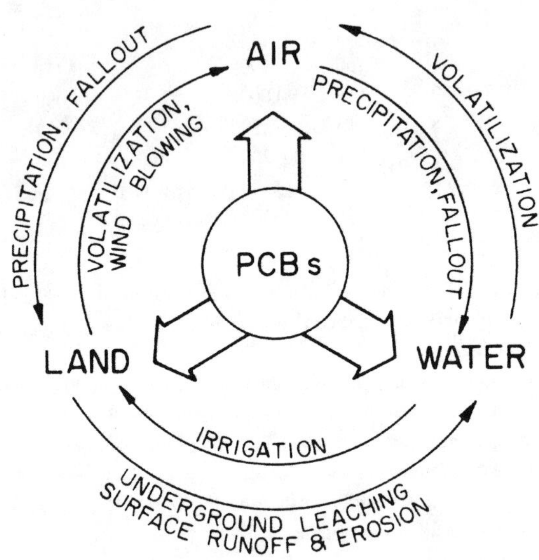

FIGURE 1.10. PCB environmental cycle. (From Lauber, J. D., in *PCBs and the Environment,* Vol. 3, Waid, J. S. (Ed.), CRC Press, Boca Raton, FL, 1986, 86. With permission.)

TABLE 1.11
Concentrations (ng/l) of DDT and Metabolites in Coastal and Nearshore Surface Waters

Region	Year	N	p,p′-DDT	p,p′-DDE	p,p′-DDD	ΣDDT
Baltic Sea						
Western Baltic Sea	1974	21	0.2 ± 0.1 (0.1–0.3)		2.4 ± 0.6 (0.8–3.4)	
Western Baltic Sea	1975	18	0.2 ± 0.1 (nd–0.4)		0.1 ± 0.1 (nd–0.4)	
Mediterranean						
Marseille	1971		(70–180)	(30–120)		
Monaco	1974	3				(0.04 ± 0.01[a] (0.04–0.05)
Northeast Atlantic						
British Isles	1974	31	+[h]	+	+	0.05 ± 0.07 (<0.01–0.25)
Brest	1977–1978					1.5 ± 2.8
Northwest Atlantic						
North Carolina	1972	4	nd[c] (<8.0)	7.9 ± 6.9 (3.5–18.1)	nd (<0.8)	
Texas (Corpus Christi Bay)	1980	8				1.32 ± 1.04 (0.2–3.1)
Southwest Atlantic						
Brazil (Santos Estuary)	1974–1975	64				(nd–40)
Argentina[d] (Blanca Bay)	1980–1981	15	37.9 ± 87.2 (nd–346.0)		nd	
North Pacific						
Hawaii Canal	1971	6	9.0 (0.4–41.0)	0.3 (0.1–0.6)	3.6 (0.1–10.0)	
Northeast Pacific						
San Francisco Bay	1969	24	5 ± 5 (1–18)	2 ± 1 (1–4)	2 ± 2 (1–7)	8 ± 5 (3–23)
Monterey Bay to San Diego	1970	4	+	+	+	2.4 ± 0.2 (2.4–2.7)
California coast	1973					(0.9–17.5)
California coast	1974	7	0.5 ± 0.6 (<0.1–1.9)	1.0 ± 1.6 (<0.1–4.6)		
Southern California Bight	1973	14	0.6 ± 0.3 (0.3–1.3)	0.10 ± 0.03 (0.06–0.13)		0.7 ± 0.3 (0.34–1.43)
Southern California	1975	20	(<0.05–0.8)	(<0.05–5.9)	(<0.03–2.2)	
North Sea						
German Bight	1974	24	0.3 ± 0.2 (0.1–0.6)		1.5 ± 0.6 (0.5–2.6)	
German Bight	1975	17	0.33 ± 0.15 (0.11–0.63)	0.11 ± 0.06 (nd–0.21)	0.20 ± 0.15 (nd–0.57)	
Hanö Bight	1975	8	0.04 ± 0.08 (nd–0.21)	1.7 ± 1.6 (0.8–5.2)		
Antarctic[e]						
Syowa Station	1981	6	+	+		0.008 ± 0.008 (0.001–0.021)

Note: See original source for particular studies.

[a] Mean ± SD (range).

[b] Present.

[c] Not detected.

[d] o,p′-DDT, 13.4 ± 24.5, (nd–94.5).

[e] o,p′-DDT, +.

Source: Harding, G. C., *Mar. Ecol. Prog. Ser.*, 33, 167, 1986. With permission.

TABLE 1.12
Concentrations (ng/l) of DDT and Metabolites in Open Ocean Surface Waters

Region	Year	N	p,p′-DDT	o,p′-DDT	p,p′-DDE	p,p′-DDD	ΣDDT
Mediterranean	1972		(2–24)		(0.5–2.0)	0.7–2.0	
Mediterranean	1977	3	1.2[a]		0.2 = ΣDDE + DDD		1.4
Mediterranean	1978	16	0.4		0.2 = ΣDDE + DDD		0.6
Mediterranean	1979	32	0.7		0.3 = ΣDDE + DDD		1.0
Mediterranean	1980	35	0.5		0.5 = ΣDDE + DDD		1.0
Subarctic Atlantic	1977	10	0.4		0.5 = ΣDDE + DDD		0.9
Subarctic Atlantic	1978	73	0.2		0.4 = ΣDDE + DDD		0.6
Subarctic Atlantic	1979	54	0.4		0.2 = ΣDDE + DDD		0.6
Subarctic Atlantic	1980	49	0.2		0.1 = ΣDDE + DDD		0.3
Subtropical Atlantic	1977	15	0.7		0.1 = ΣDDE + DDD		0.8
Subtropical Atlantic	1978	30	0.3		0.2 = ΣDDE + DDD		0.5
Subtropical Atlantic	1979	49	0.7		0.1 = ΣDDE + DDD		0.8
Subtropical Atlantic	1980	39	0.5		0.2 = ΣDDE + DDD		0.7
Tropical Atlantic	1979	11	0.3		0.4 = ΣDDE + DDD		0.7
Tropical Atlantic	1980	34	0.4		0.3 = ΣDDE + DDD		0.7
Northwest Atlantic							
Gulf Stream							
Sargasso Sea	1972	6	<8 nd[b]		2.9 ± 2.7 (<0.35–7.5)	<0.8 nd	
Sargasso Sea	1973	9	0.2 ± 0.1[c] (<0.15–0.5)	<0.05			
Northeast Atlantic							
Norway to Iceland	1973		+[d]		+	+	500 ± 90
Norway to Ireland	1975		+		+	+	500 ± 90

North Pacific (NC Pacific Gyre)	1972	2	<0.002		<0.01		
Bering Sea	1978	7	+		+		0.02 ± 0.01 (0.01–0.04)
Bering Sea	1981	2	0.0016, 0.0007		0.0011, 0.0007		
Northwest Pacific	1975	13	+		+	+	0.11 ± 0.05 (0.06–0.23)
Northwest Pacific	1976	8	+		+	+	0.90 ± 0.26 (0.52–1.35)
Northwest Pacific	1978	6	+		+	+	0.25 ± 0.03 (0.22–0.27)
Northwest Pacific	1979	8	+		+	+	0.38 ± 0.41 (0.02–1.17)
Northeast Pacific							
Mexico	1975	23	(<1–20)		(<1–<10)	(<1–<80)	
Indo-Pacific	1980–81	15	0.03 ± 0.03 (0.005–0.091)	0.01 ± 0.01 (0.001–0.018)	0.004 ± 0.003 (0.001–0.013)		0.04 ± 0.04 (0.007–0.13)
Indian Ocean							
Arabian Sea and Bay of Bengal	1976	6	+		+	+	0.10 ± 0.04 (0.06–0.16)
Antarctic (below Australia)		12	0.01 ± 0.01 (0.003–0.053)	0.003 ± 0.002 (0.001–0.006)	0.002 ± 0.001 (0.001–0.005)		0.02 ± 0.01 (0.005–0.058)
China Sea	1977	3	+		+	+	0.08 ± 0.02

Note: See original source for particular studies.

[a] Mean.

[b] Not detected.

[c] Mean ± SD (range).

[d] Present.

Source: Harding, G. C., *Mar. Ecol. Prog. Ser.*, 33, 167, 1986. With permission.

APPENDIX 2. ORGANOCHLORINE COMPOUNDS IN ESTUARINE AND MARINE SEDIMENTS

TABLES 2.1
Chlorinated Hydrocarbon Concentrations (ng/g dry weight) in Surface Sediments from Coastal and Nearshore Areas

Location	ΣPCB	ΣDDT
Northwest Atlantic		
Gulf of Maine	<100,[e] 40–340, trace–130	
New Bedford Bay, MA	8400	
Escambia Bay, FL	<30–480,000	
New York Bight	0.5–2200	
Gulf of Mexico, U.S.	0.2–35	<0.03
Gulf of Mexico, Caribbean (Mexico)		0.3–2.27
Chesapeake Bay	4–400	
Hudson-Raritan Estuary	286–1950	116–739
Northeast Atlantic		
Brittany	<0.5	
Brest, France		
(subtidal)	0.4–185	<0.1–17.6
(intertidal)	3.3–2100	8.0–52.2
Seine Estuary	15 ± 8	<0.1–0.4
Loire Estuary	51 ± 34	<0.5
Irish Sea, U.K.	<2–2890	
Ivory Coast	2–213	2–997
Mediterranean		
Venice	15	1.4
Adriatic	1–17	1.3, <1–8
Northwestern	0.3–1200, 11–61, <2–250	3.6, 0.7–44
Tyrrhenian	0.6–3200	4, <20
Ionian	0.8–457	
Aegean	1.3–775	7.1–1893
Eastern	1.9–4.0	
Central and western	0.5–323	<1–28
Marseille Bay	157 ± 12	10–50
Northeast Pacific		
San Pedro Basin, CA	1–13	5–30
Santa Monica Basin, CA	0–9	30–160
Palos Verdes, CA	80–7420	1600–100,000
Puget Sound, WA	80–640	
San Francisco Bay	30–50	
Northwest Pacific		
Gulf of Thailand	nd	22–56
Osaka Bay, Japan	40–2000	
Harimanada Bay	50–400	
Southwest Pacific		
Queensland estuaries, Australia	6–350	
Port Phillip Bay, Australia	<10–390	
Bass Strait, Australia	<10	
Manukau Harbour, New Zealand	0.5–14.2	1.2–2.3

TABLES 2.1 (continued)
Chlorinated Hydrocarbon Concentrations (ng/g dry weight) in Surface Sediments from Coastal and Nearshore Areas

Location	ΣPCB	ΣDDT
Southwest Atlantic		
Brazil	nd–84	nd–25
Rio de Janeiro, Brazil	10–38	6–22
North Sea	11.5–40.5	
Dunkerque	134 ± 134	<0.1
Norway	14–28	0.16–0.36
Baltic Sea		
Sweden	40–160	
Finland	10 (<10–20)	
Kiel Bay	8.4–10.8	2.0–2.4
Eckernforde Bight	134–212	28–46
Indian Ocean		
East India (Bay of Bengal)		nd–980
West India (Arabian Sea)		43 (14–358)[a]

Note: See original source for particular studies.

[a] Factors of 2 used to convert from wet to dry weight concentrations.

Source: Fowler, S. W., *Mar. Environ. Res.*, 29, 1, 1990. With permission.

TABLE 2.2
Ranges in Concentration of PCBs (ng/g dry weight) Reported for Sediments in Areas Ranging from Relatively Uncontaminated to Highly Contaminated

Area	PCB Conc.
Mediterranean Sea	0.8–9
Gulf of Mexico	0.2–35
Chesapeake Bay	4–400
Lake Superior	5–390
Tiber estuary	28–770
Rhine-Meuse estuary	50–1000
New York Bight	0.5–2200
Palos Verdes Peninsula	30–7900
Hudson River	tr–6700
Escambia Bay	190–61,000

Note: tr = trace. See original source for particular studies.

Source: Phillips, D. J. H., in *PCBs and the Environment,* Vol. 2, Waid, J. S. (Ed.), CRC Press, Boca Raton, FL, 1986, 132. With permission.

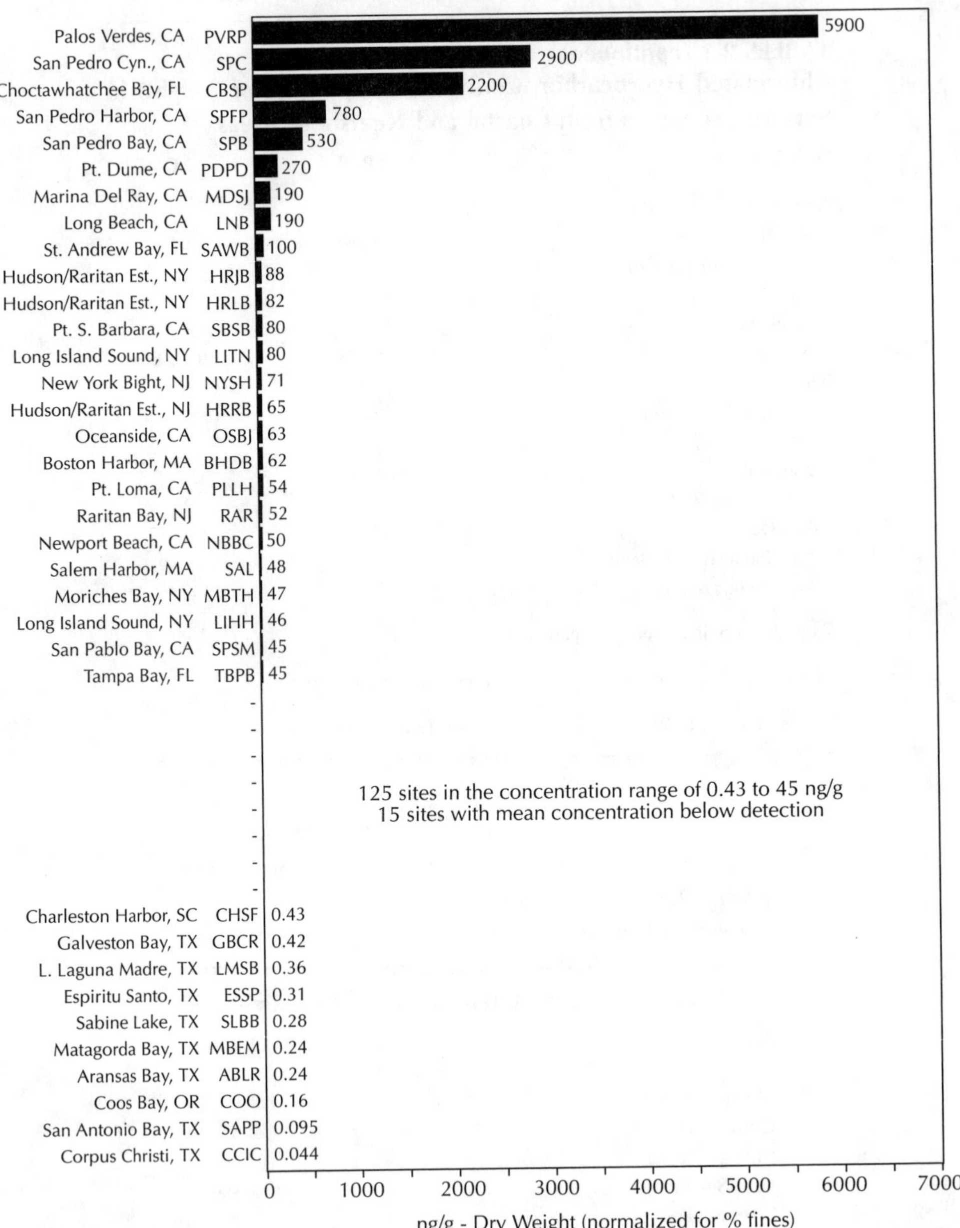

FIGURE 2.1. Total DDT in sediments of various estuarine and coastal systems of the U.S. (From NOAA, *A Summary of Data on Individual Organic Contaminants in Sediments Collected During 1984, 1985, 1986, 1987*, NOAA Tech. Mem. NOS OMA 47, National Oceanic and Atmospheric Association, Rockville, MD, 1989.)

TABLE 2.3
PCBs and PAHs in Sediments from Selected Estuaries in the U.S., 1984[a]

Estuary	Total PCBs	Total PAHs
Casco Bay, ME	95.28	7320.00
Merrimack River, MA	52.97	1730.00
Salem Harbor, MA	533.58	10,220.00
Boston Harbor, MA	17,104.86	26,440.00
Buzzard's Bay, MA	308.46	1710.00
Narragansett Bay, RI	159.96	2350.00
East Long Island Sound, NY	10.00	48,560.00
West Long Island Sound, NY	234.43	8430.00
Raritan Bay, NJ	443.89	5010.00
Delaware Bay, DE	2.50	330.00
Lower Chesapeake Bay, VA	51.00	410.00
Pamlico Sound, NC	ND	219.25
Charleston Harbor, SC	9.10	802.98
Sapelo Sound, GA	ND	22.28
St. Johns River, FL	140.00	1926.91
Charlotte Harbor, FL	ND	26.51
Tampa Bay, FL	ND	27.10
Apalachicola Bay, FL	12.00	200.25
Mobile Bay, AL	ND	96.79
Round Islands, MS	ND	52.36
Mississippi River Delta, LA	34.00	603.41
Barataria Bay, LA	ND	106.08
Galveston Bay, TX	ND	68.05
San Antonio Bay, TX	ND	8.98
Corpus Christi Bay, TX	ND	28.17
Lower Laguna Madre, TX	ND	0.00
San Diego Harbor, CA	422.10	5000.00
San Diego Bay, CA	6.74	0.00
Dana Point, CA	7.06	22.87
Seal Beach, CA	46.71	257.96
San Pedro Canyon, CA	159.56	527.00
Santa Monica Bay, CA	14.00	68.25
San Francisco Bay, CA	123.46	5976.03
Bodega Bay, CA	4.18	11.00
Coos Bay, OR	3.19	234.67
Columbia River Mouth, OR/WA	8.77	145.03
Nisqually Reach, WA	4.23	0.00
Commencement Bay, WA	20.60	1200.00
Elliott Bay, WA	329.87	4700.00
Lutak Inlet, AK	5.50	0.00
Nahku Bay, AK	6.60	100.00

Note: PCBs = polychlorinated biphenyls; PAHs = polycyclic aromatic hydrocarbons. ND = not detected.

[a] Parts per billion.

Source: NOAA, *National Status and Trends Program for Marine Environmental Quality: Progress Report and Preliminary Assessments of Findings of the Benthos Surveillance Project — 1984*, National Oceanic and Atmospheric Administration Office of Ocean Resources Conservation and Assessment, Rockville, MD, 1987.

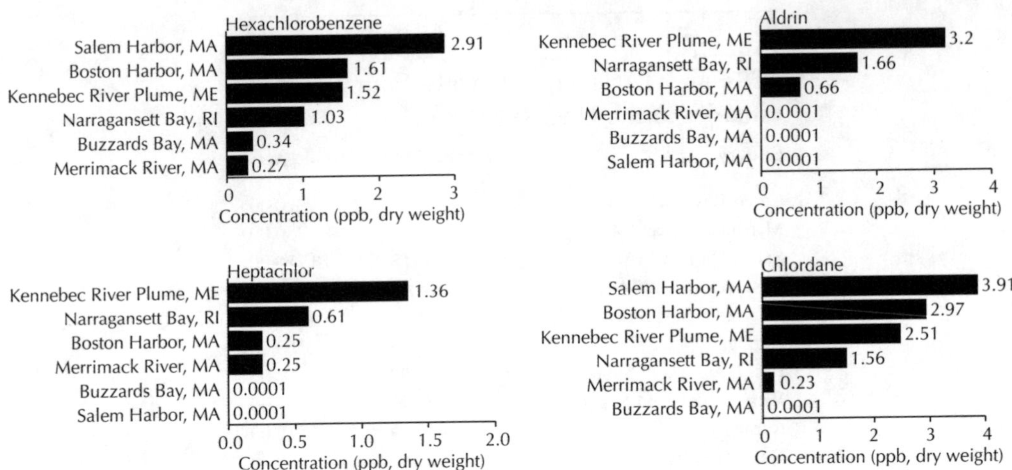

FIGURE 2.2. Concentrations of individual pesticides in New England marine sediments. (From Larsen, P. F., *Rev. Aquat. Sci.*, 6, 67, 1992. With permission.)

TABLE 2.4
Concentrations of Chlorinated Organic Contaminants (ng/g dry weight) in Bottom Sediments from San Francisco Bay

Compound	Site No.																
	25	**26**	**27**	**28**	**29**	**30**	**31**	**32**	**33**	**34**	**35**	**36**	**37**	**38**	**39**	**40**	**41**
DDE	3.9	1.5	0.7	1.1	0.7	0.5	1.3	1.4	1.8	1.8	1.5	<0.1	1.5	1.5	1.5	1.2	1.2
DDD	1.9	1.3	0.7	3.1	0.1	0.8	2.6	2.7	6.3	2.8	2.4	<0.1	2	1.7	1.9	1.5	1.7
DDT	3.2	0.2	0.1	0.2	<0.1	<0.1	3	1.5	0.7	2.5	0.5	<0.1	0.2	0.1	0.2	0.2	<0.1
Total DDT	9	3	1.5	4.4	0.8	1.3	6.9	5.6	8.8	7.1	4.4	<0.1	3.7	3.3	3.6	2.9	2.9
gamma-Chlordane	0.7	0.1	<0.1	0.1	<0.1	<0.1	<0.1	<0.1	<0.1	<0.1	0.1	<0.1	<0.1	<0.1	0.1	<0.1	<0.1
alpha-Chlordane	<0.1	<0.1	<0.1	<0.1	<0.1	<0.1	<0.1	<0.1	<0.1	<0.1	<0.1	<0.1	<0.1	<0.1	<0.1	<0.1	<0.1
trans-Nonachlor	0.5	0.1	<0.1	0.7	<0.1	<0.1	0.1	<0.1	<0.1	<0.1	0.1	<0.1	<0.1	<0.1	<0.1	<0.1	<0.1
cis-Nonachlor	0.3	<0.1	<0.1	<0.1	<0.1	<0.1	<0.1	<0.1	<0.1	<0.1	<0.1	<0.1	<0.1	<0.1	<0.1	<0.1	<0.1
Total chlordane	1.5	0.2	<0.1	0.8	<0.1	<0.1	0.1	<0.1	<0.1	<0.1	0.2	<0.1	<0.1	<0.1	0.1	<0.1	<0.1
Tetrachloro PCB	3	1.5	0.5	0.9	0.8	0.5	0.9	1	1.8	1	0.7	<0.1	0.4	0.2	0.3	0.4	1.3
Pentachloro PCB	2.3	3.2	0.8	1.3	1	0.8	2.9	1.5	2.8	1.1	1.2	<0.1	0.6	1	0.6	1.3	1.5
Hexachloro PCB	2.2	3.4	1.4	1.7	2.4	1.5	2.3	1.3	1.7	0.6	0.8	<0.1	0.4	0.3	0.4	0.3	1.7
Total PCB	7.5	8.1	2.7	3.9	4.2	2.8	6.1	3.8	6.3	2.7	2.7	<0.1	1.4	1.5	1.3	2	4.5
DCPA	<0.1	<0.1	<0.1	<0.1	<0.1	<0.1	<0.1	<0.1	<0.1	<0.1	<0.1	<0.1	<0.1	<0.1	<0.1	<0.1	<0.1

Source: Pereira, W. E., Hostettler, F. D., Cashman, J. R., and Nishioka, R. S., *Mar. Pollut. Bull.*, 28, 434, 1994. With permission.

APPENDIX 3. ORGANOCHLORINE COMPOUNDS IN ESTUARINE AND MARINE ORGANISMS

TABLE 3.1
List of Target Chemical Residues Surveyed in U.S. Coastal Fish and Shellfish

Residue	Use and Occurrence
Polychlorinated biphenyls (PCB)	Dielectric fluid in capacitors; transformer fluid; lubricants; hydraulic fluids; plasticizers; cutting oil extenders; carbonless paper; banned in 1976; total is either sum of chlorination mixtures (Aroclors®) or number
Total PCB	
Aroclor® 1016	
Aroclor® 1242	
Aroclor® 1248	
Aroclor® 1254	
Aroclor® 1260	
PCB by chlorination number (2–10)	
DDT and structurally related chemicals	
DDE (o-p and p-p′)	Insecticides; DDT metabolites
DDD (TDE: o-p and p-p′)	Insecticides; DDT metabolites
DDT (o-p and p-p′)	Insecticides; parents of DDD/DDE
Total DDT	Sum of parent and metabolites
Kelthane (Dicofol)	Acaracide; parent of DDE
Methoxychlor	Insecticide
Cyclodiene pesticides	
Technical chlordane	Insecticide; mix of constituents
Chlordane (*trans*- and *cis*-)	Insecticides; major constituents of technical chlordane
Nonachlor (*trans*- and *cis*-)	Insecticides; minor constituents of technical chlordane
Oxychlordane	Chlordane metabolite
Heptachlor	Insecticide; minor constituent of technical chlordane
Heptachlor epoxide	Metabolite of heptachlor
Endosulfan (I and II)	Insecticides; 7:3 mixture of stereoisomers
Endosulfan sulfonate	Metabolite of endosulfans
Aldrin	Insecticide
Dieldrin	Insecticide; Aldrin metabolite
Endrin	Insecticide
Hexachlorocyclohexane insecticides	
a-BHC	Constituent of BHC insecticide mix
Y-BHC (Lindane)	Insecticide; BHC constituent
Hexachlorocyclopentadiene pesticides	
Kepone	Acaricide, larvicide, fungicide, ant bait
Mirex	Insecticide (fire ant control)
Chlorinated camphenes	
Toxaphene	Insecticide (cotton)
Carboxylic acid derivatives	
2,4-D (2,4-DEP)	Weed herbicide (in cereals)
2,4,5-T	Wood plant herbicide
DCPA (dacthal)	Pre-emergence weed herbicide
Chlorinated benzenes and phenols	
HCB (hexachlorobenzene)	Fungicide
PCP (pentachlorophenol)	Wood preservative

Source: Mearns, A. J., Matta, M. B., Simecek-Beatty, D., Buckman, M. F., Shigenaka, G., and Wert, W. A., *PCB and Chlorinated Pesticide Contamination in U.S. Fish and Shellfish: A Historical Assessment Report,* NOAA Tech. Mem. NOS OMA 39, National Oceanic and Atmospheric Administration, Seattle, WA, 1988.

TABLE 3.2
Summary of Action Limits and Proposed Criteria (in ppm wet weight) for PCBs and Pesticides in Fish or Shellfish

			Predator Protection Levels	
Chemical	FDA Action Limit	NSSP Shellfish	NAS[c] Aquatic Wildlife	NAS[d] Marine Wildlife
Aldrin	0.30	0.20[a]	0.10[b]	[e]
Dieldrin	0.30	0.20[a]	0.10[b]	[e]
Endrin	0.30	0.20[a]	0.10[b]	[e]
DDT DDE (Sum) DDD (TDE)	5.0	1.5	1.0	0.05
Chlordane	0.3	0.03	0.10[b]	0.05
Heptachlor	0.3	0.20[a]	0.10[b]	0.05
Heptachlor epoxide		0.20[a]		
Lindane		0.20	0.10[b,e]	0.05
BHC (other than lindane)		0.20		
Methoxychlor		0.20		0.05
Endosulfan			0.10[b]	0.05
Mirex	0.1			0.05
Kepone	0.4 (crabs) 0.3 (fish, shellfish)			
Toxaphene	5.0		0.10[b]	0.05
HCB			0.1	0.05
2,4-D		0.50		
PCBs	2.0		0.5	0.5[f]
TCCD				

Note: FDA = U.S. Food and Drug Administration; NSSP = National Shellfish Sanitation Program; NAS = National Academy of Sciences.

[a] "Alert" level if combined value of the five pesticides exceeds 0.20 ppm wet weight; shellfish bed should be closed if combined values exceed 0.25 ppm wet weight.
[b] Singly or in combination with others listed.
[c] In whole fish.
[d] Homogenate of at least 25 fish of appropriate size and species.
[e] Sum of these four pesticides should not exceed 0.005 ppm wet weight.
[f] Add all Aroclors® for total.

Source: Mearns, A. J., Matta, M. B., Simececk-Beatty, D., Buchman, M. F., Shigenaka, G., and Wert, W. A., *PCB and Chlorinated Pesticide Contamination in U.S. Fish and Shellfish: A Historical Assessment Report,* NOAA Tech. Mem. NOS OMA 39, National Oceanic and Atmospheric Administration, Seattle, WA, 1988.

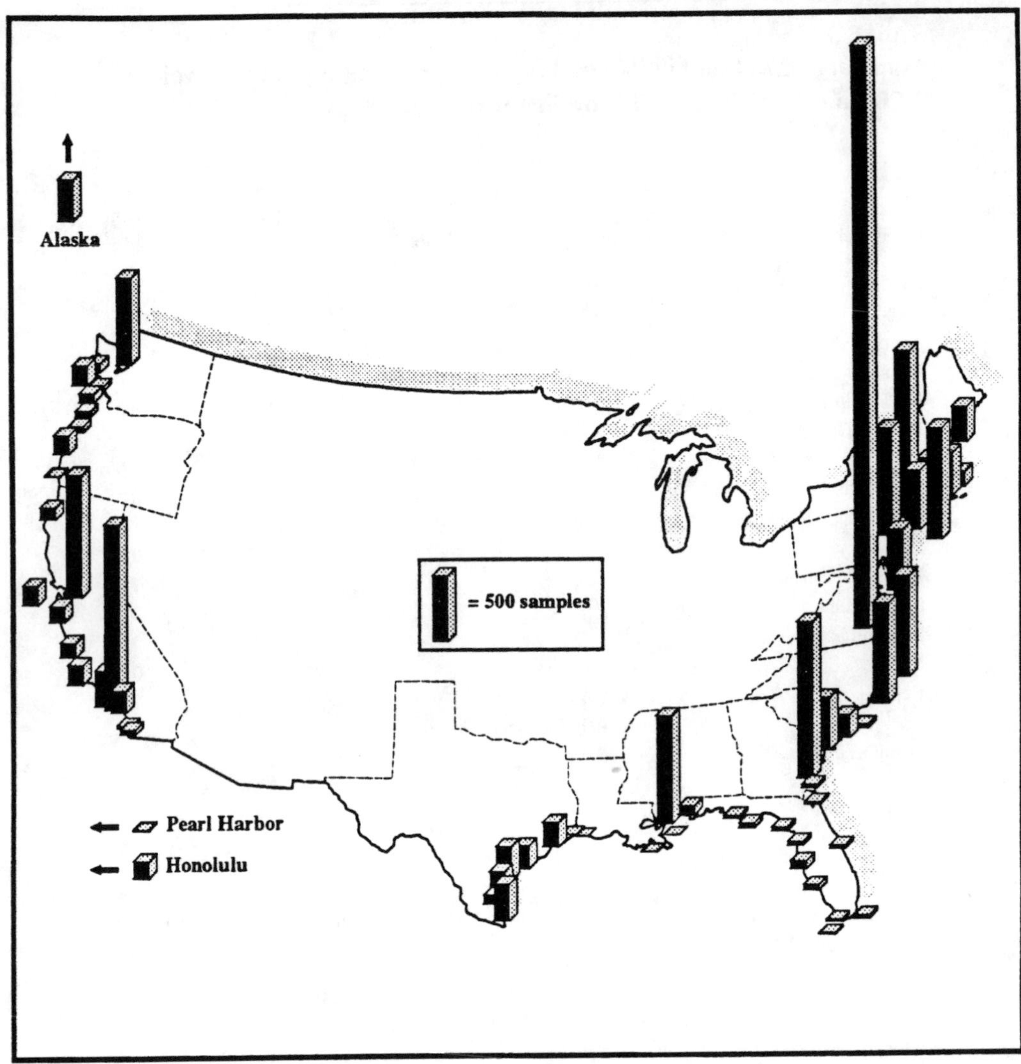

FIGURE 3.1. Geographic distribution of numbers of samples analyzed for PCBs and/or chlorinated pesticides in the U.S., 1940 to 1985. (From Mearns, A. J., Matta, M. B., Simececk-Beatty, D., Buchman, M. F., Shigenaka, G., and Wert, W. A., *PCB and Chlorinated Pesticide Contamination in U.S. Fish and Shellfish: A Historical Assessment Report,* NOAA Tech. Mem. NOS OMA 39, National Oceanic and Atmospheric Administration, Seattle, WA, 1988.)

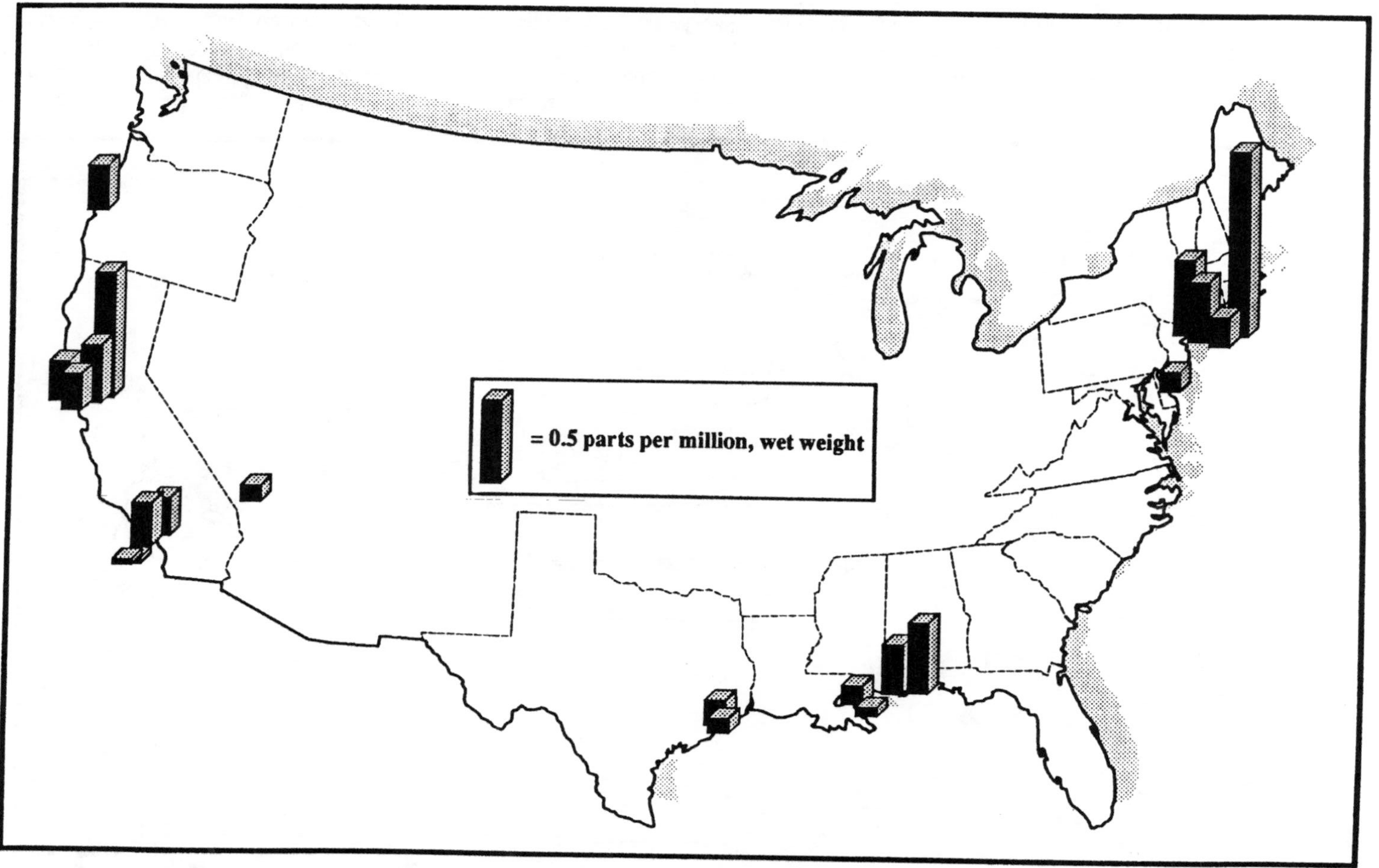

FIGURE 3.2. Total PCBs (Aroclor®) in muscle of coastal and ocean fish sampled in the U.S. from 1979 to 1981. Bar represents mean of all species means at each site. (From Mearns, A. J., Matta, M. B., Simececk-Beatty, D., Buchman, M. F., Shigenaka, G., and Wert, W. A., *PCB and Chlorinated Pesticide Contamination in U.S. Fish and Shellfish: A Historical Assessment Report,* NOAA Tech. Mem. NOS OMA 39, National Oceanic and Atmospheric Administration, Seattle, WA, 1988.)

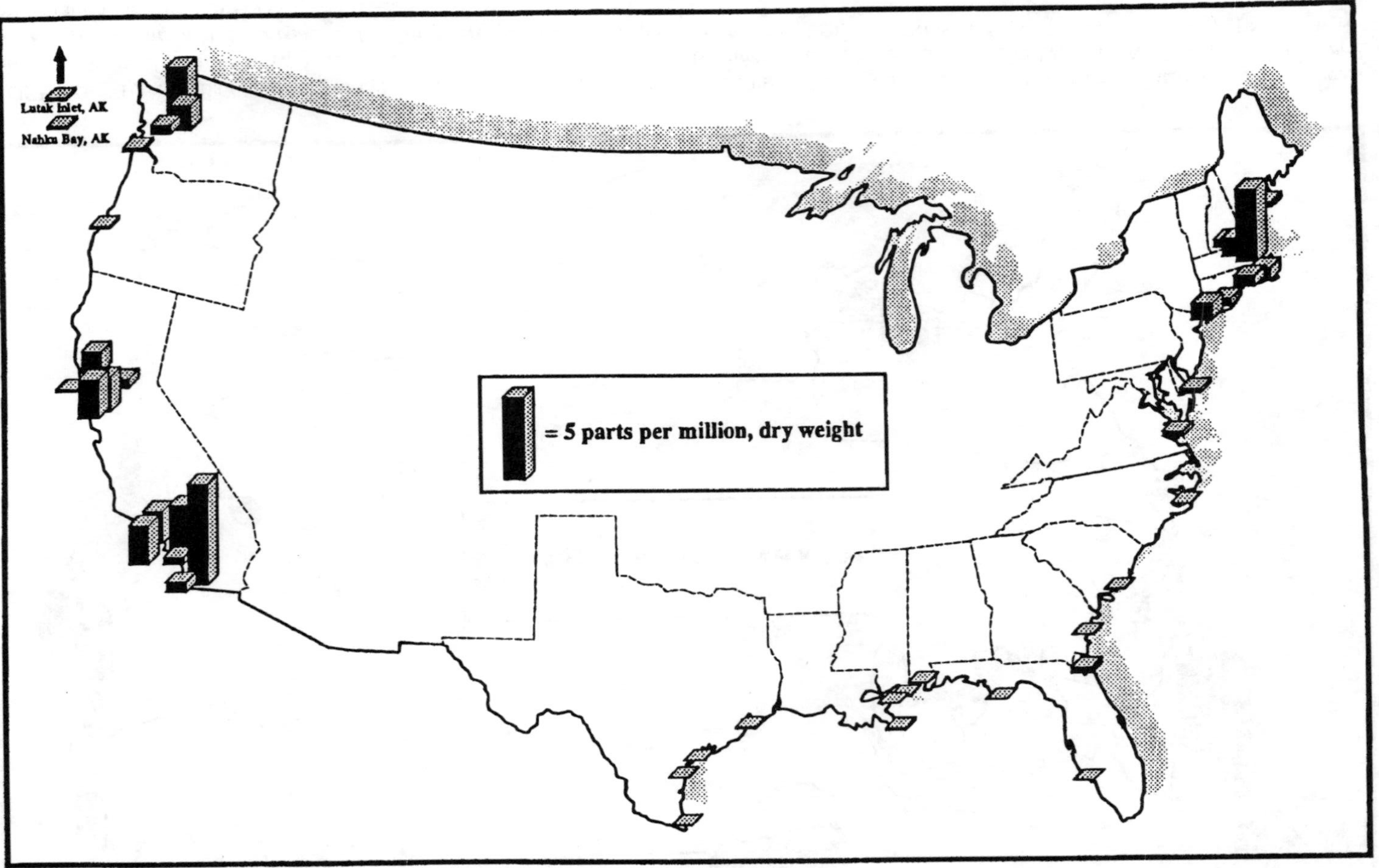

FIGURE 3.3. Total PCBs (chlorination number) in liver tissue of estuarine fish composites collected at 42 sites in the U.S. in 1984. Computed from original data for the NOAA National Status and Trends Benthic Surveillance Project. Bar represents mean of all species means at each site. (From Mearns, A. J., Matta, M. B., Simececk-Beatty, D., Buchman, M. F., Shigenaka, G., and Wert, W. A., *PCB and Chlorinated Pesticide Contamination in U.S. Fish and Shellfish: A Historical Assessment Report,* NOAA Tech. Mem. NOS OMA 39, National Oceanic and Atmospheric Administration, Seattle, WA, 1988.)

TABLE 3.3
Comparison of Average PCB Concentrations in Livers of Nearshore and Estuarine Fish from Nine Areas Occupied by National Surveys 1976–1977 and in 1984

Area	tPCB (ppm wet weight)	
	1976–1977	1984
Western Long Island Sound	0.62	0.81
Lower Chesapeake Bay, VA	0.62	0.28
Duwamish River/Elliott Bay, WA	26.7	4.23
Nisqually Reach, WA	0.31	0.49
Columbia River, OR	0.24	0.20
Coos Bay, OR	(less than) 0.20	0.15
Southern San Francisco Bay, CA	0.22	1.23–2.30
Palos Verdes/San Pedro Canyon, CA	18.63	2.27
Dana Point, CA	0.07	0.38
Median	0.31	0.49
Range:		
minimum	(less than) 0.20	0.15
maximum	26.7	4.23

Note: See original source for particular studies.

Source: Mearns, A. J., Matta, M. B., Simecek-Beatty, D., Buchman, M. F., Shigenaka, G., and Wert, W. A., *PCB and Chlorinated Pesticide Contamination in U.S. Fish and Shellfish: A Historical Assessment Report,* NOAA Tech. Mem. NOS OMA 39, National Oceanic and Atmospheric Administration, Seattle, WA, 1988.

TABLE 3.4
Median or Geometric Mean DDT Concentrations in Bivalves and Fish for Several National Survey Events

	tDDT or DDE (ppm wet weight) by Sampling Period			
Organism: Substrate	**1965–1972**	**1972–1975**	**1976–1977**	**1984–1986**
Bivalves	0.024[a]		0.01[b]/0.001[c]	0.003
Fish, whole juvenile		0.014[d]	ND	ND
Fish, muscle		0.110[e]	0.012[f]	
Fish, liver			0.220[f]	0.054[g]
Fish, whole F.W.	0.7–1.1	0.4–0.6	0.370	

Note: See original source for particular studies. ND = not determined; F.W. = fresh weight.

[a] Median of 8180-site means composited from 7839 samples.
[b] Median of 89-site means composited from 188 samples.
[c] Median of 80-site values or site means.
[d] Median of 144-site means composited from 1524 composites.
[e] Median of area or site means from samples.
[f] Median of 19-site means from samples.
[g] Median of 42-site medians from 126 composites.

Source: Mearns, A. J., Matta, M. B., Simecek-Beatty, D., Buchman, M. F., Shigenaka, G., and Wert, W. A., *PCB and Chlorinated Pesticide Contamination in U.S. Fish and Shellfish: A Historical Assessment Report,* NOAA Tech. Mem. NOS OMA 39, National Oceanic and Atmospheric Administration, Seattle, WA, 1988.

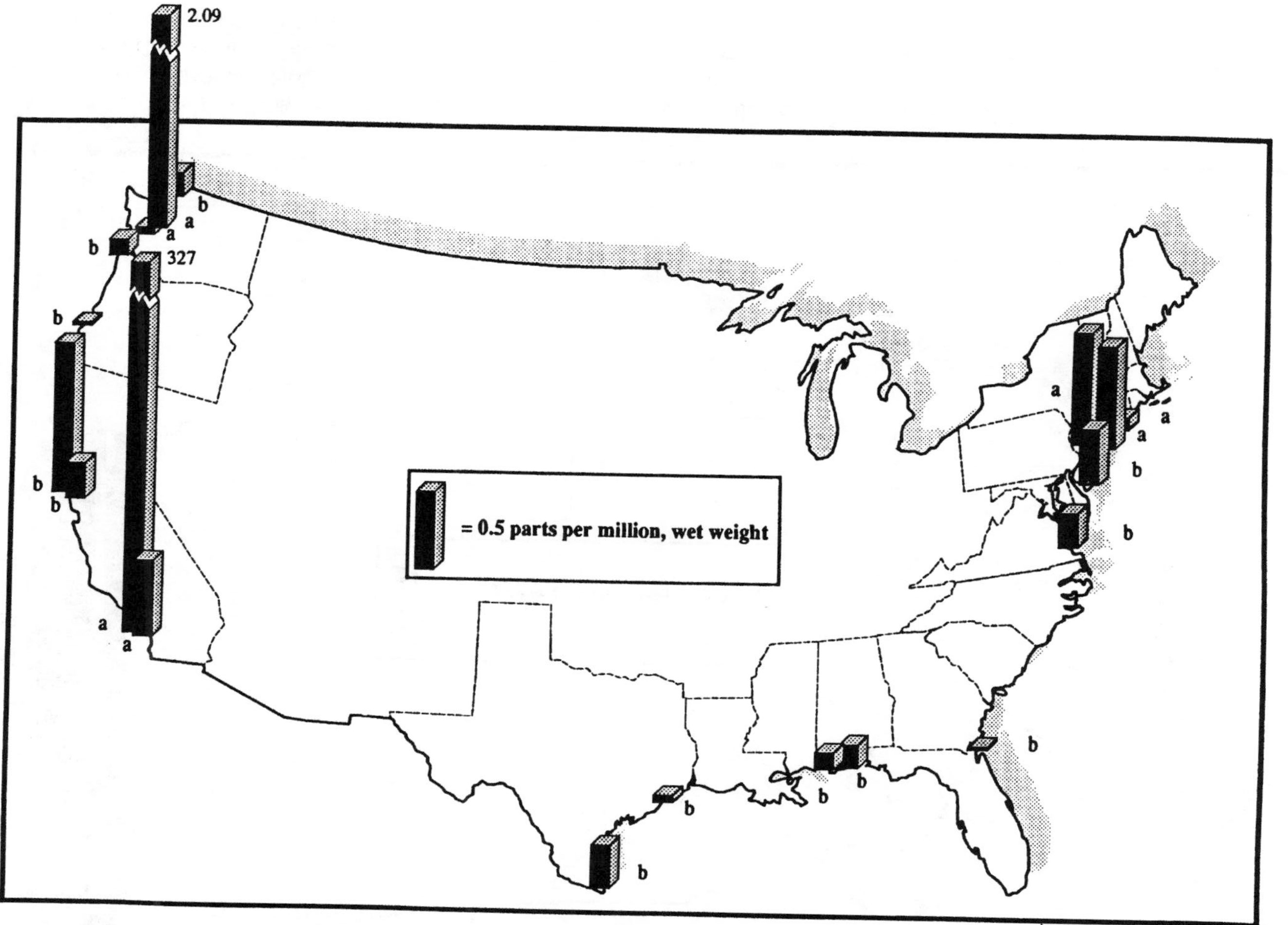

FIGURE 3.4. Total DDT in liver of coastal and estuarine fish from 19 sites sampled in the U.S. in 1976 and 1977. Bar represents mean of all individual or composite values for all species collected at a site. (From Mearns, A. J., Matta, M. B., Simececk-Beatty, D., Buchman, M. F., Shigenaka, G., and Wert, W. A., *PCB and Chlorinated Pesticide Contamination in U.S. Fish and Shellfish: A Historical Assessment Report*, NOAA Tech. Mem. NOS OMA 39, National Oceanic and Atmospheric Administration, Seattle, WA, 1988.)

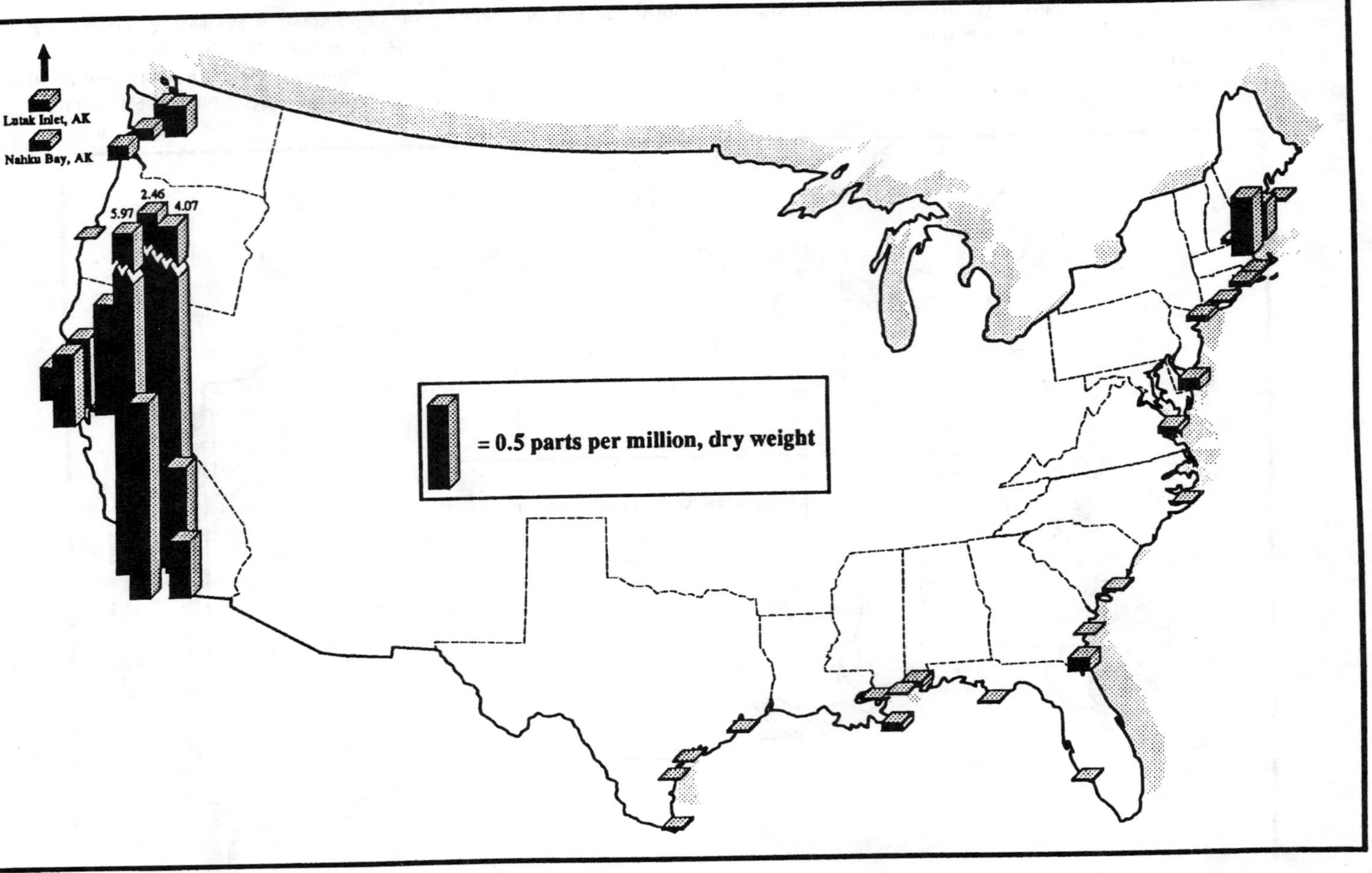

FIGURE 3.5. Total DDT in liver of estuarine fish composites collected at 42 sites in the U.S. in 1984. Computed from original data for the 1984 NOAA National Status and Trends Benthic Surveillance Project. (From Mearns, A. J., Matta, M. B., Simececk-Beatty, D., Buchman, M. F., Shigenaka, G., and Wert, W. A., *PCB and Chlorinated Pesticide Contamination in U.S. Fish and Shellfish: A Historical Assessment Report*, NOAA Tech. Mem. NOS OMA 39, National Oceanic and Atmospheric Administration, Seattle, WA, 1988.)

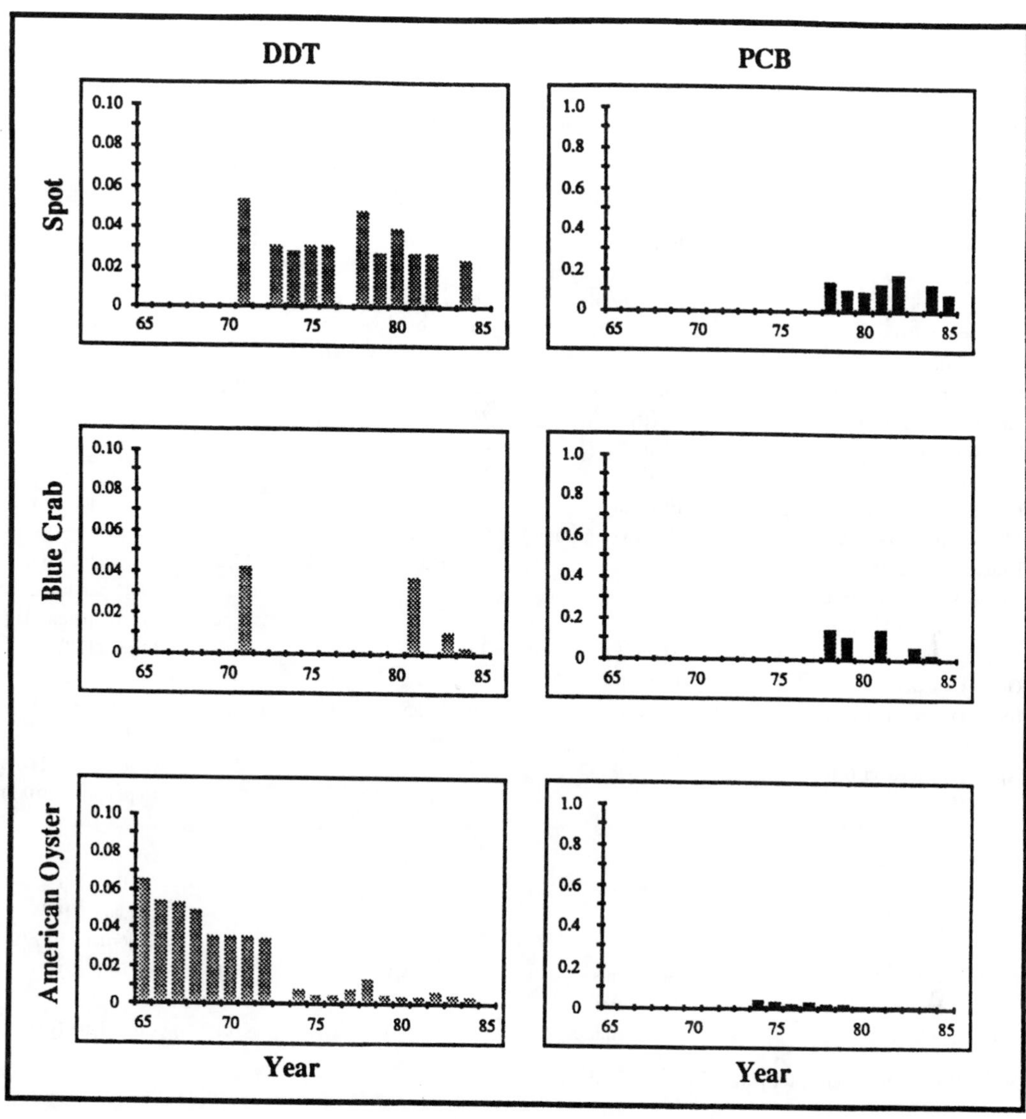

FIGURE 3.6. Comparison of annual variations of DDT and PCB contamination in all samples of spot, blue crabs, and oysters from the Chesapeake Bay, 1965 to 1985. Units in parts per million. (From Mearns, A. J., Matta, M. B., Simececk-Beatty, D., Buchman, M. F., Shigenaka, G., and Wert, W. A., *PCB and Chlorinated Pesticide Contamination in U.S. Fish and Shellfish: A Historical Assessment Report,* NOAA Tech. Mem. NOS OMA 39, National Oceanic and Atmospheric Administration, Seattle, WA, 1988.)

TABLE 3.5
Average Concentrations of Chlorinated Hydrocarbon Residues in Biota and Sediments from the Gulf of Mexico and Adjacent Estuaries and Bays

Location	Sample	N	PCBs (ng/g)	DDTs (ng/g)	Dieldrin (ng/g)	Others (ng/g)
Escambia Bay (FL)	Fish	7	11	—	—	—
Aransas Bay (TX)	Fish, crab, oysters	9	—	49	9	—
Gulf of Mexico	Fish, shrimp	46	66	62	—	—
Gulf of Mexico	Plankton	29	95	7	—	—
Gulf of Mexico	Fish	18	33	19	—	—
San Antonio Bay (TX)	Crabs	62	—	16	2.1	—
San Antonio Bay (TX)	Oysters	30	—	25	8.9	—
San Antonio Bay (TX)	Clams	43	—	20	2.9	—
San Antonio Bay (TX)	Shrimp	23	—	2	1.8	—
Mississippi Delta (LA)	Plankton	5	84	1	12	—
Mississippi Delta (LA)	Fish (mesopelagic)	27	25	10	—	—
Estuaries of Texas, Mississippi, Louisiana, Alabama, and Florida	Fish	24	203	18.2	15.2	Toxaphene (200) Ethyl parathion (75) Methyl parathion (47)
St. Louis and Mississippi Bays (MS)	Mollusks, fish	37	—	—	—	Mirex (139)
Gulf of Mexico	Fish, shrimp	27	25	10	—	—
Mexican coastal lagoons	Oysters	9	55	15	0.9	Chlordane (nd) Endrin (nd)
Apalachicola River (FL)[a]	Clams	9	21	29	2	Chlordane (21) Hep. epoxide (0.3)
Mississippi Delta (LA)	Sediments	—	18.7	4.2	—	—
Gulf of Mexico coast	Sediments	—	2	1.3	—	—
Nueces Estuarie (TX)	Sediments	—	4.7	1.5	—	HCB (0.11) Lindane (0.03) Chlordane (0.77)
Galveston Bay (TX)	Sediments	—	1.1	0.2	0.1	HCB (0.49)
Apalachicola Bay (FL)	Sediments	56	3.0	3.3	—	—
Apalachicola River (FL)[a]	Sediments	12	1.0	1.7	<0.1	Chlordane (<1)

Note: See original source for particular studies. nd = not detected.

[a] Median concentrations.

Source: Sericano, J. L., Wade, T. L., Atlas, E. A., and Brooks, J. M., *Environ. Sci. Technol.*, 24, 1541, 1990. With permission.

TABLE 3.6
Average Chlorinated Hydrocarbon Concentrations (and Ranges) in Oysters and Sediments from the Gulf of Mexico (National Programs)

Year	N	DDTs in Oysters (ng/g)	PCBs in Oysters (ng/g)	DDTs in Sediments (ng/g)	PCBs in Sediments (ng/g)
1965	58	257 ± 542[a] (<33–4730)	—	—	
1966	152	346–484[a] (<33–3890)	—	—	—
1967	155	292 ± 428[a] (<33–2790)	—	—	—
1968	136	450 ± 575[a] (<33–6490)	—	—	—
1969	142	284 ± 497[a] (<33–3530)	—	—	—
1970	144	234 ± 301[a] (<33–2200)	—	—	—
1971	140	217 ± 495[a] (<33–2840)	—	—	—
1972	60	162 ± 206[a] (<33–933)	—	—	—
1976	9	18.7 ± 12.4[b] (6.0–42.0)	71.2 ± 104 (<20–336)	—	—
1977	9	11.0 ± 9.10[b] (2.8–28.0)	83.0 ± 87.8 (16–297)	—	—
1984	11	—	—	1.83 ± 1.89 (0.1–7.0)	23.0 (nd–34.0)

Note: See original source for particular studies.

[a] Recalculated on dry weight basis.
[b] Calculated as DDE.

Source: Sericano, J. L., Wade, T. L., Atlas, E. A., and Brooks, J. M., *Environ. Sci. Technol.*, 24, 1541, 1990. With permission.

TABLE 3.7
Representative Bioconcentration Factors (BCF) of Some Pesticides

Compound	Species	Conc. in Water		Duration of Exposure	BCF
DDT	Fathead minnow	0.5–2	μg/l	56–112 d	100,000
	Golden shiner	0.3	μg/l	15 d	100,000
	Green fish	0.11–0.33	μg/l	15 d	17,500
	Pinfish and Atlantic croaker	0.1–1	μg/l	15 d	10,000 to 38,000
	Daphnia magna	8–50	μg/l	24 h	16,000 to 23,000
	Salmo gairdneri	133–176	μg/l	84 d	21,300 to 51,300
Dieldrin	Sculpin	0.017–8.6	μg/l	32 d	300 to 11,000
	Guppy	0.8–4.3	μg/l	18 d	49,307
	Bluegill (static test)	1	μg/l	48 h	2441
	Bluegill (flowthrough test)	1.5	μg/l	48 h	1727
Heptachlor	Sheepshead minnow	6.5–21	μg/l	96 h	7400 to 21,300
	Pinfish	0.32–32	μg/l	96 h	3800 to 7700
	Spot	1.2–3.7	μg/l	96 h	3000 to 13,800
Mirex	Juvenile pinfish	25–46	μg/l	3 d	3800
Kepone	Sheepshead minnow juveniles	41–780	μg/l	21 d	2600
	Adult male	41–780	μg/l	21 d	7600
	Adult female	41–780	μg/l	21 d	5700
HCH	Pinfish	18.4–31.3	μg/l	4 d	218
	Sheepshead minnow	41.9–108.3	μg/l	4 d	490
	Guppy	10–1400	μg/l	24 h	500
Technical HCH	Pinfish, edible part	1.4–36	μg/l	24 h	500
	Pinfish, offal	1.4–36	μg/l	28 d	175
Chlordane	Goldfish	3.4	μg/l	4 d	67 to 162
	Sheepshead minnow	15–51	μg/l	4 d	12,600 to 18,700
	Pinfish	5.4–15.2	μg/l	4 d	3000 to 7500
HCB	Killifish	160–380	μg/l	6.68 h	65 to 710
	Fathead minnow	—	—	—	16,200
	Green sunfish	—	—	—	21,900
	Rainbow trout	—	—	—	5500
Permethrin	Juvenile Atlantic salmon	1.4–12	μg/l	89 h	73
Cypermethrin	Juvenile Atlantic salmon	1.4–12	μg/l	96 h	3 to 7
Fenvalerate	Juvenile Atlantic salmon	1	μg/l	96 h	200
PCP	Trout	2	mg/l	24 h	100
	Killifish	57–120	μg/l	168 h	10 to 64
	Killifish	100–610	μg/l	240 h	8 to 50
Trichlorophenol	*Poecilia* female	610	μg/l	36 h	12,180
	Poecilia male	350	μg/l	36 h	7000
Leptophos	Bluegill sunfish	240	mg/l	10 d	750
Fenitrothion	Killifish	800	μg/l	10 d	53
	Coho salmon underyearling	560	μg/l	24 h	16
Diazinon	Topmouth gudgeon	10	μg/l	7 d	152
	Silver crucian carp	10	μg/l	7 d	37
	Carp	10	μg/l	7 d	65
	Guppy	10	μg/l	7 d	18
Fluridone	Fathead minnow	140	mg/l	10 weeks	64
2,4-D	Bluegill sunfish	3	mg/l	8 d	1
2,4,5-T	Bluegill sunfish	3	mg/l	8 d	1
MCPA	Trout	10–100	mg/l	10–28 d	1
Fosamine ammonium	Channel catfish	1.1	mg/l	4 weeks	1

TABLE 3.7 (continued)
Representative Bioconcentration Factors (BCF) of Some Pesticides

Compound	Species	Conc. in Water		Duration of Exposure	BCF
Hexamethyl phosphoramide	Sheepshead minnow	0.5	mg/l	28 d	1
Benthiocarb	Fathead minnow	28	μg/l	2.5 d	446 to 471
	Channel catfish	29	μg/l	3 d	120
	Bluegill	28	μg/l	5 d	91
	Longear sunfish	99	μg/l	1–5 d	280 to 300

Source: Murty, A. S., *Toxicity of Pesticides to Fish,* Vol. 1, CRC Press, Boca Raton, FL, 1986, 70. With permission.

TABLE 3.8
Relative Proportions of DDT Components in Fish and Other Organisms from North Atlantic

	Components of ΣDDT (%)		
Group	p,p′-DDE	p,p′-DDD	p,p′-DDT
Bivalves	51	36	13
Crustacea	84	12	4
Groundfish	45	14	41
Pelagic fish	45	17	38

Source: Murty, A. S., *Toxicity of Pesticides to Fish,* Vol. 1, CRC Press, Boca Raton, FL, 1986, 50. With permission.

TABLE 3.9
Recent Average Concentrations and Ranges of PCBs and Total DDT in the Mussel *Mytilus* and Closely Related Species Collected in Recent Years[a]

Region	PCBs (μg kg^{-1} dry weight)	ΣDDT (μg kg^{-1} dry weight)
Baltic Sea	(179–778)	(62–739)
North Sea	(106–362)	(15–143)
Irish Sea	(57–1070)	(92–590)
English Channel	(380–480)	(35–112)
Mediterranean		
Adriatic	(17.9–68)	(11.8–102)
Spain	(10.8–1264)	(60–288)
France	(83–1825)	
Northwest Atlantic		
U.S.	(10–6808)	(2.8–1109)
Canada	(11–258)	
Northeast Atlantic		
France	(96–1345)	
Southwest Atlantic		
Brazil	4.1 (2.5–10.5)	2.3 (1.23–5.61)
Northeast Pacific		
U.S.	(607–2052)	(5.4–1077)
Mexico		(63.5–193)
Northwest Pacific		
Hong Kong	(20–3136)	
Thailand	(11–241)	(179–235)
Japan	(3.1–364)	
	864 (17–2682)	
The Gulf	(1.7–110)[b]	(0.9–29.1)[b]
Northwest Indian Ocean Arabian Gulf		
Oman	4.9	3.6
	(0.1–69)[b]	(0.5–7.7)[b]
Pakistan	7.8	213
Southeast Pacific		
Chile		33.7
Southwest Pacific		
Australia, Port Phillip Bay	(14–879)	

[a] If weight data are not reported, a wet weight/dry weight factor of 5.6 has been applied.
[b] Rock oysters.

Source: Fowler, S. W., *Mar. Environ. Res.*, 29, 1, 1990. With permission.

TABLE 3.10
Range of Tissue Contaminant Concentrations Found in Mussel Species Around the World[a]

Country	ΣPCBs	ΣPAHs	Cd
Netherlands	25–430	30–1300	0.3–2.5
Finland		440–1200[b]	
U.S.-east	low–930	low–3600	1.0–6.0
U.S.-west	low–4250		
Norway	76–276	2200–15,400	1.5–2.7
France[c]	250–1350	4000–20,000	1.0–4.0
Hong Kong	23–4480[b]		

Note: Tissue concentrations of contaminants were measured in *M. edulis* unless stated otherwise.

[a] Concentration of ΣPCBs and ΣPAHs in μg/kg dry weight; Cd in mg/kg dry weight.
[b] Concentrations were transformed from wet weight to dry weight assuming a ratio of 8.
[c] Read: lower than–higher than.

Source: Eertman, R. H. M. and de Zwaan, A., in *Biomonitoring of Coastal Waters and Estuaries,* Kramer, K. J. M. (Ed.), CRC Press, Boca Raton, FL, 1994, 269. With permission.

TABLE 3.11
PCB Concentrations in Open Ocean Plankton

Location	Year	PCB Conc. (ng/g wet weight) N	Range	Mean
North Atlantic	1970		300–450	380
Northeast Atlantic	Before 1972	22	10–110	
North and South Atlantic	1970–1972	53		200
South Atlantic	1971	4	18–640	200
Western North Pacific	1981	1		1.1
Bering Sea	1982	3	1.0–1.6	1.3
Western South Pacific	1981	3	1.2–2.3	1.7
Antarctic (Ross Sea)	1972	1		<3
Antarctic (50–65°S, 124–126°E)	1981	3	0.2–1.0	0.5

Source: Tanabe, S. and Tatsukawa, R., in *PCBs and the Environment,* Vol. 1, Waid, J. S. (Ed.), CRC Press, Boca Raton, FL, 1986, 151. With permission.

TABLE 3.12
PCB Concentrations in Open Ocean Fishes

Location/Species	Year	N	Analyzed Portion	PCB Conc. (ng/g fresh weight) Range	Mean
North Atlantic					
Flying fish	1970–1971		Whole		50
Flying fish (*Cypselurus exsilens*)	1970–1971		Muscle		1.4
Flying fish (*Prognichthys rondeletii*)	1970–1971		Muscle		4
Trigger fish (*Canthidermis maculatus*)	1970–1971		Muscle		1.9
Western North Pacific (off Japan)					
Emmelichthys struhsakeri	1980	3	Whole	12–75	44
Tropidinis amoenus	1979	3	Whole	17–71	35
Priacanthus boops	1979	3	Whole	12–73	45
Bering Sea					
Herring (*Clupea pallasi*)	1973	1	Muscle		80
Walleye pollock (*Theragra chalcogramma*)	1973	2	Muscle	40–40	40
Flatfish (*Limanda aspera*)	1973–1974	7	Muscle	20–130	50
Chum salmon (*Oncorynchus keta*)	1980	1	Whole		5.0
Sockeye salmon (*Oncorynchus nerka*)	1980	1	Whole		15
Chum salmon (*Oncorynchus keta*)	1982	3	Whole	5.3–9.8	7.3
Walleye pollock (*Theragra chalcogramma*)	1982	4	Whole	9.8–13	11
Eastern South Pacific (off Chile)					
Cheilodactylus sp.	1978	5	Muscle	1.2–2.2	1.5
Merluccius australis	1977	5	Muscle	0.3–0.6	0.4
Brama sp.	1978	5	Muscle	0.6–1.6	0.9
Neophrynichthys marmoratus	1978	4	Muscle	0.2–0.3	0.2
Coelorhyncus fasciatus	1978	5	Muscle	0.1–0.2	0.2
Micromesistius australis	1977	3	Whole	0.11–0.33	0.19
Coelorhynchus fasciatus	1978	4	Whole	0.06–0.11	0.09
North Indian (Arabian Sea)					
Argyrops spinifer	1976	2	Whole	0.74–1.4	1.1
Thryssa vitrirostris	1976	3	Whole	0.93–2.0	1.6
South Indian (off Australia)					
Coryphaena hippurus	1980	5	Whole	0.02–0.05	0.03
Antarctic					
Two whole fish	1972				2.0
Pagothenia borchgrevinki	1981	21	Whole	0.18–0.77	0.31
Trematomus berenacchii	1981	5	Whole	0.12–0.24	0.17
Trematomus hansoni	1981	4	Whole	0.28–0.59	0.48
Trematomus newnesi	1981	2	Whole	0.08–0.33	0.21

Source: Tanabe, S. and Tatsukawa, R., in *PCBs and the Environment*, Vol. 1, Waid, J. S. (Ed.), CRC Press, Boca Raton, FL, 1986, 152. With permission.

TABLE 3.13
Organochlorine Concentrations (μg/g fat weight basis) in Cod-Liver Oil from the Southern Part of the Baltic Proper During 1971–1989

Organochlorine Compound	1971–1980		1981–1989	
	Range	Mean	Range	Mean
PCBs	9.1–18	14	7.0–14	10
DDTs	9.4–25	16	3.1–9.0	6.4
HCHs	0.29–1.1	0.60	0.50–0.63	0.56
HCB	0.29–0.46	0.37	0.17–0.34	0.26
Aldrin	0.010–0.057	0.027	0.015–0.022	0.018
Dieldrin	0.26–0.41	0.33	0.20–0.31	0.26
Chlordanes	0.17–0.35	0.28	0.19–0.42	0.31
Heptachlor	0.002–0.004	0.003	0.002–0.003	0.002
Heptachlor epoxide	0.004–0.018	0.011	0.013–0.023	0.018

Source: Kannan, K., Falandysz, J., Yamashita, N., Tanabe, S., and Tatasukawa, R., *Mar. Pollut. Bull.*, 24, 358, 1992. With permission.

TABLE 3.14
Total PCB Concentrations in Fish Livers from Nine Estuarine and Coastal Marine Systems

Area	tPCB (ppm wet weight)	
	1976–1977	1984
Western Long Island Sound	0.62	0.81
Lower Chesapeake Bay, VA	0.62	0.28
Duwamish River/Elliott Bay, WA	26.7	4.23
Nisqually Reach, WA	0.31	0.49
Columbia River, OR	0.24	0.20
Coos Bay, OR	<0.20	<0.15
Southern San Francisco Bay, CA	0.22	1.23–2.30
Palos Verdes/San Pedro Canyon, CA	18.63	2.27
Dana Point, CA	0.07	0.38
Median	0.31	0.49
Range		
minimum	<0.20	0.15
maximum	26.7	4.23

Source: Mearns, A. J., Matta, M. B., Simecek-Beatty, D., Buchman, M. F., Shigenaka, G., and Wert, W. A., *PCB and Chlorinated Pesticide Contamination in U.S. Fish and Shellfish: A Historical Assessment Report,* NOAA Tech. Mem. NOA OMA 39, National Oceanic and Atmospheric Administration, Seattle, WA, 1988.

TABLE 3.15
Concentrations of Chlorinated Organic Contaminants (ng/g wet weight) in Livers of Striped Bass (*Morone saxatilis*) in 1992

	Sample No.							Control No.				
Compound	**4**	**5**	**6**	**7**	**8**	**9**	**10**	**12**	**13**	**14**	**15**	**16**
DDE	103	131	297	172	137	104	62	3.4	<0.1	1	<0.1	1.7
DDD	35	47	86	40	39	32	34	1.5	0.2	<0.1	<0.1	<0.1
DDT	8.8	11	13	13	11	6.1	7.3	0.8	0.2	<0.1	<0.1	<0.1
Total DDT	147	189	396	225	187	142	103	5.7	0.4	1	<0.1	1.7
gamma-Chlordane	4.4	5.6	6.1	5.1	1.3	3.4	5.3	<0.1	<0.1	<0.1	<0.1	<0.1
alpha-Chlordane	5.7	9.4	7.4	7.1	2.1	5	6.2	<0.1	<0.1	<0.1	<0.1	<0.1
trans-Nonachlor	10	18	11	15	6.2	6.3	7.1	<0.1	<0.1	<0.1	<0.1	<0.1
cis-Nonachlor	4.8	9.2	4.5	7.4	3.1	3	1.4	<0.1	<0.1	<0.1	<0.1	<0.1
Total chlordane	25	42	29	35	13	18	20	<0.1	<0.1	<0.1	<0.1	<0.1
Tetrachloro PCB	13	1.9	7.4	9.3	2.9	5	24	<0.1	<0.1	<0.1	<0.1	<0.1
Pentachloro PCB	50	113	18	56	19	13	35	<0.1	<0.1	<0.1	<0.1	<0.1
Hexachloro PCB	77	155	11	84	22	17	8	<0.1	<0.1	<0.1	<0.1	<0.1
Total PCB	140	270	36	149	44	35	67	<0.1	<0.1	<0.1	<0.1	<0.1
DCPA	1.9	<0.1	8.7	5.8	0.7	3.1	2	<0.1	<0.1	<0.1	<0.1	<0.1

Source: Pereira, W. E., Hostettler, F. D., Cashman, J. R., and Nishioka, R. S., *Mar. Pollut. Bull.*, 28, 434, 1994. With permission.

TABLE 3.16
PCB Concentrations (ng/kg wet weight) in Muscle Tissue of Fish from the Mediterranean

Species	Location	Date	PCB[a]
Mullus barbatus	Ligurian Sea (Italy)	1977	188–1486
	North and Central Adriatic (Italy)	1976–1979	69–211
	Central Adriatic (Yugoslavia)	1975–1979	<1–497
	North Adriatic (Yugoslavia)	1973	3
	Sicily coast	1976–1977	17–373
	Augusta Bay, Sicily	1980	300
	North Aegean	1975–1979	703
	Saronikos Gulf	1975–1982	8–138[b]
		1975–1976	4–1100
	Eastern Turkey	~1980	2
	Israel	1975–1979	60
M. surmuletus	Israel	1975–1979	69
Mugil auratus	Eastern Turkey	~1980	10
M. cephalus	North and Central Adriatic (Italy)	1972	870
Sardina pilchardus	North and Central Adriatic (Italy)	1970	37–1060
		1972	620
	North Adriatic (Yugoslavia)	1973	2–19
	France	1975	51–309
	Spain	1970	540–6900
	August Bay, Sicily	1980	2300–6100
Engraulis encrasicholus	Ligurian Sea (Italy)	1977–1978	88–232
	Sicily coast	1976–1977	9–176
	North and Central Adriatic (Italy)	1970	510–960
		1972	370
		1976–1979	119–162
	North Adriatic (Yugoslavia)	1977	11–23
Thunnus thynnus	France	1975	95–407
		1977	6–89
	Sicily coast	1976–1977	9–44
	North and Central Adriatic (Italy)	1976–1979	344
	North Aegean	1975–1979	2613
Euthynnus alletteratus	Ligurian Sea (Italy)	1977–1978	191–1020
Sarda sarda	Ligurian Sea (Italy)	1977–1978	1133–14,020
Xiphias gladius	North Aegean	1975–1979	364
Boops boops	Israel	1975–1979	74
B. salpa	Libya	1982	2.5
Scorpaena scrofa	Libya	1982	3.9
Sprattus sprattus	North and Central Adriatic (Italy)	1970	620–920
Gobius paganellus		1972	100
Pleuronectes flesus		1972	250
Squalus acanthias		1972	720
Esox lucius		1972	350
Anguilla anguilla		1972	720
	Augusta Bay, Sicily	1980	2500
Saurida undosquamis	Israel	1975–1979	236
Merluccius merluccius		1975–1979	16
Trachurus mediterraneus		1975–1979	63
Upeneus moluccensis		1975–1979	151

TABLE 3.16 (continued)
PCB Concentrations (ng/kg wet weight) in Muscle Tissue of Fish from the Mediterranean

Species	Location	Date	PCB[a]
Pagellus acarne		1975–1979	151
P. erythrinus		1975–1979	188
Maena maena		1975–1979	91
Dentex macrophthalmus		1975–1979	195

[a] In some cases the ranges given are ranges of mean values. Single values represent either means of several determinations or a concentration based on a single measurement of a pooled sample containing tissues from several individuals.

[b] Computed from original data using 0.3 dry/wet weight ratio.

Source: Fowler, S. W., in *PCBs and the Environment,* Vol. 3, Waid, J. S. (Ed.), CRC Press, Boca Raton, FL, 1986, 224. With permission.

TABLE 3.17
PCBs (μg/g wet weight) in Eggs and Tissues of Mediterranean Seabirds

Species	Location	Tissue	No.	PCBs Range	PCBs $\bar{X}$
Cory's shear water (*Calonectris diomedea*)	Crete	Eggs	2	3.2–4.1	3.65
Manx shearwater (*Puffinus puffinus*)	Gibraltar	Muscle	2	0.99–1.1	1.02
Shag (*Phalacrocorax aristotelis*)	Gibraltar	Muscle	3	4.1–7.1	5.62
Lesser black-backed gull (*Larus fuscus*)	Gibraltar	Muscle	1		0.52
Herring gull (*Larus argentatus*)	Madeira	Muscle	1		0.4
	Gibraltar	Muscle	3	0.82–3.8	2.12
	Balearic Iss.	Egg	1		11.0
	Rhone delta	Eggs	6	16–160	52
	Crete	Egg	1		5.32
	Cyprus	Eggs	2	1.2–4.2	2.7
	Tuscan Iss.	Eggs	25		~8.4
	Po delta	Eggs	4		4.0[a]
Audouin's gull (*Larus augouini*)	Morocco	Eggs	3	3.1–4.2	3.82
	Balearic Iss.	Eggs	4	12.8–20.6	16.8
Black-headed gull (*Larus ridibundus*)	Rhone delta	Eggs	4	1.6–7.8	5.45
	Po delta	Muscle	8	0.61–3.6	1.65
	N. Adriatic	Eggs	17		3.2[a]
Little gull (*Larus minutus*)	Malta	Liver	1		3.6
Common tern (*Sterna hirundo*)	Po delta	Muscle	2	0.43–8.0	4.21
	Po delta	Eggs	22		3.2[a]
	N. Adriatic	Eggs	13		4.1[a]
Little tern (*Sterna albifrons*)	Sardinia	Eggs	12		~4.0
	N. Adriatic	Eggs	16		2.6[a]
	C. Sardinia	Eggs	6		2.8[a]
	S. Sardinia	Eggs	6		4.9[a]
Black tern (*Chlidonias nigra*)	Po delta	Muscle	1		<0.01
Coot (*Fulica atra*)	Po delta	Muscle	1		0.21
	Malta	Liver	1		0.1
	Po delta	Eggs	3		0.19[a]
Gull-billed tern (*Gelochelidon nilotica*)	S. Sardinia	Eggs	7		1.1[a]
	N. Adriatic	Eggs	15		1.7[a]
Slender-billed gull (*Larus genei*)	S. Sardinia	Eggs	33		3.5[a]
Black-winged stilt (*Himantopus himantopus*)	N. Adriatic	Eggs	5		0.14[a]
Avocet (*Recurvirostra avocetta*)	N. Adriatic	Eggs	5		0.12[a]
Little egret (*Egretta garzetta*)	Po delta	Eggs	9		2.3[a]
Night heron (*Nycticorax nycticorax*)	Po delta	Eggs	8		2.3[a]

[a] Computed from dry weight values using wet/dry ratio of 3 for terns, 4.3 for gulls, and 3.5 for others.

Source: Fowler, S. W., in *PCBs and the Environment,* Vol. 3, Waid, J. S. (Ed.), CRC Press, Boca Raton, FL, 1986, 228. With permission.

TABLE 3.18
Accumulation of PCBs in Birds from Field Experiments

Avian Species	Residue Level in Prey	Residue Level (ppm, wet weight) Whole Body	Muscle	Liver	Brain	Eggs
Great-crested grebe (*Podiceps cristatus*)	0.17 (0.16–0.18)	—	—	11	—	—
	0.12 (0.04–0.35)	32	—	—	—	—
	0.8–1.3	—	—	—	—	13
Brown pelican (*Pelecanus occidentalis*)	0.19 (0.17–0.23)	—	—	—	—	2.2
White pelican	0.05–0.11	2.3	3.1	4.5	—	1.7
Double-crested cormorant (*Phalacrocorax auritus*)		3.6	2.3	2.0	—	5.7
White-crested cormorant (*Phalacrocorax carbo*)	1.6	6.2	—	—	4.4	2.9
White-tailed sea eagle (*Haliaetus albicilla*)	0.03–0.18	—	150–240	130	29–70	—
Common guillemot (*Uria aalge*)	0.27 (0.01–1.0)	—	—	—	—	7.9–21
	0.01–2.0	3.4	—	0.4	—	—
Herring gull (*Larus argentatus*)	2.2	—	—	—	—	124–157

Source: Peakall, D. B., in *PCBs and the Environment,* Vol. 2, Waid, J. S. (Ed.), CRC Press, Boca Raton, FL, 1986, 34. With permission.

TABLE 3.19
PCB Levels in Herring Gulls and Their Eggs

Location	Tissue	Year of Collection	Residual Level (ppm, wet weight)
Norway (west coast)	Egg	1972	3.1–12.6
Baltic (Gdansk Bay)	Muscle	1975–1976	23–150
Finland	Muscle	1972–1974	0.68–38
East Scotland	Muscle	1971–1975	0.2–1.2
Camague, France	Egg	1972	16–160
New Brunswick, Canada	Egg	1969–1972	3.1–8.2
Lake Ontario	Egg	1974–1975	74–261
Maine	Egg	1977	(0–32.0)
			7.76 (30)
Virginia	Egg	1977	(0.13–16.70)
			9.06 (28)
East Frisian Island, Germany	Egg	1975	26.5
		1971	5.5
Denmark (North Sea)	Egg	1971	2.1 (1.3–2.6)
Baltic		1972	92 (21–199)

Source: Adapted from Peakall, D. B., in *PCBs and the Environment,* Vol. 2, Waid, J. S. (Ed.), CRC Press, Boca Raton, FL, 1986, 43. With permission.

TABLE 3.20
Contamination of Various Species of Pelagic Seabirds by DDT, Its Metabolites, and PCBs

Species	Place of Capture (Place of Reproduction)	Tissues Analyzed	DDT and Metabolites (ppm)	PCBs (ppm)
Fulmarus glacialis	California (Alaska)	Whole bird	7.1	2.3
Puffinus creatopus	Mexico (Chile)	Whole bird	3.0	0.4
P. griseus	California (New Zealand)	Fats	11.3	1.1
		Fats	40.9	52.6
P. gravis	New Brunswick (Southern Atlantic)	Fats	70.9	104.3
Pterodroma cahow	Bermudas (same)	Whole bird	6.4	—
Oceanodroma leuchorhoea (Leach's Petrel)	California (same)	Fats *ex ovo*	953	351
Oceanites oceanicus (Wilson petrel)	New Brunswick (Antarctica)	Fats	199	697

Source: Ramade, F., in *Aquatic Ecotoxicology: Fundamental Concepts and Methodologies,* Boudou, A. and Ribeyre, F. (Eds.), CRC Press, Boca Raton, FL, 1989, 156. With permission.

TABLE 3.21
Levels of PCBs (mg/kg) in Extractable Fat from Seal Blubber

Species	N	M	Range
Grey seal (*Halichoerus grypus*)			
Males and females			
Killed	60	110	47–330
Found dead	54	820	84–4500
Ringed seal (*Phoca hispida*)			
Males			
Killed	24	100	35–270
Found dead	12	240	38–640
Females			
Killed	61	93	27–190
Found dead	6	270	125–710
Common seal (*Phoca vitulina*)			
Pups			
Killed	30	43	18–100
Found dead	43	100	12–500

Note: The material is divided into actively killed animals and those found dead; N = number of specimens analyzed; M = mean value.

Source: Olsson, M., PCBs in the Baltic environment, in *PCBs and the Environment,* Vol. 3, Waid, J. S. (Ed.), CRC Press, Boca Raton, FL, 1986, 181. With permission.

TABLE 3.22
A Comparison of PCB and Total DDT Concentrations (Mean Values with Ranges) in Pinniped Blubber

Latitude		N	PCBs (μg g^{-1} wet weight)	ΣDDT (μg g^{-1} wet weight)	Sample
Arctic	76°N	28	$\bar{x}$ = 0.23 (0.06–1.10)	$\bar{x}$ = 0.06 (0.01–0.40)	Atlantic walrus
Temperate N	53°N	7	$\bar{x}$ = 189.42 (22.00–576.00)	$\bar{x}$ = 10.85 (0.51–25.40)	Harbor seal
	48°N	22	$\bar{x}$ = 4.00 (1.00–11.00)	$\bar{x}$ = 1.70 (0.60–3.10)	Harp seal
	48°N	8	$\bar{x}$ = 15.70 (±5.80)	$\bar{x}$ = 3.50 (±1.00)	Gray seal
	33°N	4	$\bar{x}$ = 17.10 (12.00–25.00)	$\bar{x}$ = 103.20 (51.00–203.00)	California sea lion
Temperate S	38°S	11	$\bar{x}$ = 0.69 (0.05–3.87)	$\bar{x}$ = 4.03 (0.03–12.05)	Australian fur seal
	69°S	1	0.04	0.17	Weddell seal
Antarctic	70°S	20	$\bar{x}$ = 0.09 (0.01–0.76)	$\bar{x}$ = 0.07 (nd–0.15)[a]	Ross seal

[a] nd = not detected.

Source: Smillie, R. H. and Waid, J. S., Polychlorinated biphenyls and organochlorine pesticides in the Australian fur seal, *Arctocephalus pusillus doriferus, Bull. Environ. Contam. Toxicol.*, 39, 358, 1987. With permission.

TABLE 3.23
PCB Concentrations in the Blubber of Male Pinnipeds and Cetaceans

Location	Species	Year	N	PCB Conc. (μg/g fresh weight) Range	Mean
Arctic					
Canada	Ringed seal	1972	4	0.05–1.5	0.58
	Ringed seal	1972	15	1–6	4.1
North Greenland	Atlantic walrus	1975–1977	8	0.16–1.1	0.36
North Atlantic					
Newfoundland	Harp seal	1970	1		26
Gulf of St. Lawrence	Harp seal	1971	7	6–22	13
Nova Scotia	Atlantic white-sided dolphin	1972	1		37
Rhode Island	Striped dolphin	1972	1		39
Caribbean	Long-snouted dolphin	1972	1		5.0
South Atlantic					
Uruguay	Franciscana dolphin	1974	5	3.2–18	6.8
North Pacific					
Bering Sea	Dall's porpoise	1980	4	3.5–6.8	5.2
Japan	Striped dolphin	1978	3	22–23	22
	Finless porpoise	1968–1975	2	64–96	80
	Pilot whale		1		2.0
California	Common dolphin	1974–1976	10	80–300	120
	Pilot whale		1		14
Hawaii	Rough-toothed dolphin	1976	3	7.0–14	9.4
Eastern tropical	Striped dolphin	1973–1976	3	2.6–7.6	5.7
South Pacific					
Eastern tropical	Fraser's dolphin	1973–1976	1		5.2
	Striped dolphin		1		5.0
New Zealand	Dusky dolphin	1980	1		1.4
Antarctic					
Syowa Station	Weddell seal	1981	1		0.038

Source: Tanabe, S. and Tatsukawa, R., in *PCBs and the Environment,* Vol. 1, Waid, J. S. (Ed.), CRC Press, Boca Raton, FL, 1986, 153. With permission.

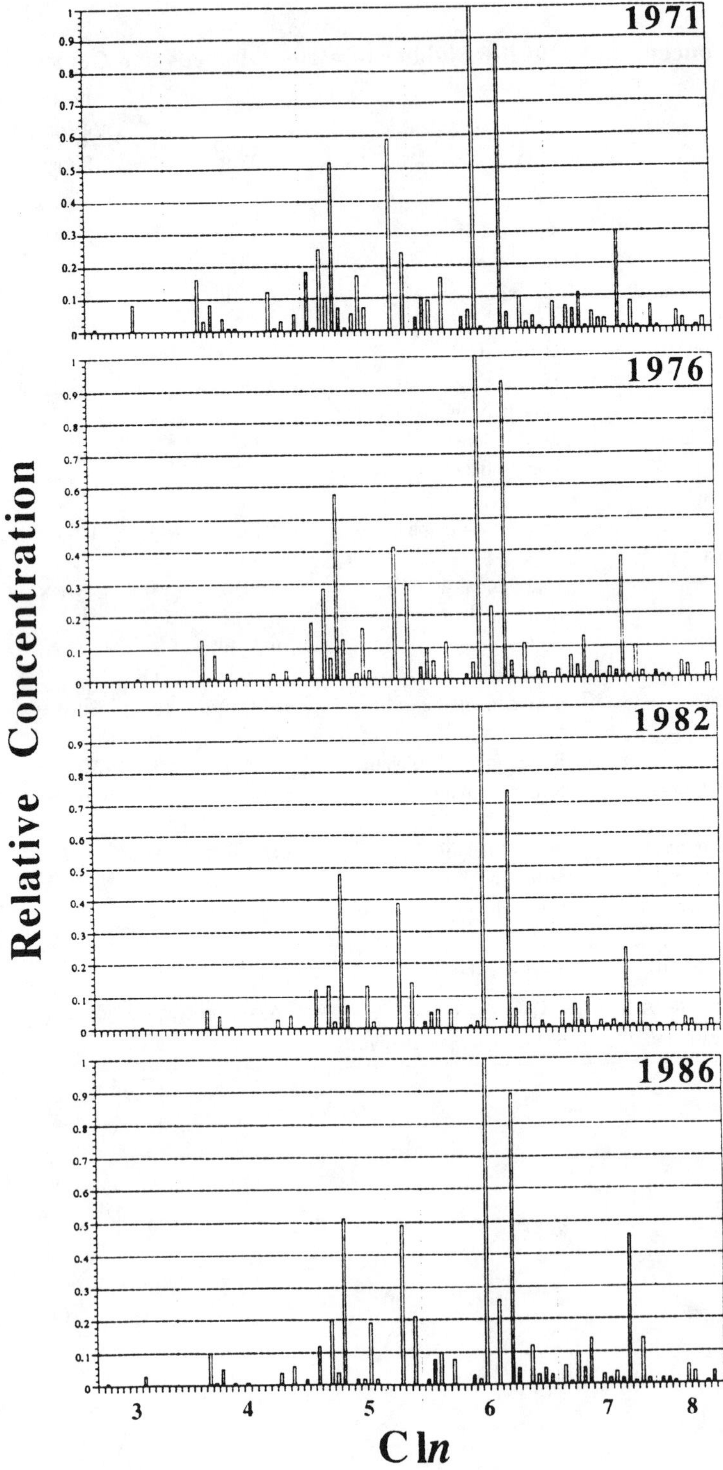

FIGURE 3.7. Temporal variation of PCB isomers and congener compositions in northern fur seals collected in 1971, 1976, 1982, and 1986. Vertical bars represent concentrations of individual PCB isomers relative to the most abundant isomer (IUPAC No. 153) present in each sample, the latter being assigned a relative concentration of 1.0. (From Tanabe, S., Sung, J.-K., Choi, D.-Y., Baba, N., Kiyota, M., Yoshida, K., and Tatsukawa, R., *Environ. Pollut.*, 85, 305, 1994. With permission.)

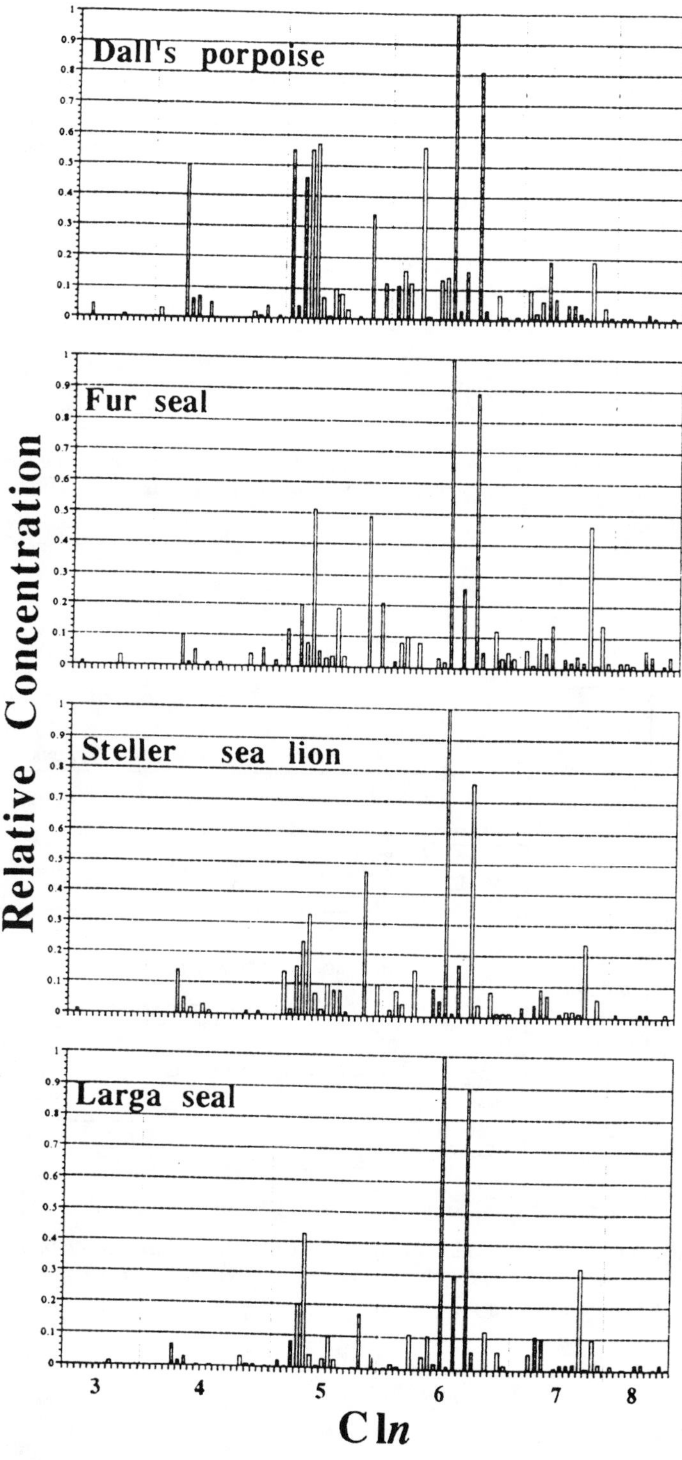

FIGURE 3.8. Comparison of PCB isomer and congener compositions in different species of marine mammals. Vertical bars represent concentrations of individual PCB isomers relative to the most abundant isomer (IUPAC No. 153) present in each sample, the latter being assigned a relative concentration of 1.0. (From Tanabe, S., Sung, J.-K., Choi, D.-Y., Baba, N., Kiyota, M., Yoshida, K., and Tatsukawa, R., *Environ. Pollut.*, 85, 305, 1994. With permission.)

TABLE 3.24
Bioaccumulation and Toxicity of Toxaphene in Estuarine Organisms

Species	96-h LC_{50} (mg/l)	96-h BCF[a] (tissue/water)
Pink shrimp (*Penaeus duorarum*)	1.4	3100–20,600
Grass shrimp (*Palaemonetes pugio*)	4.4	3100–20,600
Sheepshead minnow (*Cyprinodon variegatus*)	1.1	400–1200
Pinfish (*Lagodon rhomboides*)	0.5	400–1200
Longnose killifish (*Fundulus similis*)	0.1–1.4[b]	4200–60,000 (whole body)

[a] BCF = bioconcentration factor.
[b] 28 days.

Source: Sergeant, D. B. and Onuska, F. I., in *Analysis of Trace Organics in the Aquatic Environment,* Afghan, B. K. and Chau, A. S. Y. (Eds.), CRC Press, Boca Raton, FL, 1989, 75. With permission.

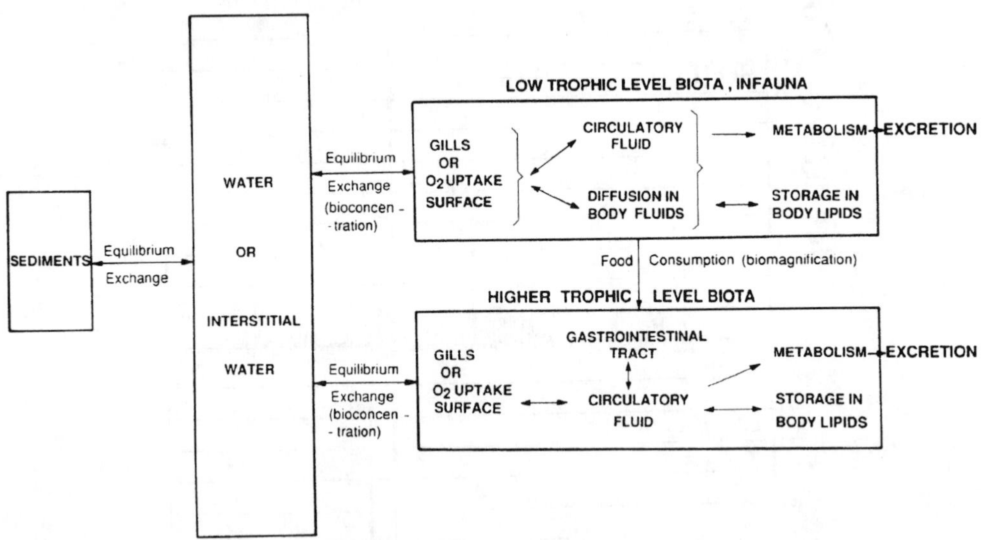

FIGURE 3.9. Diagrammatic representation of routes of uptake and clearance of lipophilic chemicals by aquatic biota. (From Connell, D. W. (Ed.), *Bioaccumulation of Xenobiotic Compounds,* CRC Press, Boca Raton, FL, 1989, 100. With permission.)

TABLE 3.25
Summaries of Summed Concentrations of Three Chlordane Compounds (Σ *cis*- and *trans*-Chlordane, *cis*- and *trans*-Nonachlor, Oxychlordane, and Heptachlor) in Fish and Shellfish Collected in the Coastal and Estuarine Environment of the U.S. 1986–1987

Matrix	N	Σ Chlordane Range	Minimum Species	Minimum Location	Maximum Species	Maximum Location
Bottom-feeding fish (whole body)	40	6.91[b]–409	Red drum	So. Brunswick R., GA	Stingray	Colorado Lagoon, CA
Predatory fish (fillet)	8	6.94[b]–30.9	Spotted seatrout	North River, GA	Atlantic croaker	Houston Ship Channel, TX
Predatory fish (whole body)	8	7.50[b]–42.5	Sheepshead	North River, GA	Bluefish	Manhasset Bay, NY
Shellfish	6	7.50[b]–11.9	Softshell clam	Grays Harbor, WA	Shellfish (unidentified)	Lake Pontchartrain, LA

[a] Values in ng/g (ppb) wet weight.

[b] All five chlordane compound concentrations below quantitation limits; value shown reflects one half specified limits.

Source: Shigenaka, G., *Chlordane in the Marine Environment of the United States: Review and Results from the National Status and Trends Program,* NOAA Tech. Mem. NOS OMA 55, National Oceanic and Atmospheric Administration, Seattle, WA, 1991.

TABLE 3.26
Ranges of Concentrations of Chlordane Compounds Measured in Tissues of Marine Mammals

Species	Tissue	Reporting Basis	Target Compounds	Range (ppb)
Harbor seal (*Phoca vitulina*)	Blubber	Wet	(not defined)	ND
Humpback whale (*Megaptera novaeangliae*)	Blubber	Wet	α-c	ND–200
Sperm whale (*Physeter catodon*)				ND–300
Dense-beaked whale (*Mesoplodon densirostris*)				100–300
Atlantic pilot whale (*Globicephala melaena*)				600–1400
Atlantic white-sided dolphin (*Lagenorhynchus acutus*)				ND
Saddleback dolphin (*Delphinus delphis*)				1200
Striped dolphin (*Stenella caeruleoalba*)				1400–2700
Pacific white-sided dolphin (*Lagenorhynchus obliquidens*)				5000
Pacific pilot whale (*Globicephala scammoni*)				ND
Grey seal (*Halichoerus gryphus*)	Blubber	Lipid	α, γ, oxy-c, *t*-n	10,000
Harbor seal (*Phoca vitulina*)	Liver	Wet	(chlordane)	<70–<100
			(heptachlor)	<20
	Brain		(chlordane)	<60
			(heptachlor)	<20
Harbor seal (*Phoca vitulina*)	Blubber	Lipid	*c*-, *t*-, oxy-c, *t*-n	500–5700
	Liver			2300
	Brain			940
White-beaked dolphin (*Lagenorhynchus albirostris*)	Blubber			5400–11,300

Note: See original source for particular studies. ND = not determined.

Source: Shigenaka, G., *Chlordane in the Marine Environment of the United States: Review and Results from the National Status and Trends Program,* NOAA Tech. Mem. NOS OMA 55, National Oceanic and Atmospheric Administration, Seattle, WA, 1991.

6 Heavy Metals

INTRODUCTION

Among the most intensely studied contaminants in estuarine and marine environments are heavy metals, a group of elements with atomic weights ranging from 63.546 to 200.590 and characterized by similar electronic distribution in the external shell (e.g., copper, zinc, cadmium, zinc).[1] Because of their toxicity to estuarine and marine organisms above a threshold availability and their persistence in the environment, heavy metals pose potentially hazardous conditions. They may be subdivided into two categories: (1) transitional metals (e.g., cobalt, copper, iron, manganese) which are essential to metabolism at low concentrations but may be toxic at high concentrations, and (2) metalloids (e.g., arsenic, cadmium, lead, mercury, selenium, tin) which generally are not required for metabolic function but are toxic at low concentrations. An approximate order of decreasing toxicity of common heavy metals in aquatic organisms is as follows: mercury, cadmium, copper, zinc, nickel, lead, chromium, aluminum, and cobalt.[2] At elevated levels, heavy metals act as enzyme inhibitors in organisms; however, the toxicity of a given metal can vary significantly from one species to another. The accumulation of heavy metals by marine organisms is a function of many factors, such as temperature, salinity, diet, spawning, and the ability of the organism to regulate metal concentrations in the body.[3]

SOURCES

Heavy metals enter the sea via several major routes, most notably riverine influx, atmospheric deposition, and anthropogenic activities. A large quantity of metals in river water originates from the weathering of rocks and leaching of soils and, hence, depends in part on the occurrence of metals and ore-bearing deposits in the drainage area. However, the metal burden is often augmented by anthropogenic sources where rivers flow through urban or industrialized centers.

The atmosphere represents the principal route of entry of certain metals to the sea. For example, lead derived largely from alkyl leaded gasolines is delivered to the ocean surface by processes in the atmosphere.[4] Mercury released by volcanic activity and aluminum derived from wind-blown dust of weathering shales and other rocks account for large atmospheric inputs of these elements to the sea. Occurring as gases or aerosols in the atmosphere, heavy metals may reach the sea surface by dry and wet deposition, as well as gaseous exchange.

The coastal supply of heavy metals results mainly from atmospheric and riverine delivery systems. In the deep sea, volcanic eruptions and hydrothermal vent emissions at ridge crest spreading centers release considerable amounts of heavy metals to oceanic waters. Aside from natural sources, a wide range of human activities mobilizes heavy metals. Common anthropogenic sources of these elements are mining, smelting, refining, electroplating, and electric generating stations. Automobile emissions, the disposal of sewage sludges, dredged spoils, ash, and the use of antifouling paints on marine vessels and structures also yield appreciable concentrations of heavy metals in the sea. The largest fraction of these contaminants is trapped in estuaries, embayments, and inner shelf areas.

CHEMICAL FORMS

Trace metals occur in the following chemical forms during transit to the sea: (1) in solution as inorganic ion and both inorganic and organic complexes, (2) adsorbed onto surfaces, (3) in solid organic particles, (4) in coatings on detrital particles after coprecipitation with and sorption onto mainly iron and manganese oxides, (5) in lattice positions of detrital crystalline material, and (6) precipitated as pure phases, possibly on detrital particles.[5] Although heavy metals exist in dissolved, colloidal, and particulate phases in seawater, the concentration of dissolved forms is low. Being particle reactive, heavy metals rapidly sorb onto suspended particulate matter as they enter riverine or estuarine waters.[6] Interaction with suspended particulates in estuaries is pronounced in the reactive turbidity maximum.[7] Ultimately, most of the metals are removed to bottom sediments in estuaries which serve as a repository for these elements. For example, the trapping efficiency of suspended material in Chesapeake Bay is 98 ± 2%; therefore, only a minor fraction of metals escapes to coastal waters from this system.[8] In general, more than 90% of particulate matter transported by rivers settles in estuarine and coastal marine areas.[9]

DISTRIBUTION

Both physical and geochemical processes control the distribution and transport of heavy metals in estuaries.[10] Sediment-sorbed heavy metals often are resuspended from the bottom during storms and other turbulent periods, with estuarine circulation patterns returning the particles upestuary. Trace metal removal during estuarine mixing is accelerated through precipitation or interactions with particle surfaces or flocculating colloids, coprecipitation with organic, iron and/or manganese hydrous oxides, increased affinity of the metals for anions in seawater, and uptake by organisms.[11] While the uptake of dissolved trace metals onto solid phases already present in estuarine water is important to the overall concentration of the elements in solution, so is the release of material into solution from particulate phases by dissolution, desorption, and autolytic respiratory biological processes.[12]

Heavy metals clearly undergo varying degrees of recycling in estuaries. This recycling involves removal of elements from solution by coprecipitation with particulate matter or adsorption onto particle surfaces and their subsequent deposition. Some heavy metals associated with particulate phases are desorbed from suspended matter under suitable conditions and remobilized in the estuary. As particulate carrier phases experience oxidation and/or dissolution, trace metals are regenerated, either in the water column or from surficial bottom sediments.[13]

Organisms also influence the cycling of heavy metals. Phytoplankton utilize heavy metals in metal-requiring and metal-activated enzyme systems which catalyze major steps in glycolysis, the tricarboxylic acid cycle, photosynthesis, and metabolism.[13] Zooplankton and filter-feeding benthic fauna facilitate the removal of heavy metals to bottom sediments by consolidating them into fecal material that settles to the sea floor. Heavy metals bound to dead plants and animals, fecal pellets, and crustacean molts generally account for more than 90% of the vertical transport of the metals in the water column.[14] Bioturbating activities of benthic fauna redistribute these metal-bound materials and may promote the release of other metals to interstitial and overlying waters. Rooted vascular plants (e.g., seagrasses) assimilate metals which otherwise would be lost indefinitely to bottom sediments, later releasing them to various environmental compartments.

Because of their large burden of trace metals, seafloor sediments exert strong control on the biogeochemical cycling of some heavy metals. Bottom sediments contain from three to five times the concentration of heavy metals found in overlying waters; thus, the bioavailability of even a small fraction of the total sediment burden assumes considerable importance. The concentrations and bioavailabilities of metals in estuarine sediments depend on many different processes, such as:[15]

1. Mobilization of metals to interstitial waters and their chemical speciation
2. Transformation of metals

3. Control exerted by major sediment components (e.g., iron oxides and organics) to which metals are preferentially bound
4. Competition between sediment metals for uptake sites in organisms
5. Influences of bioturbation, salinity, redox or pH on these processes

The flocculation and sedimentation of riverborne heavy metals in estuaries results in depleted concentrations in oceanic environments. Most of the riverborne heavy metals accumulate in the upper and middle reaches of estuaries, largely due to the occurrence of the turbidity maximum zone and settlement of metal-sorbed, fine-grained sediments. Dissolved forms of heavy metals transported by rivers may be removed quickly in some estuaries owing to adsorption and coprecipitation. Despite chemical and physical processes that tend to concentrate heavy metals in the middle and upper reaches of estuaries, advection within the benthic boundary layer displaces some heavy metals downestuary or to areas on the inner shelf beyond the estuarine mouth.

Bottom sediments are long-term integrators of metal inputs to estuarine and shallow coastal environments. Thus, information from sediment cores can establish long-term trends of trace metals in these environments.[8,16-19] Historical profiles of metal accumulation in the coastal zone typically show both a natural and an anthropogenic signal. Background values vary greatly among different source regions and with changes in grain size and mineralogy.[17] The distribution of metals in sediment cores has been used to generate historical pollution records and temporal sequences of pollution intensities for industrialized estuaries.[19] Such records have been invaluable for assessing the effectiveness of recent pollution control programs.

BIOAVAILABILITY AND BIOACCUMULATION

Estuarine and marine organisms tend to accumulate heavy metals from the environment, with both metal speciation and bioavailability being important in this process.[20,21] The accumulation of metals in biota occurs via several pathways, including the ingestion of food and suspended particulate material containing sorbed metals, the uptake of metals either directly from bottom sediments or interstitial waters, and the removal of metals from solution.[22] The major routes of metal uptake by invertebrates are solution and food.[23] The drinking of water and consumption of food are primary routes of metal uptake by fish.[24] The gills play a significant role in the entry of dissolved metals. Numerous factors influence the uptake of trace metals by estuarine and marine organisms, most notably extrinsic physico-chemical factors controlling the metal bioavailability — for example, dissolved metal concentration, temperature, salinity, presence or absence of chelating agents, presence or absence of other metals — and intra- and interspecifically variable intrinsic factors such as surface impermeability, nutritional state, stage of molt cycle, and throughput of water by osmotic flux, many of which are in turn affected by other physico-chemical factors.[23]

The literature regarding toxic effects of heavy metals on estuarine and marine organisms is extensive. Above a threshold bioavailability, all trace metals are potentially toxic.[23] Metal toxicity may be a function of the free metal ionic activity in seawater rather than the total concentration of the metal.[1] The capacity of some estuarine and marine organisms to store, remove, or detoxify metal contaminants varies considerably.[21] Certain species are able to regulate the body concentration of particular trace metals to an approximately constant level over a wide range of ambient dissolved metal availabilities.[22] While metals may be stored in the skeletal structure, concretions, or intracellular matrices of organisms,[25] the voiding of feces, eggs, and molting products tends to counter these storage effects.

Metallothioneins (low-molecular-weight, sulfhydryl-rich, metal-binding proteins), which occur in prokaryotes and eukaryotes, are known to be involved in cellular regulation, sequestration, and detoxification of metals.[26-28] Lysosomes (cellular structures involved in intracellular digestion and transport) likewise sequester trace metals and are also a factor in heavy metal homeostasis.[21,29] Metal-rich lysosomes have been detected in annelids, crustaceans, hydroids, mollusks, and algae.[1,30] Despite their potential for regulating heavy metal concentrations in organisms,[31] metallothioneins and other metal-binding proteins have a finite metal-binding capacity. When this capacity is exceeded, the toxic effects of the contaminants may be manifested. George and Olsson[32] have shown recently that the induction of metallothionein synthesis in fish by cadmium, copper, zinc, and mercury provides a valuable monitoring procedure to assess the biological availability and impact of these metals in the aquatic environment. Consequently, metallothioneins may be useful as indicators of trace metal pollution.

The exposure of estuarine and marine organisms to toxic levels of metal contaminants elicits a range of pathological responses, such as tissue inflammation and degeneration, lack of repair and regeneration of damaged tissue, neoplasm formation, and genetic derangement.[21] Growth inhibition may arise, as well as changes in physiology, reproduction, and development. Feeding behavior, respiratory metabolism, and digestive efficiency also can be adversely affected.

The accumulation of heavy metals by certain organisms has proven to be useful in the assessment of water and sediment contamination. Species employed to monitor contaminant levels in this fashion are known generically as bioindicators, bioaccumulative indicators, biomonitors, or sentinel organisms.[33-35] There are several prerequisites for a species to act as an efficient biomonitor of trace metal contaminants in aquatic environments. Biomonitors should be strong accumulators of trace contaminants and should reflect their ambient bioavailabilities. They should have widespread geographic distributions and year-round availability. In addition, these organisms should be easy to sample and to identify taxonomically. Preferably, they should be sessile and tolerant of changes in salinity and turbidity.[35] Among the most notable sentinel organisms are mussels and oysters. During the past two decades, emphasis has been placed on marine mussels as biomonitors of trace toxic substances in coastal marine waters. These bivalves form the basis of the Mussel Watch Program sponsored by the U.S. Environmental Protection Agency and the National Status and Trends Mussel Watch Program sponsored by the National Oceanic and Atmospheric Administration (NOAA).[36-40]

CONCENTRATIONS IN SEA WATER

The following treatment of metals is not intended to be exhaustive. However, the elements selected for discussion are those considered to be environmentally and toxicologically significant in estuarine and marine systems. They are representive of the following groups in the periodic table: IB, IIB, IVA, VA, and VIA.

Copper

The concentration of copper in oceanic waters amounts to about 0.1 µg/l.[41] Higher levels (~0.2 to >100 µg/l) are found in estuaries.[15] A large fraction of dissolved copper may sorb to particulate matter in estuaries and concentrate in bottom sediments. The metal binds readily to organic matter both in river water and sea water. Important anthropogenic inputs of copper in estuarine and coastal waters include sewage sludge dumpsites, municipal waste discharges, runoff from copper mines, antifouling paints, and other sources.

Zinc

While zinc concentrations in the open ocean are less than 1 µg/l,[42] significantly higher levels occur in estuaries and coastal waters. For example, Bryan and Langston[15] reported zinc concentrations as high as 23 µg/l in the Bristol Channel-Severn estuary and 70 µg/l in the North Sea. Even higher levels are found in heavily industrialized systems.

Cadmium

Open ocean concentrations of cadmium range from approximately 0.2 to 60 ng/l, with lowest values (~0.1 to 10 ng/l) generally occurring in subtropical and central gyres.[13,43] The cadmium content of open waters in enclosed seas is somewhat elevated (~10 to 60 ng/l). The level of cadmium in coastal waters removed from highly industrialized areas varies from 1 to 100 ng/l.[43] Accumulations in coastal areas impacted by industrial wastes and sewage sludge usually are much higher than in the open ocean, often leading to pronounced onshore-offshore gradients in cadmium concentrations. Bruland and Franks[41] uncovered a distinct onshore-offshore gradient in cadmium concentrations from 22 to 0.22 ng/l in the northwest Atlantic. A similar gradient exists in the North Pacific.[44] There is a close coupling of cadmium with nutrient biochemical cycles. In addition to nutrient chemistry, salinity and sediment load appear to be important factors controlling cadmium concentrations within and outside estuaries.[43] Although point sources can be significant, most cadmium enters the marine hydrosphere via atmospheric loading and riverine discharges, being derived from refining operations, copper and nickel smelting, and fuel combustion.[45]

Mercury

Natural inputs of mercury to aquatic systems originate from the weathering of mercury-bearing rocks and ores (e.g., cinnabar), the fallout of atmospheric gases from volcanoes and geothermal vents, and the emission of deep-sea hydrothermal vents. Contamination of surface waters from the atmosphere due to the combustion of fossil fuels may be significant. Important anthropogenic sources prior to 1980 were slimicides used in lumber and paper pulp industries, antifouling paints in the shipping industry, pesticides and seed dressings in agriculture, chemicals in the pharmaceutical industry, and mercury electrodes in the chloralkali industry. These sources have been largely eliminated, at least in the northern hemisphere.

The concentration of mercury in the open ocean ranges from 0.001-0.004 μg/l.[46] In coastal waters, mercury levels increase by about an order of magnitude,[15] but near estuaries concentrations vary considerably.[43] In Puget Sound, a heavily polluted system, mercury concentrations are similar to those of the open ocean, averaging 0.2 to 1.0 ng/l.[47] However, the Tagus estuary of Portugal shows signs of large-scale mercury contamination, with total dissolved concentrations as high as 80 ng/l.[48] Comparable values of mercury have been recorded in Minamata Bay, Japan (50 to 70 ng/l)[49] and the New York Bight (10 to 90 ng/l).[50] Many estuaries are effective sinks for mercury since it rapidly sorbs to suspended particulates, especially carbon compounds, and settles to the bottom. When mercury contamination occurs in estuaries, anthropogenic input of inorganic mercury is often suspected.[15] "Hot spots" of mercury contamination are evident in the Thames estuary, Liverpool Bay, Bristol Channel, and parts of the North Sea.[51]

Lead

Values of lead in surface waters of the North Atlantic and North Pacific are 5 to 50 ng/l, but they decline by a factor of 7 to 10 in deeper waters below the thermocline.[43] North Atlantic surface waters have concentrations three times greater than those of the North Pacific, reflecting the greater input of lead contamination along the East Coast of North America. The quantities of lead in the surface waters of the South Pacific are approximately 8 to 20 times less than those of the North Pacific and North Atlantic.[52] Somewhat higher concentrations are evident in the Baltic and Mediterranean seas compared to the North Atlantic due to their enclosed nature and the input from heavily industrialized areas in proximity to them. Between 1981 and 1989, lead concentrations in North Atlantic surface waters dropped by 50% owing to a decrease of lead input from the atmosphere.[4] The reduction in the use of leaded gasoline in North America during the past two decades is considered to be largely responsible for this decline.

The concentrations of lead in estuaries and coastal waters generally exceed those in the open ocean by a factor of 10 or more, with the greatest fraction associated with particulates.[53] These waters are commonly impacted directly by anthropogenic sources of lead. Most of this lead contamination originates from the mining, smelting, and refining of lead and other metal ores; vehicle emissions; and industrial emissions originating from the production, use, recycling, and disposal of lead-containing products.[54]

Tin

Values of inorganic tin in open ocean waters, present mainly as $SnO(OH_3)^-$, are between 0.003 and 0.008 μg/l, with concentrations being about 10 to 100 times greater in inshore and estuarine waters. "Hot spots" of dissolved inorganic tin occur in some estuaries that receive industrial discharges, such as Poole Harbor, England, where levels are as high as 48 μg/l.[15] Important anthropogenic sources of the metal are tanning, tinplate manufacturing, and smelting. Inorganic tin sorbs rapidly to particulate matter as it enters estuarine waters and is removed to bottom sediments. Particulate fluxes dominate the transport of tin to coastal oceanic waters.[53]

Organic forms of tin have gained increasing attention in recent years because of their potential impacts on biotic communities.[55-60] Organotin compounds are potent toxins that have been used as biocides, stabilizers in polyvinyl chloride, wood preservatives, preservatives for mildew control in textiles, and as catalysts in the production of polyurethane foam in room-temperature vulcanization of silicone elastomers.[61,62] Of all organotin compounds, tributyltin (TBT) is particularly toxic to a wide range of aquatic organisms.[63] As a component of marine antifouling paints, timber prservatives, and slimicides, TBT has become a widespread contaminant in estuarine and marine waters. In estuarine waters, TBT levels range from less than 1 ng/l in boat-free areas to levels above 600 ng/l near marinas.[15] It tends to concentrate in the air-seawater surface microlayer, on particles suspended in the water column, and in shallow water sediments in and around harbors, boatyards, and marinas. TBT degrades to dibutyltin, monobutyltin, and inorganic tin mediated by microorganisms and UV irradiation.[59,64] Estuarine particulates scavenge TBT, with significant concentrations of the contaminant (0.005 to 0.5 μg/g) removed to bottom sediments.[15]

Arsenic

Dissolved arsenic exists in a variety of chemical forms, principally arsenite, As(III); arsenate, As(V); and the methylated arsenicals (monomethylarsonate and dimethylarsinate).[53] In the open ocean, arsenic concentrations are about 1 to 2 μg/l. Although similar levels of arsenic are observed in many estuaries, values occasionally exceed 100 μg/l in waters receiving mine wastes or industrial discharges.

Selenium

Selenium occurs in multiple oxidation states and different chemical forms within an oxidation state (e.g., organic and inorganic) in natural waters, with three dissolved species — selenate(VI), selenite(IV), and organo-Se — widely recognized in seawater.[65] The levels of selenium in oceanic water vary from 0.06 to 0.12 μg/l.[66] Similar values are documented in estuarine and coastal marine waters. Elevated measurements are most evident in the immediate vicinity of industrial discharges. For example, Langston[67] reported selenium concentrations up to 29 μg/l in Poole Harbor, England. Combustion of fossil fuels, fly ash disposal, and industrial chemical discharges can increase selenium concentrations, but they are usually localized.[15]

CONCENTRATIONS IN SEDIMENTS

Sediments in estuaries and coastal environments serve as a repository of heavy metals. They may release these metals to surrounding waters in three ways: (1) by desorption from suspended particles upon contact with seawater, (2) by desorption from bottom sediments, and (3) by diffusion from interstitial water subsequent to diagenetic alteration of sediments.[5] The heavy metal content of estuarine bottom sediments is a function of their chemical and mineralogical composition which has been related to the grain size of the particles. Most heavy metals are associated with the fine-grained fraction of sediments. Muddy sediments contain substances (e.g., clay, chloride ion, humic acid) that complex or chelate metal ions.[68] The complexation of trace metals to organic matter, primarily sedimentary humic compounds, is an important mechanism for their transport.

Copper, Zinc

Levels of copper in estuarine sediments range from approximately 10 μg/g dry weight in pristine areas to 2000 μg/g dry weight at impacted sites. By comparison, zinc sediment concentrations in estuaries vary from baseline levels of less than 100 μg/g dry weight to 3000 μg/g dry weight. Interstitial waters of estuarine surface sediments have concentrations between a few micrograms per liter and about 100 μg/l, with zinc concentrations being as high as 396 μg/l.[15]

Cadmium

Cadmium levels in deep-sea sediments generally are less than 0.5 μg/g dry weight, but reach values as high as 60 μg/g dry weight in upwelling areas near Antarctica diatomaceous oozes.[43,69] Values in nearshore surface sediments range from near 0 to a maximum of 200 μg/g dry weight in contaminated areas of the Bay of Naples.[70] Two "hot spots" of cadmium contamination in U.S. coastal sediments are the New York Bight and Southern California Bight, where highest cadmium levels amount to 9.6 and 140 μg/g dry weight, respectively.[43] Cadmium concentrations in estuarine sediments typically range from about 0.2 to 10 μg/g dry weight.[15]

Mercury

The content of mercury in deep-sea sediments of the North Atlantic span two orders of magnitude from 0.008 to 0.6 μg/g dry weight. Values generally are greater in estuarine and coastal marine sediments. For example, Fowler[43] reported mercury concentrations in coastal systems from less than 0.1 μg/g dry weight at noncontaminated sites to 5–32 μg/g dry weight in heavily impacted areas, such as Minamata Bay and Raritan Bay. However, Bryan and Langston[15] documented somewhat lower mercury levels (0.03 to 6 μg/g dry weight) in UK estuarine sediments.

Lead

Values of lead in surface sediments of the deep sea are on the order of 8 to 80 μg/g dry weight.[71,72] Higher concentrations are found in clays (47 to 80 μg/g dry weight) than in oozes (13 to 17 μg/g dry weight). The range of lead concentrations in nearshore sediments is higher, typically between 10 and 100 μg/g dry weight.[43] However, "hot spots" receiving municipal or industrial wastes exhibit considerably higher levels. For instance, highest levels in surface sediments of the Southern California Bight, New York Bight, and heavily polluted areas of the Mediterranean Sea amount to 540, 270, and 280 μg/g dry weight, respectively.[73-75] Levels in estuarine sediments may be even greater; for example, Bryan and Langston[15] listed values of inorganic lead in estuarine bottom sediments of 25 to 2700 μg/g dry weight.

Tin

Localized high levels of tin occur in estuarine sediments receiving industrial discharges. In certain estuaries of southwest England affected by metal-mining operations (e.g., the Hayle and Helford estuaries), total concentrations of tin in bottom sediments exceed 1000 μg/g dry weight.[15] In more pristine systems, concentrations are usually two to three orders of magnitude lower. Levels of organotin are likewise elevated in estuarine sediments affected by major anthropogenic sources. Hence, due to high organotin releases from shipyards into the Sado estuary, Spain, the concentration of total butyltin in bottom sediments ranges from 235 to 12,200 ng/g dry weight.[76] In U.S. coastal sediments, the range of butyltin concentrations is <5 to 282 ng/g dry weight, with an average value of 36 ng/g dry weight.[56] Tributyltin (TBT) in surficial sediments of selected east coast estuaries in England displays high spatial variability ranging from <3 to 3935 ng/g dry weight.[77] A lower spread of TBT values has been recorded in coastal sediments of Bahrain (128 to 1930 ng/g dry weight), with a mean of 732 ng/g dry weight.[78] Typical surface sediment values of TBT in estuaries range from 0.005 to 0.5 μg/g dry weight.[15]

Arsenic

Arsenic measurements in bottom sediments of 19 U.K. estuaries are about 5 to 1740 μg/g dry weight.[15] The degree of contamination depends on proximity to areas of mine wastes and inputs from industrial outfalls, particularly smelting and metal refining operations. Estuarine sediments act as a sink for arsenic since dissolved inorganic forms sorb rapidly onto precipitating amorphous iron oxyhydroxides and manganese oxides. However, there may be significant post-depositional remobilization and release of dissolved species to the water column.

Selenium

Concentrations of selenium in estuarine sediments are generally less than 1.5 μg/g dry weight.[15] However, selenium levels may exceed 10 μg/g dry weight in contaminated sediments affected by anthropogenic chemical discharges. The highest concentrations are observed in the immediate vicinity of the discharge sources.

CONCENTRATIONS IN BIOTA

Although a number of trace metals are essential elements for growth and survival of estuarine and marine organisms (e.g., cobalt, copper, iron, zinc), excessive amounts can adversely affect physiological function, reproduction, and development, oftentimes culminating in death. Other metals (e.g., cadmium, chromium, lead, mercury) have no known biological function and are more toxic than the aforementioned essential elements.[79] Even at sublethal levels of these toxicants, the behavior of an organism can be modified, such that it may be less capable of dealing successfully with changes in the environment.[80] Toxicological modification of feeding behavior can alter predator-prey interactions, thereby influencing trophic relationships. A major concern over the years has been the potential of metal biomagnification with increasing trophic levels along food chains. Excluding methylmercury, however, little metal biomagnification has been demonstrated among estuarine and marine organisms.[15,68]

In more recent years, emphasis has been placed on the use of biomonitors to study contaminated aquatic environments. Assessment of entire biotic communities in polluted coastal environments is too time-consuming, expensive, and impractical. Hence, the generally accepted approach is to employ biomonitors, which provide accurate results.[81] Two groups of organisms are especially valuable in biomonitoring studies: seaweeds, which are employed to assess the availability of contaminants in solution, and filter-feeding mollusks (e.g., mussels, oysters), which primarily accumulate heavy metals from suspended particulate matter.[82]

Copper

Laboratory studies have established that dissolved concentrations of copper between 1 and 10 μg/l can significantly impact a large number of estuarine and coastal marine organisms.[83] For example, concentrations within this range increase the mortality of bay scallops (*Argopecten irradians*), Atlantic surf clams (*Spisula solidissima*), and isopods (*Idotea balthica*).[84,85] Other species, such as the blue mussel *Mytilus edulis*, have increased tolerance to copper and other heavy metals owing to the induction of copper-binding metallothioneins.[86]

Investigations in the field indicate a positive correlation between copper concentrations in sediments and those observed in seagrasses and polychaetes.[87,88] Rygg,[89] working in Norwegian fjords, found an absence of the most sensitive benthic species in bottom sediments with copper levels above 200 μg/g dry weight. The values of copper in organismal tissue are usually substantially higher in specimens collected from known areas of contamination. Copper concentrations in benthic species inhabiting contaminated estuaries may be one to two orders of magnitude greater than those inhabiting noncontaminated systems.[15] A comparison of copper concentrations in mussels and oysters collected from U.S. coastal waters during a mussel watch program in the late 1970s with those during a mussel watch program in the late 1980s indicates that the levels of the metal have generally increased nationwide.[36]

Concentrations of copper in liver and kidney tissues of adult seabirds from Gough Island in the South Atlantic Ocean are mostly in the range 3 to 8 μg/g wet weight. Comparable liver copper levels (5 to 10 μg/g wet weight) occur in seabirds of the New Zealand region.[91] Somewhat higher zinc levels (30 to 70 μg/g wet weight) are recorded in these avifauna, however.

Zinc

Concentrations of zinc in bivalves of British estuaries often exceed 1000 μg/g dry weight, but may be greater than 4000 μg/g dry weight in contaminated systems (e.g., Fal and Gannel estuaries).[15] Oysters, in particular, frequently contain elevated levels of zinc. For example, in the lower reaches of Restronguet Creek, England, the mean concentrations of zinc in oysters are more than 10,000 μg/g dry weight.

While a positive correlation often exists between zinc levels in sediments and those in the roots, rhizomes, and leaves of seagrasses (*Zostera marina*),[87] the body zinc concentrations of a number of benthic invertebrates (e.g., *Nereis diversicolor, Perinereis cultrifera*) are relatively independent of ambient concentrations. In these animals, the tissue concentrations of zinc appear to be internally regulated. Consequently, their zinc concentrations are less likely than the concentrations of other metals to reflect environmental changes.[15] For some species (e.g., the shrimp *Penaeus vannamei*), zinc concentrations are size dependent and may signal different metabolic requirements of young and old individuals.[92] Zinc loading in the mussel *Mytilus edulis* takes place rapidly, with individuals exposed to either particulate (elemental) zinc or soluble ($ZnCl_2$) zinc at levels of 10 μg/g experiencing considerable mortality after 14 days.[93]

Cadmium

Mean concentrations of cadmium in mussels (*Mytilus* spp.) from several coastal regions worldwide range from approximately 1 to 5 μg/g dry weight.[43] Some oceanic fauna, such as pelagic cephalopods and euphausiids, have relatively high natural levels of cadmium.[94] For example, maximum cadmium concentrations in euphausiids of the northeast Pacific amount to about 5.5 μg/g dry weight.[95] The oceanic amphipod *Themisto gaudichaudii* may contain levels of the metal in excess of 50 μg/g dry weight, and this species appears to be a significant source of cadmium for pelagic seabirds.[96] Fish feces represent an important vector for the transfer of cadmium between planktonic and benthic organisms, as demonstrated by the planktivorous reef fish *Chromis punctipinnis*.[97]

Cadmium can be deleterious to a wide variety of marine organisms. For instance, Kayser and Sperling[98] observed growth inhibition in the phytoplankter *Prorocentrum micans* at cadmium levels of 1 μg/l. Paffenhofer and Knowles[99] documented lower rates of reproduction in the copepod *Psuedodiaptomus coronatus* at 5 μg/l. This concentration also caused a decrease in population abundance of the isopod *Idotea balthica*,[85] as well as depressed growth of juvenile plaice *Pleuronectes platessa*.[100] For nauplii and adults of the copepod *Tisbe battaglai*, cadmium toxicity tests have revealed 96-h LC_{50} values of 0.46 and 0.34 mg/l, respectively. By comparison, the 96-h LC_{50} value of cadmium for larvae of the finfish *Cyprinodon variegatus* amounts to 1.23 mg/l.[101]

Several factors may confound cadmium toxicity. In crustaceans and fish, for example, the toxicity of cadmium appears to change as a function of ambient salinity. Calcium has been shown to exert a sparing effect on cadmium toxicity in the bay mysid *Mysidopsis bahia*, although the effect is less pronounced than effects attributed to cadmium-salinity interactions.[102]

Mercury

Mercury-tainted seafood is a major societal health concern, as evidenced by the effects of Minamata disease which proved fatal for more than 100 Japanese. In estuarine and coastal marine environments, sediments are the principal source of mercury accumulation in biota. While inorganic mercury is the dominant form in bottom sediments, a significant amount of mercury accumulating in the tissues of benthic invertebrates and fish is methylmercury. Thus, an estimated 40 to 90% of the mercury occurring in shellfish and more than 90% of that in finfish is methylmercury.[14] Unlike most metals, methylmercury undergoes biological magnification as mentioned above. Moving up the food chain, therefore, the concentration of methylmercury increases by several orders of magnitude.[15]

Localized contamination of estuaries and embayments due to anthropogenic activities has been coupled to elevated levels of mercury in many species of organisms.[103] In contrast, mercury concentrations in oceanic biota, primarily pelagic fauna, have not been linked unequivocally to anthropogenic sources. The levels of mercury in euphausiids from different oceanic regions range from 0.026 to 0.497 µg/g dry weight. Most oceanic species of fish contain 150 µg/kg of mercury in muscle tissue. Oceanic species that tend to have the highest mercury concentrations — large pelagics commonly contain 1000 to 5000 µg/kg of mercury in muscle — include marlin (*Makaira indica*), tuna (*Thunnus* spp.), and swordfish (*Xiphias gladius*). High levels of mercury in these species result from bioaccumulation and biomagnification.[104,105] Bioaccumulation occurs because mercury, when methylated, is very effectively absorbed by aquatic organisms.[45,106]

Mean values of mercury in mussels (*Mytilus* spp.) from various coastal regions worldwide are about 0.1 to 0.4 µg/g dry weight. Mytilids from some regions (e.g., northern Mediterranean and southwest Pacific) have very high levels of the element. In the northern Mediterranean, mussels contain mean mercury concentrations as high as 7 µg/g wet weight, and elevated concentrations of the metal have been recorded in other taxa from this area as well, such as crustaceans, fish, marine mammals, and seabirds. Mercury levels in mussels from the southwest Pacific average as much as 2.7 µg/g dry weight.[43] The U.S. Mussel Watch Program has detected "hot spots" of mercury in Tampa Bay, FL, Copano Bay, TX, and the Hudson Bay.[107]

Aside from mercury contamination of invertebrates and fish, high incidences of abnormalities in some seabirds have been correlated with mercury residues in tissues. Mercury tends to accumulate in the liver and feathers of birds.[104] Becker et al.,[108] investigating mercury in eggs, feathers, and livers of the common tern (*Sterna hirundo*), herring gull (*Larus argentatus*) and black-headed gull (*Larus ridibundus*), from the German North Sea coast, noted highest contamination levels in terns and lowest levels in black-headed gulls. Mercury concentrations in tern tissues were about four times those in the gulls. Honda et al.[109] likewise obtained high mercury readings in pelagic seabirds from the North Pacific, with peak values recorded in the black-footed albatross *Diomedea nigripes*. Gerlach[110] reported that high mercury concentrations (1 to 5 mg/kg, occasionally up to 20 mg/kg) are found regularly in the following marine birds: goosander (*Mergus merganser*), guillemot (*Uria aalge*), kittiwakes (*Rissa tridactyla*), fulmars (*Fulmaris glacialis*), skuas (*Stercorarius* spp.), and eiderducks (*Somateria mollissima*).

Although marine mammals accumulate large quantities of mercury, they rarely display any harmful effects of the contaminant. The liver is a critically important mercury accumulator organ in marine mammals.[111-113] Law et al.[112] measured mercury levels of 0.26 to 430 µg/g wet weight in eight species of seals, porpoises, and dolphins from waters around the British Isles. In an earlier study, mercury concentrations in the liver of six species of marine mammals collected along the coast of Wales and the Irish Sea varied between 0.5 and 280 µg/g wet weight.[113] These values are high enough to be a cause of concern.

Lead

The toxicity of lead to estuarine and marine organisms is a function in part of its chemical form. Inorganic lead compounds, for example, are generally less toxic than organolead compounds.[54] Apart from acute toxic effects which may arise at high lead concentrations, a variety of sublethal responses can occur in organisms at much lower levels of the contaminant. These responses may be manifested as anemia, depressed growth, reduced abundance, diminished egg-hatching success, fin degeneration, behavioral abnormalities, and other conspicuous changes.

Compared with other metals, such as cadmium and mercury, lead in the sea is not particularly toxic.[105] Some animals accumulate high concentrations of lead without any apparent harm. For example, the bivalve *Scrobicularia plana* in the Gannel estuary, England, has accumulated more than 990 μg/g dry weight of lead in its tissues, about 100 times above normal, with no demonstrable adverse effects.[114] The level of lead in the limpet *Acmaea digitalis* from the California coast near San Francisco is also elevated (~100 μg/g dry weight), yet it fails to show any effects ascribable to the metal. In contaminant waters of the Sorfjord in Norway, the concentration of lead in mussels amounts to approximately 3000 μg/g dry weight.[105] By contrast, the mean concentrations of lead in mussels from coastal regions worldwide range from 1 to 16 μg/g dry weight.[43]

Along the Danish marine coast, lead concentrations in plaice (*Pleuronectes platessa*) and flounder (*Platichthys flesus*) generally lie between 0.05 and 0.6 μg/g dry weight. Levels of the metal declined in these two species between 1985 and 1989 compared with the early 1980s, concurrently with a substantial decline in lead emissions from gasoline.[115] Lead in cod (*Gadus morhua*) from the northwest Atlantic is well below the level permissible in food products.[116] Benthopelagic rattail fish from the deep North Atlantic and North Pacific have mean concentrations of lead totaling 0.012 to 0.016 μg/g dry weight.[117]

The effects of lead on birds vary with the form of the contaminant ingested. Following the consumption of shot, experimental mallards (*Anas platyrhynchos*) concentrated the metal in the liver or kidney at levels of 6 to 20 μg/g wet weight.[118] Other mallards fed inorganic lead (as chloride salts) at a concentration of 100 μg/g exhibited elevated lead in the liver amounting to 7.2 μg/g dry weight.[119] Custer and Hohman[120] demonstrated that a significant number of canvasbacks (*Aythya valisineria*) wintering in Louisiana incurred high levels of lead (>6.7 μg/g dry weight) in their liver. Wilson et al.[121] ascribed the mortality of birds in the Mersey estuary (i.e., dunlin, *Calidris alpina*; blackheaded gulls, *Larus rudibundes*; redshank, *Tringa totanus*; mallard, *Anas platyhynchos*) to organolead compounds, specifically trialkyl lead. Dunlin, which were most strongly affected by the contaminants, had mean alkyl lead levels of about 10 μg/g wet weight in the liver.

Concentrations of lead in marine mammals (i.e., seals, porpoises, dolphins, and one whale) from waters around the British Isles were found to be 0.05 to 7.0 μg/g wet weight.[113] Levels of lead in Antarctic fur seals (*Arctocephalus gazella*) from Bird Island, South Georgia (54.5°S, 38°W) averaged 0.1 mg/kg dry weight.[122] Contamination of fur seals in an area remote from industrial activity underscores the importance of atmospheric transport in the global distribution of some heavy metals.

Tin

The sublethal and lethal effects of organotin compounds on estuarine and marine organisms are the subjects of ongoing investigations. Many of these studies have focused on various molluscan species — especially oysters, clams, mussels, and scallops — which tend to accumulate the compounds and consequently have been used as effective biomonitors of the contaminants. Some sublethal effects of tributyltin (TBT) on mollusks include oyster shell thickening, reduced growth rates in mussels, imposex in stenoglossan gastropods, and breakdown of sexual differentiation, oogenesis, and egg production in *Ostrea edulis*. Lee[123] suggested that these effects are related to the binding of TBT metabolites to cellular proteins and the inhibition of detoxifying enzyme systems (e.g., cytochrome P-450 systems and glutathione S-transferases) by TBT.

Davies et al.[124] detailed how the Pacific oyster (*Crassostrea gigas*) accumulated up to 1.41 mg/kg of tin and 0.87 mg/kg TBT from antifouling paints over a period of 41 weeks, but lost 90% of this burden during depuration. Young Pacific oysters displayed reduced growth and other sublethal effects when exposed to TBT at concentrations as low as 0.01 μg/l.[105] In addition to reduced growth, Alzieu[125] observed decreased settlement and increased shell abnormalities in *C. gigas* cultured near marinas where TBT levels were significant. Dyrynda[126] described an increased incidence of abnormal shell thickening in this species due to TBT. His and Robert[127] also ascertained decreased condition factor in oysters exposed to two antifouling paints.

Minchin et al.[128] detected tributyltin levels of 0.7 μg/g wet weight in adult populations of the scallop *Pecten maximus* in Mulroy Bay on the northern coast of Ireland, and attributed this contamination to the introduction and subsequent use of organotin dip-nets on salmonid farms in the bay. In a study of the cellular and biochemical responses of the American oyster *Crassostrea virginica* and blue mussel *Mytilus edulis* to 0.7 μg/l TBT during a 90-d test period (60-d exposure, 30-d recovery), Pickwell and Steinert[129] discovered that both bivalves accumulated tin in their digestive glands to levels comparable to flowing seawater of the test tanks. Whereas the oysters suffered virtually no deaths during the 60-d test period, approximately 50% of the mussels died within this interval of time. Widdows and Page[130] determined that the clearance (= feeding) rate of *M. edulis* was significantly reduced above a threshold TBT concentration of 3 to 4 μg/g, with severe inhibition of growth taking place above 4 μg/g.

Wade et al.[56] recorded mean bivalve butyltin concentrations 18 times greater than those in bottom sediments sampled from U.S. coastal areas. Kure and Depledge[131] determined that the clam *Mya arenaria* concentrated TBT from Danish coastal waters by a factor of 57,000 to 220,000. Stewart and Thompson[60] discerned highest concentrations of TBT (314 ng/g dry weight) in *Mytilus edulis* samples collected along the southeastern coast of British Columbia, Canada.

An unusually damaging sublethal effect of TBT contamination in neogastropod dogwhelks is an apparent alteration in the hormonal system which triggers the imposition of male characters on females, a condition known as imposex. Smith and McVeagh[63] disclosed widespread imposex in dogwhelks (*Lepsiella* spp.) in New England coastal waters, especially in proximity to permanent mooring sites and locations of high seasonal pleasure craft activity. Imposex is a potentially serious condition for a species because the afflicted population ultimately may be eliminated from broad expanses of the sea floor due to a failure of reproduction. Imposex is now recognized as a problem affecting gastropods other than dogwhelks (e.g., *Cronia*, *Drupella*, *Ilyanassa*, *Naquetia*, *Nucella*, *Thais*, etc.) from coastal waters worldwide.[105]

Organotin compounds are magnified to relatively high levels in benthic invertebrates, even though they usually constitute less than 5% of the total tin in sediments. For instance, they comprise from 60% of the total tin in *Nereis diversicolor* to 97% in deposit-feeding bivalves such as *Mya arenaria*. There is little evidence that removal of TBT to bottom sediments reduces the bioavailability of organotin compounds.[15]

Marine mammals also accumulate organotin. For instance, Iwata et al.[62] measured butyltin compound (BTC) residues in the blubber of eight species of marine mammals caught between 1981 and 1993 in seas surrounding Japan and in the Indian, North Pacific, and Antarctic oceans. The highest residue levels were registered on a finless porpoise from the Seto-Inland Sea, Japan, with a BTC concentration of 770 ng/g wet weight.

Mammals, as well as crustaceans and fish, appear to possess the necessary enzymes to break down TBT. The liver of seabirds also may play a role in organotin detoxification. Hence, food chain magnification of TBT does not seem to develop among the upper trophic levels.[132]

Arsenic

Although arsenic is present largely in inorganic form in water and sediments of estuarine and marine environments, it is usually organically bound in organisms inhabiting these systems. Inorganic forms of arsenic pose a significant toxicity hazard to aquatic biota;[133] however, marine biota, unlike terrestrial organisms, can convert inorganic forms of the metal into organic arsenic compounds.[134] When organically bound in organisms, arsenic is relatively nontoxic to consumers and readily excretable.[15] This is important in the assessment of potential seafood products available for consumers.[135] The metal bioaccumulates in organisms, but does not biomagnify in marine food chains.[134]

While most estuarine and coastal marine organisms probably obtain a portion of their arsenic burden from surrounding waters, bottom sediments provide the major source of the metal for benthic communities. Concentrations in some benthic species (e.g., *Nereis diversicolor*, *Macoma balthica*,

Scrobicularia plana) reflect total arsenic loads in bottom sediments.[136,137] Therefore, some species may serve as suitable biological indicators of the availability of sediment-bound arsenic.[15]

Estuarine and marine plants are potentially important agents in the biotransformation of arsenic. Experimental studies have shown that macroalgae (i.e., *Fucus visiculosus*) are quantitatively significant biotransformers of arsenic.[138] Vascular macrophytes (i.e., *Spartina alterniflora*) also appear to be prime agents in total arsenic turnover, as is evident from investigations of salt marsh systems.[139] Both algae and vascular plants may contribute substantially to the overall organoarsenic production in estuaries and its transfer through higher trophic levels.

Bryan[14] cited arsenic levels in several pelagic species from the Atlantic Ocean, including a tunicate (1.5 μg/g dry weight), two species of teleost fish (2.5 to 2.7 μg/g dry weight), a coelenterate (11 μg/g dry weight), and six species of crustaceans (14 to 42 μg/g dry weight). Values of arsenic in seabirds typically range from about 0.1 to 10 μg/g wet weight, with higher levels being coupled to arsenic poisoning.[140] Arsenic concentrations of 5 to 23 μg/l (as inorganic As^{3+}/As^{5+}) purportedly hinder the growth and succession of phytoplankton.[141] Hemoglobin production in fish is also inhibited by the ingestion of as little as 10 μg/g of arsenic present in the diet as As^{3+}.[142] Additional work is needed to assess the toxicity of arsenic on other groups of estuarine and marine organisms.

Selenium

The biotic reactivity and toxicity of selenium are a function of its chemical form.[143] In estuarine and marine organisms, most selenium is organically bound,[144] with sediments providing a significant fraction of the biologically-available substance. Diet appears to be the primary source of selenium accumulation in estuarine and marine organisms.[15] There is some indication of selenium biomagnification in marine foods chains.[145] Selenium concentrations tend to be magnified in the kidney and liver of fish, birds, and mammals but not in muscle tissue, where selenium levels are not much higher than in their food supply.[15]

Anthropogenic activities can generate localized "hot spots" of selenium contamination in biotic inhabitants of coastal environments. For example, subsurface agricultural drainage waters tainted with selenium derived from tile drain wastewaters have affected aquatic food chains of the San Joaquin Valley, CA. Fish subjected to the agricultural wastewaters in the valley have elevated levels of selenium nearly 100 times greater than those in fish from control areas.[146] Striped bass (*Morone saxatilis*) from the San Francisco estuary contain selenium in concentrations that are only 25 to 50% of those measured in the most contaminated fish from the San Joaquin River, with levels of 7.9 μg/g dry weight.[147]

Seabirds are very sensitive to selenium exposure; elevated contaminant levels in this group are linked to impaired reproduction, delayed egg laying, deformed embryos, and increased histopathological lesions, as well as reduced growth and survival.[148-150] For instance, aquatic birds, waders, and wildfowl in the San Joaquin Valley have experienced serious reproduction problems. Selenium concentrations are high in these birds, ranging from 20 to 218 mg/kg dry weight.[151] Ohlendorf et al.,[152] investigating various migratory wildfowl (i.e., avocet, coot, ducks, grebe, stilt) nesting at selenium-contaminated ponds in the San Joaquin Valley, found high residue levels in livers (94 μg/g dry weight) and eggs (28 μg/g dry weight).

Other locations where high selenium concentrations have been measured in seabirds include Galveston Bay, TX (>40 μg/g selenium in kidney),[153] a chemically impacted lagoon in the Mediterranean (up to 179 μg/g in kidney),[154] and contaminated sites on the Wadden Sea (up to 32 μg/g in kidney).[155] Seabirds make ideal biomonitoring tools for contaminants such as selenium because they occupy several higher trophic levels and are visible and conspicuous, and their population levels, reproductive activity, and pathologies can be determined.[156]

Bryan[14] found that marine animals removed from sites of contamination generally contain only a few μg/g dry weight of selenium; plants (i.e., seaweeds), about 10 times less. Values of selenium reported by Bryan[14] in marine bivalves, decapod crustaceans, cephalopods, and fish are between 1 and 4 μg/g dry weight, compared with 0.1 μg/g dry weight in seaweeds. Skaare et al.[157] observed selenium levels in the liver of Arctic seals amounting to 0.8 to 3.7 μg/g wet weight. Corresponding ranges in grey seals and harbor seals average 1.0 to 23.3 μg/g wet weight. Levels are substantially lower in the kidney and brain of these higher organisms.

REFERENCES

1. Viarengo, A., Heavy metals in marine invertebrates: mechanisms of regulation and toxicity at the cellular level, *Rev. Aquat. Sci.*, 1, 295, 1989.
2. Abel, P. D., *Water Pollution Biology*, Ellis Horwood, Chichester, 1989.
3. Pastor, A., Hernandez, F., Peris, M. A., Beltran, J., Sancho, J. V., and Castillo, M. T., Levels of heavy metals in some marine organisms from the western Mediterranean area (Spain), *Mar. Pollut. Bull.*, 28, 50, 1994.
4. Helmers, E., Mart, L., Schulz-Baldes, M., and Ernst, W., Temporal and spatial variations of lead concentrations in Atlantic surface waters, *Mar. Pollut. Bull.*, 21, 515, 1990.
5. Duinker, J. C., Suspended matter in estuaries: adsorption and desorption processes, in *Chemistry and Biogeochemistry of Estuaries*, Olausson, E. and Cato, I. (Eds.), John Wiley & Sons, Chichester, 1980, 121.
6. Niencheski, L. F., Windom, H. L., and Smith, R., Distribution of particulate trace metal in Patos Lagoon estuary (Brazil), *Mar. Pollut. Bull.*, 28, 96, 1994.
7. van den Berg, C. M. G., Complex formation and the chemistry of selected trace elements in estuaries, *Estuaries*, 16, 512, 1993.
8. Sinex, S. A. and Wright, D. A., Distribution of trace metals in the sediments and biota of Chesapeake Bay, *Mar. Pollut. Bull.*, 19, 425, 1988.
9. Hanson, P. J., Evans, D. W., Colby, D. R., and Zdanowics, V. S., Assessment of elemental contamination in estuarine and coastal environments based on geochemical and statistical modeling of sediments, *Mar. Environ. Res.*, 36, 237, 1993.
10. Paulson, W. J., Feely, R. A., Curl, H. C., Jr., and Tennant, D. A., Estuarine transport of trace metals in a buoyant riverine plume, *Est. Coastal Shelf Sci.*, 28, 231, 1989.
11. Greenaway, A. M. and Rankine-Jones, A. I., Elemental concentrations in coastal sediments from Hellshire, Jamaica, *Mar. Pollut. Bull.*, 24, 390, 1992.
12. Burton, J. D., Basic properties and processes in estuarine chemistry, in *Estuarine Chemistry*, Burton, J. D. and Liss, P. S. (Eds.), Academic Press, London, 1976, 1.
13. Bruland, K. W., Trace elements in seawater, in *Chemical Oceanography*, Vol. 8, Riley, J. P. and Chester, R. (Eds.), Academic Press, London, 1983, 157.
14. Bryan, G. W., Heavy metal contamination in the sea, in *Marine Pollution*, Johnston, R. (Ed.), Academic Press, London, 1976.
15. Bryan, G. W. and Langston, W. J., Bioavailability, accumulation, and effects of heavy metals in sediments with special reference to United Kingdom estuaries: a review, *Environ. Pollut.*, 76, 89, 1992.
16. Abu-Hilal, A. H. and Badran, M. M., Effect of pollution sources on metal concentration in sediment cores from the Gulf of Aqaba (Red Sea), *Mar. Pollut. Bull.*, 21, 190, 1990.
17. Alexander, C. R., Smith, R. G., Calder, F. D., Schropp, S. J., and Windom, H. L., The historical record of metal enrichment in two Florida estuaries, *Estuaries*, 16, 627, 1993.
18. Bricker, S. B., The history of Cu, Pb, and Zn inputs to Narragansett Bay, Rhode Island, as recorded by salt-marsh sediments, *Estuaries*, 16, 589, 1993.
19. Rae, J. E. and Allen, J. R. L., The significance of organic matter degradation in the interpretation of historical pollution trends in depth profiles of estuarine sediment, *Estuaries*, 16, 678, 1993.
20. Batley, G. E. (Ed.), *Trace Element Speciation: Analytical Methods and Problems*, CRC Press, Boca Raton, FL, 1989.
21. Capuzzo, J. M., Burt, W. V., Duedall, I. W., Park, P. K., and Kester, D. R., The impact of waste disposal in nearshore environments, in *Wastes in the Ocean*, Vol. 6, *Nearshore Waste Disposal*, Ketchum, B. H., Capuzzo, J. M., Burt, W. V., Duedall, I. W., Park, P. K., and Kester, D. R. (Eds.), John Wiley & Sons, New York, 1985, 3.
22. Hawker, D. W., Bioaccumulation of metallic substances and organometallic compounds, in *Bioaccumulation of Xenobiotic Compounds*, Connell, D. W. (Ed.), CRC Press, Boca Raton, FL, 1990.

23. Rainbow, P. S., The significance of trace metal concentrations in marine invertebrates, in *Ecotoxicology of Metals in Invertebrates,* Dallinger, R. and Rainbow, P. S. (Eds.), Lewis Publishers, Boca Raton, FL, 1993, 3.
24. Heath, A. G., *Water Pollution and Fish Physiology*, CRC Press, Boca Raton, FL, 1987.
25. George, S. G., Subcellular accumulation and detoxication of metals in aquatic animals, in *Physiological Mechanisms of Marine Pollutant Toxicity*, Vernberg, W. B., Calabrese, A., Thurberg, F. P., and Vernberg, F. J. (Eds.), Academic Press, New York, 1982, 3.
26. Engel, D. W. and Brouwer, M., Crustaceans as models for metal metabolism. I. Effects of the molt cycle on blue crab metal metabolism and metallothionein, *Mar. Environ. Res.,* 35, 1, 1993.
27. Schlenk, D., Ringwood, A. H., Brouwer-Hoexum, T., and Brouwer, M., Crustaceans as models for metal metabolism. II. Induction and characterization of metallothionein isoforms from the blue crab (*Callinectes sapidus*), *Mar. Environ. Res.,* 35, 7, 1993.
28. Roesijadi, G., Behavior of metallothionein-bound metals in a natural population of an estuarine mollusc, *Mar. Environ. Res.,* 38, 147, 1994.
29. Viarengo, A., Moore, M. N., Mancinelli, G., Mazzucotelli, A., Pipe, R. K., and Farrar, S. V., Metallothioneins and lysosomes in metal toxicity and accumulation in marine mussels: the effect of cadmium in the presence and absence of phenanthrene, *Mar. Biol.,* 94, 251, 1987.
30. Pavicic, J., Raspor, B., and Martincic, D., Quantitative determination of metallothionein-like proteins in mussels: methodological approach and field evaluation, *Mar. Biol.,* 115, 83, 1993.
31. Roesijadi, G., Metallothioneins in metal regulation and toxicity in aquatic animals — review, *Aquat. Toxicol.,* 22, 81, 1992.
32. George, S. G. and Olsson, P.-E., Metallothioneins as indicators of trace metal pollution, in *Biomonitoring of Coastal Waters and Estuaries,* Kramer, K. J. M. (Ed.), CRC Press, Boca Raton, FL, 1994, 151.
33. Phillips, D. J. H. and Rainbow, P. S., Strategies of trace metal sequestration in aquatic organisms, *Mar. Environ. Res.,* 28, 207, 1989.
34. Phillips, D. J. H. and Rainbow, P. S., *Biomonitoring of Trace Aquatic Contaminants*, Elsevier, London, 1993.
35. Phillips, D. J. H., Macrophytes as biomonitors of trace metals, in *Biomonitoring of Coastal Waters and Estuaries,* Kramer, K. J. M. (Ed.), CRC Press, Boca Raton, FL, 1994, 85.
36. Lauenstein, G. G., Robertson, A., and O'Connor, T. P., Comparison of trace metal data in mussels and oysters from a mussel watch program of the 1970s with those from a 1980s program, *Mar. Pollut. Bull.,* 21, 440, 1990.
37. O'Connor, T. P., *Coastal Environmental Quality in the United States, 1990. Chemical Contamination in Sediments and Tissues.* A Special NOAA 20th Anniversary Report, National Oceanic and Atmospheric Administration Office of Ocean Resources Conservation and Assessment, Rockville, MD, 1990.
38. O'Connor, T. P. and Ehler, C. N., Results from NOAA National Status and Trends Program on distributions and effects of chemical contamination in the coastal and estuarine United States, *Environ. Monit. Assess.,* 17, 33, 1991.
39. NOAA, *National Status and Trends Program for Environmental Quality: Second Summary of Data on Chemical Contaminants in Sediments from the National Status and Trends Program,* Tech. Mem. NOS OMA 59, National Oceanic and Atmospheric Administration Office of Oceanography and Marine Assessment, Rockville, MD, 1991.
40. O'Connor, T. P., *Recent Trends in Coastal Environmental Quality: Results from the First Five Years of the National Oceanic and Atmospheric Administration Mussel Watch Project,* National Oceanic and Atmospheric Administration Office of Ocean Resources Conservation and Assessment, Rockville, MD, 1992.
41. Bruland, K. W. and Franks, R. P., Mn, Ni, Cu, Zn, and Cd in the western North Atlantic, in *Trace Metals in Sea Water*, Wong, C. S., Boyle, E., Bruland, K. W., Burton, J. D., and Goldberg, E. D. (Eds.), Plenum Press, New York, 1983, 395.
42. Bruland, K. W., Franks, R. P., Knauer, G. A., and Martin, J. H., Sampling and analytical methods for the determination of copper, cadmium, zinc and nickel at the nanogram per liter level in sea water, *Anal. Chim. Acta,* 105, 233, 1979.
43. Fowler, S. W., Critical review of selected heavy metal and chlorinated hydrocarbon concentrations in the marine environment, *Mar. Environ. Res.,* 29, 1, 1990.

44. Bruland, K. W., Knauer, G. A., and Martin, J. H., Cadmium in northeast Pacific waters, *Limnol. Oceanogr.*, 23, 618, 1978.
45. Wren, C. D., Harris, S., and Harttrup, N., Ecotoxicology of mercury and cadmium, in *Handbook of Ecotoxicology*, Hoffman, D. J., Rattner, B. A., Burton, G. A., Jr., and Cairns, J., Jr. (Eds.), Lewis Publishers, Boca Raton, FL, 1995, 392.
46. Olafsson, J., Mercury concentrations in the North Atlantic in relation to cadmium, aluminium, and oceanographic parameters, in *Trace Metals in Sea Water*, Wong, C. S., Boyle, E, Bruland, K. W., Burton, J. D., and Goldberg, E. D. (Eds.), Plenum Press, New York, 1983, 475.
47. Bloo, N. S. and Crecelius, E. A., Determination of mercury in seawater at subnanogram per liter levels, *Mar. Chem.*, 14, 49, 1983.
48. Figueres, G., Martin, J. M., Meybeck, M., and Seyler, P., A comparative study of mercury contamination in the Tagus estuary (Portugal) and major French estuaries (Gironde, Loire, Rhone), *Est. Coastal Shelf Sci.*, 20, 183, 1985.
49. Kumagai, M and Nishimura, H., Mercury distribution in seawater in Minamata Bay and the origin of particulate mercury, *J. Oceanogr. Soc. Jap.*, 34, 50, 1978.
50. Segar, D. A. and Davis, P. G., *Contamination of Populated Estuaries and Adjacent Coastal Ocean — A Global Review,* National Oceanic and Atmospheric Administration Tech. Mem. NOS OMA 59, NOAA Office of Oceanography and Marine Assessment, Rockville, MD, 1984.
51. Förstner, U., Inorganic pollutants, particularly heavy metals in estuaries, in *Chemistry and Biogeochemistry of Estuaries,* Olausson, E and Cato, I. (Eds.), John Wiley & Sons, Chichester, 1980, 307.
52. Flegal, A. R. and Patterson, C. C., Vertical concentration of profiles of lead in the central Pacific Ocean, *Earth Planet. Sci. Lett.*, 64, 19, 1983.
53. Cutter, G. A., Trace elements in estuarine and coastal waters, *Rev. Geophys. (Suppl.), Contrib. Oceanogr.,* 1991, 639.
54. Pain, D. J., Lead in the environment, in *Handbook of Ecotoxicology*, Hoffman, D. J., Rattner, B. A., Burton, G. A., Jr., and Cairns, J., Jr. (Eds.), Lewis Publishers, Boca Raton, FL, 1995, 356.
55. Hodge, V. F., Seidel, S. L., and Goldberg, E. D., Determination of tin (IV) and organotin compounds in natural waters, coastal sediments and macroalgae by atomic absorption spectrometry, *Anal. Chem.*, 51, 1256, 1979.
56. Wade, T. L., Garcia-Romero, B., and Brooks, J. M., Butyltins in sediments and bivalves from U.S. coastal areas, *Chemosphere,* 20, 647, 1990.
57. Alzieu, C., Environmental problems caused by TBT in France: assessment, regulations, prospects, *Mar. Environ. Res.*, 32, 7, 1991.
58. Ellis, D. V., New dangerous chemicals in the environment: lessions from TBT, *Mar. Pollut. Bull.*, 22, 8, 1991.
59. Miller, M. E. and Cooney, J. J., Effects of tri-, di, and monobutyltin on heterotrophic nitrifying bacteria from surficial estuarine sediments, *Arch. Environ. Contam. Toxicol.,* 27, 501, 1994.
60. Stewart, C. and Thompson, J. A. J., Extensive butyltin contamination in southwestern coastal British Columbia, Canada, *Mar. Pollut. Bull.*, 28, 601, 1994.
61. Hall, L. W., Jr., Tributyltin environmental studies in Chesapeake Bay, *Mar. Pollut. Bull.*, 19, 431, 1988.
62. Iwata, H., Tanabe, S., Miyazaki, N., and Tatsukawa, R., Detection of butyltin compound residues in the blubber of marine mammals, *Mar. Pollut. Bull.*, 28, 607, 1994.
63. Smith, J. and McVeagh, M., Widespread organotin pollution in New Zealand coastal waters as indicated by imposex in dogwhelks, *Mar. Pollut. Bull.*, 22, 409, 1991.
64. Seligman, P. F., Valkirs, A. O., and Lee, R. F., Dedgradation of tributyltin in San Diego Bay, California waters, *Environ. Sci. Technol.*, 20, 1229, 1986.
65. Hung, J.-J. and Shy, C.-P., Speciation of dissolved selenium in the Kaoping and Erhjen rivers and estuaries, southwestern Taiwan, *Estuaries,* 18, 234, 1995.
66. Burton, J. D. and Statham, P. J., Occurrence, distribution, and chemical speciation of some minor dissolved constituents in ocean waters, in *Environmental Chemistry*, Vol. 2, Bowen, H. J. M. (Ed.), Royal Society of Chemistry, London, 1982, 234.
67. Langston, W. J., Toxic effects of metals and the incidence of metal pollution in marine ecosystems, in *Heavy Metals in the Marine Environment*, Furness, R. W. and Rainbow, P. S., CRC Press, Boca Raton, FL, 1990.
68. Kennish, M. J., *Ecology of Estuaries: Anthropogenic Effects*, CRC Press, Boca Raton, FL, 1992.
69. Simpson, W. R., A critical review of cadmium in the marine environment, *Prog. Oceanogr.*, 10, 1, 1981.

70. Griggs, G. B. and Johnson, S., Bottom sediment contamination in the Bay of Naples, Italy, *Mar. Pollut. Bull.*, 9, 208, 1978.
71. Nriagu, J. O. (Ed.), *The Biogeochemistry of Lead in the Environment*, Vols. A and B, Elsevier, Amsterdam, 1978.
72. Veron, A., Lambert, C. E., Isley, A., Linet, P., and Grousset, F., Evidence of recent lead pollution in deep northeast Atlantic sediments, *Nature*, 326, 278, 1987.
73. Katz, A. and Kaplan, I. R., Heavy metals behavior in coastal sediments of Southern California: a critical review and synthesis, *Mar. Chem.*, 10, 261, 1981.
74. Carmody, D. J., Pearce, J. B., and Yasso, W. E., Trace metals in sediments of the New York Bight, *Mar. Pollut. Bull.*, 4, 132, 1973.
75. UNEP, Co-ordinated Mediterranean Pollution Monitoring and Research Program (MED POL — Phase I), Final Report 1975-1980, MAP Tech. Rep. Ser. No. 9, United Nations Environment Program, Athens, 1986.
76. Quevauviller, P., Lavigne, R., Pinel, R., and Astruc, M., Organotins in sediments and mussels from the Sado estuarine system (Portugal), *Environ. Pollut.*, 57, 149, 1989.
77. Dowson, P. H., Bubb, J. M., and Lester, J. N., Organotin distribution in sediments and waters of selected east coast estuaries in the UK, *Mar. Pollut. Bull.*, 24, 492, 1992.
78. Hasan, M. A., and Juma, H. A., Assessment of tributyltin in the marine environment of Bahrain, *Mar. Pollut. Bull.*, 24, 408, 1992.
79. Hylland, K., Kaland, T., and Andersen, T., Subcellular Cd accumulation and Cd-binding proteins in the netted dog whelk, *Nassarius reticulatus* L., *Mar. Environ. Res.*, 38, 169, 1994.
80. Weis, J. S. and Khan, A. A., Effects of mercury on the feeding behavior of the mummichog, *Fundulus heteroclitus* from a polluted habitat, *Mar. Environ. Res.*, 30, 243, 1990.
81. Shimshock, N., Sennefelder, G., Dueker, M., Thurberg, F., and Yarish, C., Patterns of metal accumulation in *Laminaria longicruris* from Long Island Sound (Connecticut), *Arch. Environ. Contam. Toxicol.*, 22, 305, 1992.
82. King, D. G. and Davies, I. M., Laboratory and field studies of the accumulation of inorganic mercury by the mussel *Mytilus edulis* (L.), *Mar. Pollut. Bull.*, 18, 40, 1987.
83. Bryan, G. W., Pollution due to heavy metals and their compounds, in *Marine Ecology*, Vol. 5, Part 3, Kinne, O. (Ed.), John Wiley & Sons, Chichester, 1984, 1289.
84. Nelson, D. A., Miller, J. E., and Calabrese, A., Effect of heavy metals on bay scallops, surf clams, and blue mussels in acute and long-term exposures, *Arch. Environ. Contam. Toxicol.*, 17, 596, 1988.
85. Giudici, M. de N. and Guarino, S. M., Effects of chronic exposure to cadmium or copper on *Idotea balthica* (Crustacea, Isopoda), *Mar. Pollut. Bull.*, 20, 69, 1989.
86. Viarengo, A., Pertica, M., Canesi, L., Mazzucotelli, A., Orunesu, M., and Bouquegneau, J. M., Purification and biochemical characterization of a lysosomal copper-rich thionein-like protein involved in metal detoxification in the digestive gland of mussels, *Comp. Biochem. Physiol.*, 93C, 389, 1989.
87. Lyngby, J. E. and Brix, H., Monitoring of heavy metal contamination in the Limfjord, Denmark, using biological indicators and sediment, *Sci. Tot. Environ.*, 64, 239, 1987.
88. Bryan, G. W. and Gibbs, P. E., Polychaetes as indicators of heavy metal availability in marine deposits, in *Oceanic Processes in Marine Pollution*, Vol. 1, *Biological Processes and Wastes in the Ocean*, Capuzzo, J. M. and Kester, D. R. (Eds.), Krieger Publishing, Melbourne, FL, 1987, 37.
89. Rygg, B., Effects of sediment copper on benthic fauna, *Mar. Ecol. Prog. Ser.*, 25, 83, 1985.
90. Muirhead, S. J. and Furness, R. W., Heavy metal concentrations in the tissues of seabirds from Gough Island, South Atlantic Ocean, *Mar. Pollut. Bull.*, 19, 278, 1988.
91. Lock, J. W., Thompson, D. R., and Furness, R. W., and Bartle, J. A., Metal concentrations in seabirds of the New Zealand region, *Environ. Pollut.*, 75, 289, 1992.
92. Paez-Osuna, F. and Ruiz-Fernandez, C., Trace metals in the Mexican shrimp *Penaeus vannamei* from estuarine and marine environments, *Environ. Pollut.*, 87, 243, 1995.
93. Burbidge, F. J., Macey, D. J., Webb, J., and Talbot, V., A comparison between particulate (elemental) zinc and soluble zinc ($ZnCl_2$) uptake and effects in the mussel, *Mytilus edulis*, *Arch. Environ. Contam. Toxicol.*, 26, 466, 1994.
94. Thompson, D. R., Metal levels in marine vertebrates, in *Heavy Metals in the Marine Environment*, Furness, R. W. and Rainbow, P. S. (Eds.), CRC Press, Boca Raton, FL, 1990, 143.
95. Fowler, S., Trace metal monitoring of pelagic organisms from the open Mediterranean Sea, *Environ. Monit. Assess.*, 7, 59, 1986.

96. Rainbow, P. S., Copper, cadmium, and zinc concentrations in ocean amphipod and euphausiid crustaceans, as a source of heavy metals to pelagic seabirds, *Mar. Biol.,* 103, 513, 1989.
97. Krause, P. R. and Bray, R. N., Transport of cadmium and zinc to rocky reef communities in feces of the blacksmith (*Chromis punctipinnis*), a planktivorous fish, *Mar. Environ. Res.,* 38, 33, 1994.
98. Kayser, H. and Sperling, K.-R., Cadmium effects and accumulation in cultures of *Prorocentrum micans* (Dinophyta), *Helgolander Meeresunters.,* 33, 89, 1980.
99. Paffenhofer, G.-A. and Knowles, S. C., Laboratory experiments on feeding, growth, and fecundity of and effects of cadmium on *Pseudodiaptomus coronatus, Bull. Mar. Sci.,* 28, 574, 1978.
100. Westernhagen, H. von, Dethlefsen, V., and Rosenthal, H., Correlation between cadmium concentration in the water and tissue residue levels in dab, *Limanda limanda* L., and plaice, *Pleuronectes platessa* L., *J. Mar. Biol. Assoc. U.K.,* 60, 45, 1980.
101. Hutchinson, T. H., Williams, T. D., and Eales, G. J., Toxicity of cadmium, hexavalent chromium and copper to marine fish larvae (*Cyprinodon variegatus*) and copepods (*Tisbe battagliai*), *Mar. Environ. Res.,* 38, 275, 1994.
102. De Lisle, P. F. and Roberts, M. H., Jr., The effect of salinity on cadmium toxicity in the estuarine mysid *Mysidopsis bahia*: roles of osmoregulation and calcium, *Mar. Environ. Res.,* 37, 47, 1994.
103. Langston, W. J., Metals in sediments and benthic organisms in the Mersey estuary, *Est. Coast. Shelf Sci.,* 23, 239, 1986.
104. Monteiro, L. R. and Lopes, H. D., Mercury content of swordfish, *Xiphias gladius*, in relation to length, weight, age, and sex, *Mar. Pollut. Bull.,* 21, 293, 1990.
105. Clark, R. B., *Marine Pollution*, 3rd ed., Clarendon Press, Oxford, 1992.
106. Birge, W. J., Black, J. A., Westerman, A. G., and Hudson, J. E., The effect of mercury on reproduction, of fish and amphibians, in *The Biogeochemistry of Mercury in the Environment*, Nriagu, J. O. (Ed.), Elsevier/North Holland, New York, 1979, 629-655.
107. NOAA, *National Status and Trends Program for Marine Environmental Quality. Progress Report. A Summary of Selected Data on Chemical Contaminants in Tissues Collected During 1984, 1985, and 1986,* Tech. Mem. NOS OMA 38, National Oceanic and Atmospheric Administration, Rockville, Md, 1987.
108. Becker, P. H., Henning, D., and Furness, R. W., Differences in mercury contamination and elimination during feather development in gull and tern broods, *Arch. Environ. Contam. Toxicol.,* 27, 162, 1994.
109. Honda, K., Marcovecchio, J. E., Kan, S., Tatsukawa, R., and Ogi, H., Metal concentrations in pelagic seabirds from the North Pacific Ocean, *Arch. Environ. Contam. Toxicol.,* 19, 704, 1990.
110. Gerlach, S. A., *Marine Pollution: Diagnosis and Therapy*, Springer-Verlag, Berlin, 1981.
111. Marcovecchio, J. E., Moreno, V. J., Bastida, R. O., Gerpe, M. S., and Rodriguez, D. H., Tissue distribution of heavy metals in small cetaceans from the southwestern Atlantic Ocean, *Mar. Pollut. Bull.,* 21, 299, 1990.
112. Law, R. J., Fileman, C. F., Hopkins, A. D., Baker, J. R., Harwood, J., Jackson, D. B., Kennedy, S., Martin, A. R., and Morris, R. J., Concentrations of trace metals in the livers of marine mammals (seals, porpoises, and dolphins) from waters around the British Isles, *Mar. Pollut. Bull.,* 22, 183, 1991.
113. Law, R. J., Jones, B. R., Baker, J. R., Kennedy, S., Milne, R., and Morris, R. J., Trace metals in the livers of marine mammals from the Welsh Coast and the Irish Sea, *Mar. Pollut. Bull.,* 24, 296, 1992.
114. Worrall, C. M. and Widdows, J., Physiological changes following transplantation of the bivalve *Scrobicularia plana* between three populations, *Mar. Ecol. Prog. Ser.,* 12, 281, 1983.
115. Jorgensen, L. A. and Pedersen, B., Trace metals in fish used for time trend analysis and as environmental indicators, *Mar. Pollut. Bull.,* 28, 24, 1994.
116. Hellou, J., Warren, W. G., Payne, J. F., Belkhode, S., and Lobel, P., Heavy metals and other elements in three tissues of cod, *Gadus morhua,* from the northwest Atlantic, *Mar. Pollut. Bull.,* 24, 452, 1992.
117. Windom, H., Stein, D., Sheldon, R., and Smith, R., Jr., Comparison of trace metal concentrations in muscle tissue of a benthopelagic fish (*Coryphaenoides armatus*) from the Atlantic and Pacific oceans, *Deep-Sea Res.,* 34, 213, 1987.
118. Chasko, G. G., Hoehn, T. R., and Howell-Heller, P., Toxicity of lead shot to wild black ducks and mallards fed natural foods, *Bull. Environ. Contam. Toxicol.,* 32, 417, 1984.
119. Di Giulio, R. T. and Scanlon, P. F., Effects of cadmium and lead ingestion on tissue concentrations of cadmium, lead, copper, and zinc in mallard ducks, *Sci. Tot. Environ.,* 39, 103, 1984.
120. Custer, T. W. and Hohman, W. L., Trace elements in canvasbacks (*Aythya valisineria*) wintering in Louisiana, USA, 1987-1988, *Environ. Pollut.,* 84, 253, 1994.

121. Wilson, K. W., Head, P. C., and Dones, P. D., Mersey estuary (UK) bird mortalities — cause, consequences, and correctives, *Water Sci. Technol.,* 18, 171, 1986.
122. Malcolm, H. M., Boyd, I. L., Osborn, D., French, M. C., and Freestone, P., Trace metals in Antarctic fur seal (*Arctocephalus gazella*) livers from Bird Island, South Georgia, *Mar. Pollut. Bull.,* 28, 375, 1994.
123. Lee, R. F., Metabolism of tributyltin by marine animals and possible linkages to effects, *Mar. Environ. Res.,* 32, 29, 1991.
124. Davies, I. M., McKie, J. C., and Paul, J. D., Accumulation of tin and tributyltin from antifouling paint by cultivated scallops (*Pecten maximus*) and Pacific oysters (*Crassostrea gigas*), *Aquaculture,* 55, 103, 1986.
125. Alzieu, C., TBT detrimental effects on oyster culture in France — evolution since antifouling paint regulation, in *Proceedings of Oceans '86 Conference Record: Organotin Symposium*, Vol. 4, Institute of Electrical and Electronics Engineers, New York, 1986, 1130.
126. Dyrynda, E. A., Incidence of abnormal shell thickening in the Pacific oyster *Crassostrea gigas* in Poole Harbor (UK), subsequent to the 1987 TBT restrictions, *Mar. Pollut. Bull.,* 24, 156, 1992.
127. His, E. and Robert, R., Comparative effects of two antifouling paints on the oyster *Crassostrea gigas*, *Mar. Biol.,* 95, 83, 1987.
128. Minchin, D., Duggan, C. B., and King, W., Possible effects of organotin on scallop recruitment, *Mar. Pollut. Bull.,* 18, 604, 1987.
129. Pickwell, G. W. and Steinert, S. A., Accumulation and effects of organotin compounds in oysters and mussels: correlation with serum biochemical and cytological factors and tissue burdens, *Mar. Environ. Res.,* 24, 215, 1988.
130. Widdows, J. and Page, D. S., Effects of tributyltin and dibutyltin on the physiological energetics of the mussel, *Mytilus edulis*, *Mar. Environ. Res.,* 35, 233, 1993.
131. Kure, L. K. and Depledge, M. H., Accumulation of organotin in *Littorina littorea* and *Mya arenaria* from Danish coastal waters, *Environ. Pollut.,* 84, 149, 1994.
132. Ward, G. S., Cramm, G. C., Parrish, P. R., Trachman, H., and Slesinger, A., Bioaccumulation and chronic toxicity of bis (tributyltin) oxide (TBTO): tests with saltwater fish, in *Aquatic Toxicity and Hazard Assessment: Fourth Conference*, Branson, D. R., and Dickson, K. L. (Eds.), ASTM STP 737, American Society for Testing and Materials, Philadelphia, 1981, 183.
133. Leah, R. T., Evans, S. J., and Johnson, M. S., Arsenic in plaice (*Pleuronectes platessa*) and whiting (*Merlangius merlangus*) from the northeast Irish Sea, *Mar. Pollut. Bull.,* 24, 544, 1992.
134. Staveland, G., Marthinsen, I., Norheim, G., and Julshamn, K., Levels of environmental pollutants in flounder (*Platichthys flesus* L.) and cod (*Gadus morhua* L.) caught in the waterway of Glomma, Norway. II. Mercury and arsenic, *Arch. Environ. Contam. Toxicol.,* 24, 187, 1993.
135. Mat, I., Arsenic and trace metals in commercially important bivalves, *Anadara granosa* and *Paphia undulata*, *Bull. Environ. Contam. Toxicol.,* 52, 833, 1994.
136. Langston, W. J., Availability of arsenic to estuarine and marine organisms: a field and laboratory investigation, *Mar. Biol.,* 80, 143, 1984.
137. Langston, W. J., Assessment of the distribution and availability of arsenic and mercury in estuaries, in *Estuarine Management and Quality Assessment*, Wilson, J. G. and Halcrow, W. (Eds.), Plenum Press, New York, 1985, 131.
138. Rosemarin, A., Notini, M., and Holmgren, K., The fate of arsenic in the Baltic Sea *Fucus visiculosus* ecosystem, *Ambio,* 14, 342, 1985.
139. Sanders, J. G. and Osman, R. W., Arsenic incorporation in a saltmarsh ecosystem, *Est. Coastal Shelf Sci.,* 20, 387, 1985.
140. Goede, A. A., Mercury, selenium, arsenic, and zinc in waders from the Dutch Wadden Sea, *Environ. Pollut.,* 37A, 287, 1985.
141. Sanders, J. G. and Vermersch, P. S., Response of marine phytoplankton to low levels of arsenate, *J. Plankton Res.,* 4, 881, 1982.
142. Oladimeji, A. A., Qadri, S. U., and de Freitas, A. S. W., Long-term effects of arsenic accumulation in rainbow trout, *Salmo gairdneri*, *Bull. Environ. Contam. Toxicol.,* 32, 732, 1984.
143. Cutter, G. A., The estuarine behavior of selenium in San Francisco Bay, *Est. Coastal Shelf Sci.,* 28, 13, 1989.
144. Maher, W. A., Distribution of selenium in marine animals: relationship to diet, *Comp. Biochem. Physiol.,* 86C, 131, 1987.

145. Liu, D. L., Yang, Y. P., Hu, M. H., Harrison, P. J., and Price, N. M., Selenium content of marine food chain organisms from the coast of China, *Mar. Environ. Res.*, 22, 151, 1987.
146. Hamilton, S. J. and Buhl, K. J., Acute toxicity of boron, molybdenum, and selenium to fry of chinook salmon and coho salmon, *Arch. Environ. Contam. Toxicol.*, 19, 366, 1990.
147. Saiki, M. K. and Palawski, D. U., Selenium and other elements in juvenile striped bass from the San Joaquin Valley and San Francisco estuary, CA, *Arch. Environ. Contam. Toxicol.*, 19, 717, 1990.
148. Heinz, G. H., Pendleton, G. W., Krynitsky, A. J., and Gold, L. G., Selenium accumulation and elimination in mallards, *Arch. Environ. Contam. Toxicol.*, 19, 374, 1990.
149. Heinz, G. H. and Fitzgerald, M. A., Reproduction of mallards following overwinter exposure to selenium, *Environ. Pollut.*, 81, 117, 1993.
150. Hoffman, D. J., Sanderson, C. J., LeCaptain, L. J., Comartie, E., and Pendleton, G. W., Interactive effects of arsenate, selenium, and dietary protein on survival, growth, and physiology in mallard ducklings, *Arch. Environ. Contam. Toxicol.*, 22, 55, 1992.
151. Goede, A. A., Wolterbeek, H. T., and Koese, M. J., Selenium concentrations in the marine invertebrates *Macoma balthica*, *Mytilus edulis*, and *Nereis diversicolor*, *Arch. Environ. Contam. Toxicol.*, 25, 85, 1993.
152. Ohlendorf, H. M., Hothem, R. L., Bunck, C. M., Aldrich, T. W., and Moore, J. F., Relationships between selenium concentrations and avian reproduction, in *Trans. 51st North Am. Wildlife Nat. Res. Conf.*, 1986, 330.
153. King, K. A. and Cromartie, E., Mercury, cadmium, lead, and selenium in three waterbird species nesting in Galveston Bay, Texas, U.S.A., *Colonial Waterbirds*, 9, 90, 1986.
154. Cottiglia, M., Focardi, S., Leonzio, C., Mascia, C., Renzoni, A. and Fossi, C., Contaminants in tissues of shorebirds from a polluted lagoon of the Island of Sardina, *VIes Journees Etud. Pollutions, Cannes*, CIESM, 1982, 293.
155. Goede, A. A., Nygard, T., de Bruin, M. and Steinnes, E., Selenium, mercury, arsenic, and cadmium in the life cycle of the dunlin, *Calidris alpina*, a migrant wader, *Sci. Tot. Environ.*, 78, 208, 1989.
156. Burger, J. and Gochfeld, M., Heavy metal and selenium levels in feathers of young egrets and herons from Hong Kong and Szechuan, China, *Arch. Environ. Contam. Toxicol.*, 25, 322, 1993.
157. Skaare, J. U., Degre, E., Aspholm, P. E., and Ugland, K. I., Mercury and selenium in Arctic and coastal seals off the coast of Norway, *Environ. Pollut.*, 85, 153, 1994.

APPENDIX 1. SOURCES AND CONCENTRATIONS

TABLE 1.1
Natural and Anthropogenic Concentrations of Some Trace Metals

	Unit	Pb	Hg	Cd	Sb	Cr	Se	As	Cu	Zn
Earth crust concentration	mg/kg	15	0.06	0.2	0.2	?	0.09	2	45	40
Seawater concentration	µg/l	0.002	0.007	0.1	0.3	0.3	0.5	2	2	3
Total amount in world oceans	10^6 t	2.8	10	140	420	420	700	2800	2800	4200
Input by erosion	10^3 t/y	150	3.5	0.5	1.3	236	7.2	72	325	720
Input through burning of fossil fuel and cement fabrication	10^3 t/y	34	2.0	0.2	?	1.5	1.1	8.2	2.1	37
Mine production	10^3 t/y	3500	9.0	15.0	70	3000	1.2	30	7500	5000
Annual mine production in percent of total amount in world oceans	%	125	0.1	0.01	0.015	0.7	0.0002	0.001	0.3	0.1

Source: The United Nations Joint Group of Experts on the Scientific Aspects of Marine Pollution, Principles for Developing Coastal Water Quality Criteria, Rep. GESAMP (5), Rome, 1976.

TABLE 1.2
Worldwide Emissions of Trace Metals to the Atmosphere (1000 t/yr)

Metal	Natural Sources	Anthropogenic Sources
Arsenic	7.8	24
Cadmium	0.96	7.3
Copper	19	56
Nickel	26	47
Lead	19	449
Selenium	0.4	1.1
Zinc	4	314

Source: Clark, R. B., *Marine Pollution,* 3rd ed., Clarendon Press, Oxford, 1992. With permission.

TABLE 1.3
Typical Concentrations of Some Metals in Airborne Particles

Metal	Urban Air (ng m^{-3})	Rural Air (ng m^{-3})
As	5–300	1–20
Cd	0.5–200	0.5–10
Ni	1–500	1–50
Pb	10–10,000	5–500
V	10–100	3–50
Zn	200–2000	5–100
Co	0.2–20	0.1–5
Cr	2–200	1–20
Cu	10–1000	2–100
Fe	100–10,000	100–10,000

Source: Harrison, R. M. (Ed.), *Pollution: Causes, Effects, and Control,* 2nd ed., Royal Society of Chemistry, Cambridge, 1990. With permission.

TABLE 1.4
Transfer of Metals from the Atmosphere to the Sea Surface (ng/cm/yr)

Element	North Sea	Western Mediterranean	South Atlantic Bight	Tropical North Atlantic	Tropical North Pacific
Aluminium	30,000	5000	2900	5000	1200
Manganese	920	—	60	70	9
Iron	25,500	5100	5900	3200	560
Nickel	260	—	390	20	—
Copper	1300	96	220	25	8.9
Zinc	8950	1080	750	130	67
Arsenic	280	54	45	—	—
Cadmium	43	13	9	5	0.35
Mercury	—	5	24	2.1	—
Lead	2650	1050	660	310	7.0

Source: Clark, R. B., *Marine Pollution,* 3rd ed., Clarendon Press, Oxford, 1992. With permission.

TABLE 1.5
Atmospheric Deposition to the Ocean of Primarily Anthropogenic Trace Metals (10^9 g/atm)

	Cd	Cu	Ni	Zn	Aa	Pb	Hg
North Pacific	0.47–0.81	3.2–10.6	4.5–5.7	9.1–46.5	1.6–0.7	17.9	0.58
South Pacific	0.06–0.10	0.4–1.3	0.6–0.7	1.1–5.7	0.2–0.09	2.2	0.33
North Atlantic	1.47–2.54	10.2–33.3	14.1–18.1	28.6–147	5.1–2.3	56.5	0.42
South Atlantic	0.11–0.19	0.8–2.5	1.1–1.3	2.1–10.9	0.4–0.17	4.2	0.09
North Indian	0.12–0.22	0.9–2.8	1.2–1.5	2.4–12.5	0.4–0.2	4.8	0.07
South Indian	0.06–0.11	0.4–1.4	0.6–0.8	1.2–6.2	0.2–0.1	2.4	0.2
Global total	2.3–4.0	16–52	22–28	44–228	7.9–3.6	88.0	1.7

Source: The United Nations Joint Group of Experts on the Scientific Aspects of Marine Pollution, Atmospheric Input of Trace Species to the World Ocean, Rep. GESAMP (38), Rome, 1989.

TABLE 1.6
Input of Four Heavy Metals to the North Sea (t/yr)

Source	Copper	Mercury	Lead	Zinc
Rivers	1290–1330	20–21	920–980	7360–7370
Dredgings	1000	17	170	8000
Atmosphere	400–1600	10–30	2600–7400	4900–11,000
Direct discharge	315	5	170	1170
Industrial dumping	160	0.2	200	450
Sewage sludge	100	0.6	100	220
Incineration	3	—	2	12

Source: Clark, R. B., *Marine Pollution,* 3rd ed., Clarendon Press, Oxford, 1992. With permission.

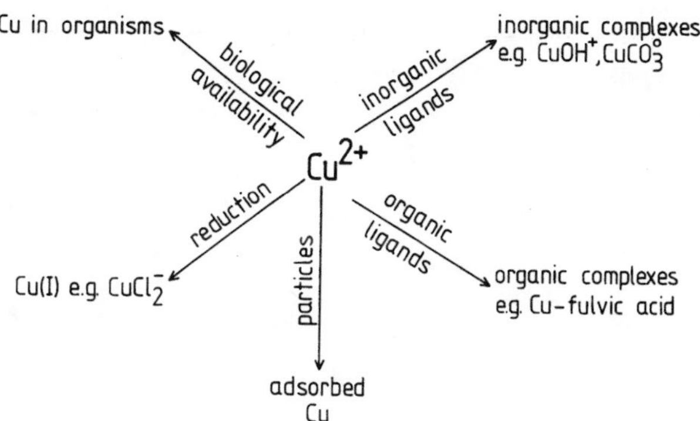

FIGURE 1.1 Major processes involving trace metals in sea water, taking copper as an example. (From Harrison, R. M. (Ed.), *Pollution: Causes, Effects, and Control,* 2nd ed., Royal Society of Chemistry, Cambridge, 1990. With permission.)

TABLE 1.7
Metal Complexation and Adsorption Models for Seawater

	Ligand Description	Competition by Other Cations	State-of-the-Art
Inorganic complexation	Small number of well-defined ligands	Can be treated readily as side reactions of the ligand	Accuracy of models is limited only by the thermodynamic data available
Organic complexation	Polydisperse range of binding sites	Empirical treatments only	Modelling is based on empirical constants which are functions of pH, etc.
Adsorption (pure phases)	Single type of binding sites in immobilized array	Treated as side reactions	Surface complexation modelling provides a successful approach
Adsorption (natural particles)	No working models are yet available; material comprises both inorganic particles and natural organic matter		Empirical K_D values provide the best available parameterization

Source: Harrison, R. M. (Ed.), *Pollution: Causes, Effects, and Control,* 2nd ed., Royal Society of Chemistry, Cambridge, 1990. With permission.

TABLE 1.8
Solubilities of Heavy Metals in Rivers

	Heavy-Metal Soluble Fraction (%)										
River	**Sb**	**Cd**	**Cr**	**Co**	**Cu**	**Fe**	**Pb**	**Mn**	**Hg**	**Ni**	**Zn**
Ruhr, F.R.G.	—	100	96	100	74	56	62	69	—	—	86
Schuylkill, U.S.	—	21	15	—	36	—	5	—	—	42	33
Thames, U.K.	72	—	22	30	—	3	—	—	23	53	31
Yarra, Australia	—	80	—	—	75	83	58	—	—	—	90
Elbe, F.R.G.	—	22	—	—	50	4	—	69	—	—	48

Source: Stephenson, T., in *Heavy Metals in Wastewater and Sludge Treatment Processes,* Vol. 2, *Treatment and Disposal,* Lester, J. N. (Ed.), CRC Press, Boca Raton, FL, 1987, 74. With permission.

TABLE 1.9
Classification of Selected Metals and Associated Ligands

Classification		
Hard Acceptor/Acid	**Intermediate**	**Soft Acceptor/Acid**
K^+, Na^+, Be^{2+}, Ca^{2+}, Mg^{2+}, Mn^{2+}, Al^{3+}, As^{3+}, Co^{3+}, Cr^{3+}, Fe^{3+}	Co^{2+}, Cu^{2+}, Fe^{2+}, Ni^{2+}, Pb^{2+}, Zn^{2+}	Ag^+, Au^+, Cu^+, Tl^+, Cd^{2+}, Hg^{2+}
Hard Donor/Base	**Intermediate**	**Soft Donor/Base**
H_2O, OH^-, F^-, Cl^-, SO_4^{2-}, CO_3^{2-}, O^{2-}, PO_4^{3-}	Br^-, NO_2^-, SO_3^{2-}	SH^-, RS^-, CN^- SCN^-, S^{2-}, CO, R_2, S, RSH
Class A Metals	**Borderline**	**Class B Metals**
K^+, Na^+, Ba^{2+}, Be^{2+}, Ca^{2+}, Mg^{2+}, Al^{3+}	Cd^{2+}, Co^{2+}, Cr^{2+}, Cu^{2+}, Fe^{2+}, Mn^{2+}, Ni^{2+}, Pb^{2+}, Sn^{2+}, V^{2+}, Zn^{2+}, Fe^{3+}, As(III), Sb(III), Sn(IV)	Ag^+, Cu^+, Tl^+, Hg^{2+}, Bi^{3+}, Tl^{3+}, Pb(IV)
Class A Ligands	**Borderline**	**Class B Ligands**
H_2O, OH^-, NO_3^-, F^-, SO_4^{2-}, CO_3^{2-}, O^{2-}, HPO_4^{2-}, PO_4^-, ROH, $RCOO^-$, ROR	Br^-, Cl^-, N_3^-, NO_2^-, O_2^-, SO_3^{2-}, O_2^{2-}, O_2, NH_3, N_2, RNH_2, R_2NH, R_3N, CONR	H^-, I^-, R^-, CN^-, RS^-, S^{2-}, CO, R_2S, R_3As
Essential	**Possibly Beneficial**	**No Apparent Metabolic Function**
Animals: Co, Cr, Cu, Fe, Mn, Mo, Ni, Se, Sn, V, Zn *Plants:* B, Cu, Fe, Mn, Mo, Se, Zn	As, Ba	Bi, Cd, Hg, Pb, Tl
Noncritical	**Toxic/Accessible**	**Toxic/Insoluble 13**
Al, Ca, Fe, K, Mg, Na	Ag, As, Be, Bi, Cd, Co, Cu, Hg, Ni, Pb, Sb, Se, Sn, Te, Tl, Zn	Ba

Note: Some elements in the original classifications have been omitted for clarity.

Source: Rudd, T., in *Heavy Metals in Wastewater and Sludge Treatment Processes,* Vol. 1, *Sources, Analysis, and Legislation,* Lester, J. N. (Ed.), CRC Press, Boca Raton, FL, 1987, 3. With permission.

TABLE 1.10
Concentration and Speciation of Elements (Including Heavy Metals) in Ocean Water

Element	Probable Main Species in Oxygenated Seawater	Range and Average Concentration at 35% Salinity[a]
Li	Li^+	25 μmol/kg
Be	$BeOH^+$, $Be(OH)_2^0$	4–30 pmol/kg; 20 pmol/kg
B	H_3BO_3	0.416 mmol/kg
C	HCO_3^-, CO_3^{2-}	2.0–2.5 mmol/kg; 2.3 mmol/kg
N	NO_3^- (also as N_2)	<0.1–45 μmol/kg; 30 μmol/kg
O	O^2 (also as H_2O)	0–300 μmol/kg
F	F^-, MgF^+	68 μmol/kg
Na	Na^+	0.468 mol/kg
Mg	Mg^{2+}	53.2 mmol/kg
Al	$Al(OH)_4^-$, $Al(OH)_3^0$	(5–40 mmol/kg; 20 mmol/kg)
Si	H_4SiO_4	<1–180 μmol/kg; 100 μmol/kg
P	HPO_4^{2-}, $NaHPO_4^-$, $MgHPO_4^0$	<1–3.5 μmol/kg; 2.3 μmol/kg
S	SO_4^{2-}, $NaSO_4^-$, $MgSO_4^0$	28.2 mmol/kg
Cl	Cl^-	0.546 mol/kg
K	K^+	10.2 mmol/kg
Ca	Ca^{2+}	10.3 mmol/kg
Sc	$Sc(OH)_3^0$	8–20 pmol/kg; 15 pmol/kg
Ti	$Ti(OH)_4^0$	(<20 nmol/kg)
V	HVO_4^{2-}, $H_2VO_4^-$, $NaHVO_4^-$	20–35 nmol/kg; 30 nmol/kg
Cr	CrO_4^{2-}, $NaCrO_4^-$	2–5 nmol/kg; 4 nmol/kg
Mn	Mn^{2+}, $MnCl^+$	0.2–3 nmol/kg; 0.5 nmol/kg
Fe	$Fe(OH)_3^0$	0.1–2.5 nmol/kg; 1 nmol/kg
Co	Co^{2+}, $CoCO_3^0$, $CoCl^+$	(0.01–0.1 nmol/kg; 0.02 nmol/kg)
Ni	Ni^{2+}, $NiCO_3^0$, $NiCl^+$	2–12 nmol/kg; 8 nmol/kg
Cu	$CuCO_3^0$, $CuOH^+$, Cu^{2+}	0.5–6 nmol/kg; 4 nmol/kg
Zn	Zn^{2+}, $ZnOH^+$, $ZnCO_3^0$, $ZnCl^+$	0.05–9 nmol/kg; 6 nmol/kg
Ga	$Ga(OH)_4^-$	(0.3 nmol/kg)
Ge	H_4GeO_4, $H_3GeO_4^-$	≤7–115 pmol/kg; 70 pmol/kg
As	$HAsO_4^{2-}$	15–25 nmol/kg; 23 nmol/kg
Se	SeO_4^{2-}, SeO_3^{2-}, $HSeO_3^-$	0.5–2.3 nmol/kg; 1.7 nmol/kg
Br	Br^-	0.84 mmol/kg
Rb	Rb^+	1.4 μmol/kg
Sr	Sr^{2+}	90 μmol/kg
Y	YCO_3^+, YOH^{2+}, Y^{3+}	(0.15 nmol/kg)
Zr	$Zr(OH)_4^0$, $Zr(OH)_5^-$	(0.3 nmol/kg)
Nb	$Nb(OH)_6^-$, $Nb(OH)_5^0$	(≤50 pmol/kg)
Mo	MoO_4^{2-}	0.11 μmol/kg
(Tc)	TcO_4^-	No stable isotope
Ru	?	?
Rh	?	?
Pd	?	?
Ag	$AgCl_2^-$	(0.5–35 pmol/kg; 25 pmol/kg)
Cd	$CdCl_2^0$	0.001–1.1 nmol/kg; 0.7 nmol/kg
In	$In(OH)_3^0$	(1 pmol/kg)
Sn	$SnO(OH)_3^-$	(1–12, ~4 pmol/kg)
Sb	$Sb(OH)_6^-$	(1.2 nmol/kg)
Te	TeO_3^{2-}, $HTeO_3^-$	?
I	IO_3^-	0.2–0.5 μmol/kg; 0.4 μmol/kg
Cs	Cs^+	2.2 nmol/kg

TABLE 1.10 (continued)
Concentration and Speciation of Elements (Including Heavy Metals) in Ocean Water

Element	Probable Main Species in Oxygenated Seawater	Range and Average Concentration at 35% Salinity[a]
Ba	Ba^{2+}	32–150 nmol/kg; 100 nmol/kg
La	La^{3+}, $LaCO_3^+$, $LaCl^{2+}$	13–37 pmol/kg; 30 pmol/kg
Ce	$CeCO_3^+$, Ce^{3+}, $CeCl^{2+}$	16–26 pmol/kg; 20 pmol/kg
Pr	$PrCO_3^+$, Pr^{3+}, $PrSO_4^+$	(4 pmol/kg)
Nd	$NdCO_3^+$, Nd^{3+}, $NdSO_4^+$	12–25 pmol/kg; 20 pmol/kg
Sm	$SmCO_3^+$, Nd^{3+}, $NdSO_4^+$	2.7–4.8 pmol/kg; 4 pmol/kg
Eu	$EuCO_3^+$, Eu^{3+}, $EuOH^{2+}$	0.6–1.0 pmol/kg; 0.9 pmol/kg
Gd	$GdCO_3^+$, Gd^{3+}	3.4–7.2 pmol/kg; 6 pmol/kg
Tb	$TbCO_3^+$, Tb^{3+}, $TbOH^{2+}$	(0.9 pmol/kg)
Dy	$DyCO_3^+$, Dy^{3+}, $DyOH^{2+}$	(4.8–6.1 pmol/kg; 6 pmol/kg)
Ho	$HoCO_3^+$, Ho^{3+}, $HoOH^{2+}$	(1.9 pmol/kg)
Er	$ErCO_3^+$, $ErOH^{2+}$, Er^{3+}	4.1–5.8 pmol/kg; 5 pmol/kg
Tm	$TmCO_3^+$, $TmOH^{2+}$, Tm^{3+}	(0.8 pmol/kg)
Yb	$YbCO_3^+$ $YbOH^{2+}$	3.5–5.4 pmol/kg; 5 pmol/kg
Lu	$LuCO_3^+$, $LuOH^{2+}$	(0.9 pmol/kg)
Hf	$Hf(OH)_4^0$, $Hf(OH)_5^-$	(<40 pmol/kg)
Ta	$Ta(OH)_5^0$	(<14 pmol/kg)
W	WO_4^{2-}	0.5 nmol/kg
Re	ReO_4^-	(14–30 pmol/kg; 20 pmol/kg)
Os	?	?
Lr	?	?
Pt	?	?
Au	$AuCl_2^-$	(25 pmol/kg)
Hg	$HgCl_4^{2-}$	(2–10 pmol/kg; 5 pmol/kg)
Tl	Tl^+, $TlCl^0$, or $Tl(OH)_3^0$	60 pmol/kg
Pb	$PbCO_3^0$, $Pb(CO_3)_2^{2-}$, $PbCl^+$	5–175 pmol/kg; 10 pmol/kg
Bi	BiO^+, $Bi(OH)_2^+$	≤0.015–0.24 pmol/kg

[a] Parentheses indicate uncertainty about the accuracy or range of concentration given.

Source: Bruland, K. W., in *Chemical Oceanography,* Vol. 8, Riley, J. P. and Chester, R. (Eds.), Academic Press, London, 1983, 157. Copyright Academic Press. With permission.

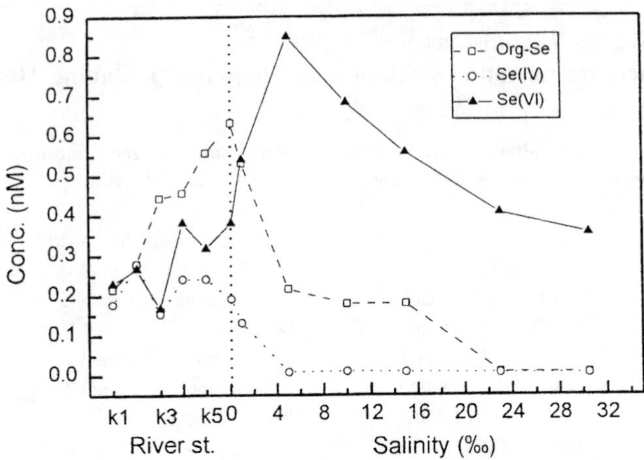

FIGURE 1.2. Selenium speciation in the Kaoping River and estuary, southwestern Taiwan. (From Hung, J.-J. and Shy, C.-P., *Estuaries,* 18, 234, 1995. With permission.)

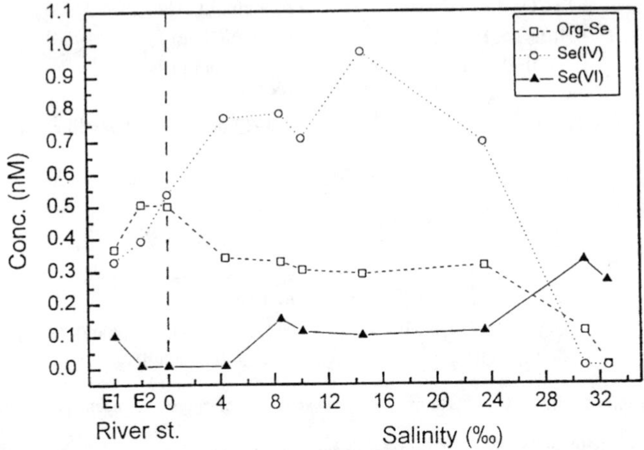

FIGURE 1.3. Selenium speciation in the Erhjen River and estuary, southwestern Taiwan. (From Hung, J.-J. and Shy, C.-P., *Estuaries,* 18, 234, 1995. With permission.)

TABLE 1.11
Concentrations of Mercury, Cadmium, and Lead in Surface Waters of the Open Ocean (ng/l)

Location	Mercury	Cadmium	Lead
Northwest Atlantic	0.7 ± 0.2, 0.5, 0.8	2.5 ± 0.9, 0.2	33, 46, 26 ± 4, 8.1
Northeast Atlantic	1.2, 0.6, 3.5	11.2, 4.8 ± 1.4	33, 33
		4.5 ± 2.9, 2, 1.1	
Northeast Pacific	0.37 ± 0.13[a]	4.5, 2.2–9, 0.16	14, 5–15
Northwest Pacific	5.0 ± 0.5[b]	14	
Southwest Atlantic			
Southeast Atlantic			
Southwest Pacific	0.42	15	4.6 ± 1.0, 3.5
Southeast Pacific	2.2	3.8 ± 2.4	16
Arctic	2.3[a]	8.1 ± 1.4, 14.5	14.8 ± 3.5
Antarctic		17–54, 13.7 ± 5.9	
Mediterranean	0.5–2.5, 3.5	8, 4.0–13.8, 14	30–150
Baltic Sea	3.0	30 ± 2.7, 34	16 ± 4.5, 41–83, 50
North Sea	1.8, 3.3	23, 59, 21 ± 6, 16	52, 40, 31, 62
Indian	4.4 ± 1.6[b]	15, 0.25	30
Arabian Sea		3.4	
Bering Sea	7.0 ± 3.3[a]		
Japan Sea	5.9 ± 1.8[a]		
East and South China Seas	5.7 ± 1.8[a]		

Note: See original source for particular studies.

[a] Reactive mercury.
[b] Total mercury.

Source: Fowler, S. W., *Mar. Environ. Res.*, 29, 1, 1990. With permission.

TABLE 1.12
Concentrations of Mercury, Cadmium, and Lead in Coastal Surface Water (ng/l)

Location	Mercury	Cadmium	Lead
Northwest Atlantic			20–125
Cape Cod			21
Long Island Sound	6 ± 1[a]		
Gulf of St. Lawrence		22	
South Atlantic Bight			25 ± 10
Canadian Arctic		59 ± 11	
New York Bight	10–90	3–270	
Northeast Atlantic			
Greenland	0.9[a]		
Irish Sea		25	
Bay of Biscaye	0.7–5.5, 3[b]	13–20	
English Channel		20–25, 11–16	
Tagus Estuary mouth	80[b]		
Scottish Coast		9–19	
Northeast Pacific			
Monterey Bay		9–23	7.6
Juan de Fuca Strait			6.5
Puget Sound	0.2–1.0[b]	70–100	20–110
Southern California Bight			25
Near Los Angeles outfall			150
Northwest Pacific			
Tokyo Bay	3–5		
Southwest Atlantic			
Brazil		1.2–12	
Southwest Pacific			
Australia		<10–60	
New Zealand (Taranaki Harbour)		2–30	
Southeast Pacific	0.1		4–11
Mediterranean	0.5–20		
Spain		7.6–9	
Rhône Delta		11 ± 6, 4, 20	77
Italy	6.3[b], 2.0[a]		
Black Sea		7.8	12
Baltic Sea	3.3(0.9–11)[b]	59	40
Bothnian Bay		34 ± 9, 39 ± 6	246 ± 66
Bothnian Sea		36 ± 8	120 ± 29
Northern Baltic Proper		39 ± 13	189 ± 24
Gulf of Finland		36 ± 5	324 ± 24
North German Coast		29 ± 6	
North Sea		46 ± 18	
Framvaren fjord, Norway	0.3–25[a]	56	73
Kattegat/Skagerrak	5 ± 2–12 ± 3	22.5	50
U.K.		10–60[b]	30–265[b]
German Bight		71	
Indian			
The Gulf	9–25[b]	16–30	60–120
Oman (Arabian Sea)	4–18[b]	5–100	50–120
Red Sea		5	
India (Arabian Sea)	30–130[b]		
India (Bay of Bengal)	<10–100[b]		

Note: See original source for particular studies.

[a] Reactive.

[b] Total dissolved.

Source: Fowler, S. W., *Mar. Environ. Res.*, 29, 1, 1990. With permission.

TABLE 1.13
Typical Background Concentrations of Some Metals in Rivers and Oceans

Metal	River (ppb)	Ocean (ppb)
Ag	0.3	0.1
Al	400	5
As	1	2.3
Cd	0.03	0.05
Co	0.2	0.02
Cr	1	0.6
Cu	5	3
Fe	670	3
Hg	0.07	0.05
Mn	5	2
Mo	1	10
Ni	0.3	2
Pb	3	0.03
Sb	1	0.2
Se	0.2	0.45
Sn	0.04	0.01
V	1	1.5
Zn	10	5

Source: Bryan, G. W., in *Marine Pollution,* Johnston, R. B. (Ed.), Academic Press, London, 1976, chap. 3. Copyright Academic Press. With permission.

TABLE 1.14
Enrichment Factors (Average or Typical Range) of Heavy Metals for Chesapeake Bay and Other Areas

Area	Cr	Mn	Ni	Cu	Zn	Pb
Chesapeake Bay (whole)	1	2	1	1	5	5
Upper (0–75 km)	1	3–6	—	2	6–8	4–7
Middle (75–200 km)	1	1	—	1	4–6	3–4
Lower (200–300 km)	1	<1	—	1	2–4	2–4
Baltimore Harbor	4	1	1	—	8	—
Hampton Roads	1	1	1	2–4	5–10	7–21
Susquehanna River	0.6	6	3	2	8	7
Offshore[a]	0.7	0.8	0.6	0.2	2	3
Delaware Bay	3	—	12	2	10	16
San Francisco Bay	—	<1	2	2	2	—

[a] Proposed disposal site on the inner Virginian continental shelf approximately 27 km off the Chesapeake Bay mouth.

Source: Sinex, S. A. and Wright, D. A., *Mar. Pollut. Bull.,* 19, 425, 1988. With permission.

TABLE 1.15
Sources of Trace Metals in San Francisco Bay, Including Municipal and Industrial Point Sources, Urban Runoff, and the Sacramento-San Joaquin River System[a]

Metal	Point Source	Urban Runoff	Rivers	Point Source Riverine[b]	Anthropogenic Riverine[c]
Ag	7.5	?	26	0.28	>0.28
As	5.7	9	37	0.15	0.39
Cd	4.0	3	27	0.15	0.26
Cr	14	15	92	0.15	0.32
Cu	31	59	203	0.15	0.44
Hg	0.8	0.2	3	0.26	0.33
Ni	29	?	82	0.35	>0.35
Pb	17	250	66	0.26	4.0
Se	2.5	?	7.4	0.33	>0.33
Zn	74	268	288	0.25	1.19

[a] Concentrations in kg/dry weight.

[b] Ratios of point source inputs and total anthropogenic input (sum of point source and urban runoff) relative to riverine input. Data from Walters, R. A. and Gartner, J. W., *Est. Coastal Shelf Sci.,* 21, 17, 1985.

[c] Data from Gunther, A. J., Davis, J. A., and Phillips, D. J. H., An Assessment of the Loading of Toxic Contaminants to the San Francisco Bay Delta, Tech. Rep., Aquatic Habitat Institute, Richmond, CA, 1987.

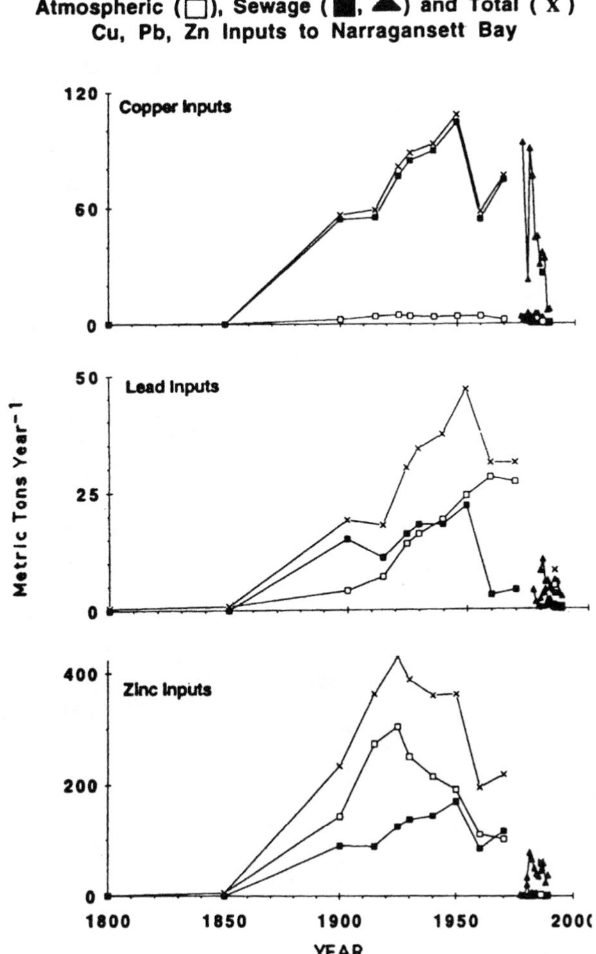

FIGURE 1.4. Long-term trends in metal loadings to Narragansett Bay, RI, from sewage treatment plants and from atmospheric emissions (fuel combustion). (From Bricker, S. B., *Estuaries,* 16, 589, 1993. With permission.)

TABLE 1.16
Concentrations of Mercury, Cadmium, and Lead in Surface Layers of Nearshore Sediments[a]

Location	Mercury	Cadmium	Lead
Baltic Sea	0.01–1.0	0.08–6.8	7–150
	0.23 (0.08–0.80)	3.2 (1.1–7.2)	54 (35–94)
Northeast Atlantic			
Irish Sea	0.07–3.3		
Tagus Estuary, Portugal	0.02–9.4		
France	0.08–0.12	0.15–0.21	31–45
Spain		1.7 ± 0.42	58 ± 22
S. Africa	0.002–1.4	0.1–2.8	2–87
Ivory Coast Lagoon	0.004–2.3		7–250
Nigeria	0.04–0.16	1.84–2.8	48–87
Northwest Atlantic			
Canada	0.04, 0.61 ± 0.33	0.16, 0.10–0.47	24
Bermuda		<0.25–0.99	6.4–230
New York Bight	0.12–4.9	<0.47–9.6	5–270
Mexico		0.1–2.4	10–91
Trinidad		0.05–4.5	6.7–29
South Carolina		0.01–0.46	0.3–30
Northeast Pacific			
Puget Sound	0.276	0.367	43.8
Southern California Bight	0.13–4.4	1.1–6.0, 0.4–140, 0.3–1.3	32–130, 10–540, 4.5–12.5
Costa Rica	0.022 ± 0.029	0.12 ± 0.08	5.3 ± 3.0
Northwest Pacific			
Korea			25–120
Seto Inland Sea		0.14–0.88	14–43
Minamata Bay	25, 25.3–32.4		
Malaysia		nd–1.25	6.5–32
Thailand		0.1–0.4, 0.005–0.11	11–18, 6.5 ± 1.5
Phillipines			36–40
China			21.6 ± 3.8
Southwest Atlantic			
Argentina		1.9–3.1, 0.24–0.44	5.3–19.9
Brazil	0.2–1.4	0.3–1.3	15–70
Southeast Pacific			
Chile	0.11–0.49	1.05–9.16	8.6–74
Southwest Pacific			
Port Phillip Bay, Australia		0.15–9.9	4.6–180
Manukau (Auckland) Harbor, New Zealand			98–247
New Zealand estuaries and harbors			43 ± 19
New Zealand fjords and sounds			35 ± 10
Mediterranean	0.01–37	0.2–49	4.5–280
Adriatic	<0.1–16.9	<0.05–5.6	5.3–96
Ligurian Sea		0.3–7.0	36–180
Sicily	0.03–2.0	2.5–4.6	4.5–17, 7.5–20
Bay of Naples	0.1–1.75	5–200	
Saronikos Gulf, Greece	0.3–10		
Israel	0.2–0.5		15–28
Spain	0.06–16.5	0.03–4.0	4.8–550
Thermaikos, Greece		0.2–5.1	18–246

TABLE 1.16 (continued)
Concentrations of Mercury, Cadmium, and Lead in Surface Layers of Nearshore Sediments[a]

Location	Mercury	Cadmium	Lead
Gulf			
Kuwait		0.75–3.0	10–40
Iraq		0.14–0.23	5.6–25.6
Oman	0.012–0.023	2.5–4.7	49–63
Indian Ocean			
Albany, Australia		0.26–7.6	13–180
Darwin, Australia		0.8–3.1	18.3–91
Bombay, India	2.5 (0.018–8), 0.038–0.08	10 ± 2	48 ± 7
Karwar, India	0.05–1.32		
Cauvery Estuary, India	0.118	1.85	38
Bay of Bengal, India	0.95–5.3		
Red Sea			
Jordan		2–18	83–225
North Sea			
Plymouth Estuary	0.02–2.6		

Note: See original source for particular studies.

[a] Concentrations expressed in μg/ dry weight. Single values are means or medians.

Source: Fowler, S. W., *Mar. Environ. Res.*, 29, 1, 1990. With permission.

TABLE 1.17
Trace Metals in Sediments from Selected Estuaries in the United States (ppm)

Estuary	Chromium	Copper	Lead	Zinc	Cadmium	Silver	Mercury
Casco Bay, ME	92.10	16.97	29.13	76.27	0.15	0.09	0.12
Merrimack River, MA	41.15	6.47	23.25	35.75	0.07	0.05	0.08
Salem Harbor, MA	2296.67	95.07	186.33	238.00	5.87	0.88	1.19
Boston Harbor, MA	223.67	148.00	123.97	291.67	1.61	2.64	1.05
Buzzards Bay, MA	73.66	25.02	30.72	97.72	0.23	0.37	0.12
Narragansett Bay, RI	93.60	78.95	60.25	144.43	0.35	0.56	0.00
East Long Island Sound, NY	37.63	11.26	22.13	58.83	0.11	0.15	0.09
West Long Island Sound, NY	131.50	111.00	69.75	243.00	0.73	0.68	0.48
Raritan Bay, NJ	181.00	181.00	181.00	433.75	2.74	2.06	2.34
Delaware Bay, DE	27.76	8.34	15.04	49.66	0.24	0.11	0.09
Lower Chesapeake Bay, VA	58.50	11.32	15.70	66.23	0.38	0.08	0.10
Pamlico Sound, NC	79.67	14.13	30.67	102.67	0.33	0.09	0.11
Sapelo Sound, GA	51.80	5.93	16.00	38.33	0.09	0.02	0.03
St. Johns River, FL	37.67	9.77	26.00	67.67	0.18	0.11	0.07
Charlotte Harbor, FL	26.47	1.17	4.33	7.20	0.08	0.01	0.02
Tampa Bay, FL	23.70	4.97	4.67	9.10	0.15	0.08	0.03
Apalachicola Bay, FL	69.17	16.93	30.67	111.67	0.05	0.06	0.06
Mobile Bay, AL	93.00	17.40	29.67	161.00	0.11	0.11	0.12
Mississippi River Delta, LA	72.27	19.40	22.67	90.00	0.47	0.17	0.06
Barataria Bay, LA	52.07	10.50	18.33	59.33	0.19	0.09	0.05
Galveston Bay, TX	41.13	8.03	18.33	33.97	0.05	0.09	0.03
San Antonio Bay, TX	39.43	5.57	11.33	32.00	0.07	0.09	0.02
Corpus Christi Bay, TX	31.43	6.63	13.00	56.00	0.19	0.07	0.04
Lower Laguna Madre, TX	24.53	5.83	11.33	36.00	0.09	0.07	0.03
San Diego Harbor, CA	178.00	218.67	50.97	327.67	0.99	0.76	1.04
San Diego Bay, CA	49.70	7.67	11.61	58.67	0.04	0.76	0.04
Dana Point, CA	39.80	10.03	18.80	53.67	0.22	0.80	0.13
Seal Beach, CA	108.33	26.00	27.37	125.00	0.17	1.27	0.59
San Pedro Canyon, CA	106.50	31.33	17.33	118.33	1.17	1.20	0.32
Santa Monica Bay, CA	53.53	10.53	33.37	46.67	0.18	0.51	0.01
San Francisco Bay, CA	1466.67	160.71	67.39	501.66	0.51	0.37	0.25
Bodega Bay, CA	246.33	0.06	2.17	38.33	0.18	1.74	0.14
Coos Bay, OR	110.30	1.47	4.65	32.00	0.62	0.31	0.11
Columbia River Mouth, OR/WA	29.53	17.00	15.90	107.67	0.86	2.14	0.25
Nisqually Reach, WA	118.07	13.33	24.57	105.33	0.68	2.62	0.32
Commencement Bay, WA	69.50	51.33	34.63	101.00	0.77	5.90	0.01
Elliott Bay, WA	114.37	96.00	20.23	166.00	0.84	1.18	0.11
Lutak Inlet, AK	58.27	26.67	15.90	180.33	0.96	0.09	0.24
Nahku Bay, AK	23.27	9.80	43.30	191.33	1.09	4.37	0.23
Charleston, SC	86.33	16.03	27.33	72.67	—	—	—

Source: Young, D. and Means, J., in *National Status and Trends Program for Marine Environmental Quality. Progress Report on Preliminary Assessment of Findings of the Benthic Surveillance Project, 1984,* National Oceanic and Atmospheric Administration, Rockville, MD, 1987; U.S. Geological Survey, National Water Summary, 1986.

TABLE 1.18
Average Concentrations of Metals in Sediments from 19 Estuaries in England Placed in Order of Decreasing Copper Concentration[a,b]

Site	Ag	As	Cd	Co	Cr	Cu	Fe	Hg	Mn	Ni	Pb	Se	Sn[c]	Zn
Restronguet Creek	3.76	*1740*	1.53	21	32	*2398*	*49,071*	0.46	485	*58*	341	—	55.9	*2821*
Fal	1.37	56	0.78	9	28	648	28,063	0.20	272	23	150	—	39.5	750
Tamar	1.22	93	0.96	21	47	330	35,124	0.83	590	44	235	1.30	8.3	452
Gannel	*4.13*[d]	174	1.35	*26*	24	150	25,420	0.08	649	38	*2753*	—	8.5	940
Tyne	1.55	24.8	*2.17*	11	46	92	28,206	0.92	395	34	187	1.23	5.4	421
Mersey	0.70	41.6	1.15	13	84	84	27,326	*3.01*	*1169*	29	124	0.30	8.3	379
Humber	0.42	49.8	0.48	16	77	54	35,203	0.55	1015	39	113	0.84	5.05	252
Medway	1.45	18.4	1.08	11	53	55	32,216	1.00	418	26	86	0.48	3.4	220
Poole	0.82	14.1	1.85	11	49	50	29,290	0.81	185	26	96	*1.51*	7.4	165
Severn	0.42	8.9	0.63	15	55	38	28,348	0.51	686	33	89	0.23	8.0	259
Hamble	0.16	18.4	0.34	10	37	31	28,132	0.43	241	19	56	0.41	3.9	105
Loughor	0.18	17.5	0.47	10	*207*	27	19,337	0.16	597	21	48	—	*161*	146
Dyfi	0.19	9.7	0.62	17	32	24	45,683	0.12	1127	33	166	0.20	0.98	212
Wyre	0.21	6.4	0.35	8	37	20	16,970	1.52	590	17	44	—	3.15	122
Avon	0.06	13.0	0.08	10	28	18	18,361	0.12	326	23	68	—	3.9	82
Teifi	0.09	10.6	0.17	10	29	13	30,280	0.04	684	23	25	0.16	1.14	87
Axe	0.13	4.8	0.17	7	27	12	14,004	0.20	248	14	26	—	1.39	76
Rother	0.17	12.4	0.13	6	29	11	15,648	0.09	259	15	20	0.18	0.62	46
Solway	0.07	6.4	0.23	6	30	7	14,816	0.03	577	17	25	0.11	0.40	59

[a] Concentrations expressed in μg/g dry weight.

[b] Data are for the <100 μm fraction of surface sediment following digestion in HNO_3.

[c] A large proportion of Sn in some samples occurs as cassiterite and is not dissolved in HNO_3 (fusion technique necessary).

[d] Figures in italics are the highest concentrations.

Source: Bryan, G. W. and Langston, W. J., *Environ. Pollut.*, 76, 89, 1992. With permission.

TABLE 1.19
Summary of Data Sources Concerning the Heavy Metal Composition of Chesapeake Bay Sediments

Area	Elements	Author(s)/Date of Publication[a]
Main stem	Al, C, Ca, Co, Cr, Cu, Fe, H, K, Mg, Mn, Na, Ni, Pb, S, Si, Ti, V	Sommer and Pyzik, 1974
	Cr, Cu, Ni, Pb	Schubel and Hirschberg, 1977
	Cd, Cu, Fe, Mn, Ni, Zn	Cronin et al., 1974
	Ag, Al, Cd, Co, Cr, Cu, Fe, Mn, Ni, Pb, V, Zn	Goldberg et al., 1978
	Al, Co, Cr, Cu, Fe, Ga, Mn, Ni, Org C, Org N, Pb, Si, Ti, V, Zn	Sinex, 1981
	As, Cd, Cu, Fe, Hg, Mn, Ni, Pb, Sn, Zn	Harris et al., 1980
Back River	Cd, Cu, Pb, Zn	Helz et al., 1975
Patapsco	Cd, Cr, Cu, Hg, Mn, Ni, Pb, Zn	Villa and Johnson, 1974
	As, Cd, Cr, Cu, Hg, Mn, Ni, Pb, Zn	Tsai et al., 1979
	Al, Co, Cr, Fe, Mn, Ni, Si, Ti, V, Zn	Sinex and Helz, 1982
Rhode River	Cd, Cr, Fe, Mn, Zn	Frazier, 1976
Patuxent	Cd, Co, Cr, Cu, Fe, Mn, Ni, Pb, Zn	Ferri, 1977
Potomac	Ag, Ba, Cd, Co, Cr, Cu, Fe, Li, Mn, Ni, Pb, Sr, V, Zn	Pheiffer, 1972
	Ba, Co, Cr, Cu, Fe, Mn, Pb, Sr, Ti, V, Zi, Zn	Mielke, 1974
Rappahannock, York	Cu, Zn	Huggett et al., 1975
Elizabeth	Al, Cd, Cr, Cu, Fe, Hg, Pb, Zn	Johnson and Villa, 1976
	Al, Co, Cr, Fe, Mn, Ni, Si, Ti, V, Zn	Helz et al., 1983
	Cd, Co, Cr, Cu, Fe, Mn, Ni, Pb, Zn	Rule, 1986

[a] See original source for particular studies.

Source: Helz, G. R. and Huggett, R. J., in *Contaminant Problems and Management of Living Chesapeake Bay Resources,* Majumdar, S. K., Hall, L. W., Jr., and Austin, H. M. (Eds.), Pennsylvania Academy of Science, Easton, 1987, 270. With permission.

TABLE 1.20
Metal Concentrations (μg/g) in Contaminated and Uncontaminated Sediments[a]

Sediment	Ag	As	Cd	Co	Cr	Cu	Fe	Hg	Mn	Ni	Pb	Sn	V	Zn
Firth of Clyde, Scotland														
Control area (mean)	<0.2	8	3.4	34	64	37	5.3×10^4	0.1	1100	50	86	19	250	160
Sewage-sludge dumpsite (max.)	5	24	7	40	310	210	6.1×10^4	2.2	1000	70	320	100	400	830
California coast														
Control area (median)	0.2	—	0.33	—	22	8.3	—	0.043	—	9.7	6.1	—	—	43
Los Angeles wastewater outfall (max.)	27	—	66	—	1500[b]	940	—	5.4	—	130	580	—	—	2900
Eastern England														
Humber Estuary TiO_2 and smelting (max.)	—	—	—	30	200	160	9.2×10^4	—	1100	63	220	—	2000[b]	430
Southwestern England estuaries														
Control (Avon) (typical)	0.1	13	0.3	10	37	19	1.9×10^4	0.12	420	28	39	28	—	98
Restronguet Creek acid mine waste (max.)	4.1	2500[b]	1.2	22	37	2500	5.8×10^4	0.22	560	32	290	1700[b]	—	3500
Tasmania														
Derwent Estuary Zn refinery and chloralkali (Hg) (max.)	—	—	860[b]	140	200	>400	16×10^4	100[b]	8900	42	1000	—	—	>10,000
Norway														
Sorfjord smelting (max.)	190[b]	—	850	—	—	12,000[b]	—	—	—	—	31,000[b]	1350	—	12,000[b]
Pacific Ocean														
Pelagic clay (>4000 m depth) (max.)	—	—	—	150[b]	—	1200	7.5×10^4	—	38,000[b]	1300[b]	—	—	—	390

[a] Reported values rounded to two significant digits.
[b] Highest value for each element.

Source: Bryan, G. W., in *Wastes in the Ocean,* Vol. 6, *Nearshore Waste Disposal,* Ketchum, B. H., Capuzzo, M. J., Burt, W. V., Duedall, I. W., Park, P. K., and Kester, D. R. (Eds.), John Wiley & Sons, New York, 1985, 41. With permission.

TABLE 1.21
Examples of Metal Concentrations (μg/g) in Contaminated Sediments[a]

Metal	Baltic Sea[b] (Various Sources)	Bristol Channel/Severn Estuary, U.K.[c] (Industry Sewage)	Mersey Estuary, U.K.[d] (Sewage, Industry Including Chlor-Alkali)	Los Angeles Outfall, California[e] (Sewage)	Derwent Estuary, Tasmania[f] (Refinery, Chlor-Alkali)	Restronguet Creek, U.K.[c] (Mining)	Port Pirie, Australia[g] (Smelter)
As		8	71			2520 (13)	151 (1.0)
Cd	8.1 (<0.01)	1.1	3.9	66 (0.3)	862	1.2 (0.3)	267 (0.5)
Cu	283 (1.0)	54	144	940 (8.3)	>400	2540 (19)	122 (3.0)
Hg	9 (0.01)	0.48	6.2	5.4 (0.04)	1130	0.22 (0.12)	8
Ni	920 (1.0)	33	44	130 (9.7)	42	32 (28)	19.4 (12)
Pb	400 (2)	88	205	580 (6.1)	>1000	400 (2)	5270 (2)
Zn	2090 (6)	255	255	2900 (43)	>10,000	2090 (6)	16,667 (11)

[a] Maximum concentrations shown together with local background values (in parentheses), where given.
[b] Data from Brugman, L., *Mar. Pollut. Bull.*, 12, 214, 1981.
[c] Data from Bryan, G. W. et al., *J. Mar. Biol. Assoc. U.K.*, Pub. 4, 1985.
[d] Data from Langston, W. J., *Estuarine Coastal Shelf Sci.*, 23, 239, 1986.
[e] Data from Mason, A. Z. and Simkiss, K., *Exp. Cell Res.*, 139, 383, 1982.
[f] Data from Kojoma, Y. and Kagi, J. H. R., *Trends Biochem. Sci.*, 3, 403, 1978.
[g] Data from Ward, T. J. et al., in *Environmental Impacts of Smelters*, Nriagu, J. O. (Ed.), John Wiley & Sons, New York, 1984, 1.

Source: Langston, W. J., in *Heavy Metals in the Marine Environment*, Furness, R. W. and Rainbow, P. S. (Eds.), CRC Press, Boca Raton, FL, 1990, 115. With permission.

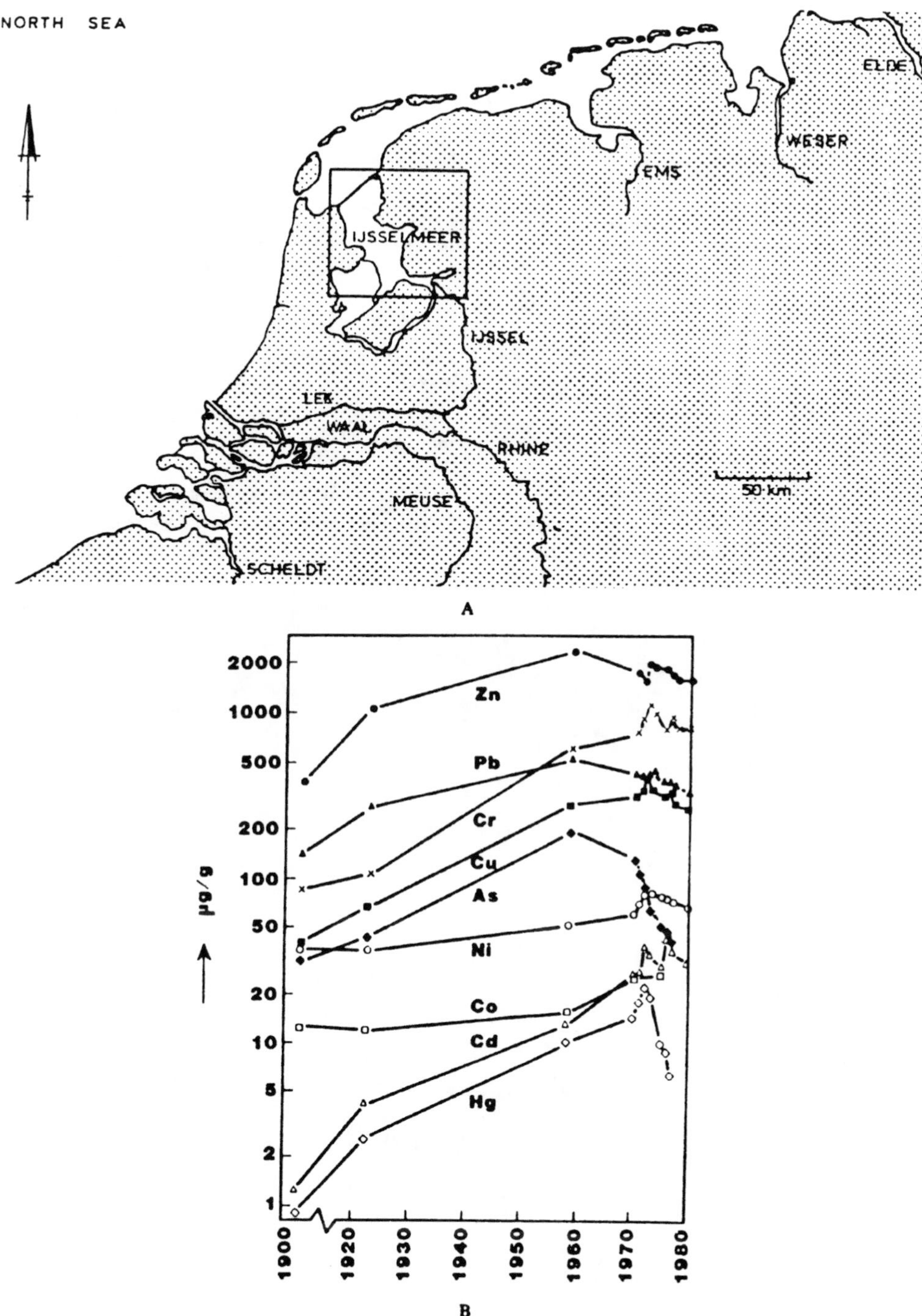

FIGURE 1.5. (A) Map showing the Ijsselmeer; (B) history of metal pollution of the Rhine River as reflected in its sediments in the Netherlands. (From Salomons, W., in *Aquatic Ecotoxicology: Fundamental Concepts and Methodologies,* Vol. 1, Boudou, A. and Ribeyre, F. (Eds.), CRC Press, Boca Raton, FL, 1989, 187. With permission.)

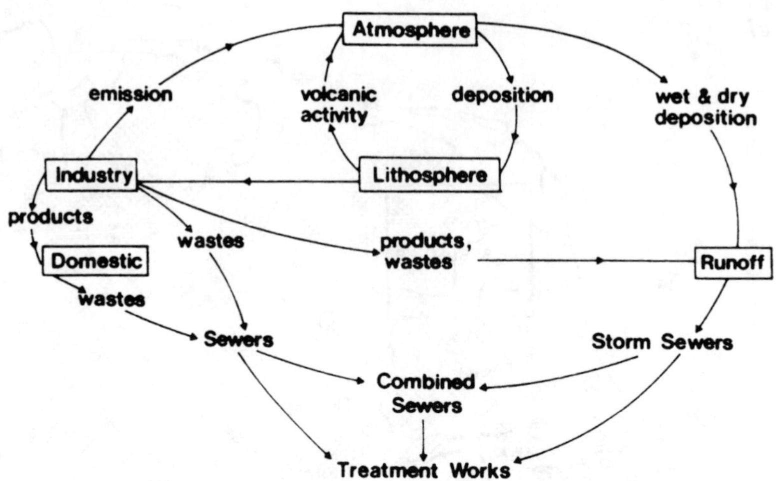

FIGURE 1.6. Sources and pathways of heavy metals entering wastewater-treatment processes. (From Stephenson, T., *Heavy Metals in Wastewater and Sludge Treatment Processes,* Vol. 1, *Sources, Analysis, and Legislation*, Lester, J. N. (Ed.), CRC Press, Boca Raton, FL, 1987, 33. With permission.)

TABLE 1.22
Sources of Heavy Metals in Domestic Wastewater by Product Type

	Al	Sb	As	Be	Bi	Cd	Cr	Co	Cu	Fe	Pb	Mn	Hg	Mo	Ni	Se	Ag	Sn	Ti	V	Zn
Automotive products	X		X	X			X	X		X	X			X					X		X
Caulking compounds	X			X			X	X		X	X								X		X
Cleaners	X						X		X	X									X		X
Cosmetics	X			X	X	X		X	X	X	X	X	X		X	X	X	X	X		X
Disinfectants													X								
Driers	X																			X	
Fillers	X																				X
Fire Extinguishers	X						X			X											X
Fuels				X					X		X										
Pesticides	X		X			X	X		X	X	X	X	X	X							X
Inks	X								X										X		X
Lubricants				X			X				X			X							X
Medicine	X	X	X		X			X	X	X											X
Oils				X					X		X										X
Ointments	X				X			X	X				X								X
Paints	X		X	X			X	X		X	X	X	X						X		X
Photography	X						X			X	X		X				X	X			
Pigments	X	X	X	X	X	X	X	X	X	X	X	X	X		X	X		X	X		X
Polish	X			X					X								X		X		X
Powders	X									X											X
Preservatives										X	X										X
Suppositories					X																X
Water treatment	X								X	X		X									X

Source: Stephenson, T., in *Heavy Metals in Wastewater and Sludge Treatment Processes*, Vol. 1, *Sources, Analysis, and Legislation*, Lester, J. N. (Ed.), CRC Press, Boca Raton, FL, 1987, 46. With permission.

TABLE 1.23
Heavy Metals in Wastewaters from Different Categories of Industries

	Al	Sb	As	Bi	Cd	Cr	Co	Cu	Fe	Pb	Mn	Hg	Mo	Ni	Se	Ag	Te	Tl	Sn
Metal industries																			
Power plants (steam generation)						X													
Foundries — ferrous		X	X	X	X	X	X	X	X	X	X	X	X	X	X		X		X
Foundries — nonferrous	X	X	X	X	X	X		X		X	X	X			X	X			
Plating	X				X	X		X				X		X		X		X	
Chemical industries																			
Cement and glass			X			X									X			X	
Organics and petrochemicals	X		X		X	X			X	X		X							X
Inorganic chemicals	X		X		X	X			X	X		X							X
Fertilizers	X		X		X	X		X	X	X				X					
Oil refining	X		X		X	X		X	X	X				X					
Others																			
Paper						X		X		X		X		X					
Leather						X													
Textiles		X				X													
Electronics		X						X				X			X		X		X

Source: Stephenson, T., in *Heavy Metals in Wastewater and Sludge Treatment Processes*, Vol. 1, *Sources, Analysis, and Legislation*, Lester, J. N. (Ed.), CRC Press, Boca Raton, FL, 1987, 51. With permission.

TABLE 1.24
Amounts of Heavy Metal Ions Removed from Sewage by Sludges

	Primary Sedimentation		Percolating Filter Treatment		Activated-Sludge Process	
Heavy Metal Ion	**Metal Concentration in Crude Sewage (mg/l)**	**Proportion Removed by Treatment (%)**	**Metal Concentration in Settled Sewage (mg/l)**	**Proportion Removed by Treatment (%)**	**Metal Concentration in Settled Sewage (mg/l)**	**Proportion Removed by Treatment (%)**
Copper					0.4	54
Copper	Up to 0.8	45	Up to 0.44	20	Up to 0.44	80
Copper					0.4–25	50–79
Copper	Up to 5	12				
Copper					28	90–93
Dichromate	Up to 1.2[a]	28	Up to 0.86[a]	32	Up to 0.86[a]	67–70
Dichromate					4.0	6.3
Dichromate					0.5–2	ca. 100
Dichromate					5	50
Dichromate					50	10
Iron (Ferric)	3–9	40	1.8–5.4	Nil	1.8–5.4	80
Lead	0.3–0.9	40	0.18–0.54	30	0.18–0.54	90
Nickel	0.1–0.3	20	0.08–10	40	0.08–0.24	30
Nickel					2.0	31
Nickel					2.5–10	30
Zinc	0.7–1.6	40	0.4–1.0	30	0.4–1.0	60
Zinc					2.5	90
Zinc					2.5	95
Zinc	Up to 5	12			7.5	100
Zinc					15	78
Zinc					20	74

[a] As Cr.

Source: Harrison, R. M. (Ed.), *Pollution: Causes, Effects, and Control,* 2nd ed., Royal Society of Chemistry, Cambridge, 1990. With permission.

APPENDIX 2. CONCENTRATIONS AND EFFECTS OF HEAVY METALS IN ESTUARINE AND MARINE ORGANISMS

TABLE 2.1
Metal Concentrations in Marine Fauna (ppm)

Metal	Geometric Mean and Location	Phytoplankton	Algae[a]	Mussels	Oysters	Gastropods	Crustaceans	Fish	Seals, Mammals[a]
Arsenic	Geometric mean	—	20	15	10	20	30	10	
	Newfoundland	—	9.8–17 (b)	1.6–5	—	4.0–11.5	3.8–7.6	0.4–0.8	
	England	—	26.0–54 (b)	1.8–15	2.6–10	8.1–38	16	1.7–8.7	
	Greenland	—	36 (b)	14–17	—	—	63–80	14.7–307	
Cadmium	Geometric mean	2	0.5	2	10	6	1	0.2	
	Spain	—	0.8–4 (b)	0.5–8	2.9–3.5	1.1–9	0.7–32	<0.4–4.3	
	England	—	0.2–53 (b)	3.7–65	6–54	3.5–1120	2.8–33	0.06–3.96+	2.2–11.6+
	Australia	—	—	4.2–83	9–174	2.8–30	—	0.05–0.4+	—
	Norway	—	1.0–13 (b)	1.9–140	—	0–51	1.9–7	<0.01–0.03+	
Copper	Geometric mean	7	15	10	100	60	70	3	
	Spain	—	5–26 (g)	6–14	120–435	5–50	110–435	<0.6–10	
	England	—	4–141 (b)	7–15	20–6480	0–1750	6–64+	0.5–14.6+	
	California	—	—	7–77	10–2100	3–177	(4–150)	(16–29.3)	14.5–386 (m)
	Norway	—	9–170 (g)	3–120	—	17–190	2–90	—	
Lead	Geometric mean	4	4	5	3	5	1	3	
	Spain	—	4–20 (g)	2–15	4–11	10–27	<1.2–11	<1.2–2.2	
	England	—	16–66 (g)	7–19	5–17	0.2–0.8	8	14–28	0–4+
	California	—	—	0.3–42	—	0.6–21	—	<0.001–5.3	0.3–34.2 (s)
	Norway	—	3–1200 (g)	2–3100	—	0–39	8.3	—	—

Mercury	Geometric mean	0.17	0.15	0.4	0.4	0.2	0.4	0.4	0.6–103 (m)
	Hawaii	—	—	—	—	0–0.03	0.03–0.12	0.02–23	0.1–700 (s)
	California	—	—	—	—	<0.01–0.07+	0.02–0.04+	0.02–0.2	
	Atlantic	0.2–5.3	<0.01–0.07+	<0.01–0.13+	0.02–180+	—	<0.05–0.6+	0.1–9.0	—
	Mediterranean	—	<0.5–0.7+	0.25–0.4	—	0.1–3.5+	0.3–4.5+	0.1–29.8+	—
	Australia	—	—	0.05–0.23+	1.5–8.2	0.32–0.65	—	0.3–16.5	0.1–106 (m)
	Norway	0.5–25.2	—	0.24–0.84	—	0.61	0.31–0.39	0.14–7.3	0.4–225 (s)
	England	—	<0.01–25.5 (g)	0.64–1.86	0.56–1.2	0.02–1.84	0.98	0.02–1.8	
Nickel	Geometric mean	3	3	3	1	2	1	1	
	England	—	4–33 (g,b)	5–12	2–174	8.8–12.3	1.1–12.3	0.5–10.6	
	California	—	—	3.3–20	—	1.8–18.5	—	—	
Silver	Geometric mean	0.2	0.2	0.3	—	1	0.4	0.1	
	California	—	—	0.7–46	—	0.4–10.7	—	0.1–1.2 (m)	
Zinc	Geometric mean	38	90	100	1700	200	80	80	
	South Africa	0.6–710	5.6+ (g)	73–113	400–886	12+	17+	3.2–7.2+	
	California	0.1–725	46–244	70–8430	1.7–288	—	—	78–875	
	Spain	—	63–345 (g)	190–370	310–920	60–120	79–330	21–220	
	Australia	—	—	170–1350	3740–38,700	56–1050	—	4–375	
	England	—	28–1240 (g)	12–779	1830–99,200	9.7–1500	36–82	2–342	
	Norway	—	20–2310 (g)	105–2370	—	87–2900	12–32	—	

Note: +, ppm wet weight; all other values in ppm dry weight.

[a] (b) = brown algae; (g) = green algae; (m) = mammals; (s) = seals.

Source: Förstner, U., in *Chemistry and Biogeochemistry of Estuaries*, Olausson, E. and Cato, I. (Eds.), John Wiley & Sons, Chichester, 1980, 307. With permission.

TABLE 2.2
Heavy Metal Levels in Marine Organisms from the Mediterranean Sea

		Cd		Pb		Hg (Org.)		Hg (Total)	
Species	**N**	**Mean**	**Range (ng/g)**	**Mean**	**Range (ng/g)**	**Mean**	**Range (ng/g)**	**Mean**	**Range (ng/g)**
Mullus barbatus-M. surmuletus	59	49	10–510	402	50–11,189	113	30–1162	139	30–1387
Sardina pilchardus	38	43	10–107	105	50–463	84	30–890	105	30–1000
Micromessistius poutassou	5	67	12–207	960	50–3238	160	30–356	191	42–406
Thunnus thynnus	7	43	10–110	163	50–350	399	40–1120	499	68–1359
Engraulis encrasicolus	18	53	10–274	543	50–8296	49	30–111	60	30–125
Lithognatus mormirus	14	19	10–58	90	50–301	87	30–293	104	30–321
Maena	6	24	10–30	74	50–125	81	30–194	99	30–238
Xiphias gladius	3	13	10–20	53	50–60	347	280–410	417	340–470
Trisopterus fuscus	2	33	10–55	108	50–166	117	114–121	142	122–163
Merluccius merluccius	9	24	10–87	114	10–414	54	30–110	66	30–140
Chamalea gallina	1	601		241		nd		nd	
Scorpaena porcus	3	11	10–12	55	50–64	59	30–90	66	30–100
Trigla lineata	3	40	10–80	139	50–450	79	40–110	100	40–130
Serranus cabrilla	2	21	10–32	383	50–715	107	30–183	129	49–210
Boops boops	1	312		215		53		75	
Citharus linguatula	2	38	30–46	63	50–77	66	30–102	71	30–112
Pagellus centrodontus-P. erithynnus	5	32	10–51	169	50–350	132	40–240	168	50–280
Naucrates ductor	4	28	10–50	210	50–450	nd		nd	
Seriola dumerili	5	46	10–170	126	50–420	34	30–40	42	30–60
Charax puntazzo	1	10		50		60		90	
Trachinus draco	2	40	10–70	70	50–90	nd		45	40–50
Diplodus sargus	2	35	30–40	2220	50–4390	95	80–110	110	90–130
Lepidorhombus	1	40		50		320		320	
Aristeomorpha antennatus-A. foliacea	23	80	20–180	157	50–790	364	31–820	460	40–1000
Nephrops norvegicus	1	nd		nd		390		480	
Macropipus depurator	12	171	10–500	582	50–1656	270	30–1414	307	30–1462
Pennaeus kerathurus	4	41	30–50	56	50–74	352	36–1110	400	42–1240
Mytilus galloprovincialis	28	222	10–1037	743	50–2830	96	30–1310	122	30–1730
Murex brandaris	14	3500	270–13,581	375	50–430	118	30–430	158	45–520
Truncularioptsis trunculus	3	759	520–1225	116	52–220	81	55–130	120	84–180
Patella caerulea	2	450	150–750	980	670–1290	nd		nd	
Donax vittatus	1	40		360		nd		nd	

Note: nd = not detected.

Source: Paster, A. et al., *Mar. Pollut. Bull.*, 28, 50, 1994. With permission.

TABLE 2.3
Cadmium Concentrations in Biota (μg/g dry weight), Sediment (μg/g dry weight), and Water (μg/l) from Three Estuaries

Feeding Type	Species	Bristol Channel-Severn Estuary	Restronguet Creek[a]	Looe Estuary[b] (mean values)	Ratio Bristol Channel: Looe Estuary
Deposit feeders	*Scrobicularia plana* (bivalve)	26.1	4.5	1.6	16:3
	Macoma balthica (bivalve)	3.9	—	0.67	5:8
	Nereis diversicolor (polychaete)	3.3	1.8	0.53	6:2
Suspension feeder	*Mytilus edulis* (bivalve)	57.0	6.6[c]	1.8	31:7
Herbivores	*Patella vulgata* (gastropod)	277	16	8.6	32:2
	Littorina littorea (gastropod)	38.7	2.7	1.4	27:6
	Littorina littoralis (gastropod)	105	5.6	7.6	14:1
Carnivore	*Nucella lapillus* (gastropod)	144	23	12.8	11:2
Brown alga	*Fucus vesiculosus* (brown seaweed)	29.8	1.4	1.3	22:9
	Sediment <100 μm	0.74	1.5	0.20	3:7
	Waters	Mean 1.25[d]	Range 0.7–38	—	—

[a] Maxima provided by Bryan, G. W. and Gibbs, P. E., *Mar. Biol. Assoc. (U.K.),* Occ. Publ., No. 2, 1983.
[b] Data primarily from Bryan, G. W. and Hummerstone, L. G., *J. Mar. Biol. Assoc. (U.K.),* 57, 75, 1977.
[c] Transplanted for 6 months.
[d] Data from Owens, M., *Mar. Pollut. Bull.,* 15, 41, 1984.

Source: Bryan, G. W. and Langston, W. J., *Environ. Pollut.,* 76, 89, 1992. With permission.

TABLE 2.4
Typical Metal Bioconcentration Factors of Selected Aquatic Organisms

	Marine Organisms					Freshwater Organisms		
Metal	**Phytoplankton**	**Zooplankton**	**Macrophytes**	**Mollusks**	**Fish**	**Macrophytes**	**Mollusks**	**Fish**
Ti	2700	—	—	—	—	—	—	—
Cr	7800	—	2880	21,800	—	—	267	10
Mn	3800	3900	—	2300	373	1450	—	23
Fe	28,300	114,600	—	14,400	—	3642	—	190
Ni	570	560	1050	4000	235	—	650	85
Co	—	—	—	—	50	1367	300	90
Cu	2800	1800	2890	3800	127	158	1500	60
Zn	5500	8800	7000	27,300	533	318	2258	228

Source: Williams, S. L., Aulenbach, D. B., and Clesceri, N. L., in *Aqueous Environmental Chemistry of Metals,* Rubin, A. J. (Ed.), 1976. Copyright Ann Arbor Science Publishers. With permission.

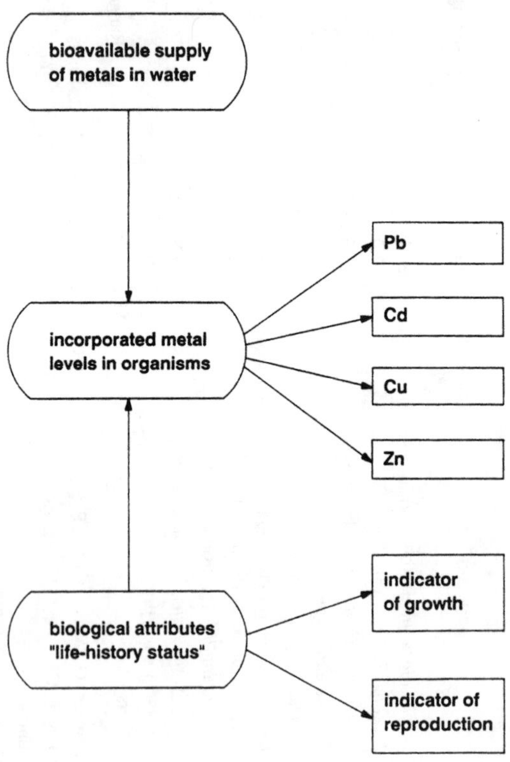

FIGURE 2.1. Conceptual model for biological monitoring of metals in organisms, showing latent variables on left-hand side and measurable indicator variables on right-hand side. (From Zauge, G.-P. and Petri, G., in *Ecotoxicology of Metals in Invertebrates,* Dallinger, R. and Rainbow, P. S. (Eds.), Lewis Publishers, Boca Raton, FL, 1993, 73. With permission.)

TABLE 2.5
Average Concentrations and Ranges of Heavy Metals in the Mussel *Mytilus* and Closely Related Species[a]

Region	Cadmium	Mercury	Lead
Baltic Sea and Kattegat	2.6 ± 2.6 (0.4–12.9)	(0.04–0.57)	10 ± 2 (1.4–20)
North Sea	1.4 ± 0.8 (0.2–11)	(0.08–1.00)	(0.9–12)
Irish Sea	2.1 ± 1.2 (0.2–5.4)	0.36 ± 0.23 (<0.09–0.96)	15.9 (5.2–30.7)
English Channel	1.4 ± 0.8 (0.5–4.7)	0.24 ± 0.12 (0.09–1.30)	5.3 (3.2–7.2)
Northeast Atlantic			
Gulf of Gascony	1.0 ± 0.8 (0.3–2.6)	(0.11–1.23)	
Clyde Estuary, Scotland	2.5 (1.7–3.8)	0.50 (0.22–0.86)	19.5 (10.0–30)
Mediterranean	1.3 ± 0.7 (0.4–3.0)	(0.04–1.52)	0.8 ± 0.8
Adriatic	(0.73–1.6)	(0.10–0.20)	(1.18–2.13)
Northwest Pacific	(0.4–5.2)	(0.04–2.1)	0.73 ± 0.51
Hong Kong	0.75 (0.15–1.09)		9.6 (6.9–13.4)
Southwest Pacific	(0.4–18.5)	(0.10–2.7)	
Southeastern Australia		(0.06–0.4)	
New Zealand	<1.0		(1.4–30.5) 0.67 (0.1–2.0)
Northeast Pacific			
U.S. and Canada	4.8 ± 2.6 (0.5–16.2)	0.24 ± 0.08 (0.09–0.49)	2.47 ± 4.0 (0.42–23.3)
Mexico	(1.8–2.7)	(0.27–0.44)	(1.53–1.87)
Southeast Pacific	3.3		(13.6–18.4)
Northwest Atlantic	1.9 ± 0.7 (0.9–4.2)	0.16 ± 0.10 (0.05–0.47)	4.6 ± 3.2 (1.4–14)
Gulf of Saint Lawrence	2.0 ± 0.6 (1.0–4.2)	0.16 ± 0.08 (0.05–0.40)	
Southwest Atlantic			
Argentina	(3.6–6.0)		(0.8–1.9)
Brazil		(0.12–0.17)	
Northwest Indian Ocean			
Arabian Gulf			
Oman	16.4 ± 8.6 (8.3–25.5) 7.6 (5.7–21.3)	0.082 ± 0.006 (0.076–0.088)	1.32 ± 0.99 (0.4–2.33)
Pakistan	(0.24–0.61)	(0.064–0.070)	(2.11–2.57)
GOA, India	7.7	0.50	7.3

Note: If weight data are not reported, concentrations given in the original reference as wet weight have been multiplied by 5.6 to obtain dry weight values. See original source for particular studies.

[a] Expressed in μg/g dry weight.

Source: Fowler, S. W., *Mar. Environ. Res.*, 29, 1, 1990. With permission.

TABLE 2.6
Concentrations (ppm) of Heavy Metals in Above-Ground Portions of *Spartina alterniflora*

Location	Cu	Pb	Hg	Mn	Zn	Cd
Massachusetts	3	2–26		48	11–31	0.06–0.21
Rhode Island	12–16	21–22			42–69	0.2–0.3
New Jersey	2–8	0.3–3	0.01–0.03	8–92	5–100	0.00–0.15
Virginia	2–5			35–140	12–23	
North Carolina	0.1–5	0.01–2	0.0–0.16	6–95	4–70	0.0–0.33
South Carolina	0.6–5	0.1–3	0.0–0.07	5–180	1–101	0.0–0.58
Georgia	0.8–8	0.4–55	0.01–1.13	30–77	0.4–64	0.0–1.14
Florida	1	1.5	0.01	39	14	0.06–0.018

Source: Nixon, S. W. and Lee, V., Wetlands and Water Quality: A Regional Review of Recent Research in the United States on the Role of Freshwater and Saltwater Wetlands as Sources, Sinks, and Transformers of Nitrogen, Phosphorus, and Various Heavy Metals, Tech. Rep. Y-86-2, U.S. Army Engineer Waterways Experimental Station, Vicksburg, MS, 1986.

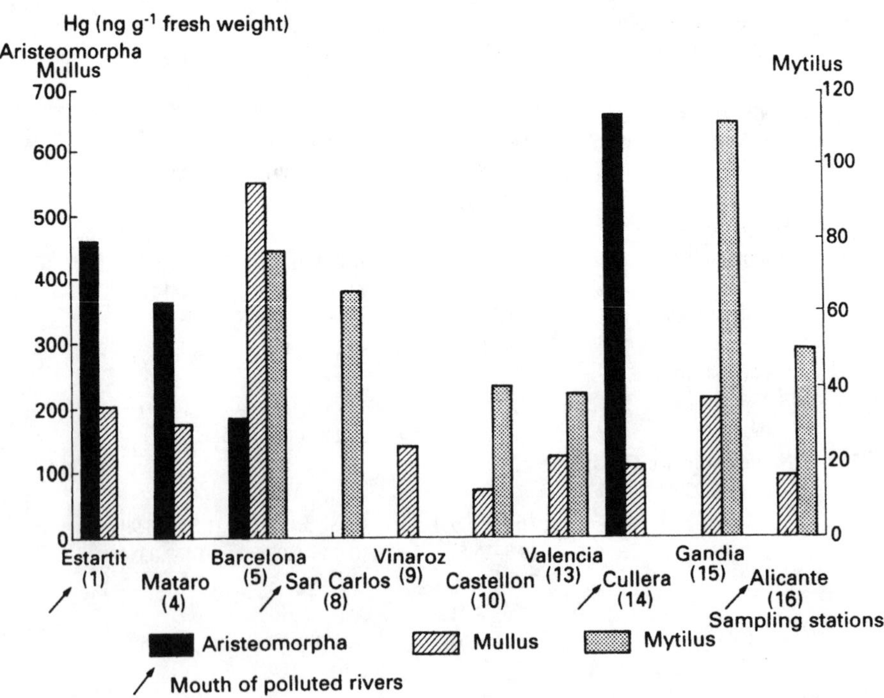

FIGURE 2.2. Concentrations of mercury in three species from the Mediterranean Sea. From Pastor, A., Hernandez, F., Peris, M. A., Beltran, J., Sancho, J. V., and Castillo, M. T., *Mar. Pollut. Bull.*, 28, 50, 1994. With permission.)

TABLE 2.7
Rankings, on Four Geographical Scales, of Metal Concentrations in Fish Livers from the Kennebec River Plume

Region (No. Sites)	Pb	Ag	Zn	Cu	Sn	Cd	Ni	Cr	Hg
Country (43)	1	3	4	10	15	18	18	18	22
East coast (14)	1	1	1	1	4	3	5	5	7
New England (6)	1	1	1	1	1	3	1	3	3
Gulf of Marine (4)	1	1	1	1	1	1	1	3	2

Source: Larsen, P. F., *Rev. Aquat. Sci.*, 6, 67, 1992. With permission.

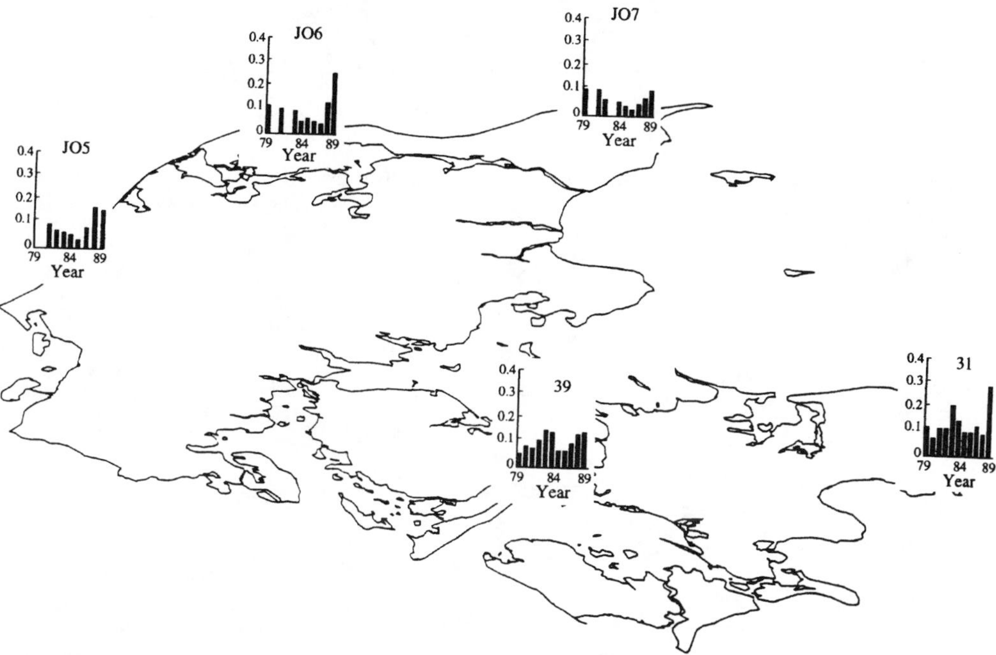

FIGURE 2.3. Cadmium concentrations (ppm wet weight) in fish liver from the Danish marine area, flounder in the Belts, and plaice in the North Sea. (From Jorgensen, L. A. and Pedersen, B., *Mar. Pollut. Bull.*, 28, 24, 1994. With permission.)

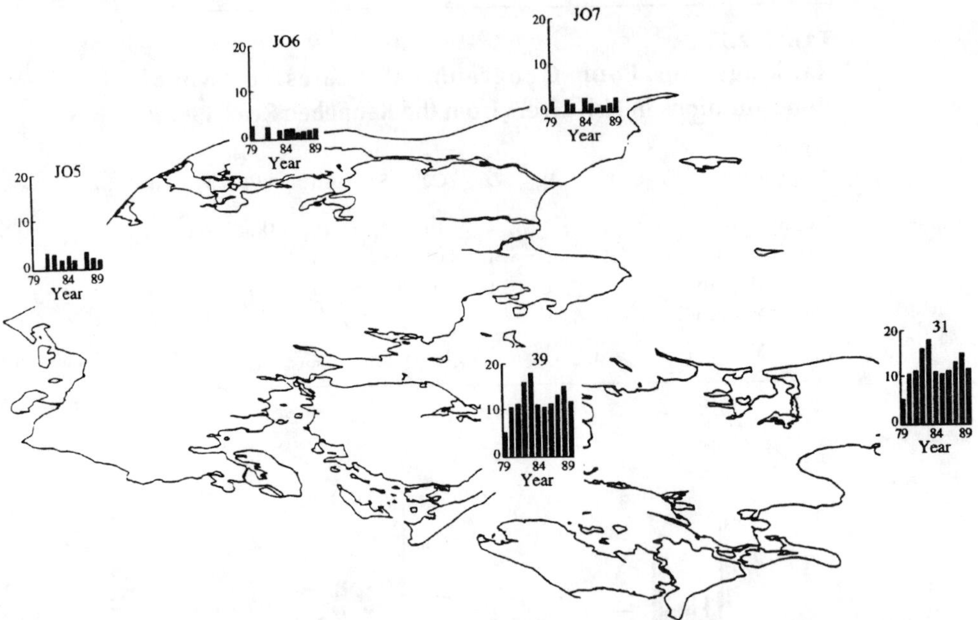

FIGURE 2.4. Copper concentrations (ppm wet weight) in fish liver from the Danish marine area, flounder in the Belts, and plaice in the North Sea. (From Jorgensen, L. A. and Pedersen, B., *Mar. Pollut. Bull.*, 28, 24, 1994. With permission.)

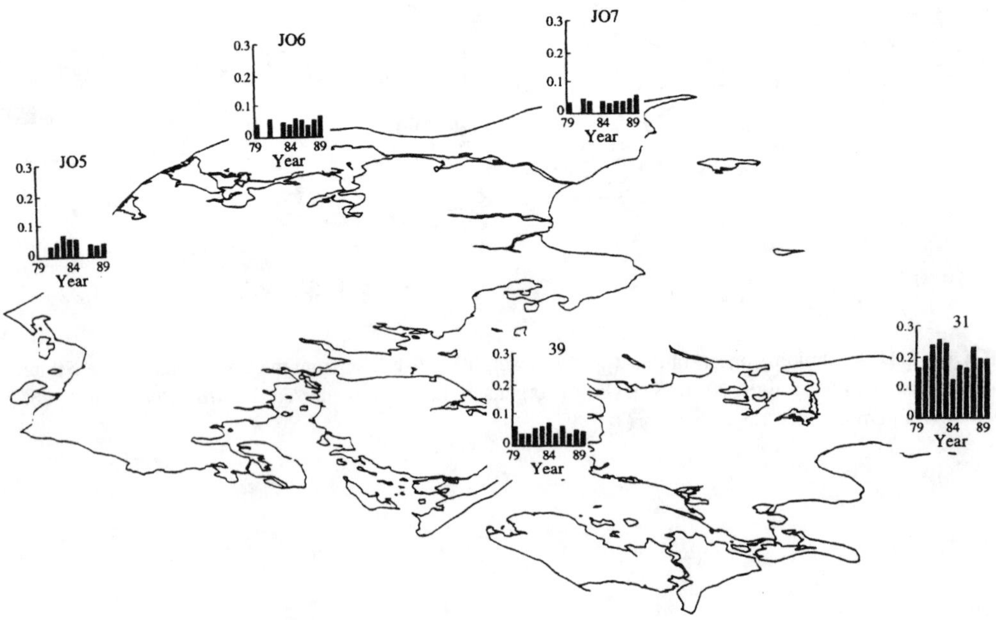

FIGURE 2.5. Mercury concentrations (ppm wet weight) in fish muscle from the Danish marine area, flounder in the Belts, and plaice in the North Sea. (From Jorgensen, L. A. and Pedersen, B., *Mar. Pollut. Bull.*, 28, 24, 1994. With permission.)

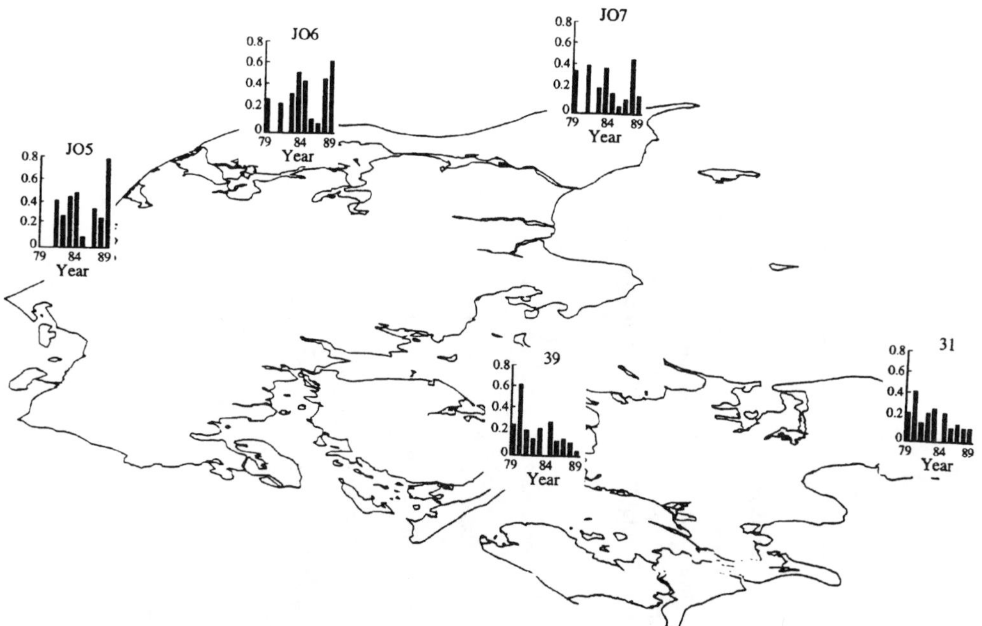

FIGURE 2.6. Lead concentrations (ppm dry weight) in fish liver from the Danish marine area, flounder in the Belts, and plaice in the North Sea. (From Jorgensen, L. A. and Pedersen, B., *Mar. Pollut. Bull.*, 28, 24, 1994. With permission.)

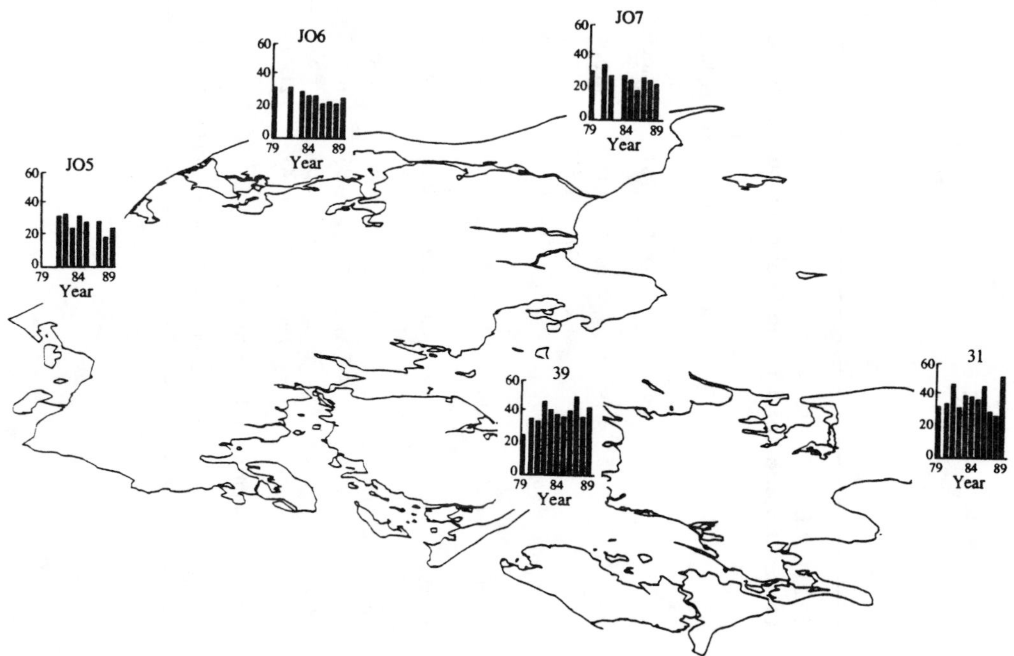

FIGURE 2.7. Zinc concentrations (ppm wet weight) in fish liver from the Danish marine area, flounder in the Belts, and plaice in the North Sea. (From Jorgensen, L. A. and Pedersen, B., *Mar. Pollut. Bull.*, 28, 24, 1994. With permission.)

TABLE 2.8
Organotin and Total Tin in Sediments and Benthic Organisms from Poole Harbor

	Sites (N)		Tin (μg/g dry weight)				Organotin (%, mean ± SE)		Concentration Factor (Mean ± SE) Relative to Sediment
			DBT	TBT	Total Organotin	Total Tin	TBT/Total Organotin	Organotin/Total Tin	
Sediments	16	Min	0.007	0.009	0.019	3.12	48.2 ± 2.08	2.95 ± 0.4	—
		Max	0.573	0.517	1.05	26.1			
Fucus vesiculosus	15	Min	0.02	0.01	0.04	0.11	40.0 ± 4.0	27.8 ± 4.3	1.0 ± 0.4
		Max	0.29	0.22	0.44	1.77			
Nereis diversicolor	16	Min	0.15	0.03	0.18	0.27	24.2 ± 3.0	60.3 ± 4.5	6.7 ± 1.1
		Max	5.65	1.59	6.49	7.03			
Littorina littorea	9	Min	0.23	0.10	0.33	0.33	35.0 ± 3.3	80.4 ± 12.0	7.7 ± 1.2
		Max	3.22	1.12	4.34	8.81			
Scrobicularia plana	12	Min	0.27	0.26	0.53	1.11	36.6 ± 3.6	76.3 ± 7.2	11.6 ± 2.0
		Max	7.29	3.05	10.40	11.53			
Mya arenaria	8	Min	1.78	5.55	6.83	7.62	82.6 ± 5.9	96.5 ± 3.5	52.1 ± 21.9
		Max	9.95	11.45	21.40	21.40			

Note: DBT = dibutylin; TBT = tributylin.

Source: Bryan, G. W. and Langston, W. J., *Environ. Pollut.*, 76, 89, 1992. With permission.

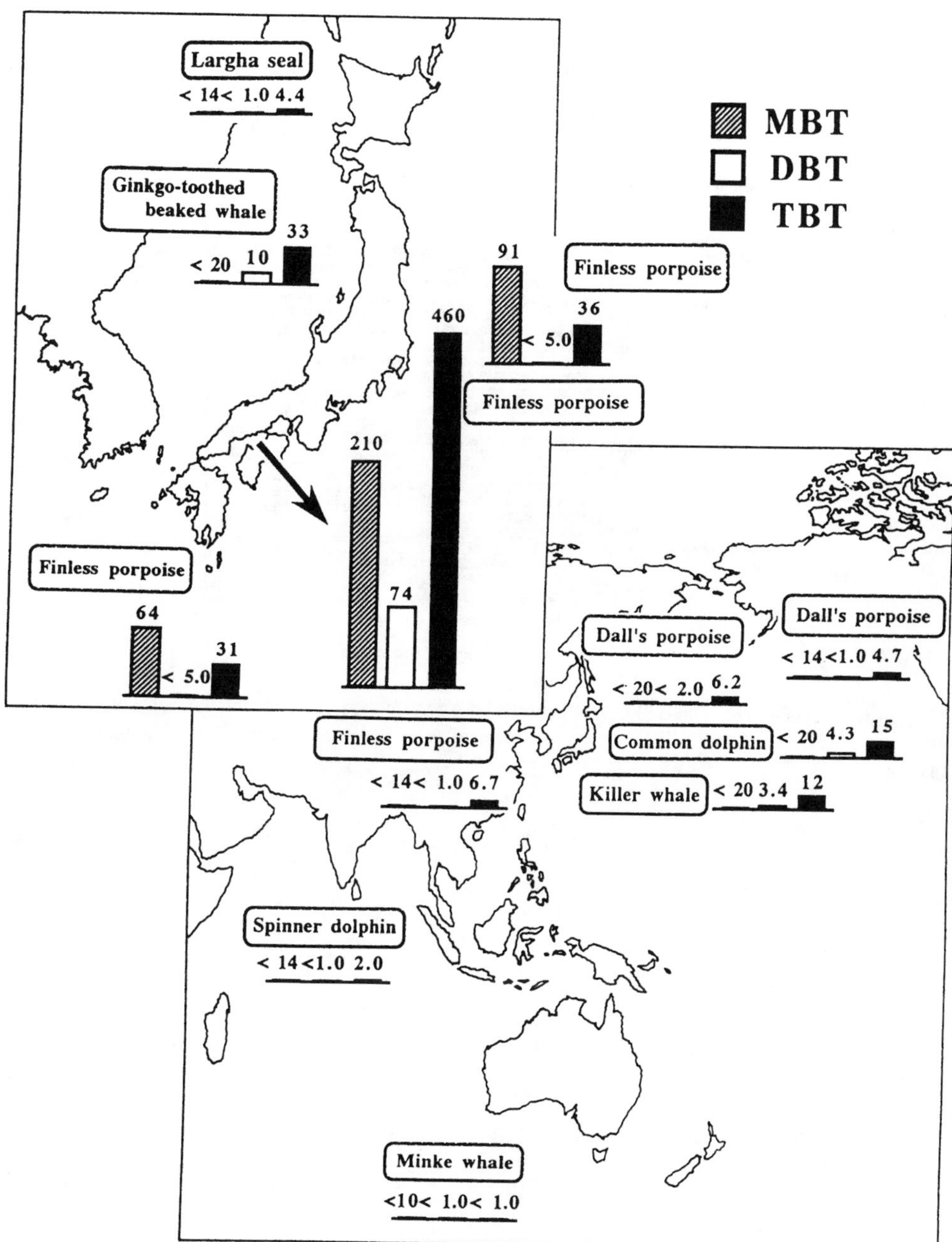

FIGURE 2.8. Concentrations (ng/g wet weight) of butyltin compounds in the blubber of marine mammals. (From Iwata, H., Tanabe, S., Miyazaki, N., and Tatsukawa, R., *Mar. Pollut. Bull.*, 28, 607, 1994. With permission.)

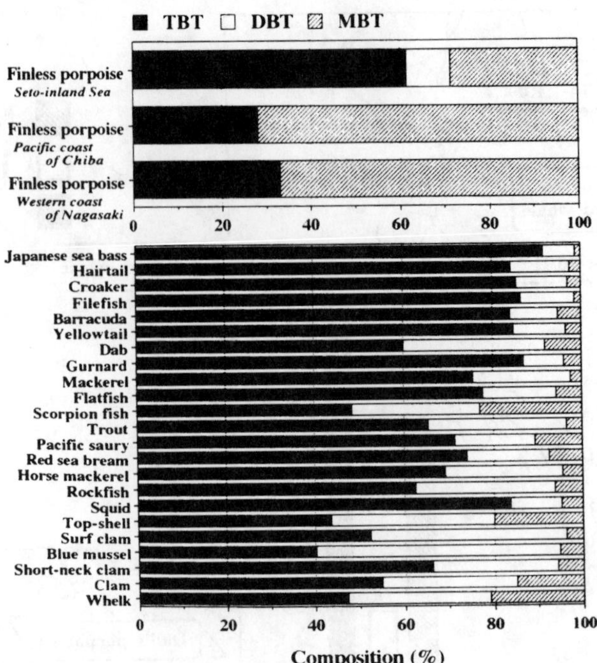

FIGURE 2.9. Comparison of butyltin compositions between finless porpoises and various aquatic organisms. (From Iwata, H., Tanabe, S., Miyazaki, N., and Tatsukawa, R., *Mar. Pollut. Bull.*, 28, 607, 1994. With permission.)

TABLE 2.9
Effects of Tributyltin on Estuarine and Marine Organisms

Species	Concentration TBT in water (ng/l as SN)	Effect
Dogwhelk		
Nucella lapillus	<1–5	Imposex, impaired reproduction
Mudsnails		
Ilyanassa obsoleta	1–10	Imposex
Hinia (= *Nassarius*) *reticulatus*	1+	Imposex
Oysters		
Crassostrea gigas	1	Shell thickening initiated
(spat)	8	Significant shell thickening
(larvae)	20	Reduced growth and viability
Clam larvae		
Mercenaria mercenaria	4	Reduced growth
Venerupis decussata	40	Reduced growth and viability
Mussel larvae		
Mytilus edulis	40	Reduced viability
Amphipod larvae		
Gammarus oceanicus	120	Reduced growth
Lobster larvae		
Homarus americanus	400	Reduced viability
Juvenile fish		
Cyprinodon variegatus	100	Mortality
Microalgae		
Dunaliella tertiolecta		
Pavlova lutheri	40–400	Reduced growth
Skeletonema costatum		

Note: See original source for particular studies. TBT = tributyltin.

Source: Bryan, G. W. and Langston, W. J., *Environ. Pollut.*, 76, 89, 1992. With permission.

TABLE 2.10
Classification of Essential Elements

Element Group	Subgroup	Number	Members	Tissue Conc.
Major	Bulk elements	5	C, H, N, O, S	g/kg
	Major ions	6	Na, K, Ca, Mg, P, Cl	g/kg
Trace	Essential	9	Co, Cu, I, Fe, Mn, Mo, Se, Si, and Zn	μg–mg/kg
	Desirable	6	As, Cr, F, Ni, Sn, V	μg/kg

Source: Langston, W. J., in *Heavy Metals in the Marine Environment,* Furness, R. W. and Rainbow, P. S. (Eds.), CRC Press, Boca Raton, FL, 1990, 125. With permission.

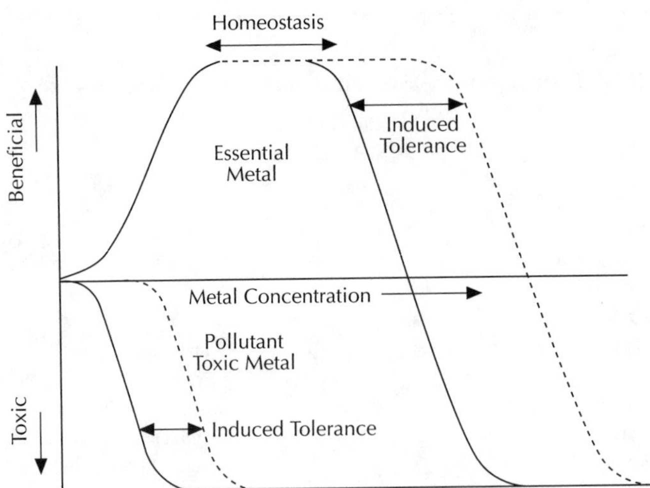

FIGURE 2.10. Effects of essential and pollutant metals on cellular metabolism. Intracellular concentrations of essential trace metals are controlled by homeostatic mechanisms. Once this capacity is exceeded, enzyme inhibition often occurs and the metal becomes toxic. Most nonessential (pollutant) metals are inhibitory. Inducible tolerance (detoxification systems) raises the inhibitory thresholds. (From George, S. G., in *Heavy Metals in the Marine Environment,* Furness, R. W. and Rainbow, P. S. (Eds.), CRC Press, Boca Raton, FL, 1990, 125. With permission.)

TABLE 2.11
Toxic Metals of Importance in Marine Pollution Based on Potential Supply and Toxicity, Listed in Order of Decreasing Toxicity[a]

	Rate of Mobilization (10^9 g/year)				Relative Critical Index (km³/year)	
	A	B	C	D		
Element	(Man) Fossil Fuels	(Natural) River Flow	Total	Toxicity (μg/l)	A/D	C/D
Mercury	1.6	2.5	4.1	0.1	16,000	41,000
Cadmium	0.35	?	3.0	0.2	1750	15,000
Silver	0.07	11	11.1	1	70	11,100
Nickel	3.7	169	164	2	1350	82,000
Selenium	0.45	7.2	7.7	5	90	1540
Lead	3.6	110	113.6	10	360	11,360
Copper	2.1	250	252.1	10	210	25,210
Chromium	1.5	200	201.5	10	150	20,150
Arsenic	0.7	72	72.7	10	70	7270
Zinc	7	720	727	20	330	36,350
Manganese	7.0	250	257	20	350	12,850

[a] Equals the volume of seawater that would be contaminated annually to the indicated level of toxicity by the specified rates of addition, both by natural processes and anthropogenic activities.

Source: Ketchum, B. H., in *Oceanography: The Past,* Sears, M. and Merriman, D. (Eds.), Springer-Verlag, New York, 1980, 397. With permission.

TABLE 2.12
Acute Toxicity (48- to 96-h LC_{50} or EC_{50}) Data for Certain Phyla Commonly Used as Marine Test Organisms

Test Organism	Copper	Mercury	Lead	Zinc
Arthropoda (crustaceans)	50–10,000	4–400	700–3000	200–5000
Annelida	100–500	10–90	—	800–50,000
Mollusca	200–8000	4–30,000	800–30,000	100–40,000
Vertebrata (Salmonidae)	30–500	—	—	20,000–70,000
Algae				
Chlorophyta	—	<5–400	—	50–7000
Chrysophyta (diatoms)	5–50	0.1–10	—	200–500

Note: Total concentration ranges (μg/l) given for the four trace elements reflect variations in experimental design (i.e., chemical, physical, and biological characteristics) as well as inherent randomness in organism response. Concentration ranges for algae represent concentrations causing significant growth inhibition. Dashes indicate that sufficient data were not obtained to determine an LC_{50} or EC_{50} range, generally because tests in this category had emphasized sublethal toxicity and bioaccumulation.

Source: Leland, H. V. and Kuwabara, J. S., in *Fundamentals of Aquatic Toxicology: Methods and Applications,* Rand, G. M. and Petrocelli, S. R. (Eds.), Hemisphere Publishing, New York, 1985, 374. With permission.

TABLE 2.13
Several Effluents and Chemicals (Including Heavy Metals) Reported To Be More Toxic to Aquatic Plants Than to Faunal Species

Acrylates	Aldrin
Anionic surfactant	Dieldrin
Cationic surfactants	Endrin
Sodium salts	Chlordane
Acridine	2,4-D
Nickel	Potassium chlorate
Cadmium	Sodium tetraborate
Copper	Disodium hydrogen
Zinc	Hydrazine hydroxide
Arsenate	Industrial effluents
Chromium	Paper mill effluents
Glyphosate	Textile plant effluents
Organotin	Refinery plant effluents
Soil elutriates	Herbicide plant effluent
Potassium dichromate	Monochloroacetic acid
River water extracts	Chloramine
Polycationic polymers	Chloroacetaldehyde
Monobromoacetic acid	Dinitrotoluene
Phosphoric acid tributyl ester	Phenol
Phthalic acid dialyl ester	

Source: Klaine, S. J. and Lewis, M. A., in *Handbook of Ecotoxicology,* Hoffman, D. J., Bruton, G. A., Jr., and Cairns, J., Jr. (Eds.), Lewis Publishers, Boca Raton, FL, 1995, 198. With permission.

TABLE 2.14
Sublethal Effects of Metals — Growth

Species	Metal Concentration (μg/l)								Response
	Ag	As	Cd	Cu	Hg	Ni	Pb	Zn	
Phytoplankton									
Natural assemblage				0.3					Reduced ^{14}C fixation
Natural assemblage			1.0	1.0	1.0				Reduced ^{14}C fixation
Natural assemblage				10.0					Reduced ^{14}C fixation
Natural assemblage		23	112	6.4	<6.0	60[a]	20	20	Reduced growth
Monochrysis lutheri				21.6 (0.07 as Cu^{2+})					Reduced division
Natural assemblage					0.8				Reduced growth
Natural assemblage					1.0				Reduced productivity
Natural assemblage								15	Reduced photosynthesis
Natural assemblage		5							Reduced growth
Macroalgae									
Laminaria saccharina									
Sporeling				10	0.5 (0.5)[b]			100	Reduced growth
Sporophyte				50	50 (5.0)[b]			1000	Reduced growth
Hydroids									
Campanularia flexuosa			195	14.3	1.6			740	Reduced growth
Mollusks									
Mytilus edulis			10	3	0.3	>200[a]	>200[a]	10	Reduced shell growth
Mercenaria mercenaria	32			16	15	5700		195	Reduced growth[c]
Crassostrea virginica	25			33	12	1200			Reduced growth[c]
Fish									
Pleuronectes platessa			5	10					Reduced growth

[a] No effect observed at this concentration.
[b] Methylmercury.
[c] For larvae: concentrations also represent LC_{50} values (8 to 12 d).

Source: Langston, W. J., in *Heavy Metals in the Marine Environment*, Furness, R. W. and Rainbow, P. S. (Eds.), CRC Press, Boca Raton, FL, 1990, 107. With permission.

TABLE 2.15
Sublethal Effects of Metals — Morphology

Species	Metal Concentration (µg/l)										Response
	Ag	As	Cd	Cr	Cu	Hg	Ni	Pb	Se	Zn	
Hydroids											
Eirene viridula		300	100		10	3		300	3000	1500	Altered hydranth morphology[a]
Mollusks											
Crassostrea gigas	22	326	611	4538	5	7	349	476	>10,000	199	50% abnormal larvae
Mytilus edulis	14	>3000	1200	4469	6	6	891	758	>10,000	175	50% abnormal larvae
Crassostrea virginica	24					11				206	50% abnormal larvae
Fish											
Myoxocephalus quadricornis		32	0.5		0.8	0.1		1.2		5.3	Increase in vertebral deformities[b]

[a] Lowest reported threshold concentration.
[b] Metals applied as a mixture.

Source: Langston, W. J., in *Heavy Metals in the Marine Environment,* Furness, R. W. and Rainbow, P. S. (Eds.), CRC Press, Boca Raton, FL, 1990, 109. With permission.

TABLE 2.16
Sublethal Effects of Metals — Reproduction and Development

Species	Metal Concentration (μg/l)									Response
	Ag	As	Cd	Cr	Cu	Hg	Ni	Pb	Zn	
Macroalgae										
Champia parvula		60								Inhibition of sexual reproduction
Hydroids										
Campanularia flexuosa					0.05	0.01			500	Increase in gonozooid frequency[a]
Polychaetes										
Neanthes arenaceodentata			1000	50				3100	320	Reductions in reproduction
Capitella capitata			560	100				200	560	Reductions in reproduction
Ctenodrilus serratus			2500	50	100	50	500	1000	500	Reductions in reproduction
Bivalves										
Mytilus edulis					50				200	Development of oocytes suppressed
Spisula solidissima (germ cells)	9.5									Impaired embryogenesis
Crustaceans										
Pontoporeia affinis			5.5					4.9		Fecundity reduced
Rhithropanopeus harrisii			50						25	Hatch time increased
Tigriopus japonicus			44		6.4					Generation time doubled
Echinoderms										
Sea urchin eggs (various spp.)		1500	600	1000	10	10	600	1000	30	Fertilization and development arrested[a]
Fish										
Spring-spawning herring			5		10				10	Reduced fertilization
Fundulus heteroclitus						10				Reduced hatch
Leiostomus xanthurus (eggs)					0.064[b]					Reduced hatch
Menidia menidia (eggs)					0.025[b]					Reduced hatch

[a] Lowest reported threshold concentration.
[b] Calculated as free ion (Cu^{2+}).

Source: Langston, W. J., in *Heavy Metals in the Marine Environment,* Furness, R. W. and Rainbow, P. S. (Eds.), CRC Press, Boca Raton, FL, 1990, 111. With permission.

TABLE 2.17
Changes in Enzyme Activity Following Exposure of Fish to Heavy Metals

Chemical	Enzyme	Effect	Species	Duration of Exposure	Concentration	Organ
Pb, Hg	Alkaline phosphatase	↑	*Fundulus heteroclitus*	96 h	96-h LC_{50}	Liver
Cu, Cd	Alkaline phosphatase	NS	*F. heteroclitus*	96 h	96-h LC_{50}	Liver
Ag	Alkaline phosphatase	↓	*F. heteroclitus*	96 h	96-h LC_{50}	Liver
Pb, Cd, Cu, Hg	Acid phosphatase	↓	*F. heteroclitus*	96 h	96-h LC_{50}	Liver
Pb, Hg, Ag, Cd	Xanthine oxidase	↓	*F. heteroclitus*	96 h	96-h LC_{50}	Liver
Cu	Xanthine oxidase	↑	*F. heteroclitus*	96 h	96-h LC_{50}	Liver
Ag, Pb, Cu, Hg, Cd	Catalase	↓	*F. heteroclitus*	96 h	96-h LC_{50}	Liver
Cd, Pb, Cu, Ag, Hg	RNAase	↓	*F. heteroclitus*	96 h	96-h LC_{50}	Liver
Cr	Na/K-ATPase	↓	*Salmo gairdneri*	48 h	2.5 ppm	Kidney, intestine
Cr	Na/K-ATPase	NS	*S. gairdneri*	48 h	2.5 ppm	Gill, liver
Cr	Cyto, oxidase, succinic reductase, G-6-PO_4 dehydrogenase	NS	*S. gairdneri*	48 h	2.5 ppm	Several
Cd	Aspartate aminotransferase	↓	*Tautogolabrus*	96 h	3 and 24 ppm	Liver
Cd	Aspartate aminotransferase	↓	*Tautogolabrus*	30 d	50–100 ppb	Liver
Cd	Glucose-6-PO_4 dehydrogenase	↑	*Tautogolabrus*	30 d	50–100 ppb	Liver
Cd	Leucine aminopeptidase	↓	*Pseudopleuronectes*	60 d	5–10 ppb	Kidney
Cd	Carbonic anhydrase	↑	*Pseudopleuronectes*	60 d	5–10 ppb	Kidney
Cd	Aspartate aminotransferase	↑↓	*Mugil cephalus*	1–4 d	96-h LC_{50}	Liver
Cd	Aspartate aminotransferase	↑	*M. cephalus*	1–4 d	96-h LC_{50}	Heart, gill
Cd	Alanine aminotransferase	↑	*M. cephalus*	1–4 d	96-h LC_{50}	Heart, gill
Cd	Alanine aminotransferase	↓	*M. cephalus*	1–4 d	96-h LC_{50}	Liver
Cd	Acid phosphatase	↓	*M. cephalus*	1–4 d	96-h LC_{50}	Heart, gill, liver
Cd	Lactic dehydrogenase	↓	*M. cephalus*	1–4 d	96-h LC_{50}	Heart, gill, liver
Cd	Ten enzymes	NS	*S. gairdneri* and *S. trutta*	2 weeks or 3 months		Several
Cd	Malic dehydrogenase and propionyl-CoA-carboxylase	↓	Pacific Herring eggs		10 ppm	
Hg	Several	↓↑	*Channa punctatus*	15–60 d	3 ppb	Several
Hg	Na/K-ATPase, Mg-ATPase	↓	*Notopterus notopterus*	30 d	17–88 ppb	Gill, liver, kidney, brain

TABLE 2.17 (continued)
Changes in Enzyme Activity Following Exposure of Fish to Heavy Metals

Chemical	Enzyme	Effect	Species	Duration of Exposure	Concentration	Organ
Hg	Na/K-ATPase	↓	*Channa punctatus*	30 d	17–88 ppb	Brain
Hg	Succinic, lactic, and pyruvate dehydrogenases	↓	*Ophiocephalus punctatus* and *Heteropneustes fossilis*	30 d	0.3 ppm	Liver, stomach, intestine and caeca
Hg	Alkaline phosphatase	↑	*O. punctatus*	30 d	0.3 ppm	Brain
Hg	Acid phosphatase	↓	*O. punctatus*	30 d	0.3 ppm	Brain
Hg	Glucose-6-phosphatase, succinic, lactic and pyruvate dehydrogenase	↓	*O. punctatus*	30 d	0.3 ppm	Brain
Pb, Hg	Delta aminolevulinic acid dehydrase	↓	*F. heteroclitus*	4–14 d	0.8–10 ppm	Liver and kidney
Ag, Zn	Delta aminolevulinic acid dehydrase	↑	*F. heteroclitus*	14 d	0.02–2.2 ppm	Liver
Pb	Delta aminolevulinic acid dehydrase	↓	*S. gairdneri*	30 d	10–300 ppb	Erythrocytes, spleen, and distal kidney
Cu	Alkaline phosphatase	↑	*O. punctatus*	24 h	5–7.5 ppm	Kidney
Cu	Alkaline phosphatase	↓	*O. punctatus*	24 h	7.5 ppm	Liver
Cu	Acid phosphatase	↑	*O. punctatus*	24 h	7.5	Liver and kidney

Note: When more than one chemical is listed in sequence, they are arranged in order of decreasing effect; NS = no significant change.

Source: Heath, A. G., *Water Pollution and Fish Physiology,* CRC Press, Boca Raton, FL, 1987. With permission.

TABLE 2.18
Concentration of Copper, Zinc, and Cadmium in Subcellular Fractions from Hepatopancreas of *Nassarius reticulatus* Exposed to Copper, Zinc, and Cadmium in Sea Water

Fraction	Control (mean ± sd)	Cu-Exposed (mean ± sd)	Zn-exposed (mean ± sd)	Cd-exposed (mean ± sd)
Cu				
Mitochondrial	32.5 ± 11.0	52.2 ± 6.2[a]	66.1 ± 15.4[a]	8.4 ± 0.8[a]
Microsomal	10.3 ± 2.9	10.7 ± 3.5	13.3 ± 5.7	2.2 ± 0.3[a]
Cytosolic	5.1 ± 1.7	64.7 ± 9.3[a]	6.1 ± 1.9	2.5 ± 2.0
Zn				
Mitochondrial	109.0 ± 31.5	67.4 ± 15.5	188.2 ± 37.0[a]	18.0 ± 2.4[a]
Microsomal	17.2 ± 1.8	14.4 ± 6.1	26.9 ± 1.5	2.8 ± 1.0[a]
Cytosolic	747.3 ± 48.8	591.9 ± 34.9	1122.5 ± 162.5[a]	146.9 ± 30.2[a]
Cd				
Mitochondrial	0.22 ± 0.05	0.19 ± 0.04	0.21 ± 0.06	4.60 ± 1.38[a]
Microsomal	0.09 ± 0.03	0.05 ± 0.02	0.06 ± 0.01	1.10 ± 0.52[a]
Cytosolic	2.75 ± 0.32	2.41 ± 0.60	2.75 ± 0.95	36.35 ± 8.52[a]

Note: Each value is the mean of three pooled samples of eight individuals. All values in μg/g wet weight of original hepatopancreas.

[a] Values significantly different from control ($p < 0.05$, Dunnett's test).

Source: Kaland, T., Andersen, T., and Hylland, K., in *Ecotoxicology of Metals in Invertebrates,* Dallinger, R. and Rainbow, P. S. (Eds.), Lewis Publishers, Boca Raton, FL, 1993, 37. With permission.

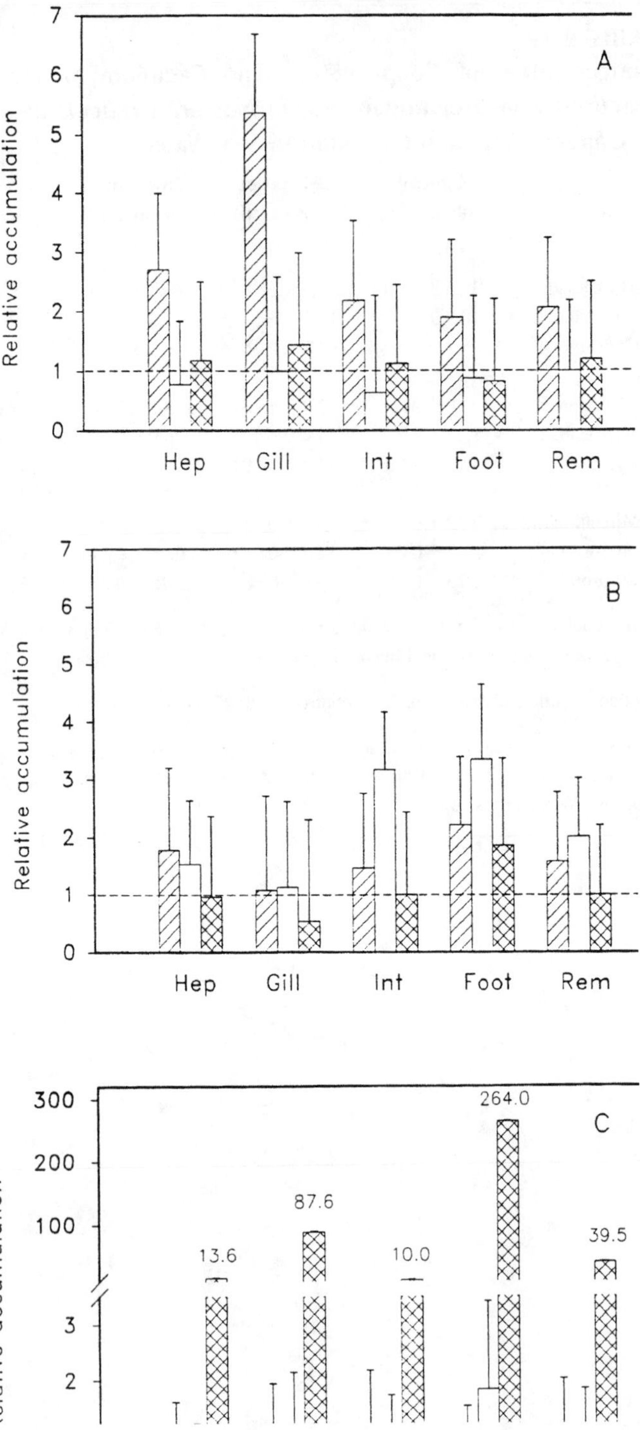

FIGURE 2.11. Accumulation of copper (diagonally hatched), zinc (blank), and cadmium (cross-hatched) in different organs of the marine gastropod *Nassarius reticulatus,* relative to control levels. The given values are summed concentrations from the subcellular fractions of each tissue. Hep = hepatopancreas; Int = intestine; Rem = remainder. (A) Gastropods exposed to 50 µg/l Cu. (B) Gastropods exposed to 1.5 mg/l Zn. (C) Gastropods exposed to 0.9 mg/l Cd. Vertical lines denote one standard deviation. (From Kaland, T., Andersen, T., and Hylland, K., in *Ecotoxicology of Metals in Invertebrates,* Dallinger, R. and Rainbow, P. S. (Eds.), Lewis Publishers, Boca Raton, FL, 1993, 37. With permission.)

TABLE 2.19
Ranges of Typical Accumulated Concentrations of Zinc in Crustaceans and Bivalve Mollusks

Crustaceans	Body Concentration Range
Barnacles	
Semibalanus balanoides	1220–113,250
Elminius modestus	4900–11,700
Balanus amphitrite	2726–11,990
Caridean Decapods	
Palaemon elegans	76–138
Pandalus montagui	119–214
Crangon crangon	56–174
Bivalve Mollusks	**Soft Tissue Concentration Range**
Oysters	
Ostrea edulis	660–3280
Crassostrea virginica	322–12,675
Saccostrea cucullata	430–8629
Mussels	
Mytilus edulis	14–500
Perna viridis	77–164
Septifer virgatus	74–116

[a] Expressed in μg/g dry weight.

Source: Rainbow, P. S., in *Ecotoxicology of Metals in Invertebrates,* Dallinger, R. and Rainbow, P. S., (Eds.), Lewis Publishers, Boca Raton, FL, 1993, 3. With permission.

TABLE 2.20
Metal Concentrations (µg/g) of Euphausiids (*Meganyctiphanes norvegica*) Collected from the Northeast Atlantic (July 1985) and the Firth of Clyde, Scotland (October 1984)

	Northeast Atlantic	Firth of Clyde
Zinc		
Mean	102	43.0
SD	54.3	7.7
Range	40.1–281	26.9–62.5
N	29	30
Copper		
Conc. (0.05 g)	57.5	35.8
95% CL	39.6–83.3	31.2–41.3
Range	8.8–67.2	30.8–72.6
N	29	30

Note: Figures are quoted either as the mean with standard deviation in the absence of a size effect on metal concentration (zinc) or as the metal concentration of a 0.05-g euphausiid estimated from the best fit regression line of log concentration vs. log dry weight with a 95% confidence limit (CL) when there is a significant size effect in at least one population (copper).

Source: Rainbow, P. S., Moore, P. G., and Watson, D., *Est. Coastal Shelf Sci.*, 28, 567, 1989. With permission.

TABLE 2.21
Spatial and Temporal Variability of Heavy Metals in the Isopod *Ceratoserolis trilobitoides* from the Antarctic Ocean[a]

	Gould Bay 1985	Elephant Island 1987	Test Equality of: Variances Levene		Means ANOVA		Means Welch	
			F	p	F	p	F	p
Pb								
Mean	0.49	0.25	0.75	0.39	14.7	0.00	14.8	0.00
SD	0.36	0.26						
N	78	88						
Cd								
Mean	1.9	2.1	15.3	0.00	1.83	0.18	2.01	0.16
SD	0.7	1.5						
N	78	90						
Cu								
Mean	38	46	4.71	0.03	15.2	0.00	15.5	0.00
SD	11	14						
N	78	85						
Zn								
Mean	40	40	0.54	0.47	0.01	0.92	0.01	0.92
SD	16	19						
N	60	90						

Note: N = number of determinations regarding 12 to 18 individual organisms analyzed in five parallel samples in different series of determinations, respectively. ANOVA: assuming equality of variances. Welch test: not assuming equality of variance. Null hypotheses are to be rejected if $p < 0.05$.

[a] Mean concentrations in mg/kg dry weight.

Source: Zauge, G.-P. and Petri, G., in *Ecotoxicology of Metals in Invertebrates,* Dallinger, R. and Rainbow, P. S. (Eds.), Lewis Publishers, Boca Raton, FL, 1993, 73. With permission.

TABLE 2.22
Variability of Heavy Metals in Males and Females of the Isopod *Ceratoserolis trilobitoides* from the Antarctic Ocean[a]

			Tests Equality of					
			Variances Levene		Means ANOVA		Means Welch	
	Males 1987	Females 1987	F	p	F	p	F	p
Pb								
Mean	0.24	0.30	1.97	0.16	0.47	0.49	0.60	0.44
SD	0.24	0.44						
N	34	55						
Cd								
Mean	1.4	2.6	0.51	0.48	16.3	0.00	18.4	0.00
SD	1.1	1.5						
N	35	55						
Cu								
Mean	36	51	0.08	0.78	36.7	0.00	34.5	0.00
SD	13	11						
N	32	53						
Zn								
Mean	44	37	24.4	0.00	2.70	0.10	2.04	0.16
SD	25	12						
N	35	55						

Note: N = number of determinations regarding 7 to 11 individual organisms analyzed in five parallel samples in different series of determinations, respectively. ANOVA: assuming equality of variances. Welch test: not assuming equality of variances. Null hypotheses are to be rejected if $p < 0.05$.

[a] Mean concentrations in mg/kg dry weight.

Source: Zauge, G.-P. and Petri, G., in *Ecotoxicology of Metals in Invertebrates,* Dallinger, R. and Rainbow, P. S. (Eds.), Lewis Publishers, Boca Raton, FL, 1993, 73. With permission.

TABLE 2.23
Variability of Heavy Metals in Males and Females of the Amphipod *Eusirus propeperdentatus* from the Antarctic Ocean[a]

	Males	Females	Tests for Equality of					
			Variances Levene		Means ANOVA		Means Welch	
			F	p	F	p	F	p
Pb								
Mean	0.62	0.49	0.17	0.68	0.37	0.55	0.45	0.51
SD	0.56	0.64						
N	10	65						
Cd								
Mean	8.6	10.9	1.07	0.31	1.37	0.25	1.22	0.29
SD	6.2	5.7						
N	10	65						
Cu								
Mean	118	115	8.75	0.00	0.05	0.83	0.12	0.73
SD	17	36						
N	9	59						
Zn								
Mean	57	41	4.56	0.04	9.18	0.00	24.8	0.00
SD	8	16						
N	10	64						

Note: N = number of determinations regarding 3 to 13 individual organisms analyzed in 3 to 5 parallel samples in different series of determinations, respectively. ANOVA: assuming equality of variances. Welch test: not assuming equality of variances. Null hypotheses are to be rejected if $p < 0.05$.

[a] Mean concentrations in mg/kg dry weight.

Source: Zauge, G.-P. and Petri, G., in *Ecotoxicology of Metals in Invertebrates*, Dallinger, R. and Rainbow, P. S. (Eds.), Lewis Publishers, Boca Raton, FL, 1993, 73. With permission.

TABLE 2.24
Molar Ratios of Trace Metals (Relative to Copper = 1000) in Crustaceans from the Antarctic Ocean[a]

Species	Taxon[b]	Molar Ratios Pb	Cd	Zn	Cu	Cu (mg/kg)	Class
Antarctic Ocean waters		20	150	1100	1000	—	1
Maxilliphimedia longipes	A	15	330	33,000	1000	5.5	1
Aega antarctica	I	10	62	14,800	1000	7.7	2
Gondogeneia antarctica	A	1	14	790	1000	37.1	2
Serolis bouvieri	I	5	36	660	1000	37.6	2
Ceratoserolis trilobitoides	I	3	27	925	1000	41.8	2
Bovallia gigantea	A	7	19	840	1000	53.0	2
Paraceradocus gibber	A	8	42	1200	1000	53.0	2
Eurymera monticulosa	A	2	18	650	1000	63.7	3
Euphausia superba	E	1	29	490	1000	65.7	3
Notocrangon antarcticus	D	4	110	650	1000	67.2	3
Natatolana spp.	I	4	54	1230	1000	68.2	3
Waldeckia obesa	A	1	7	3600	1000	80.8	3
Cheirimedon femoratus	A	1	6	550	1000	85.0	3
Chorismus antarcticus	D	5	80	460	1000	92.7	3
Eusirus propeperdentatus	A	1	50	450	1000	107	4
Serolis pagenstecheri	I	5	7	378	1000	123	4
Glyptonotus antarcticus	I	5	5	415	1000	149	4

[a] Mean concentrations in mg/kg dry weight.
[b] A (Amphipoda); E (Euphausiacea); D (Decapoda); I (Isopoda).

Source: Zauge, G.-P. and Petri, G., in *Ecotoxicology of Metals in Invertebrates,* Dallinger, R. and Rainbow, P. S. (Eds.), Lewis Publishers, Boca Raton, FL, 1993, 73. With permission.

TABLE 2.25
Metal Bioconcentration Factors Calculated for Antarctic Crustaceans on Basis of Dry Weight

Species	Taxon[a]	BCFs × 10³ Pb	Cd	Cu	Zn
Antarctic Ocean waters[b]		13	54	200	230
Maxilliphimedia longipes	A	19	54	28	740
Gondogeneia antarctica	A	11	18	186	132
Bovallia gigantea	A	93	34	265	200
Paraceradocus gibber	A	109	70	265	272
Eurymera monticulosa	A	32	36	318	188
Waldeckia obesa	A	20	20	404	1309
Cheirimedon femoratus	A	15	17	425	210
Eusirus propeperdentatus	A	34	176	535	212
Notocrangon antarcticus	D	60	248	336	198
Chorismus antarcticus	D	122	246	464	190
Euphausia superba	E	19	64	329	145
Aega antarctica	I	21	17	39	526
Serolis bouvieri	I	43	49	188	111
Serolis pagenstecheri	I	142	29	615	203
Ceratoserolis trilobitoides	I	29	38	209	173
Natatolana spp.	I	70	123	341	372
Glyptonotus antarcticus	I	195	28	745	277

[a] A (Amphipoda); E (Euphausiacea); D (Decapoda); I (Isopoda).
[b] Expressed in ng/kg.

Source: Zauge, G.-P. and Petri, G., in *Ecotoxicology of Metals in Invertebrates,* Dallinger, R. and Rainbow, P. S. (Eds.), Lewis Publishers, Boca Raton, FL, 1993, 73. With permission.

TABLE 2.26
Mean Concentrations of Cadmium in the Total Body and Hepatopancreas of *Idotea baltica* Males After a 10-d Exposure to 1000 μg/l Cadmium

Treatment	Cd μg/g dry weight: Total Body $\bar{X}$ SE	Hepatopancreas $\bar{X}$ SE
Control	0.68 ± 0.34	1.37 ± 0.34
Cadmium-exposed	264 ± 112	650 ± 218

Source: de Nicola, M., Cardellicchio, N., Cambardella, N., Guarino, S. M., and Marra, C., in *Ecotoxicology of Metals in Invertebrates,* Dallinger, R. and Rainbow, P. S. (Eds.), Lewis Publishers, Boca Raton, FL, 1993, 103. With permission.

7 National Monitoring Surveys

INTRODUCTION

To assess the effects of chemical contaminants on environmental quality of estuarine and coastal marine systems, several national monitoring programs have been conducted during the past three decades. These national surveys commenced with the Estuarine Mollusk Project of the National Pesticide Monitoring Program (NPMP), which spanned 7 years (1965 to 1972) and measured pesticides monthly in clams, oysters, and mussels from at least 180 sites in 15 coastal states.[1] From 1972 to 1976, the NPMP monitored pesticides in juvenile estuarine fish at 144 sites nationwide.[2] The U.S. Environmental Protection Agency (USEPA) subsequently developed the U.S. Mussel Watch consisting of two annual nationwide surveys (1976 to 1978) of PCB and DDE contamination at about 100 sites.[3] The National Oceanic and Atmospheric Administration's (NOAA) National Marine Fisheries Service completed one nationwide survey for PCBs in finfish over the period 1979 to 1981.[4] The Benthic Surveillance Project and the Mussel Watch Project of NOAA's National Status and Trends Program began in 1984 and 1986, respectively.[5] Both are multiyear, nationwide monitoring surveys of chemical contaminants in estuarine and coastal marine environments. Since 1990, the USEPA has undertaken estuarine and coastal monitoring as a component of its Environmental Monitoring and Assessment Program (EMAP). Unlike NOAA's National Status and Trends Program, EMAP does not analyze chemical contaminants in mussels or oysters, although it has measured the same chemicals as the National Status and Trends Program in sediments. EMAP also does not sample at fixed points; site selection is a vigorously random procedure. When its sampling strategy is complete, EMAP expects to annually sample and analyze sediments from approximately 800 sites, a number which represents more than three times the number of sites sampled by the National Status and Trends Program. For future assessments of national coastal contamination, joint NOAA/USEPA efforts are planned using data collected from both national monitoring programs.[6]

The aforementioned long-term, large-scale monitoring programs have provided valuable data on the environmental quality of the nation's estuarine and coastal marine waters by assessing the status of contaminants in select media (i.e., sediments and organisms). By measuring contaminant levels in bivalves and marine fish, these programs also effectively monitor conditions of biological resources nationwide. Large-scale data sets derived from the monitoring programs are of value for comparing site values to regional or national averages but have more limited application for assessing site-specific problems which have been addressed by other programs, such as the National Estuary Program of the USEPA.[7] Monitoring stations are representative of relatively large areas in order to delineate chemical distributions over regional and national scales. To avoid sampling bias by contaminant sources, no sites are knowingly selected near waste discharge points or poorly flushed industrialized waterways. Furthermore, no mollusk samples are collected from artificial substrates (e.g., pilings which are often chemically treated).[6]

Existing long-term monitoring data collected for an estuary, embayment, or coastal area can be used to identify reference sites, if a subsequent spill or other anthropogenic impact occurs nearby. Similarly, existing long-term, regional monitoring can shed light on natural variability and trends, thereby establishing a scientifically meaningful database from which future changes in environmental quality may be determined.[1] Sampling sites of national monitoring programs also may overlap in space with those of local monitoring programs sponsored by state, municipal, or other agencies and, therefore, can provide points of comparison and a basis for extrapolating local results to larger scales.[6] While not all chemicals can be tracked by these national monitoring programs, major classes of contaminants (principally priority pollutants) are targeted, including organochlorine compounds (e.g., DDT, chlordane, PCBs), polycyclic aromatic hydrocarbons (e.g., benzo(a)pyrene, fluorene, pyrene), and trace metals (e.g., arsenic, lead, mercury).[7]

NATIONAL STATUS AND TRENDS PROGRAM

The National Oceanic and Atmospheric Administration's National Status and Trends Program, the most comprehensive monitoring survey of U.S. coastal waters, consists of four major components: (1) Benthic Surveillance Project, (2) Mussel Watch Project, (3) Biological Effects Surveys and Research, and (4) Historical Trends Assessment. The following discussion draws heavily from reports of the Status and Trends Program of NOAA's Coastal and Estuarine Assessment Branch in Rockville, MD. The goals of the Benthic Surveillance Project are to measure concentrations of chemical contaminants (aromatic hydrocarbons and their metabolites, chlorinated hydrocarbons, and trace metals) in sediments and in bottom-dwelling finfish at selected sites in urban and nonurban embayments, to determine the prevalences of diseases (e.g., neoplasia, fin erosion, etc.) in these species, and to evaluate temporal trends in the aforementioned parameters.[8-11] The strategy of the Mussel Watch Project is to collect mussels or oysters annually from indigenous populations within a target size range at the same sites and times and to search for trends in chemical contamination. The Mussel Watch Project, like the Benthic Surveillance Project, also delineates chemical contamination in sediments.[12-14] Bivalve mollusk (i.e., mussels and oysters) samples in the Mussel Watch Project are examined not only for the same contaminants as analyzed in bottom fish samples of the Benthic Surveillance Project, but also for radionuclides. Bottom sediments are monitored for pollutants in both the Benthic Surveillance and Mussel Watch Projects because they serve as repositories of chemical contaminants. The exposure of bottom-dwelling fish to chemical pollutants is closely linked to seafloor sediments; hence, these organisms are considered to be reliable indicators of local pollution. Therefore, they are important target organisms in the Benthic Surveillance Project. Since bivalve mollusks have been used successfully as sentinel organisms to detect aquatic contamination, mussels and oysters have been selected in the Mussel Watch Project as indicators of environmental quality.

In 1986, the National Status and Trends Program initiated Biological Effects Surveys and Research to further investigate those regions where laboratory analyses of samples indicated a potential for substantial environmental degradation and biological impacts of the contaminants. Most of this effort has focused on living marine resources. Studies have targeted a number of key subject areas, such as genetic damage of organisms, reproductive impairment, evaluation of new indicators of contamination (DNA damage and enzymatic activity in fish livers), sediment toxicity, and the relation of these effects to contaminant concentration gradients.

Historical Trends Assessment studies involve a closer examination of the environmental conditions in individual regions of the U.S. The available data and ancillary information pertaining to the status and trends of toxic contaminants and their effects in regions of concern are used to assess the present state of knowledge on the magnitude and extent of degradation to living resources and their habitat. More recently, the National Status and Trends Program added sediment coring to better evaluate the trends of chemical contaminants.

The National Status and Trends Program of NOAA assesses the levels of more than 70 contaminants and certain associated effects in biota and sediments as part of a nationwide monitoring effort. It provides data for making spatial and temporal comparisons of contaminant levels to determine which regions around the country are of greatest concern with regard to existing or developing potential for environmental degradation. It includes measurements of concentrations of 24 polycyclic aromatic hydrocarbons, 20 congeners of polychlorinated biphenyls (PCBs), DDT and its breakdown derivatives (DDD and DDE) plus 9 other chlorinated pesticides, butyltins, 12 toxic trace elements, and 4 major elements in sediments, mussels, and oysters at a network of more than 240 regionally representative sites of the Mussel Watch Project. Additionally, the Benthic Surveillance Project determines the levels and effects of the same chemicals in the livers of bottom-dwelling fish and associated sediments at about 75 sites. The frequency of external disease and internal tumors is documented in the fish studied. Data from all monitored sites are stored on national NOAA databases, analyzed, and made available to estuarine managers and the public in numerous reports.

National Benthic Surveillance Project

Initiated in 1984 by NOAA as a component of the National Status and Trends Program, the National Benthic Surveillance Project (NBSP) is a cooperative effort between the National Marine Fisheries Service (NMFS) and the Office of Oceanography and Marine Assessment of the National Ocean Service. The initial phase of the NBSP involved a cooperative effort between three NMFS centers (i.e., Northeast, Southeast, and Northwest centers) using similar protocols and analytical instrumentation. Between 1984 and 1988, 60 sites in embayments along the Atlantic, Gulf, and Pacific coasts were sampled on an annual basis, with levels of chemical contaminants measured in sediments and selected fish species and chemical pollution-related pathological conditions registered on representative bottom fish.[8]

Chemical contaminants measured in sediments of the NBSP include PCBs, aromatic hydrocarbons, organochlorine pesticides, and selected trace metals.[9-11] The concentrations of these xenobiotics are also recorded in the tissues of target fish. In addition, pathological conditions in the fish (e.g., prevalences of gross external lesions and microscopic lesions in the liver and kidney) are registered. Information obtained from these analyses enables investigators to determine which chemicals or classes of chemicals are associated with sediment particulates and which ones are bioaccumulated by the biota. Because fish are highly mobile and tend to integrate contaminants from relatively broad geographic areas, the levels of xenobiotics observed in benthic fish are more representative of the overall degree of contamination in an area than concentrations found in a limited number of sediment samples or in sedentary organisms. Consequently, the concentrations of chemical contaminants in sediments and fish, together with the cause-and-effect relationships established between the contaminants and diseases generated in the organisms, have become useful indicators of environmental degradation.

Due to the broad geographical areas covered by the NBSP, only a limited number of samples have been collected at each location (e.g., a major embayment). NOAA has adapted several site selection criteria for sediment and biotic sampling to meet project goals. These are summarized by Hanson et al.[15] as follows:

1. Broad geographic coverage of the coastline is targeted to establish a national baseline representative of the diversity of habitats and pollutant impacts.
2. The assemblage of sampling locations should represent a range of anthropogenic impacts from relatively pristine to urban-industrial.
3. A collection site is selected to be subtidal, not intertidal; a depositional zone for sediments; integrative of contaminant accumulations within the location; an area that has not been or will not be dredged or scoured and will not undergo slumping; located outside the zone of initial dilution of a point source discharge and outside the zone of an authorized dumpsite.
4. Each collection site contains three stations within an area 2 km in diameter.

In urban embayments, the sampling sites have been selected in areas that integrate waste inputs from multiple sources and have not been selected directly adjacent to any known sources of contaminants or near established dredge disposal areas.[9,16] This sampling strategy avoids the most highly contaminated and less representative areas (i.e., "hot spots") within locations.[17]

Samples from the Benthic Surveillance Project have been analyzed at the following NOAA laboratories:

1. Woods Hole, MA
2. Gloucester, MA
3. Beaufort, NC
4. Charleston, SC
5. Seattle, WA

In addition, Mussel Watch samples are analyzed at the following locations:

1. Battelle Laboratories in Duxbury, MA
2. Battelle Laboratories in Sequim, WA
3. Texas A & M University Geochemical and Environmental Research Group laboratories in College Station, TX
4. La Jolla, CA Laboratory of Scientific Applications International Corporation

Data generated in all these laboratories identify the areas around the U.S. coastline having the highest levels of contamination. Once discovered, these highly contaminated areas become the subject of more intensive surveys of biological and chemical conditions to discern whether the observed levels of contamination are degrading biological communities and habitats. These more intensive studies are performed by NOAA's National Marine Fisheries Laboratories in Beaufort, NC and Seattle, WA. Gradients of contamination and measures of biological properties (e.g., reproductive impairment, genetic damage, and sediment toxicity) typically are carried out in these laboratories.

Results of NBSP sampling indicate that the most contaminated sites along the Pacific Coast are those in highly urbanized and industrialized areas, such as Elliott Bay (Seattle), Commencement Bay (Tacoma), San Pedro Bay (Los Angeles/Long Beach), and San Diego Bay. Most other urban sites have intermediate levels of contamination (e.g., San Francisco Bay, Santa Monica Bay). Nonurban sites and sites in Alaska and Oregon exhibit the lowest levels of contamination.[8,10] Similar findings are reported for the Atlantic and Gulf Coasts sampled as part of the NBSP.[15] For example, trace metal contamination in sediments is highest in areas subjected to the most intense human activity (e.g., Boston Harbor, Raritan Bay, Baltimore Harbor). Lower levels of contamination are seen in Great Bay, Narragansett Bay, and Buzzards Bay. Spatial trends in contamination reveal that the most heavily impacted sites are located along the northeast and central Atlantic Coast with fewer sites along the southeast Atlantic and Gulf Coasts.[17] The concentrations of contaminants in sediments generally are comparable at most polluted sites sampled on both the Atlantic and Pacific coasts, with the exception of Boston Harbor, which has extremely high contaminant concentrations. Tissue levels of contaminants and prevalences of pollution-related liver lesions in fish also appear to be similar for sites on both the Atlantic and Pacific coasts.[18]

Data collected by the NBSP have been used to generate a comprehensive overview of the present status of environmental quality in the nation's coastal waters.[6,9-12,19] The collection of additional data must be continued, however, to effectively evaluate the long-term temporal and spatial trends in environmental quality, as reflected by the concentrations of key contaminants and biological indicators of effects on living resources in coastal and estuarine areas. This is particularly true for urban and industrialized areas, where a relatively high variability of many of the measured parameters exists due to natural variation or patchiness of contaminant concentrations.[8]

Mussel Watch Project

Since 1986, NOAA's National Status and Trends Mussel Watch Project has measured chemical contaminants in surface sediments and tissues of mussels and oysters collected from estuarine and coastal marine waters nationwide. This project is designed to monitor both the current status and long-term trends of a selected suite of organic and inorganic contaminants (e.g., chlorinated pesticides, PCBs, PAHs, and trace metals) by sampling abiotic and biotic media from more than 240 sites along the Atlantic, Gulf, and Pacific coasts, including the Hawaiian Islands. Sampling sites are not uniformly distributed. They lie about 20 km apart in estuaries and embayments and 70 km apart along open coastlines.[6,16] Nearly half of the sampling sites occur in urban estuaries within 20 km of population centers in excess of 100,000 people, where chemical contamination is assumed to be higher and biological effects greater than in nonurban systems.[20] However, known "hot spots" of contamination are avoided.

The Mussel Watch Project has evolved since its inception in 1986. For example, some sites have been added — increasing the total from 145 in 1986 to 234 in 1990 — with further additions made to test the representations of earlier sampling sites.[14] Furthermore, sampling continues on an annual basis for bivalves but not for sediment, and replicate samples are no longer collected. Some chemicals have been added and others removed from the group of target contaminants, most of which are also found on the list of 127 priority pollutants created by the USEPA in the late 1970s.[6]

Mussels and oysters were selected as sentinel organisms for the project because they are fairly cosmopolitan, hardy, and sessile and have a limited capacity for metabolizing the contaminants. They also are easily sampled. The chemical concentrations in these organisms reflect the current contaminant loading of an ecosystem.[21,22] Molluscan species sampled by the Mussel Watch Project differ along U.S. regional coastlines. The mussel *Mytilus edulis* is collected at sites from Maine to Delaware Bay, and the oyster *Crassostrea edulis* from Delaware Bay south through Florida and around the Gulf of Mexico. The mussels *M. edulis* and *M. californianus* are sampled on the west coast, with the oyster *Ostrea sandivicensis* gathered in Hawaii. Chemical body-burden in these bivalves is a function of reproductive state. Hence, sampling is restricted to the period from November through March prior to spawning since this activity alters contaminant concentrations in the animals.

Sediment samples of the Mussel Watch Project are used principally to document the status of chemical contamination, that is, its geographic or spatial distribution. Periodic analyses of surface sediments, however, do not uncover temporal trends of the contaminants.[20] These trends are determined by monitoring chemical concentrations in mussels and oysters.

Sediment sampling since 1988 has been conducted only at Mussel Watch sites not already sampled in prior years due to the lack of data on sedimentation rates and bioturbation of the sediment. The protocol is to collect samples of the upper 2 cm of sediment, with the distance between sediment sampling stations being approximately 0.5 km. At present, it is not possible to delineate the time scale represented by the upper 2 cm of sediment at a sampling site because of insufficient knowledge of the rates of particle deposition and rates of sediment mixing.

Analyses of sediment contamination are essentially comparisons among sampling sites. While a strong species effect for some elements precludes valid comparisons of molluscan data between sites with mussels and those with oysters, there is no species effect on sediment concentrations. However, chemical concentrations vary with sediment grain size such that sites dominated by muddy sediments invariably are the most contaminated. Comparisons take into account the effect of grain size by simply dividing all concentrations by the fraction of sediment in a sample that was silt or clay (i.e., the fine fraction particles $< 63\ \mu m$). All nonsandy sample data from each site, regardless of year, are utilized to calculate mean concentrations for that site.[6]

Mussel Watch data through 1989 and Benthic Surveillance data through 1986 have been combined to form a set of sediment data on chemical contamination. Examination of this data set shows that the highest chemical contamination in the estuarine and coastal U.S. occurs near the urban centers of Boston, New York, San Diego, Los Angeles, and Seattle. A comparison of trace element contaminants in sediments from NOAA's National Status and Trends Program with data from throughout the world indicates similar mean values, although the worldwide data contain higher high concentrations and lower low concentrations owing to the inclusion of samples from "hot spots" and from sandy sediments.[23]

As noted above, temporal trends in chemical contamination of estuarine and coastal marine environments are monitored via annual sampling and chemical analysis of mussels and oysters. The sampling strategy entails the collection of these sentinel organisms from indigenous populations within a target size range at the same sites and times.[14] Two nonparametric statistical tests have been applied to the data in search of temporal trends: the Sign Test and the Spearman Rank Correlation. The Sign Test is based on the fact that there are many sites with which to examine year-to-year changes, whereas the Spearman Rank Correlation tests the correlation between the ranks of concentration and year. The Spearman Test is sensitive only to the direction of change (i.e., increasing or decreasing) rather than its magnitude.[6] Two types of trends may become evident. First, a year-to-year change in chemical concentration occurs in the same direction at a vast majority of sites. Second, a statistically meaningful relationship exists between concentration and time.[20]

Analysis of trends in molluscan concentrations during the first 5 years of the Mussel Watch Project (1986 through 1990) using the aforementioned applications reveals no statistical change between years on a national scale and no strong trends in contaminant concentrations at individual sites. However, when a direction was indicated, it was generally a decrease. In addition, when 1990 was compared with 1986, the only changes in contaminant concentrations were decreases.[12] A more recent report comparing chemical concentrations in mussels and oysters over the period from 1986 to 1993 shows many more decreases than increases in contaminant concentrations along coastal regions.[24] Based on these findings, the levels of contamination in coastal U.S. waters appear to be declining, most probably in response to environmental regulations that have banned or curtailed many toxic substances. For example, all of the chlorinated hydrocarbons monitored by the Mussel Watch Project have been banned for use in the U.S., and the use of other toxic chemicals (e.g., arsenic, cadmium, tributyltin, etc.) has been markedly curtailed.

Lauenstein et al.[25] investigated longer-term decadal trends in trace metal contamination by comparing National Status and Trends data from 1986 through 1988 with data from an earlier mussel watch program sponsored by the USEPA from 1976 through 1978.[3] Because differences in chemical contamination occurred at many sites over a 10-year period rather than at a single site, changes in concentrations are considered trends rather than random variations. Results of this analysis showed that lead concentrations decreased between the decades, although they remained highest in bivalves from urban areas during both sampling periods. For other metals (i.e., cadmium, silver, copper, nickel, and zinc) differences in concentrations were not predominantly in one direction or the other, and no national decadal difference could be detected. Despite different seasons and decades of sampling, therefore, the most common finding was that the concentrations of these metals had not changed over the interval.[14] Nevertheless, the assessment of trends over decadal and longer time scales indicated that the levels of most contaminants in NOAA's Status and Trends Program may be declining.[20]

O'Connor[20] documented an 18-year record of declining tPCB levels in mussels at Royal Palms Park on the Palos Verdes coast of Los Angeles commencing in 1971. Sericano et al.[26] likewise chronicled a dramatic historical decline in tDDT concentrations in oysters from the Gulf of Mexico, with major decreases occurring in the early 1970s. These long-term reductions in chemical contaminants support the view that the imposition of control technologies and other actions to limit or preclude contaminant releases to the environment have generally stemmed the rising levels of chemical contaminants previously observed in estuarine and coastal waters.[12]

CONCLUSIONS

Results of 8 years of National Status and Trends sampling throughout the coastal U.S. reveal that higher levels of chemical contamination of surface sediments are characteristic of urbanized estuaries. This is clearly evident for urbanized areas of the northeastern states and near San Diego, Los Angeles, and Seattle on the west coast; however, except at a few sites, high levels are relatively rare in the southeastern states and along the coastline of the Gulf of Mexico. The high levels of contaminants in urbanized estuaries, though, are lower than those expected to cause sediment toxicity. Liver tumors in fish and sediment toxicity, both extreme responses to contamination, have been found infrequently. Analyses of mussels and oysters are beginning to uncover temporal trends in contaminant levels at National Status and Trends sites. Trends observed over longer time scales (e.g., 1986 to 1993) indicate that the concentrations of most contaminants measured in the National Status and Trends Program may be declining.

REFERENCES

1. Butler, P. A., Organochlorine residues in estuarine mollusks, 1965-72, National Pesticide Monitoring Program, *Pest. Monit. J.*, 6, 238, 1973.
2. Butler, P. A. and Schutzmann, R. L., Residues of pesticides and PCBs in estuarine fish, National Pesticide Monitoring Program, *Pest. Monit. J.*, 12, 51, 1978.
3. Goldberg, E. D., Koide, M., Hodge, V., Flegal, A. R., and Martin, J., U.S. Mussel Watch: 1977-1978 results on trace metals and radionuclides, *Est. Coastal Shelf Sci.*, 16, 69, 1983.
4. Gadbois, D. F. and Maney, R. S., Survey of polychlorinated biphenyls in selected finfish species from United States coastal waters, *Fish. Bull., U.S.*, 81, 389, 1983.
5. NOAA, *National Status and Trends Program for Marine Environmental Quality: Progress Report — A Summary of Selected Data on Chemical Contaminants in Tissues Collected during 1984, 1985, and 1986,* NOAA Tech. Mem. OMA 38, OAD/NOS/NOAA, National Oceanic and Atmospheric Administration, Department of Commerce, Rockville, MD, 1987.
6. O'Connor, T. P., The National Oceanic and Atmospheric Administration's (NOAA) National Status and Trends Mussel Watch Program: national monitoring of chemical contamination in the coastal United States, in *Environmental Statistics, Assessment, and Forecasting*, Cothern, C. R. and Ross, N. P. (Eds.), Lewis Publishers, Boca Raton, FL, 1993, 331.
7. Breckenridge, R. P. and Olson, G. L., Identification and use of biomonitoring data, in *Handbook of Ecotoxicology*, Hoffman, D. J., Rattner, B. A., Burton, G. A., Jr., and Cairns, J., Jr. (Eds.), Lewis Publishers, Boca Raton, FL, 1995, 198.
8. Varanasi, U., Chan, S-L., McCain, B. B., Schiewe, M. H., Clark, R. C., Brown, D. W., Myers, M. S., Landahl, J. T., Krahn, M. M., Gronlund, W. D., and MacLeod, W. D., Jr., *National Benthic Surveillance Project: Pacific Coast. Part I. Summary and Overview of Results for Cycles I to III (1984–1986),* NOAA Tech. Mem., NMFS FNWC-156, National Oceanic and Atmospheric Administration, Rockville, MD, 1988.
9. Johnson, L. L., Stehr, C. M., Olson, O. P., Myers, M. S., McCain, B. B., Pierce, S. M., and Varanasi, U., National Status and Trends Program, *National Benthic Surveillance Project: Northeast Coast — Fish Histopathology and Relationships Between Lesions and Chemical Contaminants (1987–1989),* NOAA Tech. Mem. NMFS/NWFSC-4, National Oceanic and Atmospheric Administration, Seattle, WA, 1992.
10. Myers, M. S., Stehr, C. M., Olson, O. P., Johnson, L. L., McCain, B. B., Chan, S.-L., and Varanasi, U., National Status and Trends Program, *National Benthic Surveillance Project: Pacific Coast — Fish Histopathology and Relationships Between Toxicopathic Lesions and Exposure to Chemical Contaminants for Cycles I to V (1984–1988),* NOAA Tech. Mem. NMFS/NWSFSC-6, National Oceanic and Atmospheric Administration, Seattle, WA, 1993.
11. Turgeon, D. D., Gottholm, B. W., Wolfe, D. A., and Robertson, A., The National Status and Trends Program National Benthic Surveillance Project — Contaminants in Fish Tissues, in *Proceedings Coastal Zone '93, the 8th Symposium on Coastal and Ocean Management*, Magoon, Q. T., Wilson, S., Converse, H., and Tobin, L. T. (Eds.), American Society of Civil Engineers, New York, 1993, 3, 3474.
12. O'Connor, T. P., *Recent Trends in Coastal Environmental Quality: Results from the First Five Years of NOAA Mussel Watch Project,* National Oceanic and Atmospheric Administration Office of Ocean Resources Conservation and Assessment, Rockville, MD, 1992.
13. Turgeon, D. D., Bricker, S. B., and O'Connor, T. P., National Status and Trends Program: chemical and biological monitoring of U.S. coastal waters, in *Ecological Indicators*, McKenzie, D. H., Hyatt, D. E., and McDonald, V. J. (Eds.), Elsevier Applied Science, London, 1992. 425.

14. O'Connor, T. P., Cantillo, A. Y., and Lauenstein, G. G., Monitoring of temporal trends in chemical contamination by the NOAA National Status and Trends Mussel Watch Project, in *Biomonitoring of Coastal Waters and Estuaries,* Kramer, K. J. M. (Ed.), CRC Press, Boca Raton, FL, 1994, 29.
15. Hanson, P. J., Evans, D. W., and Colby, D. R., Assessment of elemental contamination in estuarine and coastal environments based on geochemical and statistical modeling of sediments, *Mar. Environ. Res.,* 36, 237, 1993.
16. Lauenstein, G. G., Harmon, M. R., and Gottholm, B. W., *National Status and Trends Program: Monitoring Sites Descriptions (1984–1990) for the National Mussel Watch and Benthic Surveillance Projects,* NOAA Tech. Mem. 70, NOAA/NOS/ORCA, N/ORCA21, National Oceanic and Atmospheric Administration, Silver Spring, MD, 1993.
17. McCain, B. B., Brown, D. W., Krahn, M. M., Myers, M. S., Clark, R. C., Jr., Chan, S.-L., and Malins, D. C., Marine pollution problems, North American west coast, *Aquat. Toxicol.,* 11, 143, 1988.
18. Varanasi, U., Chan, S.-L., McCain, B. B., Schiewe, M. H., Clark, R. C., Brown, D. W., Myers, M. S., Landahl, J. T., Krahn, M. M., Gronlund, W. P., and MacLeod, W. D., Jr., *National Benthic Surveillance Project: Pacific Coast. Part II. Summary and Overview of the Results for Cycles I to III (1984–1986),* NOAA Tech. Mem., NMFS F/NWC-6, National Oceanic and Atmospheric Administration, Rockville, MD, 1989.
19. NOAA, *National Status and Trends Program for Marine Environmental Quality: Progress Report and Preliminary Assessments of Findings of the Benthic Surveillance Project* — 1984, OAD/NOS/NOAA, Department of Commerce, Rockville, MD, 1987.
20. O'Connor, T. P., *Coastal Environmental Quality in the United States, Chemical Contamination in Sediments and Tissues, A Special NOAA 20th Anniversary Report,* National Oceanic and Atmospheric Administration Office of Ocean Resources Conservation and Assessment, Rockville, MD, 1990.
21. Farrington, J. W., Risebrough, R. W., Parker, P. L., Davis, A. C., de Lapps, B., Winters, J. K., Boatwright, D., and Frew, N. M., *Hydrocarbons, Polychlorinated Biphenyls, and DDE in Mussels and Oysters from the U.S. Coast, 1976–1978,* The Mussel Watch Technical Report, WHOI-82-42, Woods Hole Oceanographic Institution, Woods Hole, MA, 1982.
22. Farrington, J. W., Goldberg, E. D., Risebrough, R. W., Martin, J. H., and Bowen, V. T., U.S. "Mussel Watch" 1976–1978: an overview of the trace metal, DDE, PCB, hydrocarbon, and artificial radionuclide data, *Environ. Sci. Technol.,* 17, 490, 1983.
23. Cantillo, A. Y. and O'Connor, T. P., Trace element contaminants in sediments from the NOAA National Status and Trends Program compared to data from throughout the world, *Chem. Ecol.,* 7, 31, 1992.
24. O'Connor, T. P. and Beliaeff, B., *Recent Trends in Coastal Environmental Quality: Results from the Mussel Watch Project,* NOAA Tech. Rep., Department of Commerce, Rockville, MD, 1995.
25. Lauenstein, G. G., Robertson, A., and O'Connor, T. P., Comparisons of trace metal data in mussels and oysters from a mussel watch program of the 1970s with those from a 1980s program, *Mar. Pollut. Bull.,* 21, 440, 1990.
26. Sericano, J. L., Wade, T. L., Atlas, E. A., and Brooks, J. M., Historical perspective on the environmental bioavailability of DDT and its derivatives to Gulf of Mexico oysters, *Environ. Sci. Technol.,* 24, 1541, 1990.

APPENDIX 1. NATIONAL SURVEYS

TABLE 1.1
Chemicals Measured in the NOAA National Status and Trends Program

DDT and Its Metabolites

2,4′-DDD
4,4′-DDD
2,4′-DDE
4,4′-DDE
2,4′-DDT
4,4′-DDT

Chlorinated Pesticides other than DDT

Aldrin
cis-Chlordane
trans-Nonachlor
Dieldrin
Heptachlor
Heptachlor epoxide
Hexachlorobenzene
Lindane (γ-HCH)
Mirex

Polychlorinated Biphenyls

PCB congeners 8, 18, 28, 44, 56, 66, 101, 105, 118, 128, 138, 153, 179, 180, 187, 195, 206, 209

Polycyclic Aromatic Hydrocarbons

2-ring
Biphenyl
Naphthalene
1-Methylnaphthalene
2-Methylnaphthalene
2,6-Dimethylnaphthalene
1,6,7-Trimethylnaphthalene

3-ring
Fluorene
Phenanthrene
1-Methylphenanthrene
Anthracene
Acenaphthene
Acenaphthylene

4-ring
Fluoranthene
Pyrene
Benz(*a*)anthracene
Chrysene

5-ring
Benzo(*a*)pyrene
Benzo(*e*)pyrene
Perylene
Dibenz(*a,h*)anthracene
Benzo(*b*)fluoranthene
Benzo(*k*)fluoranthene

6-ring
Benzo(*ghi*)perylene
Indeno(1,2,3-*cd*)pyrene

Major Elements

Al	Aluminum
Fe	Iron
Mn	Manganese

Trace Elements

As	Arsenic
Cd	Cadmium
Cr	Chromium
Cu	Copper
Pb	Lead
Hg	Mercury
Ni	Nickel
Se	Selenium
Ag	Silver
Sn	Tin
An	Zinc

Tri-, di- and mono-butyltin

Source: O'Connor, T. P., Cantillo, A. Y., and Lauenstein, G. G., in *Biomonitoring of Coastal Waters and Estuaries,* Kramer, K. J. M. (Ed.), CRC Press, Boca Raton, FL, 1994, 34. With permission.

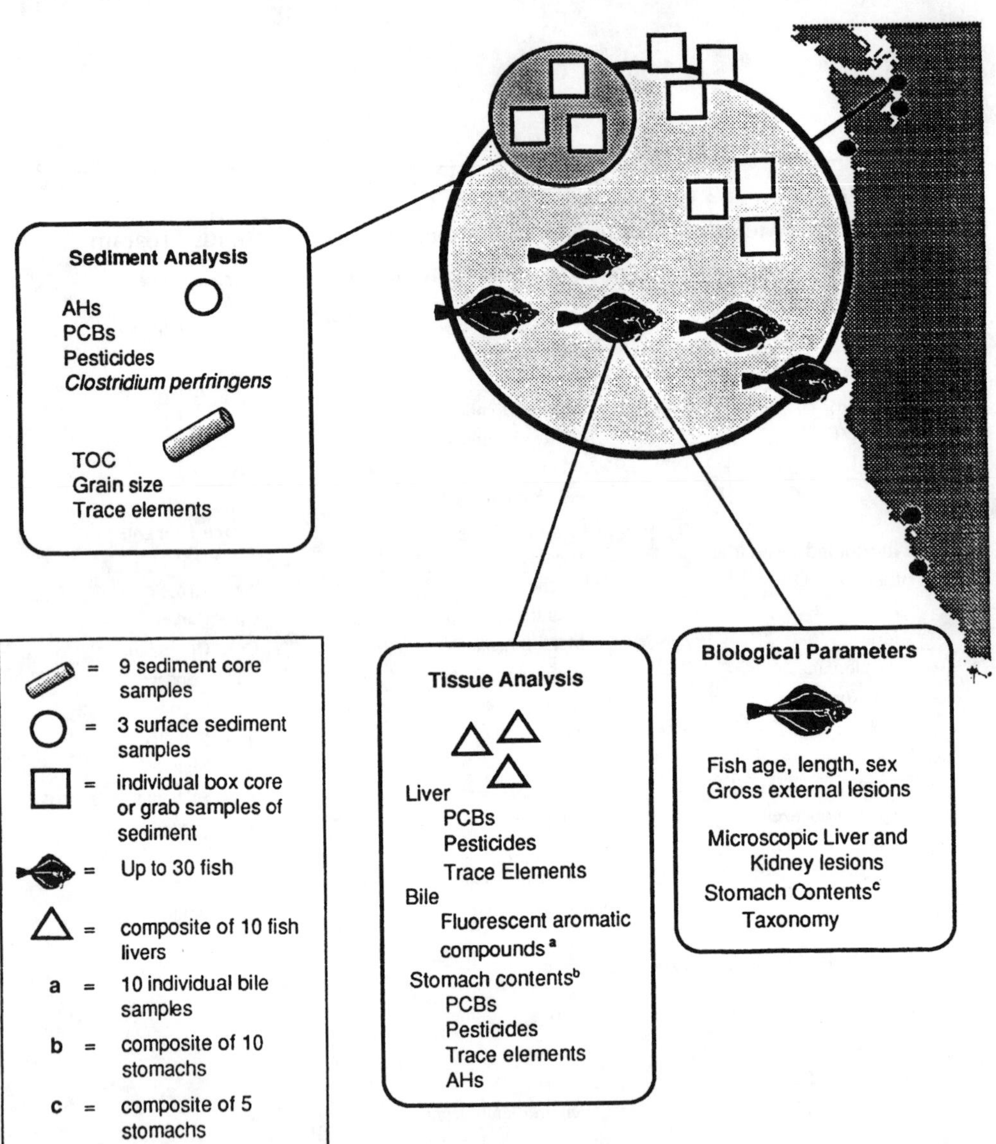

FIGURE 1.1. The sampling strategy of the National Benthic Surveillance Project. Each sampling site consists of three stations at which grab samples of sediment were taken. (From Varanasi, U., Chan, S.-L., McCain, B. B., Schiewe, M. H., Clark, R. C., Brown, D. W., Myers, M. S., Landahl, J. T., Krahn, M. M., Gronlund, W. D., and Macleod, W. D., Jr., *National Benthic Surveillance Project: Pacific Coast. Part I. Summary and Overview of the Results for Cycles I to III 1984–1986,* NOAA Tech. Rep., National Oceanic and Atmospheric Administration, Seattle, WA, 1988.)

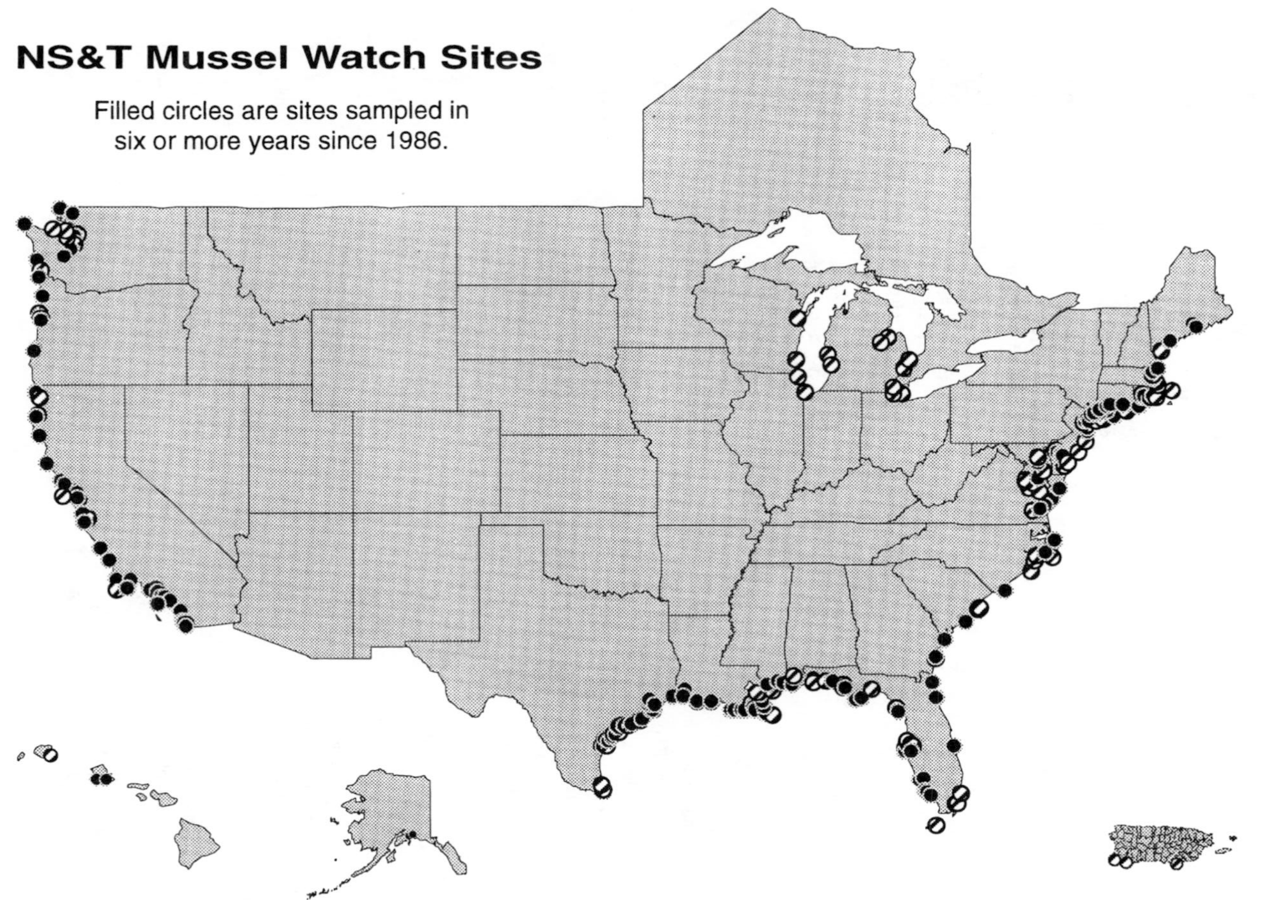

FIGURE 1.2. Map showing sampling sites of the National Status and Trends Mussels Watch Project. (From O'Connor, T. P. and Beliaeff, B., *Recent Trends in Coastal Environmental Quality: Results from the Mussel Watch Project*, NOAA Tech. Rep., Department of Commerce, Rockville, MD, 1995.)

TABLE 1.2
National Status and Trends Sites Where Contaminant Concentrations in Mollusks Were Increasing or Decreasing Over the Period 1986–1988[a]

Site	Location	State
Arsenic Decreasing		
LICR[a]	Long Island Sound	CT
LINH[a]	Long Island	CT
CBMP[a]	Chesapeake Bay	MD
CBCI[a]	Chincoteague Bay	VA
TBMK[a]	Tampa Bay	FL
ABOB	Atchafalaya Bay	LA
VBSP	Vermillion Bay	LA
YBOP[a]	Yaquina Bay	OR
Arsenic Increasing		
BBRH[a]	Buzzards Bay	MA
MRCB[a]	Matanzas River	FL
SJCB	St. Johns River	FL
LJLJ[a]	LaJolla	CA
NBWJ	Newport Beach	CA
SFSM	San Francisco Bay	CA
Cadmium Decreasing		
NBDI	Narragansett Bay	RI
LICR[a]	Long Island Sound	CT
LINH[a]	Long Island Sound	CT
LIHR	Long Island Sound	CT
LIMR[a]	Long Island Sound	NY
LITN	Long Island Sound	NY
DBFE	Delaware Bay	DE
DBBD[a]	Delaware Bay	DE
DBAP[a]	Delaware Bay	DE
CBHG[a]	Chesapeake Bay	MD
CBDP	Chesapeake Bay	VA
CHFJ[a]	Charleston Harbor	SC
CHSF	Charleston Harbor	SC
SSSI[a]	Sapelo Sound	GA
MBEM[a]	Matagorda Bay	TX
UISB[a]	Unakwit Inlet	AK
HHKL	Honolulu Harbor	HI
Cadmium Increasing		
PBPI	Penobscot Bay	ME
TBLB	Terrebonne Bay	LA
SPFP[a]	San Pedro Harbor	CA
PGLP	Pacific Grove	CA
Chromium Decreasing		
TBMK	Tampa Bay	FL
OSBJ[a]	Oceanside	CA
MDSJ	Marina Del Rey	CA
PDPD	Pt. Dume	CA
PALH[a]	St. Arena	CA
SSBI[a]	S. Puget Sound	WA
BBSM[a]	Bellingham Bay	WA
Chromium Increasing		
BBGN	Buzzards Bay	MA
LIHU[a]	Long Island Sound	NY
LIPJ[a]	Long Island Sound	NY
HRLB	Hudson/Raritan Estuary	NY
NYSH	New York Bight	NJ
NYLB[a]	New York Bight	NJ
SRTI	Savannah River Estuary	GA
MRCB	Matanzas River	FL
Copper Decreasing		
LIHR[a]	Long Island Sound	CT
LIPJ[a]	Long Island Sound	NY
CBHG	Chesapeake Bay	MD
SSSI[a]	Sapelo Sound	GA
TBMK	Tampa Bay	FL
APDB[a]	Apalachicola Bay	FL
SAWB	St. Andrew Bay	FL
OSBJ[a]	Oceanside	CA
PDPD[a]	Pt. Dume	CA
PCPC[a]	Pt. Conception	CA
TBSR	Tomales Bay	CA
Copper Increasing		
HRJB[a]	Hudson/Raritan Estuary	NY
HRUB[a]	Hudson/Raritan Estuary	NY
HRLB[a]	Hudson/Raritan Estuary	NY
NYSH[a]	New York Bight	NJ
NYSR[a]	New York Bight	NJ
CKBP	Cedar Key	FL
BBSD	Barataria Bay	LA
TBLB	Terrebonne Bay	LA
MGBP[a]	Matagorda Bay	TX
GHWJ[a]	Gray's Harbor	WA

TABLE 1.2 (continued)
National Status and Trends Sites Where Contaminant Concentrations in Mollusks Were Increasing or Decreasing Over the Period 1986–1988[a]

Lead Decreasing		
BPBP	Barber's Point	HI
Lead Increasing		
CBCC	Chesapeake Bay	VA
CBSP	Choctawhatchee Bay	FL
CCNB[a]	Corpus Christi	TX
LMSB	L. Laguna Madre	TX
BBSM[a]	Bellingham Bay	WA
Mercury Decreasing		
LIHR[a]	Long Island Sound	CT
CBCC[a]	Chesapeake Bay	VA
EVFU[a]	Everglades	FL
BPBP	Barber's Point	HI
Mercury Increasing		
LICR[a]	Long Island Sound	CT
HRUB[a]	Hudson/Raritan Estuary	NY
NYSH[a]	New York Bight	NJ
NYLB[a]	New York Bight	NJ
CBHP	Chesapeake Bay	MD
QIUB[a]	Quinby Inlet	VA
RSJC[a]	Roanoke Sound	NC
CFBI	Cape Fear	NC
MRCB	Matanzas River	FL
TBLB[a]	Terrebonne Bay	LA
SSSS	San Simeon Point	CA
Nickel Decreasing		
LIHR[a]	Long Island Sound	CT
SLBB	Sabine Lake	TX
MBGP	Matagorda Bay	TX
ABLR	Aransas Bay	TX
PLLH[a]	Pt. Loma	CA
MBVB	Mission Bay	CA
MDSJ	Marina Del Rey	CA
BBBE	Bodega Bay	CA
TBSR	Tomales Bay	CA
Nickel Increasing		
HRJB	Hudson/Raritan Estuary	NY
HRUB[a]	Hudson/Raritan Estuary	NY
NYSH	New York Bight	NJ
CBHP[a]	Chesapeake Bay	MD
PSWB[a]	Pamlico Sound	NC
SRTI	Savannah River Estuary	GA
ABOB[a]	Atchafalaya Bay	LA
Selenium Decreasing		
CBTP	Comm. Bay	WA
HHKL	Honolulu Harbor	HI
Selenium Increasing		
TBPB	Tampa Bay	FL
APCP[a]	Apalachicola Bay	FL
APDB	Apalachicola Bay	FL
CBSP	Choctawhatchee Bay	FL
MSBB	Mississippi Sound	MS
BBSD	Barataria Bay	LA
BBMB	Barataria Bay	LA
CLCL[a]	Caillou Lake	LA
ABOB[a]	Atchafalaya Bay	LA
VBSP	Vermillion Bay	LA
MBGP[a]	Matagorda Bay	TX
SFSM	San Francisco Bay	CA
Silver Decreasing		
BHDB	Boston Harbor	MA
BHDI	Boston Harbor	MA
LIHR	Long Island Sound	CT
LISI	Long Island Sound	CT
DBBD	Delaware Bay	DE
CBMP	Chesapeake Bay	MD
CBHG	Chesapeake Bay	MD
CBDP	Chesapeake Bay	VA
RSJC	Roanoke Sound	NC
CFBI	Cape Fear	NC
TBMK	Tampa Bay	FL
CCNB[a]	Corpus Christi	TX
LMSB	L. Laguna Madre	TX
Silver Increasing		
BBGN	Buzzards Bay	MA
CKBP	Cedar Key	FL
BBSD	Barataria Bay	LA
TBLB	Terrebonne Bay	LA
ABOB	Atchafalaya Bay	LA
MBGP	Matagorda Bay	TX
SFSM	San Francisco Bay	CA
JFCF	San Juan de Fuca	WA
BBSM	Bellingham Bay	WA
BPBP	Barber's Point	HI
PVMC[a]	Port Valdez	AK

TABLE 1.2 (continued)
National Status and Trends Sites Where Contaminant Concentrations in Mollusks Were Increasing or Decreasing Over the Period 1986–1988[a]

Zinc Decreasing		
LIHR[a]	Long Island Sound	CT
LIHU[a]	Long Island Sound	CT
LIMR[a]	Long Island Sound	NY
CBHG[a]	Chesapeake Bay	MD
MRCB	Matanzas River	FL
TBMK[a]	Tampa Bay	FL
SAWB[a]	St. Andrew Bay	FL
SLBB[a]	Sabine Lake	TX
IBNJ[a]	Imperial Beach	CA
ABWJ[a]	Anaheim Bay	CA
Zinc Increasing		
CKBP	Cedar Key	FL
BBSD	Barataria Bay	LA
BBMB	Barataria Bay	LA
JHJH[a]	Joseph Harbor Bayou	LA
SFSM[a]	San Francisco Bay	CA
UISB[a]	Unakwit Inlet	AK
HHKL	Honolulu Harbor	HI
PCB Decreasing		
BHDI	Boston Harbor	MA
DBKI	Delaware Bay	DE
HRLB	Hudson/Raritan Estuary	NY
PLLH	Pt. Loma	CA
SCBR	S. Catalina Island	CA
SCFP	Santa Cruz Island	CA
SBSB	Pt. Santa Barbara	CA
PCPC	Pt. Conception	CA
PDSC	Pt. Delgada	CA
GHWJ	Gray's Harbor	WA
EBFR	Elliott Bay	WA
SIWP[a]	Sinclair Inlet	WA
PRPR	Pt. Roberts	WA
PCB Increasing		
None		
DDT Decreasing		
BBAR	Buzzards Bay	MA
HRLB	Hudson/Raritan Estuary	NY
CBHG	Chesapeake Bay	MD
CBCC	Chesapeake Bay	VA
LMSB	L. Laguna Madre	TX
CBTP	Comm. Bay	WA
DDT Increasing		
MSPB	Mississippi Sound	MS
OSBJ	Oceanside	CA
MDSJ	Marina Del Rey	CA
SSBI[a]	S. Puget Sound	WA
Chlordane Decreasing		
BHDI	Boston Harbor	MA
DBFE	Delaware Bay	DE
LINH	Long Island Sound	CT
LISI	Long Island Sound	CT
LITN[a]	Long Island Sound	NY
HRLB	Hudson/Raritan Estuary	NY
Chlordane Increasing		
EVFU	Everglades	FL
APDB	Apalachicola Bay	FL
MSPB	Mississippi Sound	MS
Dieldrin Decreasing		
PBSI	Penobscot Bay	ME
BHDI	Boston Harbor	MA
HRJB	Hudson/Raritan Estuary	NY
HRLB	Hudson/Raritan Estuary	NY
NYSH	New York Bight	NJ
DBFE	Delaware Bay	DE
SAWB	St. Andrew Bay	FL
JHJH	J. Harbor Bayou	LA
Dieldrin Increasing		
MBTH	Moriches Bay	NY
YHYH	Yaquina Head	OR
Lindane Decreasing		
LISI	Long Island Sound	CT
LIPJ	Long Island Sound	NY
HRLB	Hudson/Raritan Estuary	NY
NYLB	New York Bight	NJ
Lindane Increasing		
NBDI	Narragansett Bay	RI
VBSP	Vermillion Bay	LA

TABLE 1.2 (continued)
National Status and Trends Sites Where Contaminant Concentrations in Mollusks Were Increasing or Decreasing Over the Period 1986–1988[a]

LMWpah Decreasing		
HRUB	Hudson/Raritan Estuary	NY
NYLB	New York Bight	NJ
TBMK	Tampa Bay	FL
GBTD	Galveston Bay	TX
SSSS	San Simeon Point	CA
SGSG	Pt. St. George	OR
LMWpah Increasing		
BBAR	Buzzards Bay	MA
CBMP	Chesapeake Bay	MD
MSPB	Mississippi Sound	MS
MBTP	Matagorda Bay	TX
CCNB	Corpus Christi	TX
HMWpah Decreasing		
HRUB	Hudson/Raritan Estuary	NY
TBMK	Tampa Bay	FL
SAWB	St. Andrew Bay	FL
SIWP	Sinclair Inlet	WA
HMWpah Increasing		
LINH	Long Island Sound	CT
CBMP	Chesapeake Bay	MD
CBSR	Choctawhatchee Bay	FL

[a] Less than a twofold change.

Source: NOAA, *A Summary of Data on Tissue Contamination from the First Three Years (1986–1988) of the Mussel Watch Project,* NOAA Tech., Mem. NOS OMA 49, National Oceanic and Atmospheric Administration, Rockville, MD, 1989.

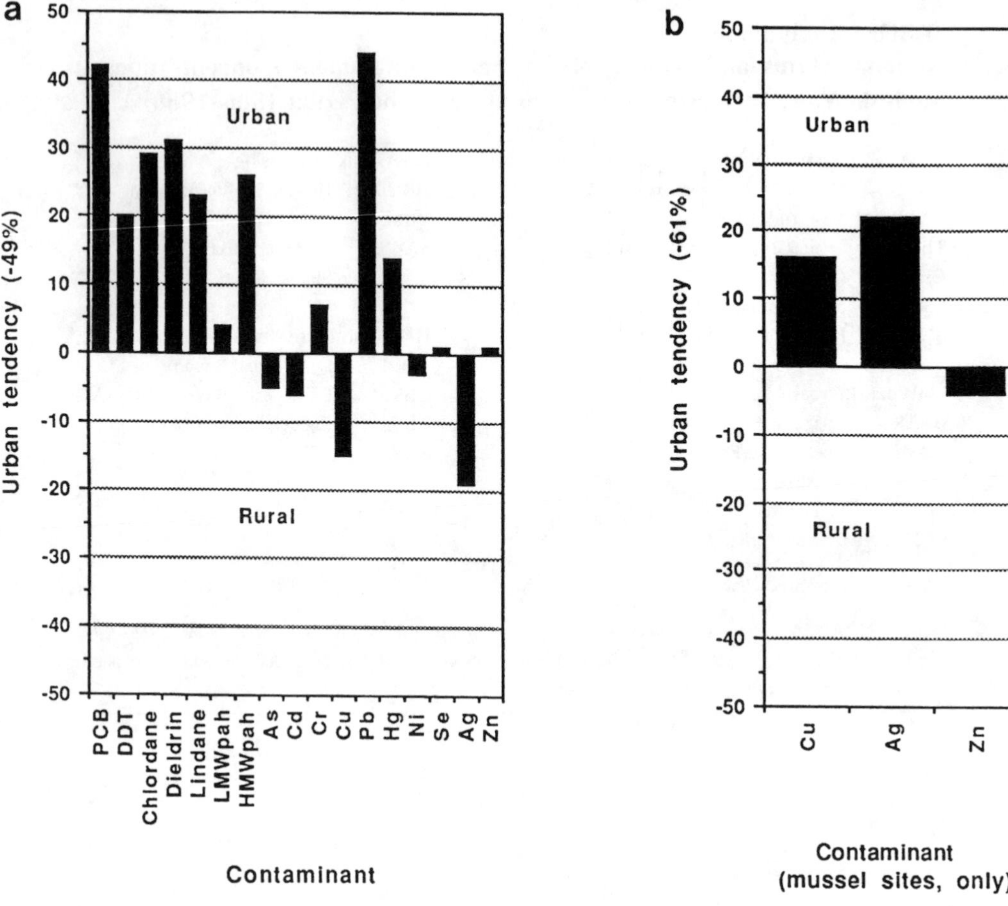

FIGURE 1.3. Tendencies for highest 20 contaminant concentrations to be in mollusks at urban sites. Among all 177 sites sampled during 1986 through 1988, 49% were in urban areas. (a) The values plotted are the percentages among all sites (minus 49%) that were urban and at which concentrations were ever among the highest 20. For example, of the 32 sites at which PCB concentrations were among the highest 20 in any year, 91% of those sites were urban and the plotted value is 91 – 49 = 42%. (b) Three chemicals, Cu, Ag, and Zn, are heavily preferred by oysters, and the urban tendencies are plotted only for those 81 sites where mussels were collected. In that case 61% of the sites were urban. (From NOAA, *A Summary of Data on Tissue Contamination from the First Three Years (1986–1988) of the Mussel Watch Project,* NOAA Tech. Mem. NOS OMA 49, National Oceanic and Atmospheric Administration, Rockville, MD, 1989.)

TABLE 1.3
Results of a Sign Test Applied to Mean Chemical Concentrations at 141 NOAA Mussel Watch Sites Sampled in at Least Four of the Years Between 1986 and 1990[a]

Chemical	86–87	87–88	88–89	89–90	86–90
Ag	—	—	—	Inc	—
As	Dec	—	Dec	Inc	—
Cd	Dec	—	—	—	Dec
Cr	Dec	—	—	—	—
Cu	—	—	—	—	—
Hg	—	—	—	—	—
Ni	—	Dec	Dec	—	Dec
Pb	Inc	—	Dec	Inc	—
Se	—	Inc	Dec	—	—
Zn	Dec	—	—	Inc	—
ΣPCB	Dec	—	—	Dec	Dec
ΣDDT	—	Dec	—	—	—
ΣCdane	Inc	Dec	—	—	Dec
ΣPAH	—[b]	Inc	Dec	—	—[b]
ΣBT	—	—	—	Dec	—

[a] Comparisons made on a year-to-year basis and between 1986 and 1990. A statistically significant (0.05 level) proportion of changes in the increasing (Inc) or decreasing (Dec) direction are indicated, as are cases with no significant direction of change (—). With 141 samples, a direction of change is significant if it is common to at least 83 samples. Since it was not routinely measured in earlier years, ΣBT comparisons have been made for the 149 sites common to 1989 and 1990.

[b] ΣPAH data for 1986 were not used. The national geometric mean for that year was 430 ng $\cdot$ g^{-1} which is 1.4 times the next highest annual geometric mean. Its high value may have been due to the analytical method used in that year for samples from the east and west coasts. The last comparison for ΣPAH is an 87–90 comparison.

Source: O'Connor, T. P., Cantillo, A. Y., and Laurenstein, G. G., in *Biomonitoring of Coastal Waters and Estuaries,* Kramer, K. J. M. (Ed.), CRC Press, Boca Raton, FL, 1994, 29. With permission.

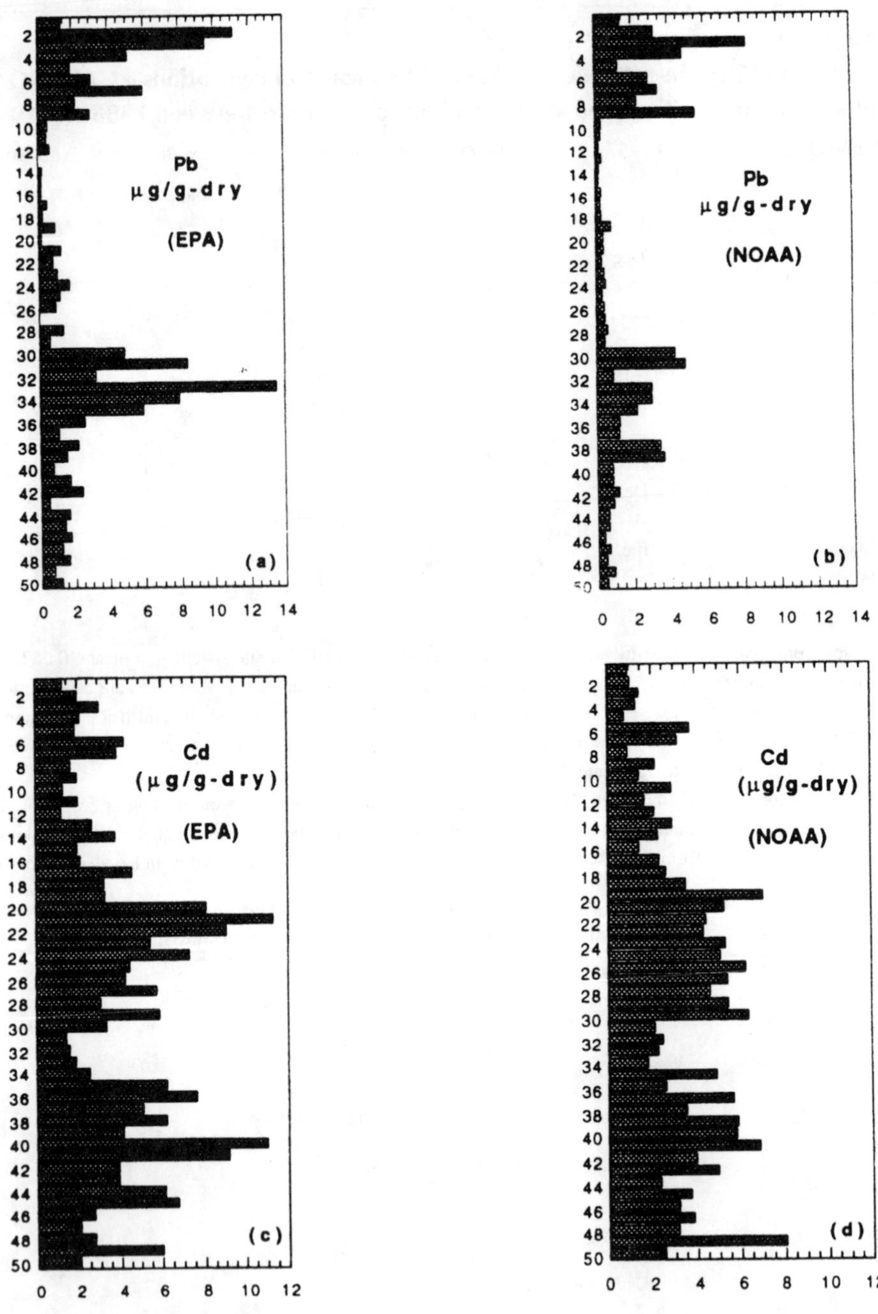

FIGURE 1.4. Mean concentrations of lead, (a) and (b), and cadmium, (c) and (d), in whole soft parts of mussels collected at 50 sites by the USEPA-sponsored program from 1976 to 1978, (a) and (c), and by the NOAA National Status and Trends Program from 1986 to 1988, (b) and (d). Sites are listed clockwise around the U.S. Sites 1 to 16 are from Maine to the east coast of Florida; sites 17 to 29 extend from the west coast of Florida to western Texas; sites 30 to 50 are from southern California to Puget Sound, WA. Oysters were collected at sites from 10 (Chesapeake Bay) through 29. Mean lead concentrations are higher in the 1970s at all but the following 14 sites: 1, 8, 9, 13, 16, 18, 20, 27, 37, 38, 39, 40, 43, and 49. Cadmium is higher in the 1970s only 26 times out of 50 and cannot be considered different. Cadmium concentrations are statistically different (Kruskall-Wallis Test, .05 level) in 12 cases; lead concentrations are statistically different in 7. (From O'Connor, T. P., Cantillo, A. Y., and Lauenstein, G. G., in *Biomonitoring of Coastal Waters and Estuaries*, Kramer, K. J. M. (Ed.), CRC Press, Boca Raton, FL, 1994, 41. With permission.)

TABLE 1.4
National Status and Trends Sites Ranking Among the Highest 20 in 1986, 1987, and 1988 and Overall for Total Polychlorinated Biphenyls (PCBs) in Mussels (*Mytilus edulis*) and Oysters (*Crassostrea virginica*)

Code	Location	State	Species	Ranking				Significant Difference?	Trend
				1986	1987	1988	Overall		
BBAR	Buzzards Bay	MA	me	1	1t	1	1	no	no
HRLB	Hudson/Raritan Estuary	NY	me	2	3	3t	2	yes	d
NYSR	New York Bight	NJ	me	3	17t	11	6	no	no
HRUB	Hudson/Raritan Estuary	NY	me	4	5t	2	3	no	no
SDHI	San Diego Bay	CA	me	5	4	5t	5	no	no
GBYC	Galveston Bay	TX	cv	6	8t	16	10t	no	no
NYSH	New York Bight	NJ	me	7	12	5t	7t	yes	no
BHDB	Boston Harbor	MA	me	8	5t	<	10t	no	no
BBGN	Buzzards Bay	MA	me	9	<	<	14	no	no
BBRH	Buzzards Bay	MA	me	10	1t	3t	4	no	no
BHDI	Boston Harbor	MA	me	11t	19	<	13	yes	d
LITN	Long Island Sound	NY	me	11t	8t	7t	9	no	no
EBFR	Elliott Bay	WA	me	13	<	<	19t	yes	d
BHHB	Boston Harbor	MA	me	14	13	<	18	no	no
HRJB	Hudson/Raritan Estuary	NY	me	15	<	10	15t	yes	no
LINH	Long Island Sound	CT	me	16	<	19t	<	no	no
LIMR	Long Island Sound	NY	me	17	<	14t	<	no	no
SAWB	St. Andrew Bay	FL	cv	18	<	17t	<	no	no
LISI	Long Island Sound	CT	me	19	<	<	<	no	no
NYLB	New York Bight	NJ	me	20	15	17t	19t	no	no
SFEM	San Francisco Bay	CA	me	—	5t	7t	7t	—	—
LIHH	Long Island Sound	NY	me	<	8t	<	17	yes	no
LIHR	Long Island Sound	CT	me	<	11	<	15t	yes	no
CBSP	Choctawhatchee Bay	FL	cv	<	14	<	<	no	no
MDSJ	Marina Del Rey	CA	me	<	16	<	<	yes	no
BHBI	Boston Harbor	MA	me	<	17t	<	<	yes	no
PBIB	Pensacola Bay	FL	cv	<	20	<	<	no	no
GBSC	Galveston Bay	TX	cv	—	—	9	12	—	—
APDB	Apalachicola Bay	FL	cv	<	<	12	<	yes	no
ABWJ	Anaheim Bay	CA	mc	<	<	13	<	no	no
SFSM	San Francisco Bay	CA	me	<	<	19t	<	no	no
LICR	Long Island Sound	CT	me	<	<	19t	<	no	no

Note: me = *Mytilus edulis;* cv = *Crassostrea virginica;* t = two or more concentrations were equal; d = decreasing trends.

Source: NOAA, *A Summary of Data on Tissue Contamination from the First Three Years (1986–1988) of the Mussel Watch Project,* NOAA Tech., Mem. NOS OMA 49, National Oceanic and Atmospheric Administration, Rockville, MD, 1989.

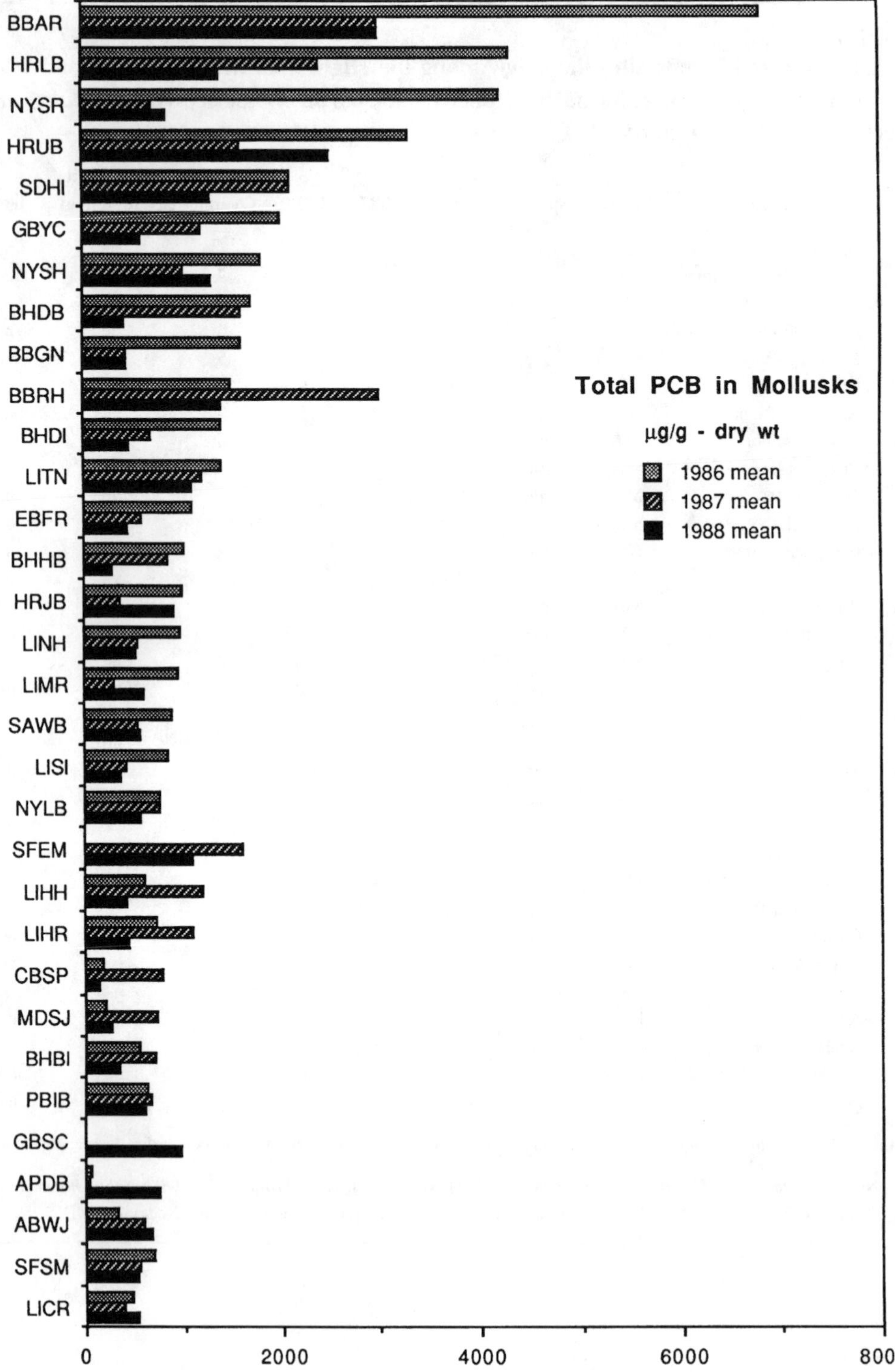

FIGURE 1.5. Total PCB concentrations in mollusks (mussels, *Mytilus edulis,* and oysters, *Crassostrea virginica*) at various National Status and Trends sites during the 1986 to 1988 survey period. See Table 1.4 for identification of estuary or coastal system. (From NOAA, *A Summary of Data on Tissue Contamination from the First Three Years (1986–1988) of the Mussel Watch Project,* NOAA Tech. Mem. NOS OMA 49, National Oceanic and Atmospheric Administration, Rockville, MD, 1989.)

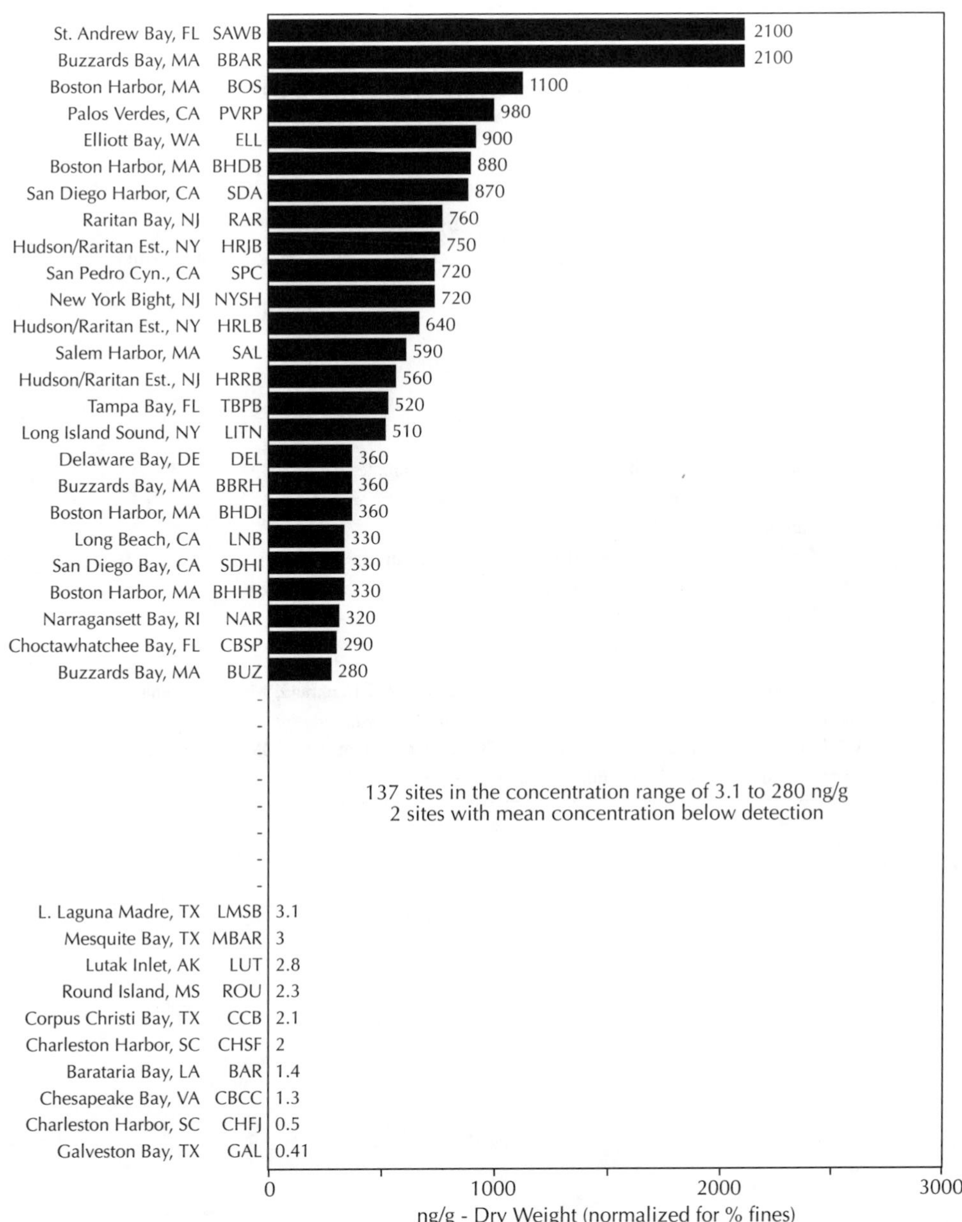

FIGURE 1.6. Total PCB concentrations in sediments at various National Status and Trends sites during the 1984 to 1987 survey period. (From NOAA, *A Summary of Data on Individual Organic Contaminants in Sediments Collected during 1984, 1985, 1986, and 1987,* NOAA Tech. Mem. NOS OMA 47, National Oceanic and Atmospheric Administration, Rockville, MD, 1989.)

TABLE 1.5
Median or Geometric Mean DDT Concentrations for Several National Surveys

Organism	tDDT or DDE (ppm wet weight) by Sampling Period			
	1965–1972	1972–1975	1976–1977	1984–1986
Bivalves	0.024[a]		0.01[b]/0.001[c]	.003
Fish, whole juvenile		0.014[d]	ND	ND
Fish, muscle		0.110[e]	0.012[f]	
Fish, liver			0.220[f]	0.054[g]
Fish, whole fresh weight	0.7–1.1	0.4–0.6	0.370	

[a] Median of 8180 site means composited from 7839 samples.
[b] Median of 89 site means composited from 188 samples.
[c] Median of 80 site values or site means.
[d] Median of 144 site means composited from 1524 composites.
[e] Median of area or site means from samples.
[f] Median of 19 site means from samples.
[g] Median of 42 site medians from 126 composites.

Source: Mearns, A. J., Matta, M. B., Simecek-Beatty, D., Buchman, M. F., Shigenaka, G., and Wert, W. A., *PCB and Chlorinated Pesticide Contamination in U.S. Fish and Shellfish: A Historical Assessment Report,* NOAA Tech. Mem. NOS OMA 39, National Oceanic and Atmospheric Administration, Seattle, WA, 1988.

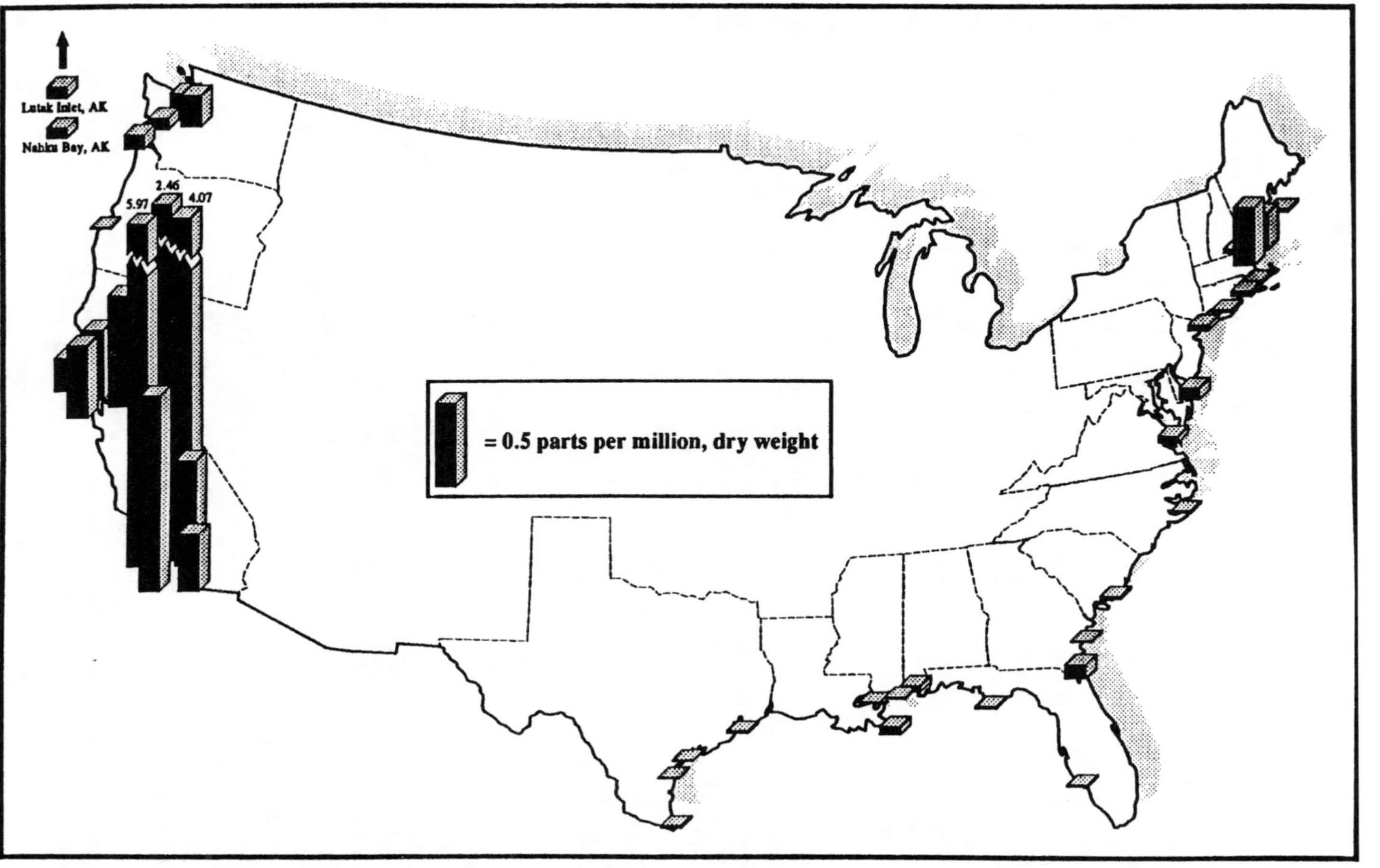

FIGURE 1.7. Total DDT in liver of estuarine fish composites collected at 42 sites in the U.S. during 1984. Data from the NOAA National Status and Trends Benthic Surveillance Project. (From Mearns, A. J., Matta, M. B., Simecek-Beatty, D., Buchman, M. F., Shigenaka, G., and Wert, W. A., *PCB and Chlordinated Pesticide Contamination in U.S. Fish and Shellfish: A Historical Assessment Report*, NOAA Tech. Mem. NOS OMA 39, National Oceanic and Atmospheric Administration, Seattle, WA, 1988.)

TABLE 1.6
National Status and Trends Sites Ranking Among the Highest 20 in 1986, 1987, and 1988 and Overall for Total DDT in Mussels (*Mytilus edulis, Mytilus californianus*) and Oysters (*Crassostrea virginica*)

Code	Location	State	Species	Ranking 1986	1987	1988	Overall	Significant Difference?	Trend
HRLB	Hudson/Raritan Estuary	NY	me	1t	7	10	5	yes	d
SPFP	San Pedro Harbor	CA	me	1t	4	1	2	no	no
PVRP	Palos Verdes	CA	mc	1t	1	2	1	no	no
ABWJ	Anaheim Bay	CA	mc	4	3	4	4	yes	no
SFDB	San Francisco Bay	CA	me	5	11t	<	16	no	no
CBSP	Choctawhatchee Bay	FL	cv	6	2	3	3	no	no
BBAR	Buzzards Bay	MA	me	7t	<	<	<	yes	d
NYSH	New York Bight	NJ	me	7t	19t	16t	14t	no	no
IBNJ	Imperial Beach	CA	mc	9	10	7	10	no	no
NYSR	New York Bight	NJ	me	10	<	<	<	no	no
LITN	Long Island Sound	NY	me	11	11t	<	17	yes	no
HRUB	Hudson/Raritan Estuary	NY	me	12t	14t	8	11	no	no
CBCC	Chesapeake Bay	VA	cv	12t	<	<	<	yes	d
SAWB	St. Andrew Bay	FL	cv	12t	<	<	<	no	no
SLSL	San Luis Ob. Bay	CA	mc	12t	<	<	<	no	no
HRJB	Hudson/Raritan Estuary	NY	me	16t	<	13	20	yes	no
DBBD	Delaware Bay	DE	cv	16t	8	<	12	no	no
PDPD	Pt. Dume	CA	mc	18	6	12	9	no	no
LIMR	Long Island Sound	NY	me	19t	<	<	<	no	no
MBCP	Mobile Bay	AL	cv	19t	<	18t	<	no	no
GBYC	Galveston Bay	TX	cv	19t	<	<	<	no	no
MBSC	Monterey Bay	CA	mc	19t	<	<	<	no	no
SFEM	San Francisco Bay	CA	me	—	5	9	7	—	—
MDSJ	Marina Del Rey	CA	me	<	9	5	8	yes	i
DBAP	Delaware Bay	DE	cv	<	13	<	<	no	no
BHDB	Boston Harbor	MA	me	<	14t	<	<	no	no
NBWJ	Newport Beach	CA	mc	<	14t	<	<	no	no
SFSM	San Francisco Bay	CA	me	<	14t	<	<	no	no
SDHI	San Diego Bay	CA	me	<	15t	<	<	no	no
PGLP	Pacific Grove	CA	mc	<	15t	<	<	no	no
DBKI	Delaware Bay	DE	cv	<	19t	<	<	no	no
MBHI	Mobile Bay	AL	cv	—	—	6	6	—	—
OSBJ	Oceanside	CA	me	<	<	11	<	yes	i
MRPL	Mississippi River	LA	cv	—	—	14	13	—	—
GBSC	Galveston Bay	TX	cv	—	—	15	14t	—	—
APDB	Apalachicola Bay	FL	cv	<	<	16t	<	no	no
MSPB	Mississippi Sound	MS	cv	<	<	18t	<	yes	i
BRFS	Brazos River	TX	cv	—	—	18t	18t	—	—
SANM	San Miguel Island	CA	mc	—	—	18t	18t	—	—

Note: me = *Mytilus edulis;* mc = *Mytilus californianus;* cv = *Crassostrea virginica;* t = two or more concentrations were equal; d = decreasing trends; i = increasing trends.

Source: NOAA, *A Summary of Data on Tissue Contamination from the First Three Years (1986–1988) of the Mussel Watch Project,* NOAA Tech. Mem. NOS OMA 49, National Oceanic and Atmospheric Administration, Rockville, MD, 1989.

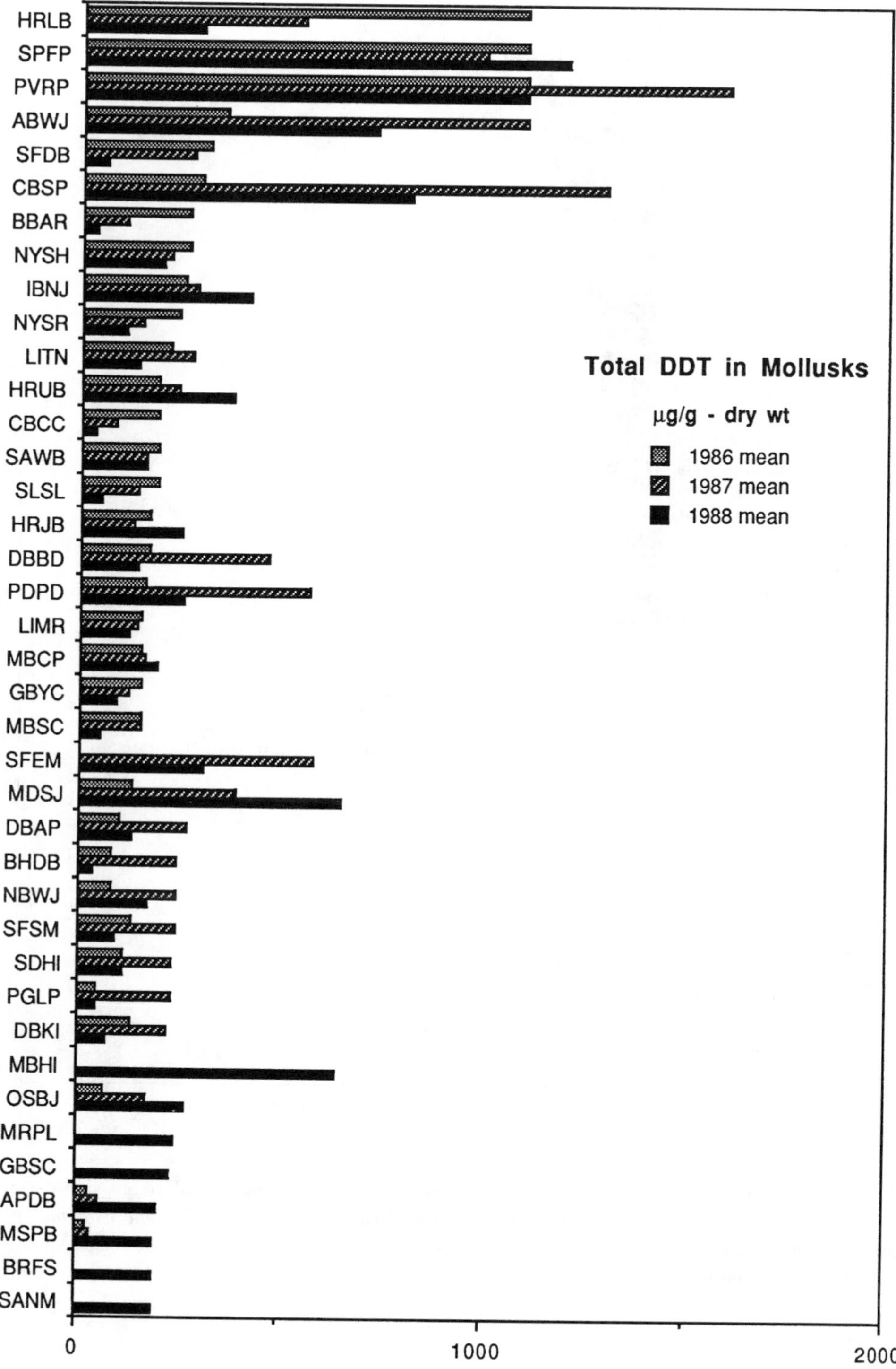

FIGURE 1.8. Total PCB concentrations in mollusks (mussels, *Mytilus edulis,* and oysters, *Crassostrea virginica*) at various National Status and Trends sites during the 1986 to 1988 survey period. See Table 1.6 for identification of estuary or coastal system. (From NOAA, *A Summary of Data on Tissue Contamination from the First Three Years (1986–1988) of the Mussel Watch Project,* NOAA Tech. Mem. NOS OMA 49, National Oceanic and Atmospheric Administration, Rockville, MD, 1989.)

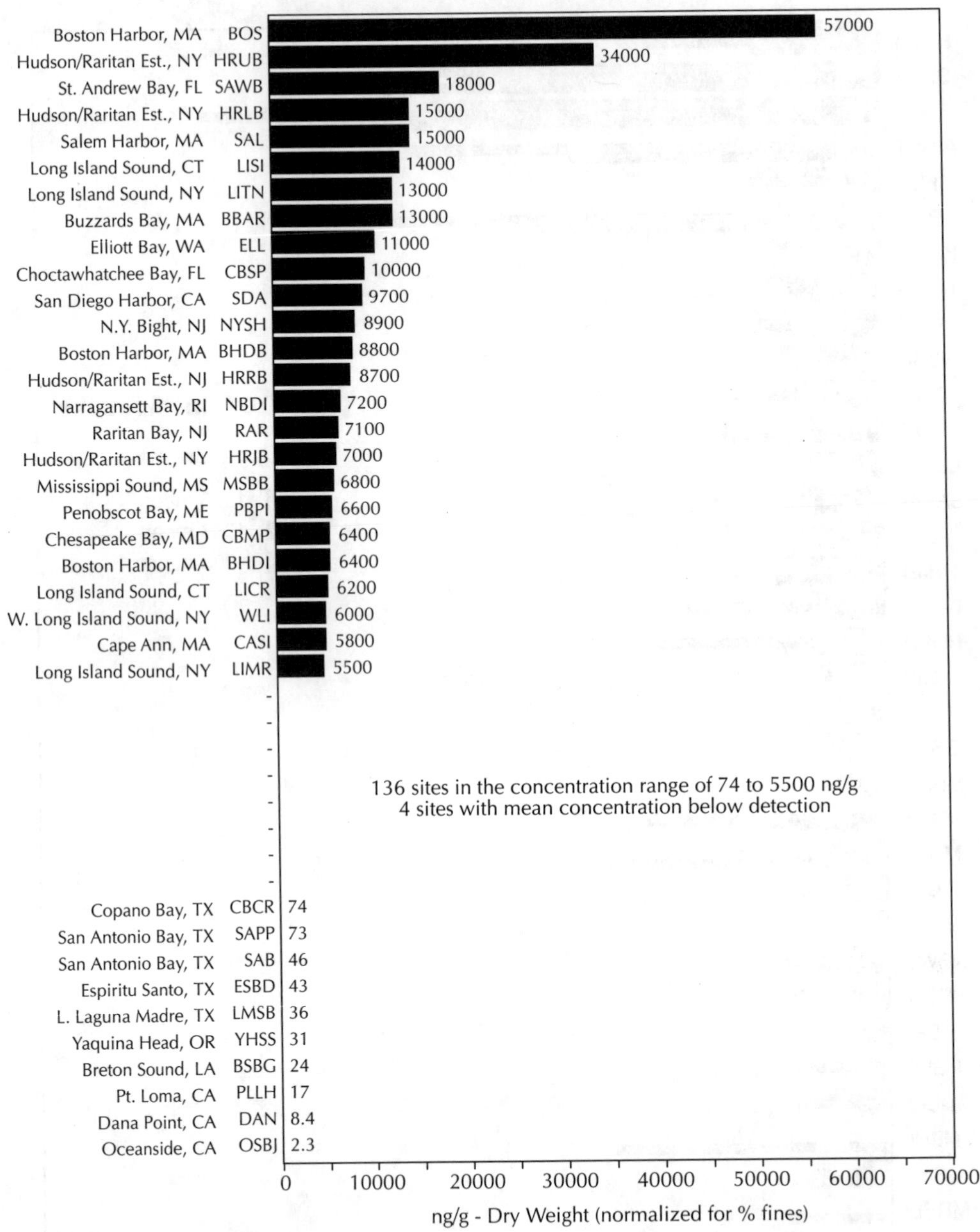

FIGURE 1.9. Total PAH concentrations in sediments at various National Status and Trends sites during the 1984 to 1987 survey period. (From NOAA, *A Summary of Data on Individual Organic Contaminants in Sediments Collected during 1984, 1985, 1986, 1987*, NOAA Tech. Mem. NOS OMA 47, National Oceanic and Atmospheric Administration, Rockville, MD, 1989.)

TABLE 1.7
National Status and Trends Sites Ranking Among the Highest 20 in 1986, 1987, and 1988 and Overall for Total PAH in Mussels (*Mytilus edulis, Mytilus californianus*) and Oysters (*Crassostrea virginica, Ostrea sandivicensis*)

Code	Location	State	Species	Ranking				Significant Difference?	Trend
				1986	1987	1988	Overall		
SAWB	St. Andrew Bay	FL	cv	1	17t	<	3	no	no
EBFR	Elliott Bay	WA	me	2	13t	1	1	yes	no
HRUB	Hudson/Raritan Estuary	NY	me	3	2	<	4	yes	d
CBRP	Coos Bay	OR	me	4	<	<	9	yes	no
BHDB	Boston Harbor	MA	me	5	8	6	6	no	no
LITN	Long Island Sound	NY	me	6	10	15t	8	no	no
SGSG	Pt. St. George	OR	mc	7	<	<	16	yes	d
BHDI	Boston Harbor	MA	me	8t	4	3	5	yes	no
CBCH	Coos Bay	OR	mc	8t	<	<	13t	yes	no
BHHB	Boston Harbor	MA	me	10t	<	<	15	yes	no
PLLH	Pt. Loma	CA	mc	10t	—	<	<	yes	no
SIWP	Sinclair Inlet	WA	me	10t	<	<	19	yes	no
SCFP	Santa Cruz Island	CA	mc	13t	<	<	<	yes	no
SSSS	San Simeon Point	CA	mc	13t	<	<	<	yes	d
SCBR	S. Catalina Island	CA	mc	15	<	<	<	no	no
BPBP	Barber's Point	HI	os	16	<	<	<	no	no
HRJB	Hudson/Raritan Estuary	NY	me	17	3	4	7	no	no
NYSH	New York Bight	NJ	me	18	<	<	<	yes	no
NYLB	New York Bight	NJ	me	19t	15	<	<	yes	d
CFBI	Cape Fear	NC	cv	19t	<	<	<	no	no
CHSF	Charleston Harbor	SC	cv	19t	<	<	<	no	no
SDHI	San Diego Bay	CA	me	<	1	<	10	yes	no
MSBB	Mississippi Sound	MS	cv	<	5t	<	18	no	no
HHKL	Honolulu Harbor	HI	os	<	5t	<	<	yes	no
SFEM	San Francisco Bay	CA	me	—	7	13	11t	—	—
BBMB	Barataria Bay	LA	cv	<	9	<	<	no	no
CBSR	Choctawhatchee Bay	FL	cv	<	11	<	<	yes	no
BHBI	Boston Harbor	MA	me	<	12	5	13t	no	no
NYSR	New York Bight	NJ	me	<	13t	<	<	no	no
CBTP	Comm. Bay	WA	me	<	16	9t	<	no	no
BBPC	Biscayne Bay	FL	cv	—	17t	—	<	no	no
BBAR	Buzzards Bay	MA	me	<	19	17	<	yes	i
RSJC	Roanoke Sound	VA	cv	<	20	<	<	no	no
PCMP	Panama City	FL	cv	—	—	2	2	no	no
NMML	North Miami	FL	cv	—	—	7t	11t	no	no
APDB	Apalachicola Bay	FL	cv	<	<	7t	<	no	no
CBCI	Chincotgue. Bay	VA	cv	<	<	9t	<	no	no
LICR	Long Island Sound	CT	me	<	<	9t	<	no	no
BBSM	Bellingham Bay	WA	me	<	<	9t	<	yes	i
DBAP	Delaware Bay	DE	cv	<	<	14	<	no	no
IRSR	Indian River	FL	cv	—	—	15t	17	—	—
UISB	Unakwit Inlet	AK	me	<	<	18	<	yes	i
LINH	Long Island Sound	CT	me	<	<	19	<	yes	i
SSBI	South Puget Sound	WA	me	<	<	20	<	yes	no
AIAC	Absecon Inlet	NJ	me	—	—	<	20	—	—

Note: cv = *Crassostrea virginica;* me = *Mytilus edulis;* mc = *Mytilus californianus;* os = *Ostrea sandivicensis*; t = two or more concentrations were equal; d = decreasing trends; i = increasing trends.

Source: NOAA, *A Summary of Data on Tissue Contamination from the First Three Years (1986–1988) of the Mussel Watch Project,* NOAA Tech. Mem. NOS OMA 49, National Oceanic and Atmospheric Administration, Rockville, MD, 1989.

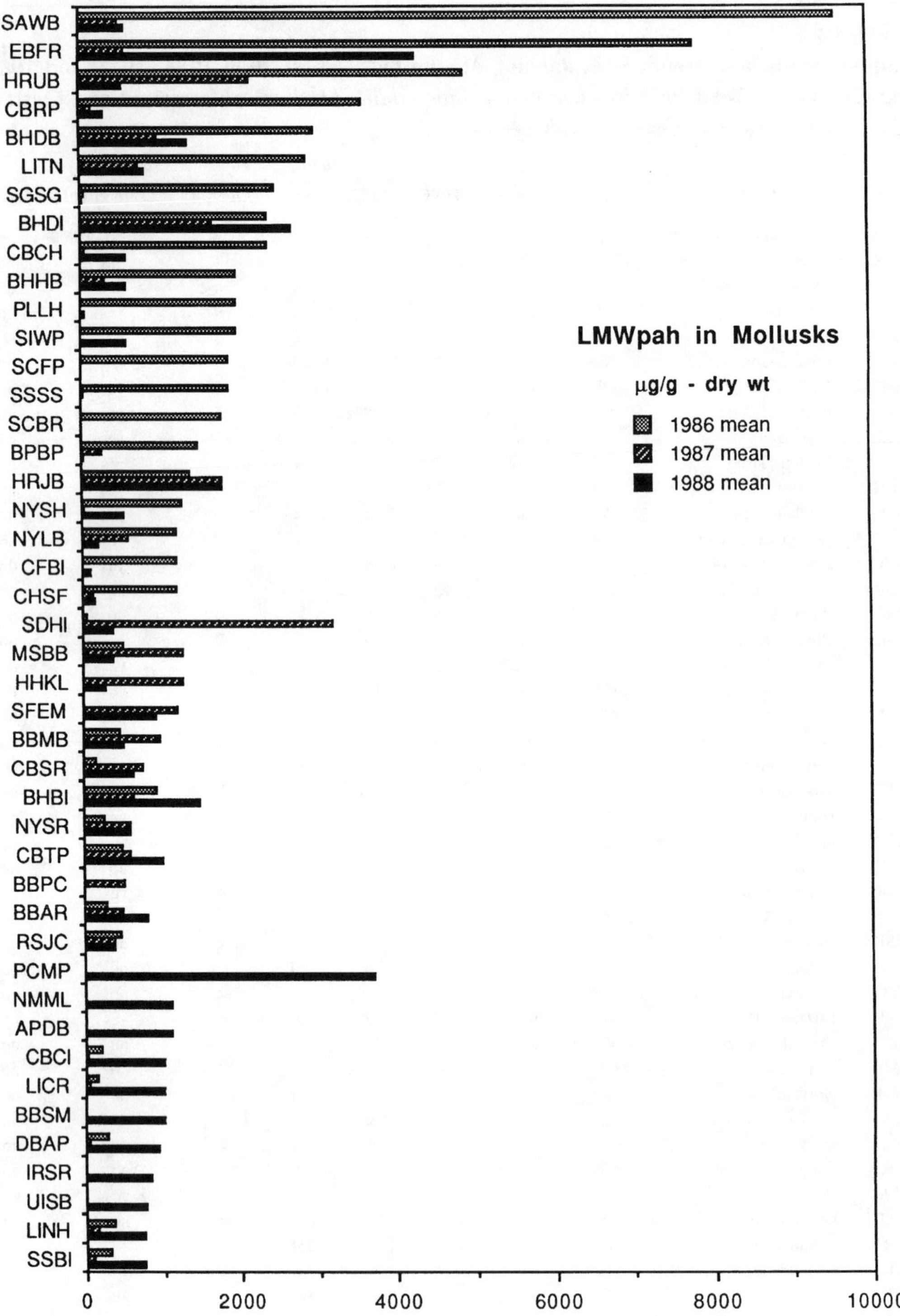

FIGURE 1.10. Low-molecular-weight PAH concentrations in mollusks from various National Status and Trends sites during the 1986 to 1988 survey period. See Table 1.7 for identification of estuary or coastal systems. (From NOAA, *A Summary of Data on Tissue Contamination from the First Three Years (1986–1988) of the Mussel Watch Project,* NOAA Tech. Mem. NOS OMA 49, National Oceanic and Atmospheric Administration, Rockville, MD, 1989.)

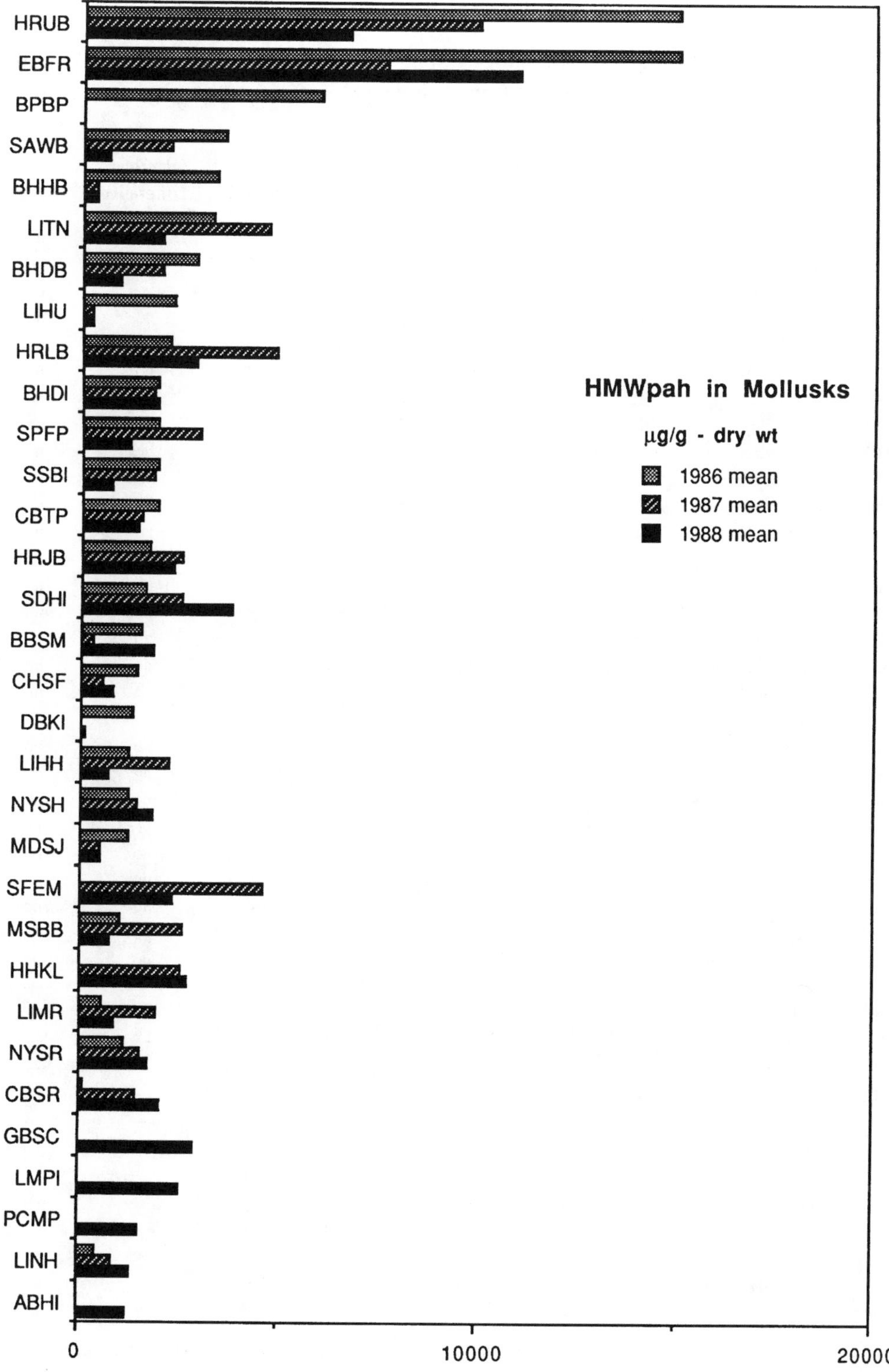

FIGURE 1.11. High-molecular-weight PAH concentrations in mollusks from various National Status and Trends sites during the 1986 to 1988 survey period. See Table 1.7 for identification of estuary or coastal systems. (From NOAA, *A Summary of Data on Tissue Contamination from the First Three Years (1986–1988) of the Mussel Watch Project,* NOAA Tech. Mem. NOS OMA 49, National Oceanic and Atmospheric Administration, Rockville, MD, 1989.)

TABLE 1.8
National Status and Trends Sites Ranking Among the Highest 20 in 1986, 1987, and 1988 and Overall for Concentrations of Copper in Oysters (*Crassostrea virginica, Ostrea sandivicensis*)

				Ranking					
Code	**Location**	**State**	**Species**	**1986**	**1987**	**1988**	**Overall**	**Significant Difference?**	**Trend**
BPBP	Barber's Point	HI	os	1	2	3	2	no	no
DBAP	Delaware Bay	DE	cv	2	5	5t	5	yes	no
SAWB	St. Andrew Bay	FL	cv	3	8	<	14t	yes	d
DBKI	Delaware Bay	DE	cv	4	13t	<	14t	no	no
VBSP	Vermillion Bay	LA	cv	5	3	15t	6t	yes	no
DBBD	Delaware Bay	DE	cv	6t	10t	<	14t	no	no
CBMP	Chesapeake Bay	MD	cv	6t	12	16t	17t	yes	no
NBNB	Naples Bay	FL	cv	8	7	8t	8	no	no
LBMP	Lake Borgne	LA	cv	9	<	16t	<	yes	no
CBCR	Copano Bay	TX	cv	10	<	7	17t	yes	no
SAMP	San Antonio Bay	TX	cv	11	<	—	<	—	—
HHKL	Honolulu Harbor	HI	os	12	1	1	1	no	no
DBFE	Delaware Bay	DE	cv	13	<	<	<	no	no
CBHP	Chesapeake Bay	MD	cv	14	<	<	<	no	no
QIUB	Quinby Inlet	VA	cv	15	<	<	<	no	no
MBEM	Matagorda Bay	TX	cv	16	<	<	<	no	no
CLSJ	Calcasieu Lake	LA	cv	17	13t	<	<	no	no
JHJH	J. Harbor Bayou	LA	cv	18t	13t	<	<	no	no
ESSP	Espiritu Santo	TX	cv	18t	<	—	<	—	—
ABLR	Aransas Bay	TX	cv	18t	<	<	<	no	no
CCNB	Corpus Christi	TX	cv	18t	17t	<	<	no	no
MSBB	Mississippi Sound	MS	cv	<	19t	<	<	no	no
SLBB	Sabine Lake	TX	cv	<	4	14	11t	yes	no
GBYC	Galveston Bay	TX	cv	<	6	<	<	no	no
ABOB	Atchafalaya Bay	LA	cv	<	9	15t	<	no	no
BSSI	Breton Sound	LA	cv	<	10t	<	<	no	no
RBHC	Rookery Bay	FL	cv	<	16	<	<	yes	no
MSPC	Mississippi Sound	MS	cv	<	17t	<	<	no	no
CBBI	Charlotte Harbor	FL	cv	<	19t	—	<	—	—
NMML	North Miami	FL	cv	—	—	2	3	—	—
KAUI	Kauai	HI	os	—	—	4	4	—	—
CLLC	Calcasieu Lake	LA	cv	—	—	5t	6t	—	—
PCMP	Panama City	FL	cv	—	—	8t	9t	—	—
LBNO	Lake Borgne	LA	cv	—	—	8t	9t	—	—
BBSD	Barataria Bay	LA	cv	<	<	11	<	yes	i
ABHI	Aransas Bay	TX	cv	—	—	12t	11t	—	—
LMPI	Laguna Madre	TX	cv	—	—	12t	11t	—	—
CCBH	Corpus Christi	TX	cv	—	—	15t	17t	—	—
MBTP	Matagorda Bay	TX	cv	<	<	16t	<	yes	no
MRPL	Mississippi River	LA	cv	—	—	16t	20t	—	—
BRFS	Brazos River	TX	cv	—	—	16t	20t	—	—

Note: os = *Ostrea sandivicensis;* cv = *Crassostrea virginica;* t = two or more concentrations were equal; d = decreasing trends; i = increasing trends.

Source: NOAA, *A Summary of Data on Tissue Contamination from the First Three Years (1986–1988) of the Mussel Watch Project,* NOAA Tech. Mem. NOS OMA 49, National Oceanic and Atmospheric Administration, Rockville, MD, 1989.

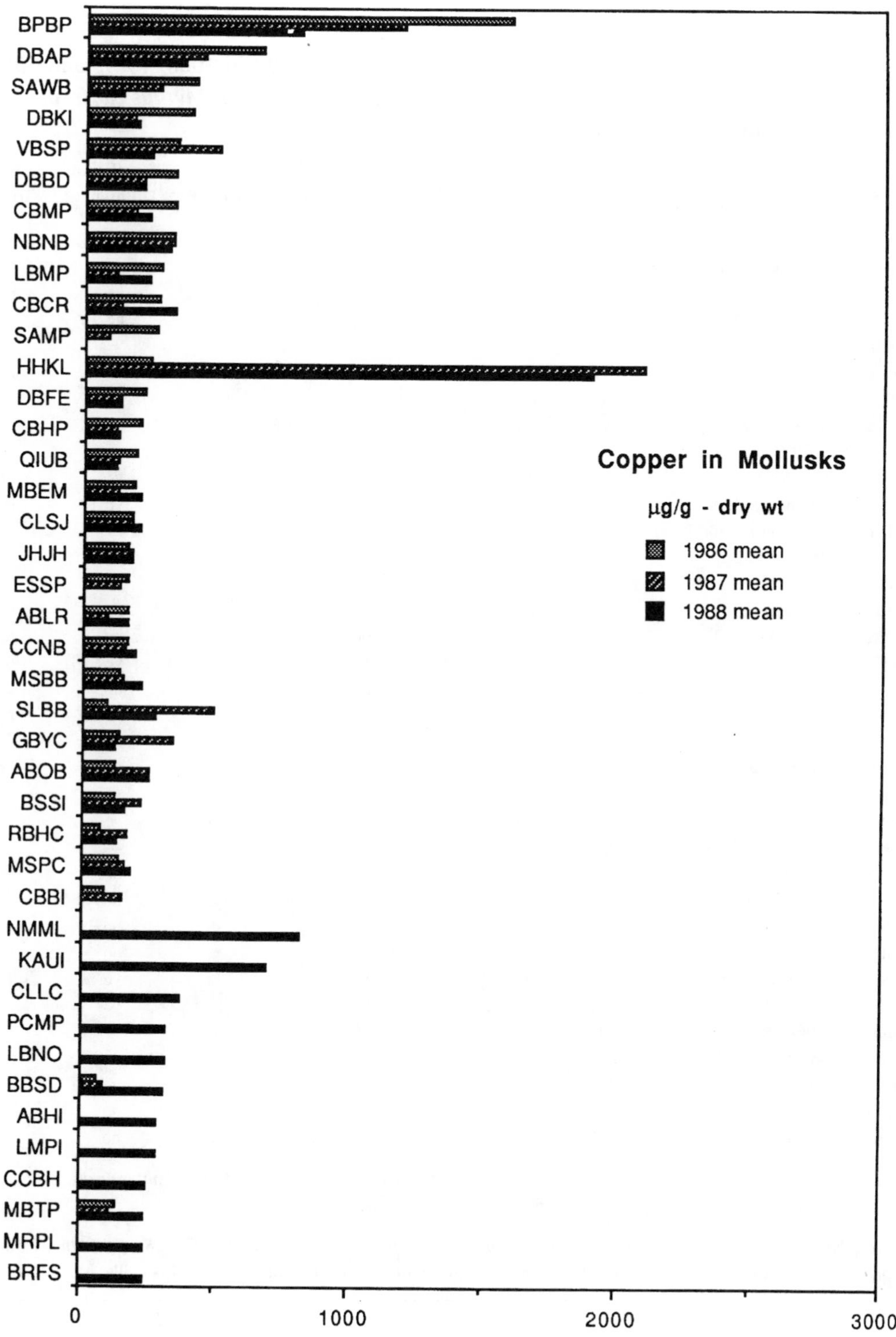

FIGURE 1.12. Concentrations of copper in oysters (*Crassostrea virginica, Ostrea sandivicensis*) from various National Status and Trends sites during the 1986 to 1988 survey period. See Table 1.8 for location of sites. (From NOAA, *A Summary of Data on Tissue Contamination from the First Three Years (1986–1988) of the Mussel Watch Project,* NOAA Tech. Mem. NOS OMA 49, National Oceanic and Atmospheric Administration, Rockville, MD, 1989.)

TABLE 1.9
National Status and Trends Sites Ranking Among the Highest 20 in 1986, 1987, and 1988 and Overall for Concentrations of Mercury in Mussels (*Mytilus edulis, Mytilus californianus*) and Oysters (*Crassostrea virginica, Ostrea sandivicensis*)

				Ranking					
Code	**Location**	**State**	**Species**	**1986**	**1987**	**1988**	**Overall**	**Significant Difference?**	**Trend**
TBPB	Tampa Bay	FL	cv	1	12t	5	5	no	no
HRUB	Hudson/Raritan Estuary	NY	me	2	1	1	2	yes	i
BPBP	Barber's Point	HI	os	3	<	<	<	yes	i
MBGP	Matagorda Bay	TX	cv	4	8	6	6	no	no
BHDI	Boston Harbor	MA	me	5	2t	<	7t	yes	no
MBTH	Moriches Bay	NY	me	6	19	7t	7t	no	no
CBBI	Charlotte Harbor	FL	cv	7	12t	—	12t	—	—
HRLB	Hudson/Raritan Estuary	NY	me	8t	12t	7t	7t	no	no
SFDB	San Francisco Bay	CA	me	8t	9	17t	15t	no	no
TBSR	Tomales Bay	CA	me	8t	<	<	<	yes	no
BBSM	Bellingham Bay	WA	me	8t	<	<	<	yes	no
HHKL	Honolulu Harbor	HI	os	8t	2t	18t	7t	no	no
PBSI	Penobscott Bay	ME	me	13t	<	17t	20t	no	no
CBSP	Choctawhatchee Bay	FL	cv	13t	<	<	<	no	no
PLLH	Pt. Loma	CA	mc	13t	<	11t	17t	no	no
SFSM	San Francisco Bay	CA	me	13t	10t	10	11	no	no
EVFU	Everglades	FL	cv	17t	<	<	<	yes	d
CBSR	Choctawhatchee Bay	FL	cv	17t	2t	<	12t	no	no
NYSH	New York Bight	NJ	me	19t	20t	11t	17t	yes	i
NYSR	New York Bight	NJ	me	19t	<	<	<	no	no
TBCB	Tampa Bay	FL	cv	19t	6t	17t	12t	no	no
PBIB	Pensacola Bay	FL	cv	19t	<	<	<	yes	no
BHBI	Boston Harbor	MA	me	<	5	18t	15t	yes	no
SFEM	San Francisco Bay	CA	me	—	6t	4	4	—	—
CBCR	Copano Bay	TX	cv	<	10t	<	<	no	no
HRJB	Hudson/Raritan Estuary	NY	me	<	15t	15t	19	no	no
BHDB	Boston Harbor	MA	me	<	15t	<	<	no	no
RBHC	Rookery Bay	FL	cv	<	15t	<	<	no	no
SSSS	San Simeon Point	CA	mc	<	18	9	20t	yes	i
GBHR	Galveston Bay	TX	cv	<	20t	<	<	yes	no
TBOT	Tampa Bay	FL	cv	—	—	2	1	—	—
SANM	San Miguel Island	CA	mc	—	—	3	3	—	—
LJLJ	La Jolla	CA	mc	<	<	13t	<	yes	no
MBSC	Monterey Bay	CA	mc	<	<	15t	<	no	no
PBPI	Penobscot Bay	ME	me	<	<	13t	<	no	no

Note: cv = *Crassostrea virginica;* me = *Mytilus edulis;* os = *Ostrea sandivicensis;* mc = *Mytilus californianus;* t = two or more concentrations were equal; d = decreasing trends; i = increasing trends.

Source: NOAA, *A Summary of Data on Tissue Contamination from the First Three Years (1986–1988) of the Mussel Watch Project,* NOAA Tech. Mem. NOS OMA 49, National Oceanic and Atmospheric Administration, Rockville, MD, 1989.

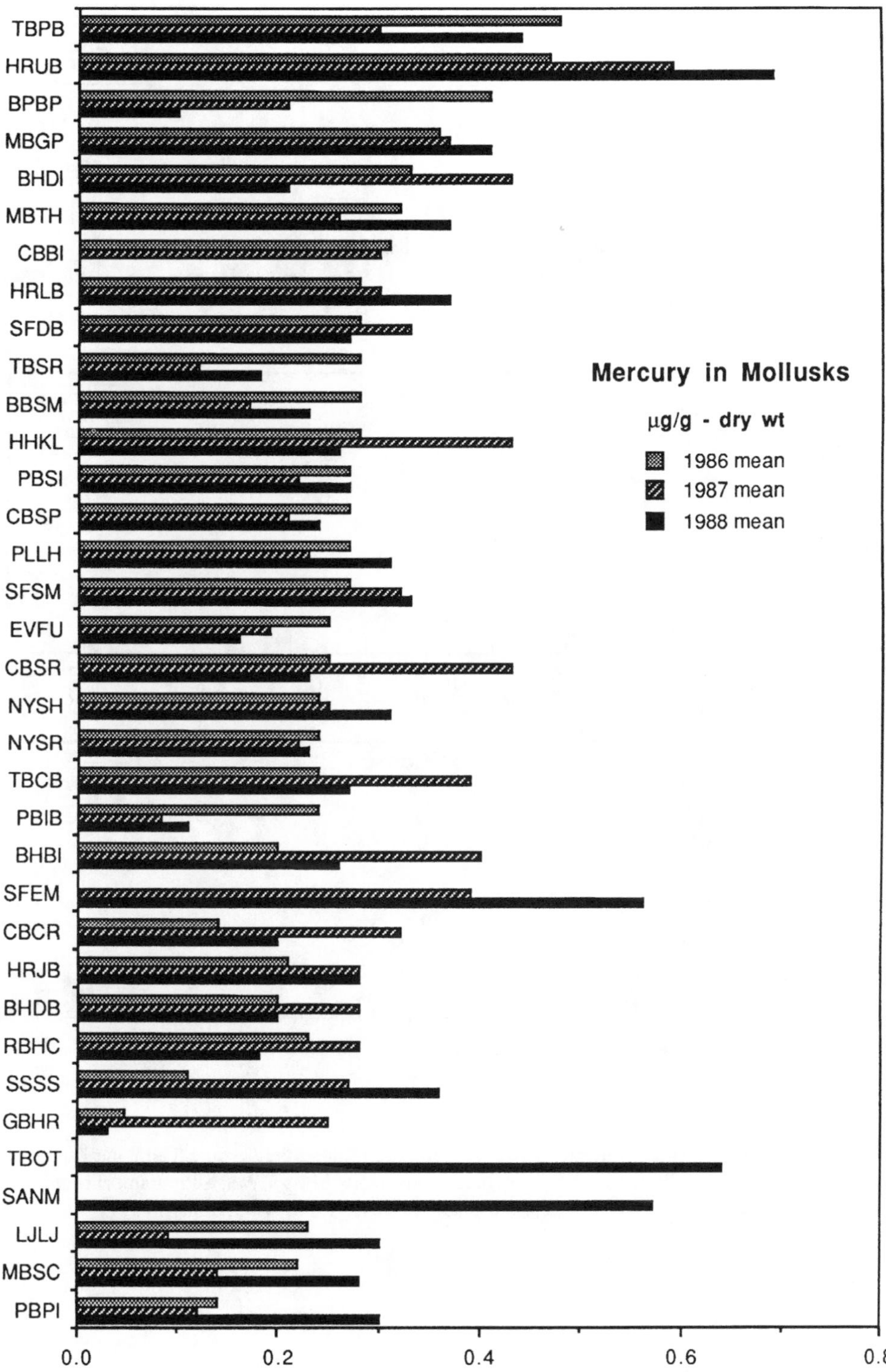

FIGURE 1.13. Concentrations of mercury in mussels (*Mytilus edulis, Mytilus californianus*) and oysters (*Crassostrea virginica, Ostrea sandivicensis*) from various National Status and Trends sites during the 1986 to 1988 survey period. See Table 1.9 for identification of estuary or coastal system. (From NOAA, *A Summary of Data on Tissue Contamination from the First Three Years (1986–1988) of the Mussel Watch Project,* NOAA Tech. Mem. NOS OMA 49, National Oceanic and Atmospheric Administration, Rockville, MD, 1989.)

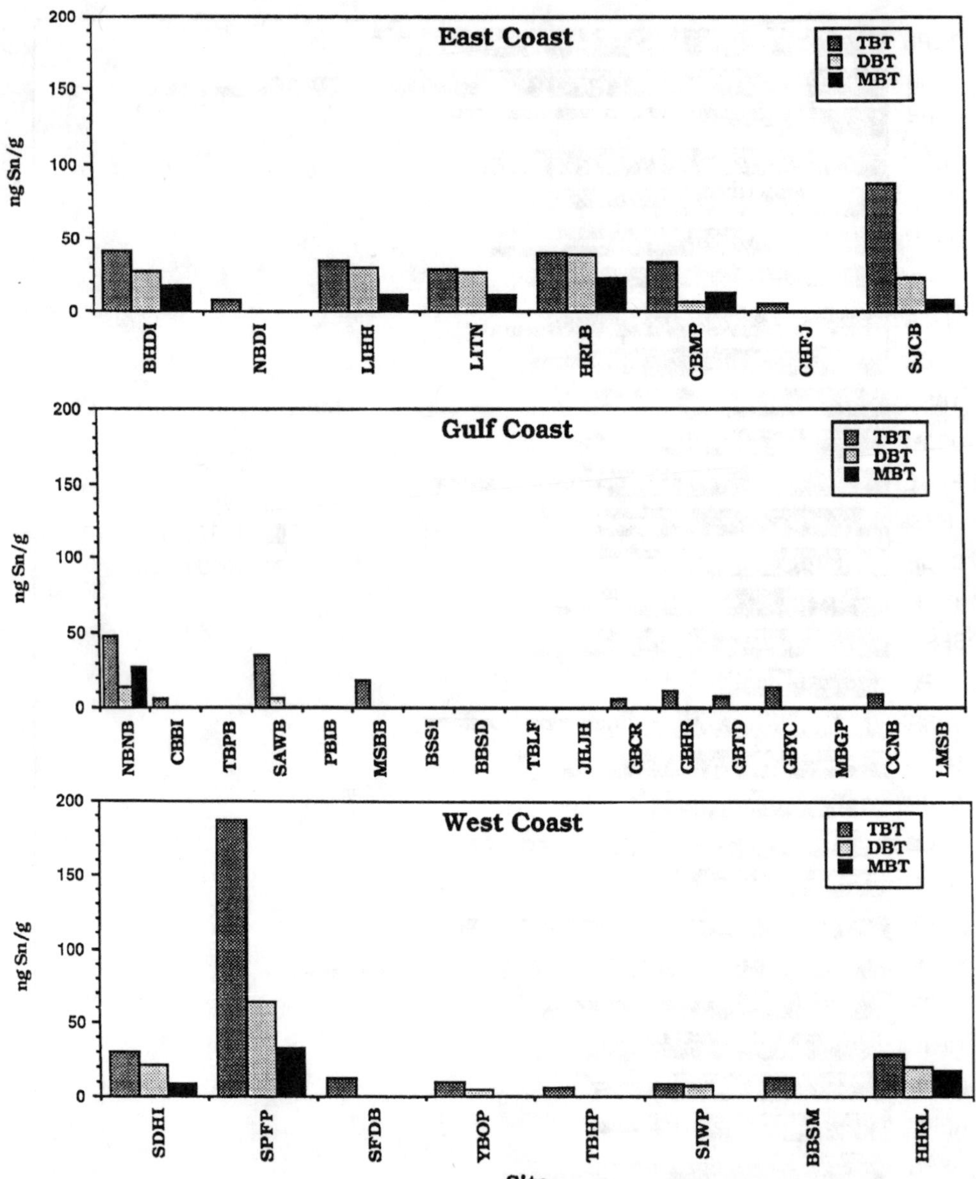

FIGURE 1.14. Comparison of sediment butyltin concentrations along the Atlantic, Gulf, and Pacific coasts of the U.S. based on data collected by NOAA's Status and Trends Mussel Watch Program. (From Wade, T. L., Garcia-Romero, B., and Brooks, J. M., *Chemosphere,* 20, 647, 1990. With permission.)

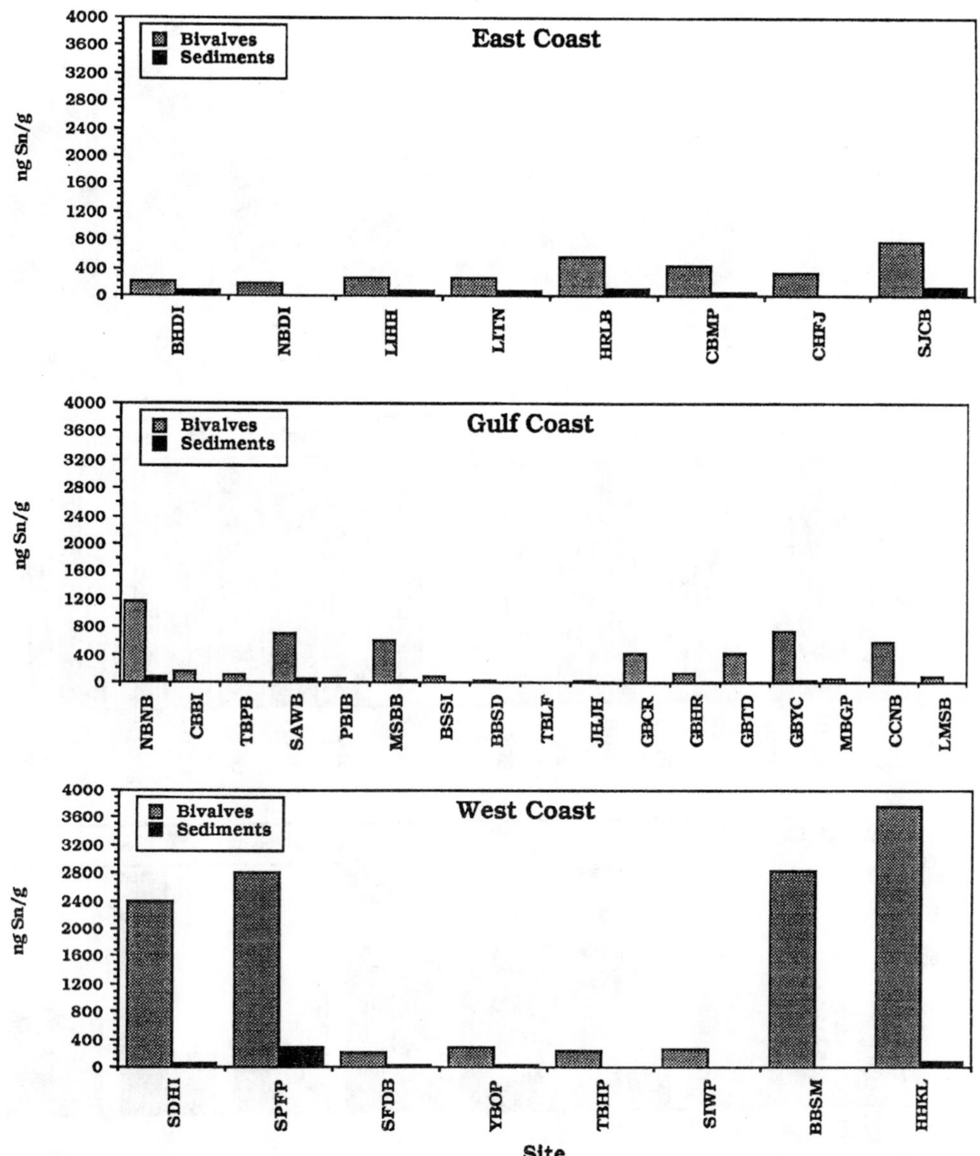

FIGURE 1.15. Comparison of total butyltin concentrations in sediments and bivalves along the Atlantic, Gulf, and Pacific coasts of the U.S. based on data collected by NOAA's Status and Trends Mussel Watch Program. (From Wade, T. L., Garcia-Romero, B., and Brooks, J. M., *Chemosphere,* 20, 647, 1990. With permission.)

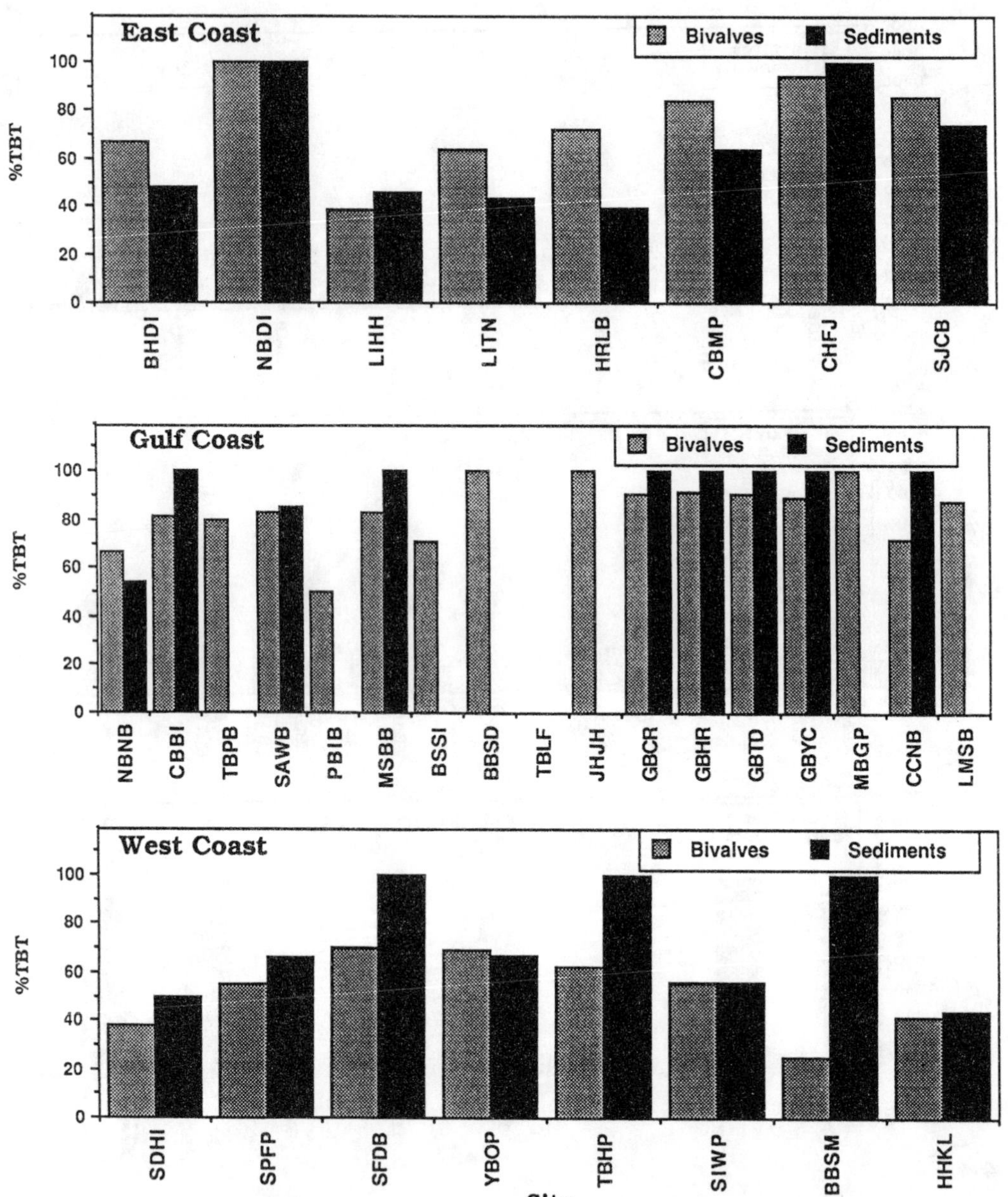

FIGURE 1.16. Comparison of tributyltin percentages in sediments and bivalves along the Atlantic, Gulf, and Pacific coasts of the U.S. based on data collected by NOAA's Status and Trends Mussel Watch Program. (From Wade, T. L., Garcia-Romero, B., and Brooks, J. M., *Chemosphere,* 20, 647, 1990. With permission.)

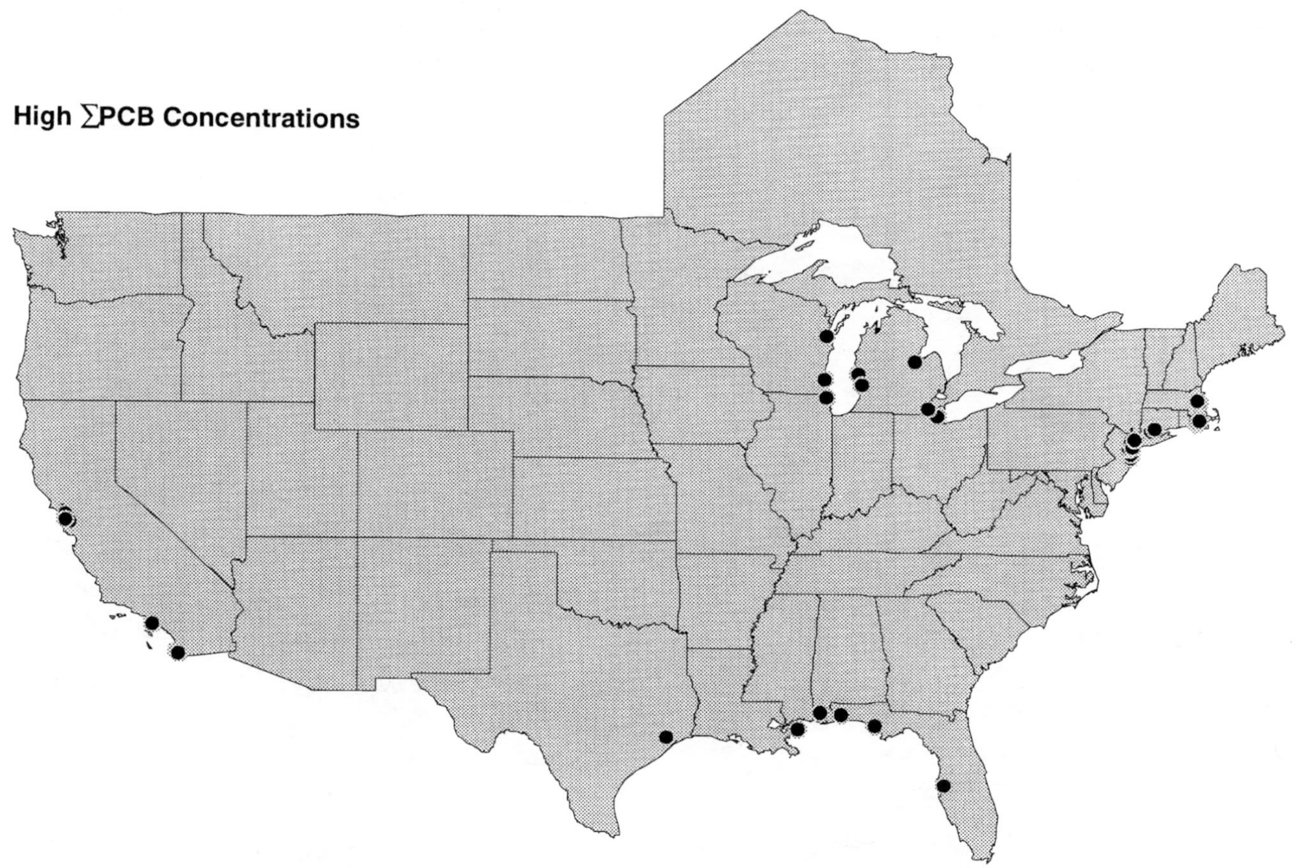

FIGURE 1.17. Locations of high PCB concentrations recorded by the National Status and Trends Mussel Watch Project. (From O'Connor, T. P. and Beliaeff, B., *Recent Trends in Coastal Environmental Quality: Results from the Mussel Watch Project*, NOAA Tech. Rep., Department of Commerce, Rockville, MD, 1995.)

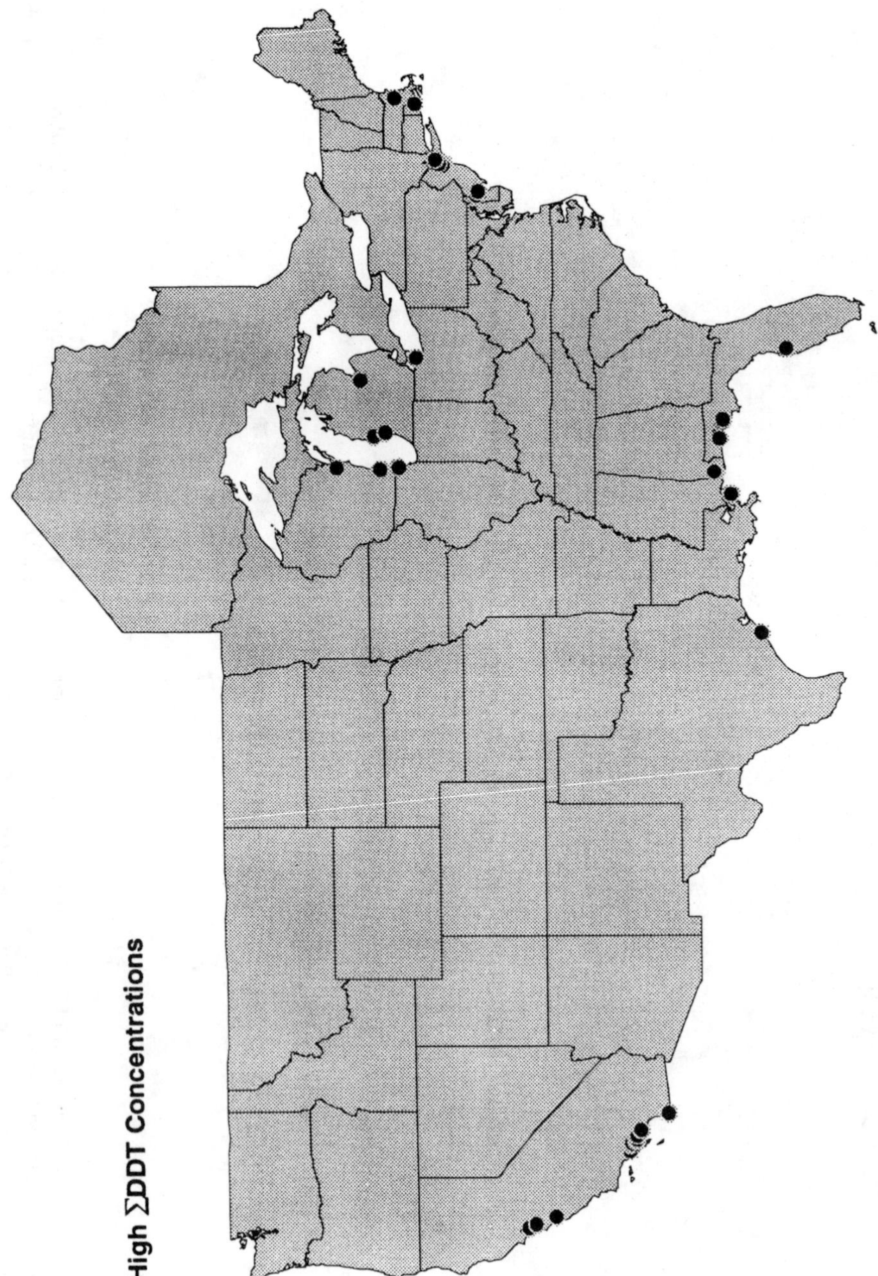

FIGURE 1.18. Locations of high DDT concentrations recorded by the National Status and Trends Mussel Watch Project. (From O'Connor, T. P. and Beliaeff, B., *Recent Trends in Coastal Environmental Quality: Results from the Mussel Watch Project*, NOAA Tech. Rep., Department of Commerce, Rockville, MD, 1995.)

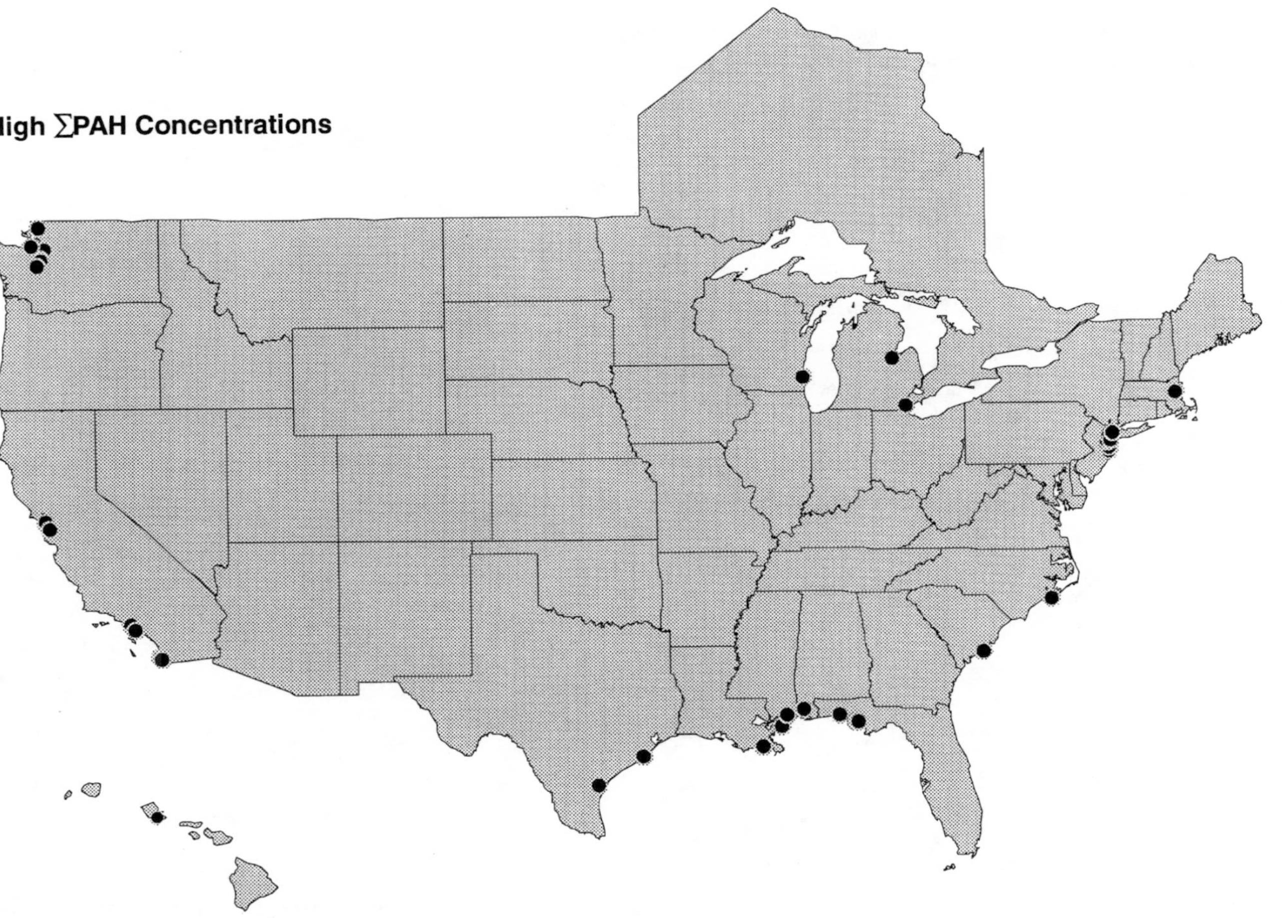

FIGURE 1.19. Locations of high PAH concentrations recorded by the National Status and Trends Mussel Watch Project. (From O'Connor, T. P. and Beliaeff, B., *Recent Trends in Coastal Environmental Quality: Results from the Mussel Watch Project,* NOAA Tech. Rep., Department of Commerce, Rockville, MD, 1995.)

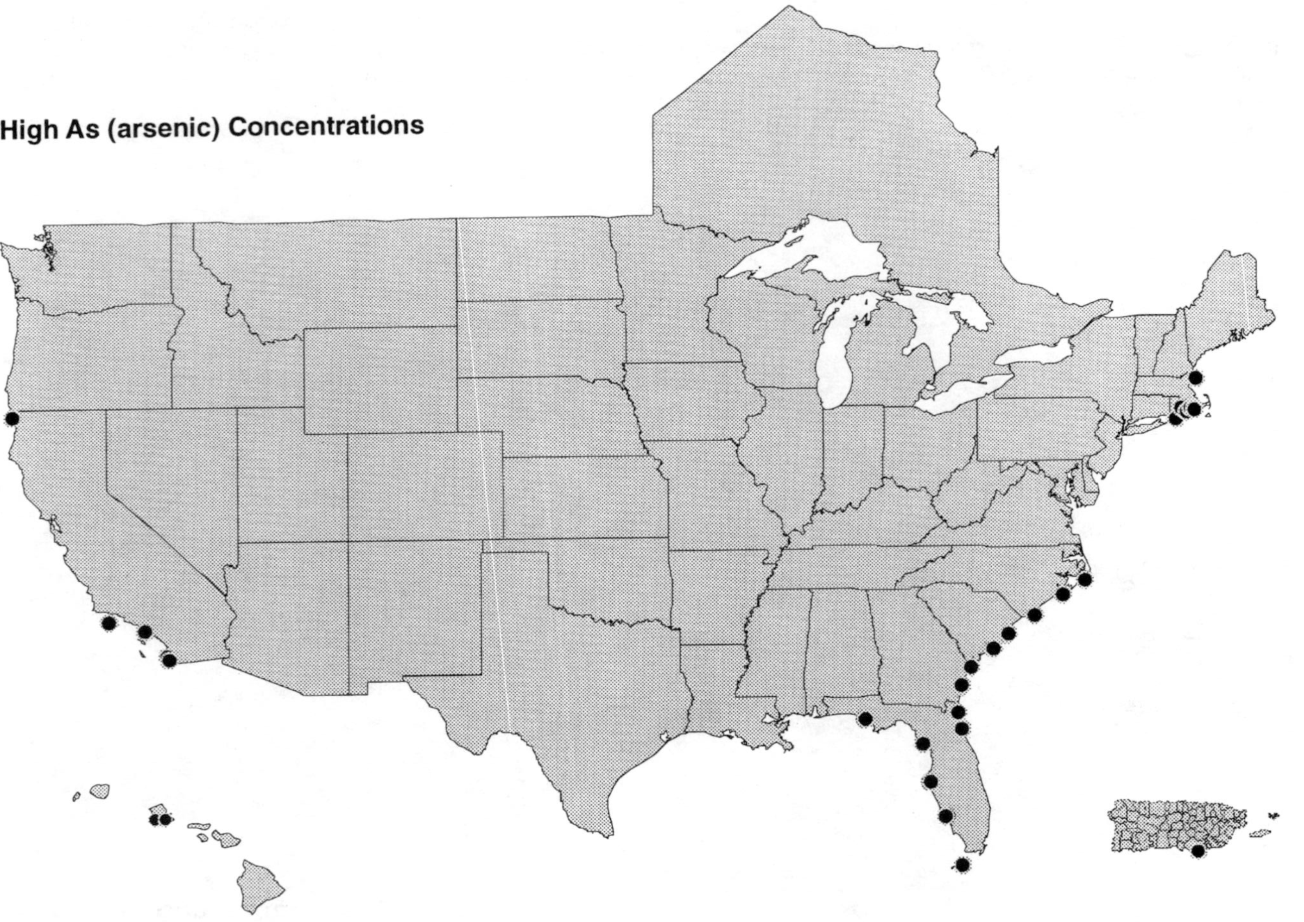

FIGURE 1.20. Locations of high arsenic concentrations recorded by the National Status and Trends Mussel Watch Project. (From O'Connor, T. P. and Beliaeff, B., *Recent Trends in Coastal Environmental Quality: Results from the Mussel Watch Project*, NOAA Tech. Rep., Department of Commerce, Rockville, MD, 1995.)

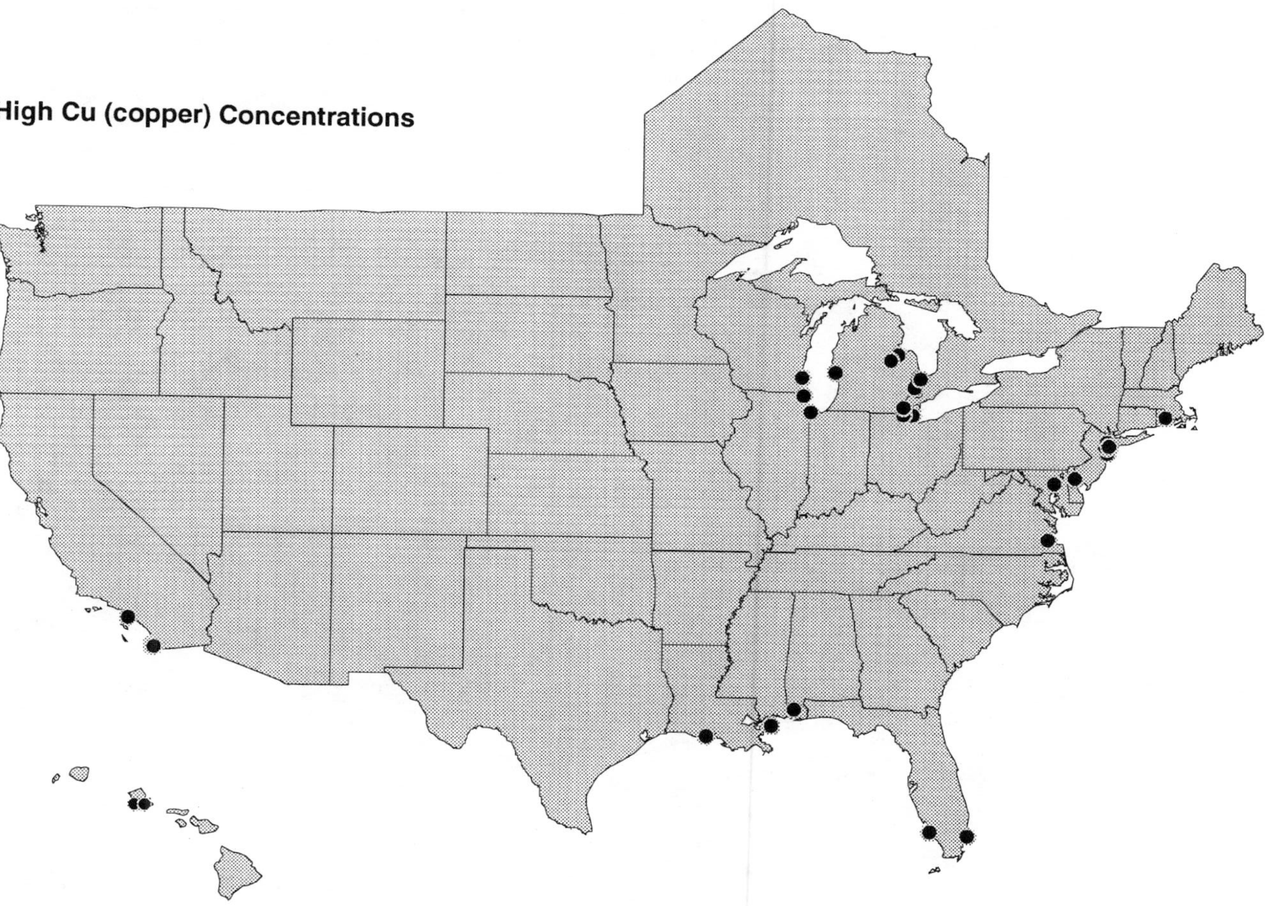

FIGURE 1.21. Locations of high copper concentrations recorded by the National Status and Trends Mussel Watch Project. (From O'Connor, T. P. and Beliaeff, B., *Recent Trends in Coastal Environmental Quality: Results from the Mussel Watch Project*, NOAA Tech. Rep., Department of Commerce, Rockville, MD, 1995.)

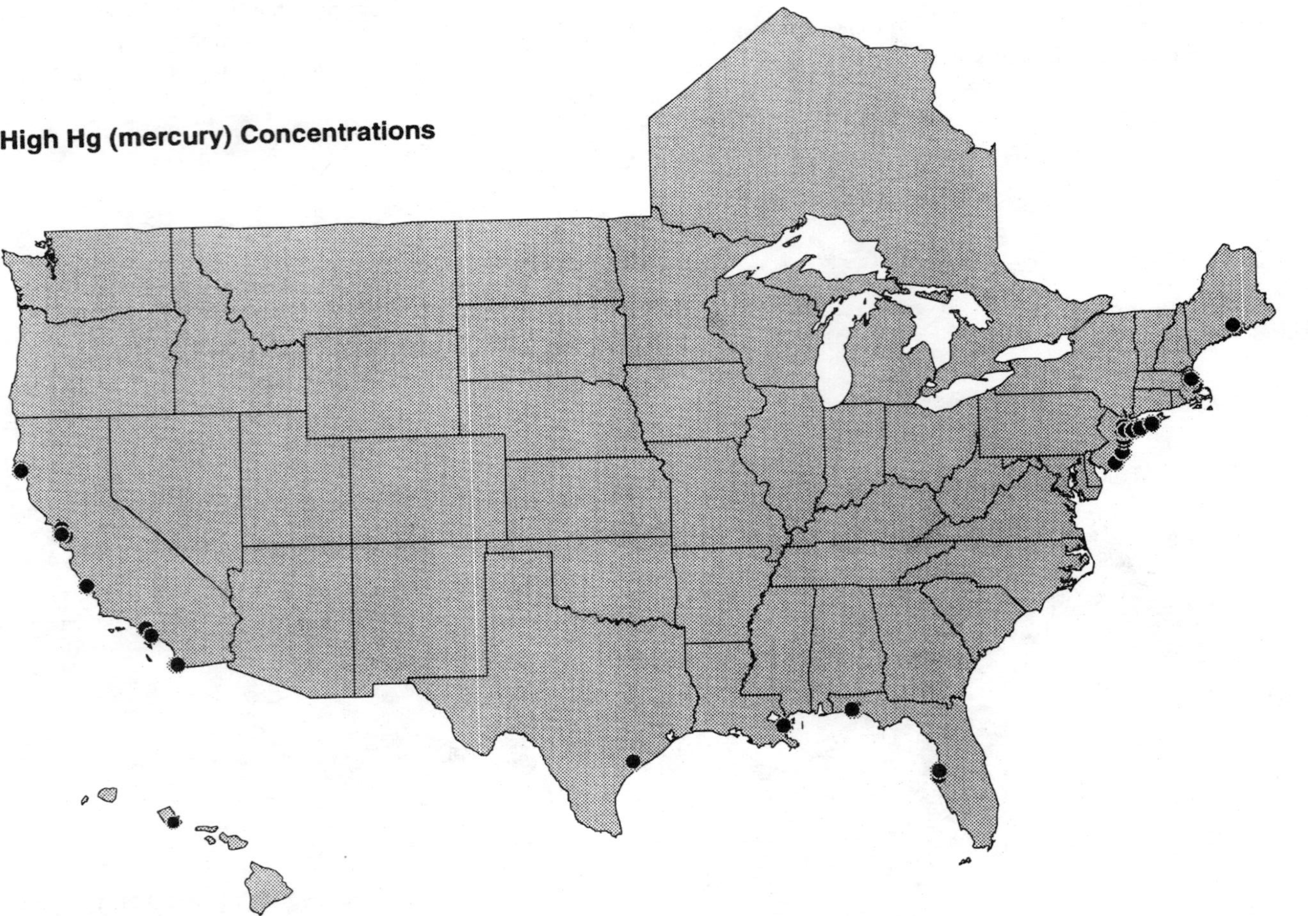

FIGURE 1.22. Locations of high mercury concentrations recorded by the National Status and Trends Mussel Watch Project. (From O'Connor, T. P. and Beliaeff, B., *Recent Trends in Coastal Environmental Quality: Results from the Mussel Watch Project*, NOAA Tech. Rep., Department of Commerce, Rockville, MD, 1995.)

APPENDIX 2. ATLANTIC COAST SURVEYS

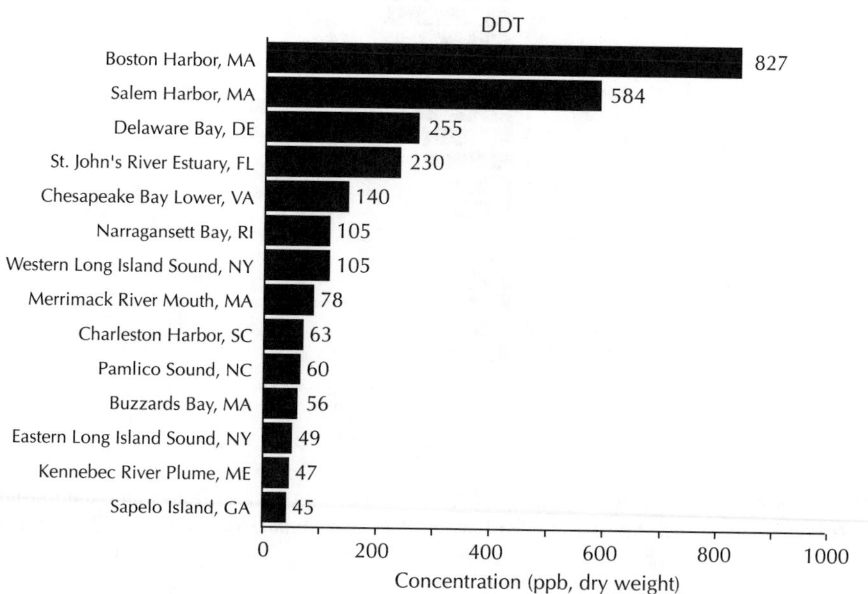

FIGURE 2.1. Concentrations of DDT in Atlantic Coast fish liver tissue sampled by the NOAA Status and Trends Program. (From Larsen, P. F., *Rev. Aquat. Sci.*, 6, 67, 1992. With permission.)

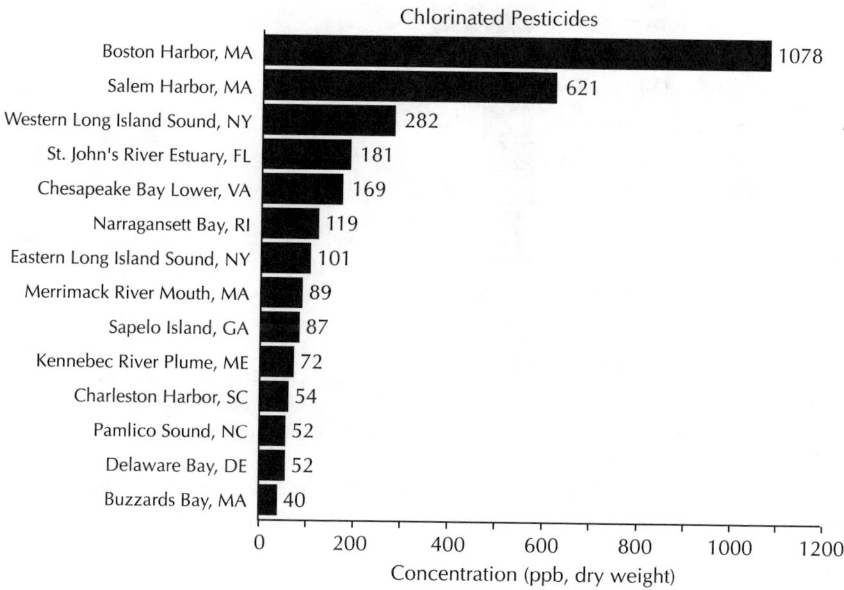

FIGURE 2.2. Concentrations of total chlorinated pesticides in Atlantic Coast fish liver tissue sampled by the NOAA Status and Trends Program. (From Larsen, P. F., *Rev. Aquat. Sci.*, 6, 67, 1992. With permission.)

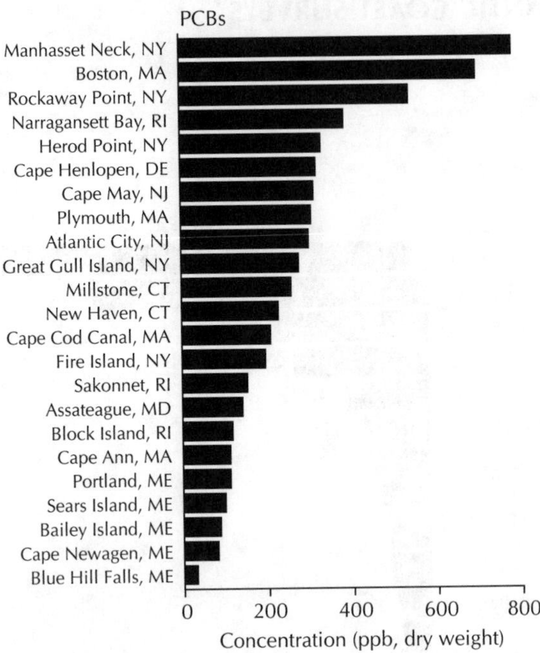

FIGURE 2.3. Concentrations of PCBs in mussels sampled at various sites from Maryland to Maine as part of the USEPA Mussel Watch Program. (From Larsen, P. F., *Rev. Aquat. Sci.*, 6, 67, 1992. With permission.)

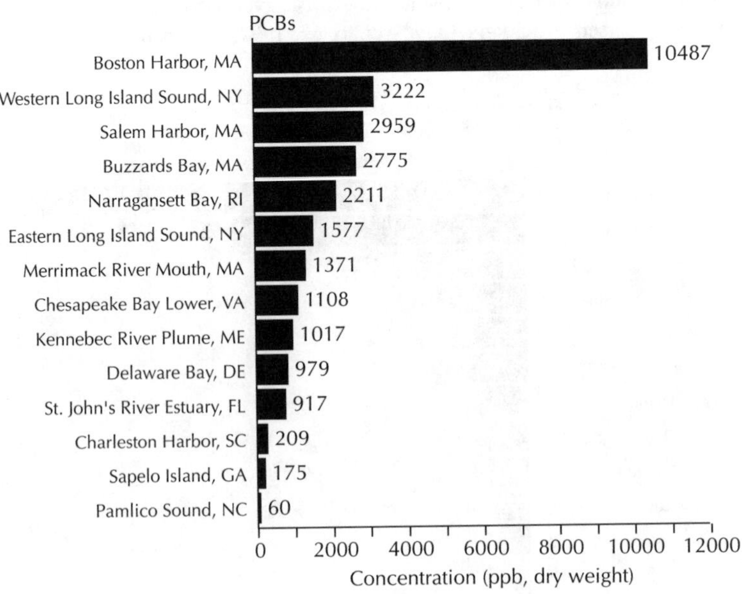

FIGURE 2.4. Concentrations of PCBs in Atlantic Coast fish liver tissue sampled by the NOAA Status and Trends Program. (From Larsen, P. F., *Rev. Aquat. Sci.*, 6, 67, 1992. With permission.)

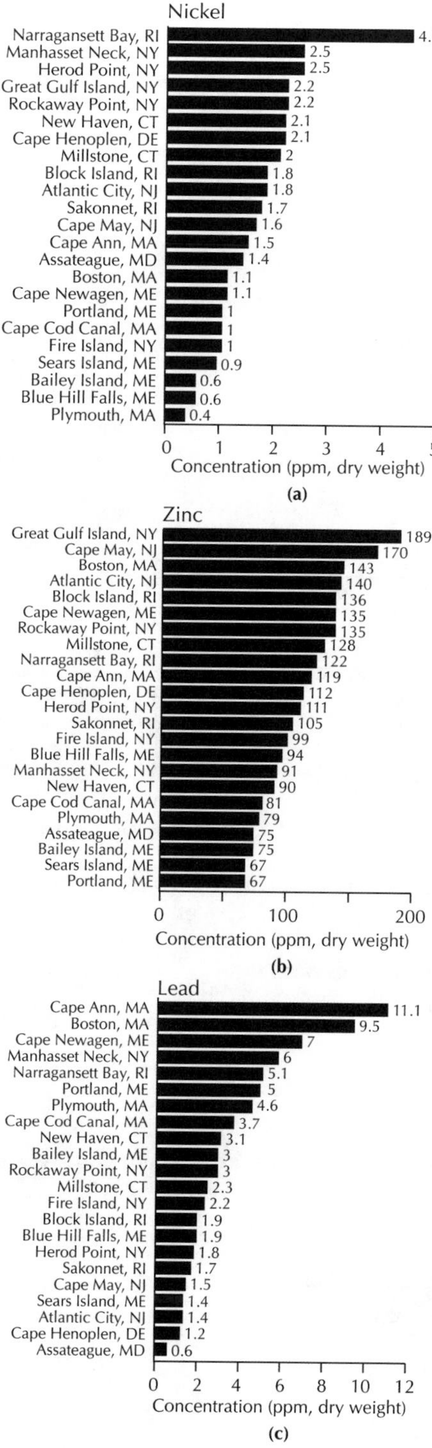

FIGURE 2.5. Concentrations of metals in mussels sampled in the northeastern U.S. by the USEPA Mussel Watch Program. (From Larsen, P. F., *Rev. Aquat. Sci.*, 6, 67, 1992. With permission.)

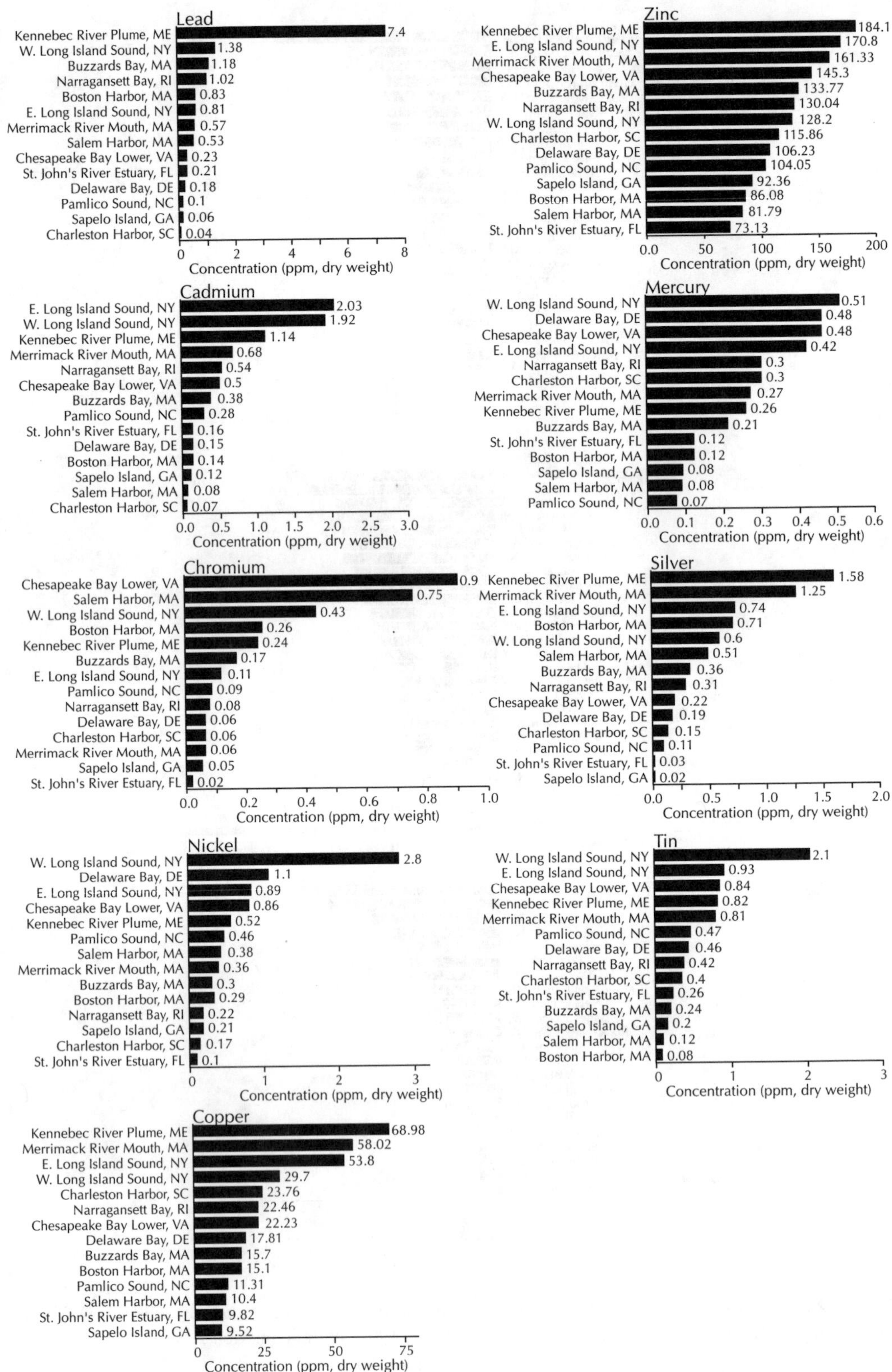

FIGURE 2.6. Concentrations of metals in Atlantic Coast fish liver tissues sampled by the NOAA Status and Trends Program. (From Larsen, P. F., *Rev. Aquat. Sci.*, 6, 67, 1992. With permission.)

TABLE 2.1
Rank Score Analysis of the NOAA Status and Trends Atlantic Coast Fish Liver Data

Rank	Site	Total Points	Zn	Cu	Ag	Sn	Cd	Cr	Ni	Pb	Hg	PCBs	DDT	Total Chlorinated Pesticides
1	Western Long Island Sound, NY	143	8	11	10	14	13	12	14	13	14	13	9	12
2	Eastern Long Island Sound, NY	124	13	12	12	13	14	8	12	9	11	9	3	8
3	Kennebec River Plume, ME	119	14	14	14	11	12	10	10	14	7	6	2	5
4	Chesapeake Bay Lower, VA	116	11	11	8	5	12	9	14	11	6	13	7	10
5	Merrimack River Mouth, MA	109	12	13	13	10	11	5	7	8	8	8	7	7
6	Narragansett Bay, RI	99	9	9	7	7	10	6	4	11	9	10	8	9
7	Boston Harbor, MA	98	3	5	11	2	4	11	5	10	5	14	14	14
8	Salem Harbor, MA	86	2	3	9	1	2	13	8	7	3	12	13	13
9	Buzzards Bay, MA	85	10	6	8	4	8	9	6	12	6	11	4	1
10	Delaware Bay, DE	84	6	7	6	8	5	3	13	4	12	5	12	3
11	Charleston Harbor, SC	58	7	10	4	6	1	4	2	1	10	3	6	4
12	Pamlico Sound, NC	56	5	4	3	9	7	7	9	3	1	1	5	2
13	St. John's River Estuary, FL	53	1	2	2	5	6	1	1	5	4	4	11	11
14	Sapelo Island, GA	30	4	1	1	3	3	2	3	2	2	2	1	6

Source: Larsen, P. F., *Rev. Aquat. Sci.*, 6, 67, 1992. With permission.

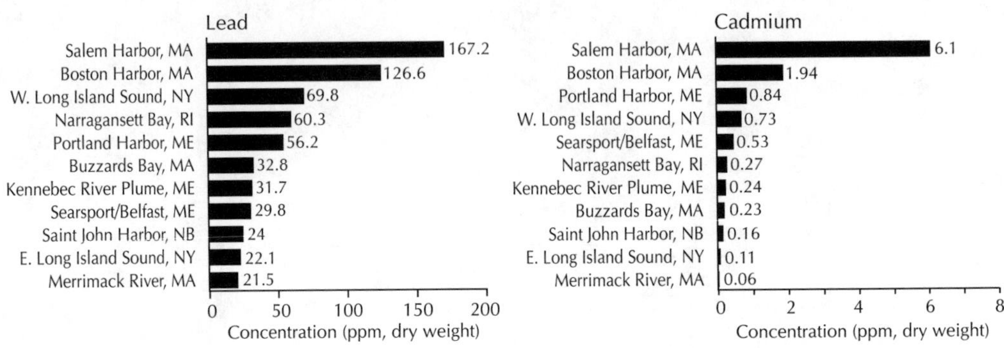

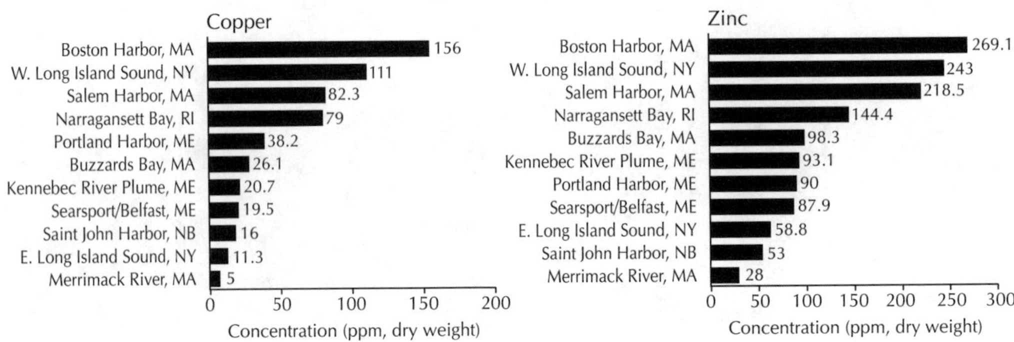

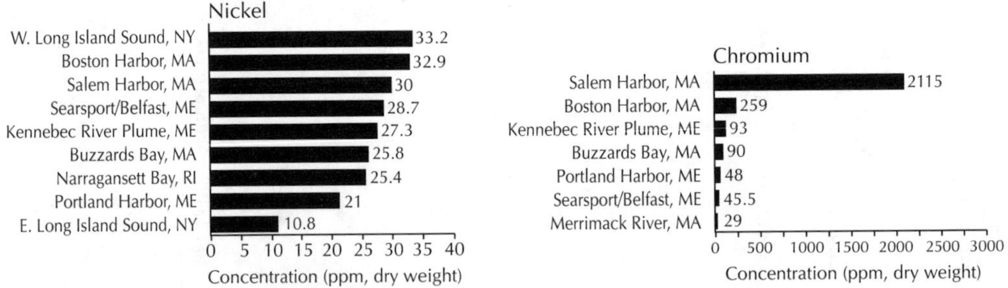

FIGURE 2.7. Concentrations of metals in northeastern U.S. coastal sediments as determined by the NOAA Status and Trends Program. (From Larsen, P. F., *Rev. Aquat. Sci.*, 6, 67, 1992. With permission.)

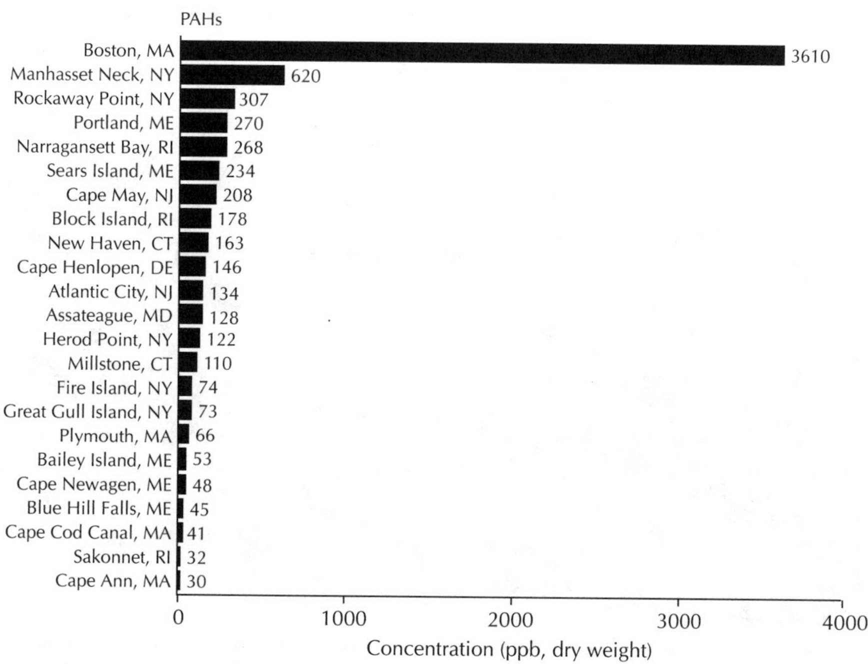

FIGURE 2.8. Concentrations of PAHs in mussels sampled in the northeastern U.S. by the USEPA Mussel Watch Program. (From Larsen, P. F., *Rev. Aquat. Sci.*, 6, 67, 1992. With permission.)

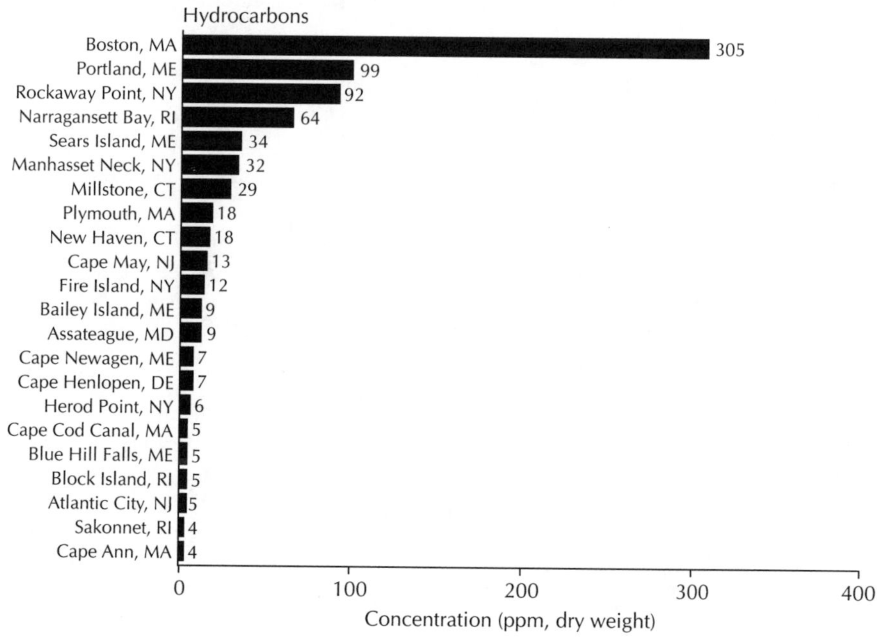

FIGURE 2.9. Concentrations of unresolved complex mixture hydrocarbons in mussels sampled in the northeastern U.S. by the USEPA Mussel Watch Program. (From Larsen, P. F., *Rev. Aquat. Sci.*, 6, 67, 1992. With permission.)

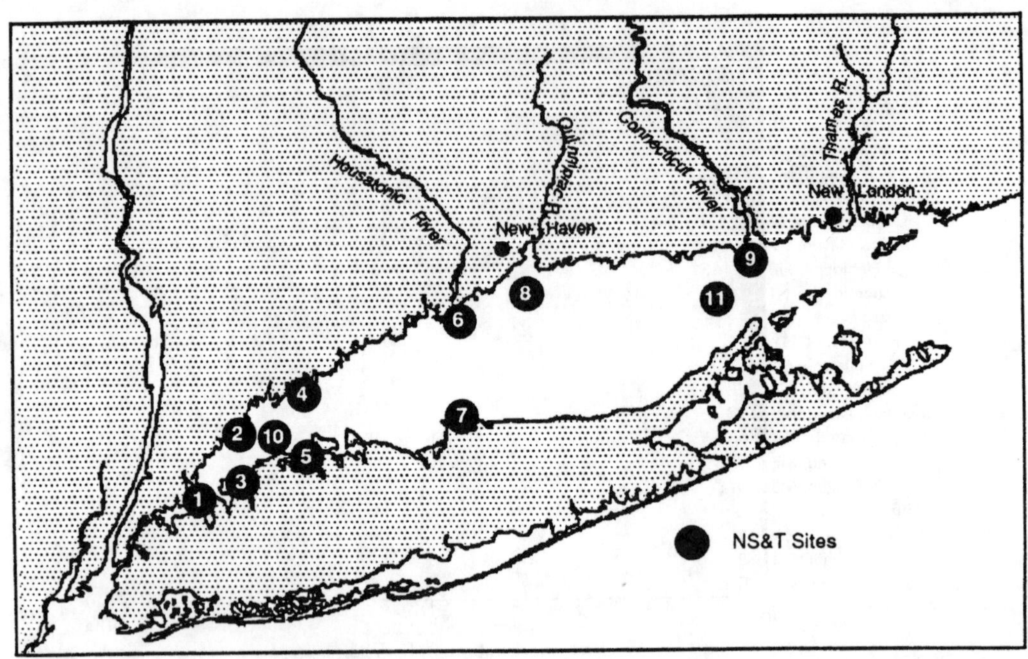

FIGURE 2.10. Locations of National Status and Trends monitoring sites in Long Island Sound. Sites 1 to 9 are Mussel Watch sites at Throgs Neck (1), Mamaroneck (2), Hempstead Harbor (3), Sheffield Island (4), Huntington Harbor (5), Housatonic River (6), Port Jefferson (7), New Haven (8), and the Connecticut River (9). Sites 10 and 11 are Benthic Surveillance sites at western (10) and eastern (11) ends of Long Island Sound. (From Turgeon, D. D. and O'Connor, T. P., *Estuaries,* 14, 279, 1991. With permission.)

TABLE 2.2
Mean Concentrations of Contaminants in Sediments and Mussels From the National Status and Trends Mussel Watch Site at Throgs Neck, New York

	Sediment		Mussel Tissue	
Contaminant	Mean Conc.[a]	(±SD)	Mean Conc.[a]	(±SD)
tDDT	77	(19)	210	(71)
tPCB	460	(92)	1300	(300)
tPAH	14,000	(3300)	4900	(3000)
tChlordane	13	(3.3)	86	(19)
Dieldrin	2.5	(3.6)	20	(20)
Antimony	3.3	(2.2)	—	—
Arsenic	9.4	(6.0)	5.6	(0.9)
Cadmium	1.9	(0.59)	3.6	(1.0)
Chromium	200	(42)	2	(0.26)
Copper	190	(48)	15	(1.8)
Lead	190	(36)	8.8	(3.7)
Mercury	1.2	(0.32)	.13	(0.03)
Nickel	56	(18)	3.1	(0.75)
Selenium	0.68	(0.36)	1.9	(0.4)
Silver	5.2	(2.3)	.32	(0.04)
Zinc	310	(68)	120	(20)

[a] Concentrations of organic contaminants in units of ng per g dry weight and concentrations of elements in units of μg per g dry weight.

Source: Turgeon, D. D. and O'Connor, T. P., *Estuaries,* 14, 279, 1991. With permission.

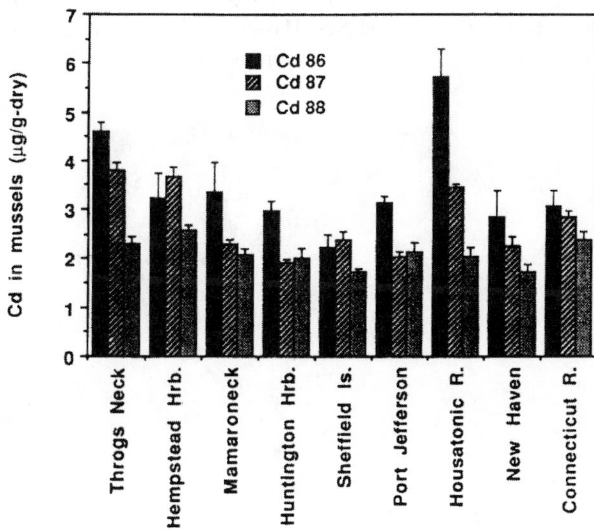

FIGURE 2.11. Cadmium concentrations in mussels at National Status and Trends sites in Long Island Sound in 1986, 1987, and 1988. (From Turgeon, D. D. and O'Connor, T. P., *Estuaries,* 14, 279, 1991. With permission.)

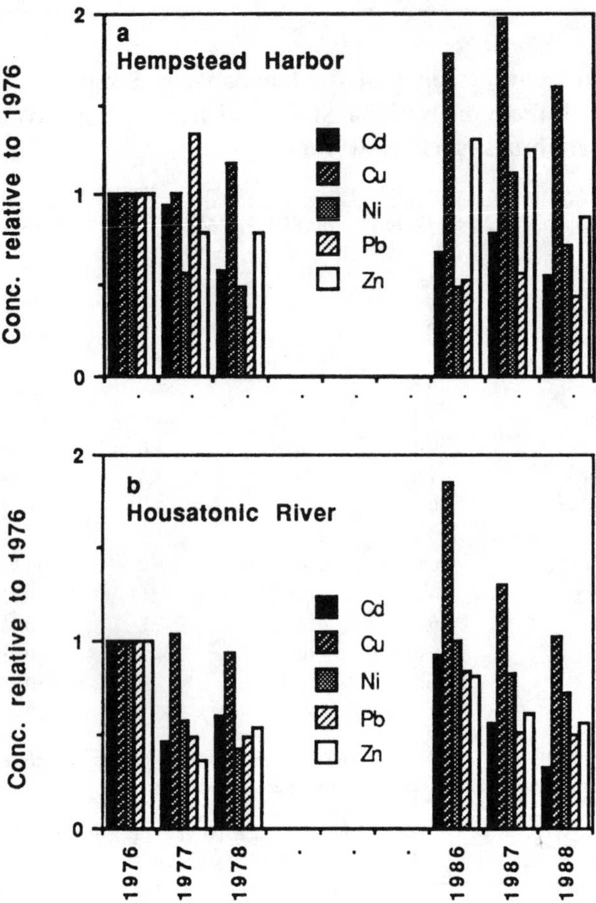

FIGURE 2.12. Comparison of National Status and Trends data (1986–1988) for Cu, Cd, Ni, Pb, and Zn in mussels with USEPA Mussel Watch data (1977–1978) for two sites in Long Island Sound. All data are shown as ratios to the 1976 concentration. The Hempstead Harbor and Housatonic River National Status and Trends sites are compared with the Manhasset Neck and New Haven USEPA sites, respectively. (From Turgeon, D. D. and O'Connor, T. P., *Estuaries,* 14, 279, 1991. With permission.)

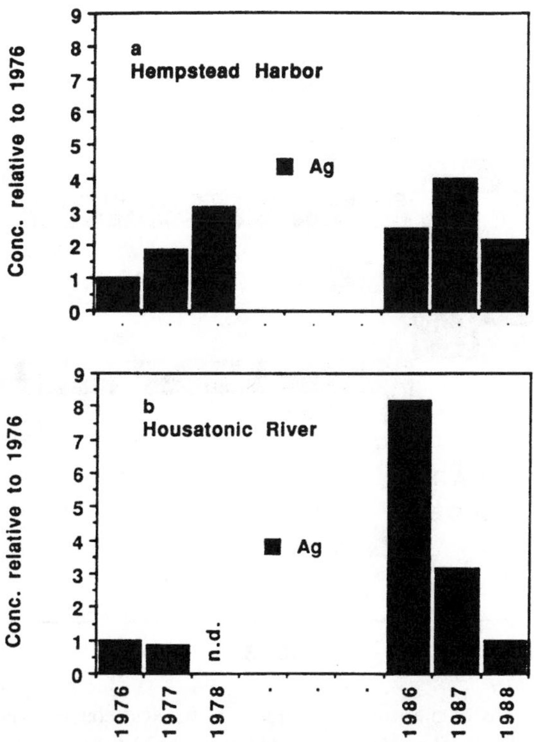

FIGURE 2.13. Comparison of National Status and Trends data (1986–1988) for Ag in mussels with USEPA Mussel Watch data (1977–1978) for two sites in Long Island Sound. All data are shown as ratios to the 1976 concentration. The Hempstead Harbor and Housatonic River National Status and Trends sites are compared with the Manhasset Neck and New Haven USEPA sites, respectively. (From Turgeon, D. D. and O'Connor, T. P., *Estuaries,* 14, 279, 1991. With permission.)

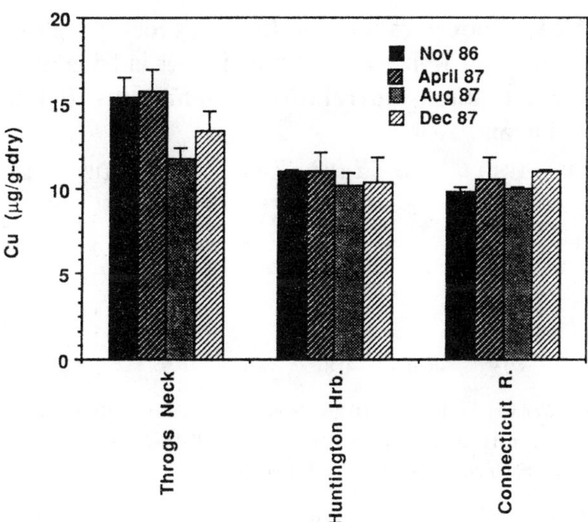

FIGURE 2.14. Copper concentrations in mussels collected at three Long Island Sound sites on four occasions between November 1986 and December 1987. In all cases three composite samples were analyzed. (From Turgeon, D. D. and O'Connor, T. P., *Estuaries,* 14, 279, 1991. With permission.)

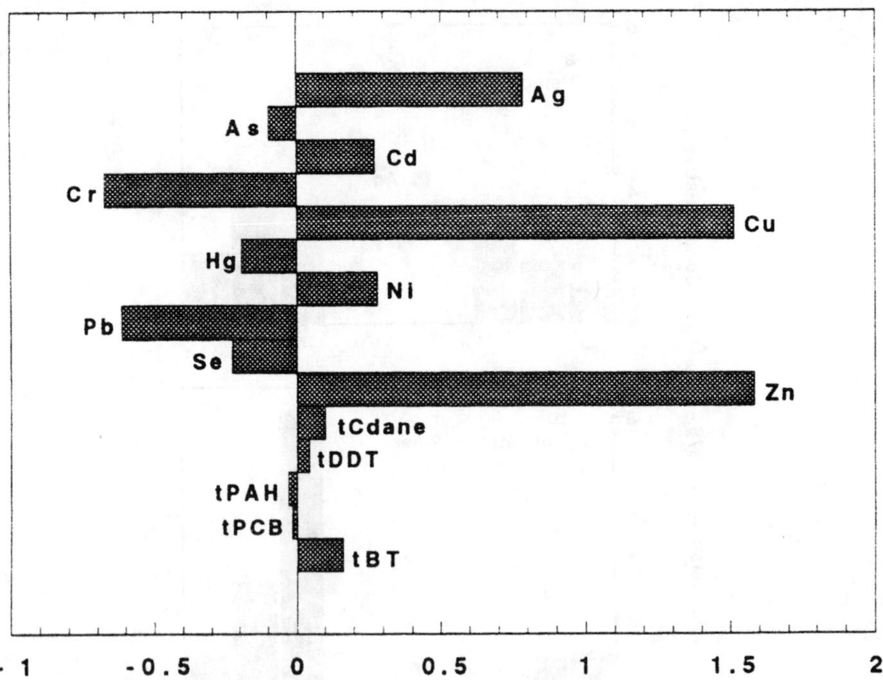

FIGURE 2.15. Logs of ratios of mean concentrations of chemical contaminants in oysters (*Crassostrea virginica*) to those in mussels (*Mytilus edulis*) collected at the NOAA Status and Trends Housatonic River site in Long Island Sound in 1989. (From O'Connor, T. P., Cantillo, A. Y., and Lauenstein, G. G., in *Biomonitoring of Coastal Waters and Estuaries,* Kramer, K. J. M. (Ed.), CRC Press, Boca Raton, FL, 1994, 35. With permission.)

TABLE 2.3
Annual Mean Concentrations of Cd (μg/g dry weight) at Nine National Status and Trends Mussel Watch Sites in Long Island Sound and Spearman Rank Correlation Coefficients (r_s) Between Concentration and Year

Year	LICR	LIHH	LIHR	LIHU	LIMR	LINH	LIPJ	LISI	LITN
1986	3.1	3.2	5.7	3.0	3.4	2.8	3.1	2.2	4.6
1987	2.8	3.7	3.5	1.9	2.3	2.2	2.0	2.4	3.5
1988	2.4	2.6	2.0	2.0	2.0	1.7	2.1	1.7	2.3
1989	2.1	1.9	2.2	1.4	1.9	1.1	1.4	1.5	1.3
1990	2.1	1.7	2.3	1.7	1.3	1.6	1.5	1.5	1.2
r_s	–0.9	–0.9	–0.6	–0.8	–1.0	–0.9	–0.8	–0.9	–1.0

Note: LICR = Connecticut River; LIHH = Hempstead Harbor; LIHR = Housatonic River; LIHU = Huntington Harbor; LIMR = Mamaroneck; LINH = New Haven; LIPJ = Port Jefferson; LISI = Sheffield Island; LITN = Throgs Neck.

Source: O'Connor, T. P., Cantillo, A. Y., and Lauenstein, G. G., in *Biomonitoring of Coastal Waters and Estuaries,* Kramer, K. J. M. (Ed.), CRC Press, Boca Raton, FL, 1994, 42. With permission.

APPENDIX 3. GULF COAST SURVEYS

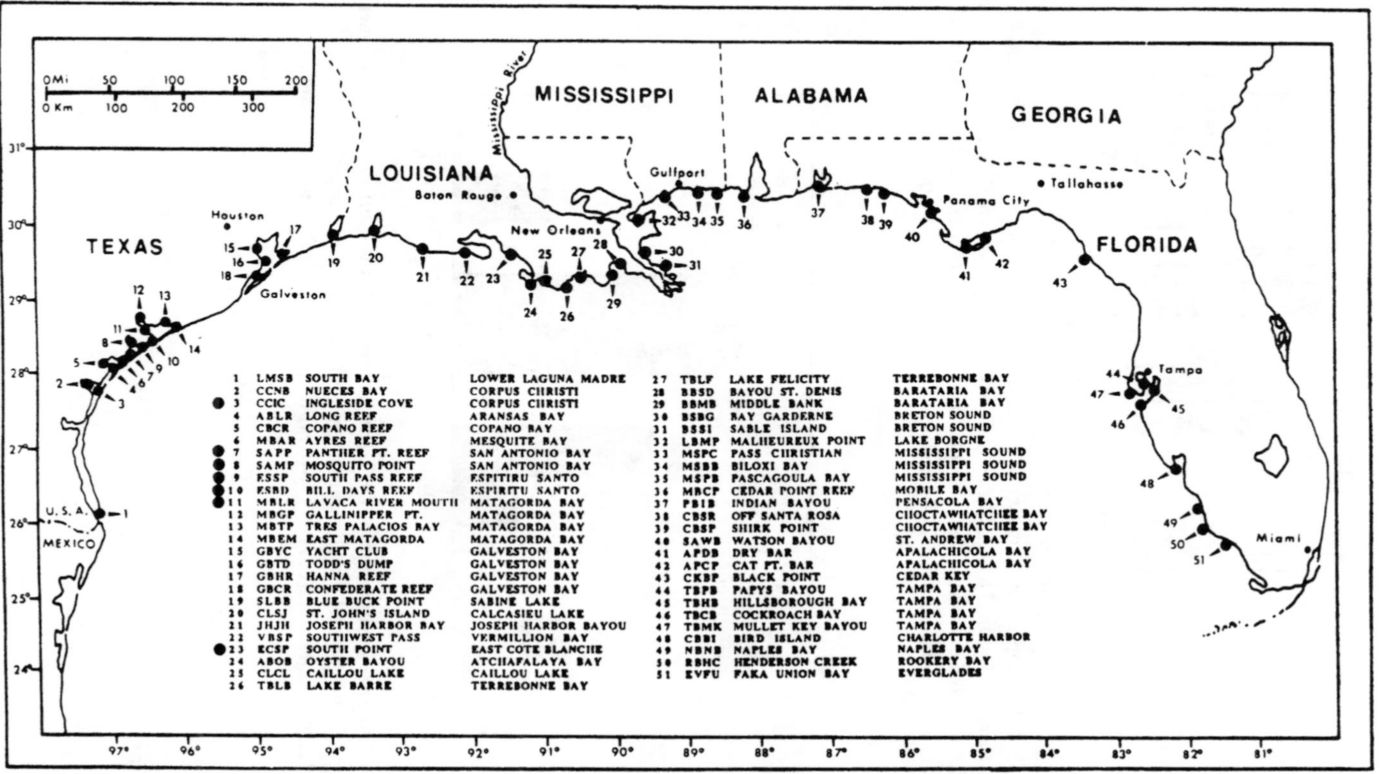

FIGURE 3.1. NOAA Status and Trends Mussel Watch sampling locations in the Gulf of Mexico during the 1986 to 1988 survey period. (From Wade, T. L. and Sericano, J. L., *Trends in Organic Contaminant Distributions in Oysters from the Gulf of Mexico, Oceans '89,* National Oceanic and Atmospheric Administration, Seattle, WA, 1989.)

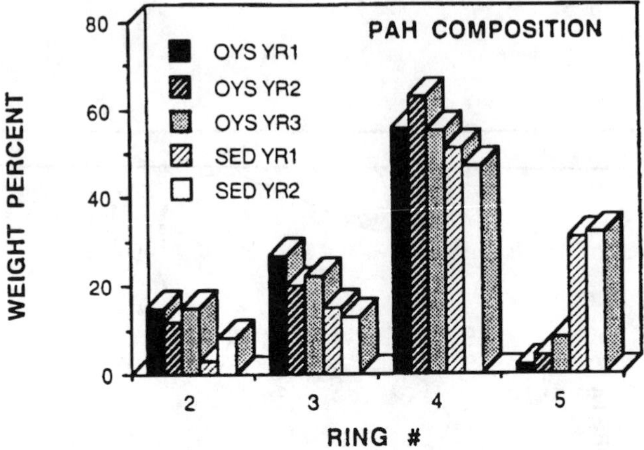

FIGURE 3.2. PAH composition in oysters (Year I, II, and III) and sediments (Year I and II) from the Gulf of Mexico determined by the NOAA Status and Trends Mussel Watch Program. (From Wade, T. L. and Sericano, J. L., *Trends in Organic Contaminant Distributions in Oysters from the Gulf of Mexico, Oceans '89,* National Oceanic and Atmospheric Administration, Seattle, WA, 1989.

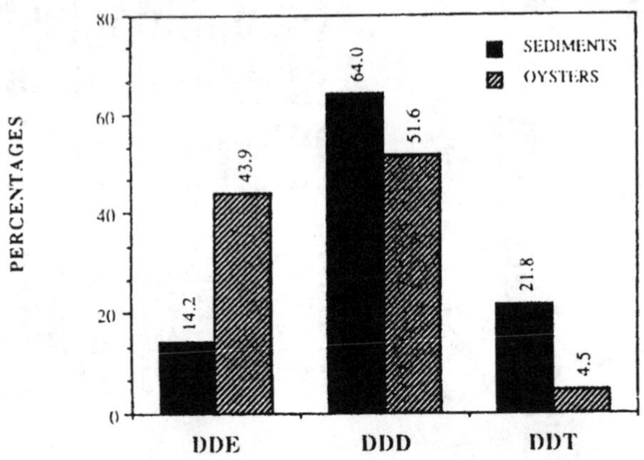

FIGURE 3.3. Average composition of DDT and its metabolites in Gulf of Mexico sediments and oysters from Status and Trends Mussel Watch sampling sites in the Gulf of Mexico. (From Sericano, J. L., Atlas, E. L., Wade, T. L., and Brooks, J. M., *Mar. Environ. Res.*, 29, 161, 1990. With permission.)

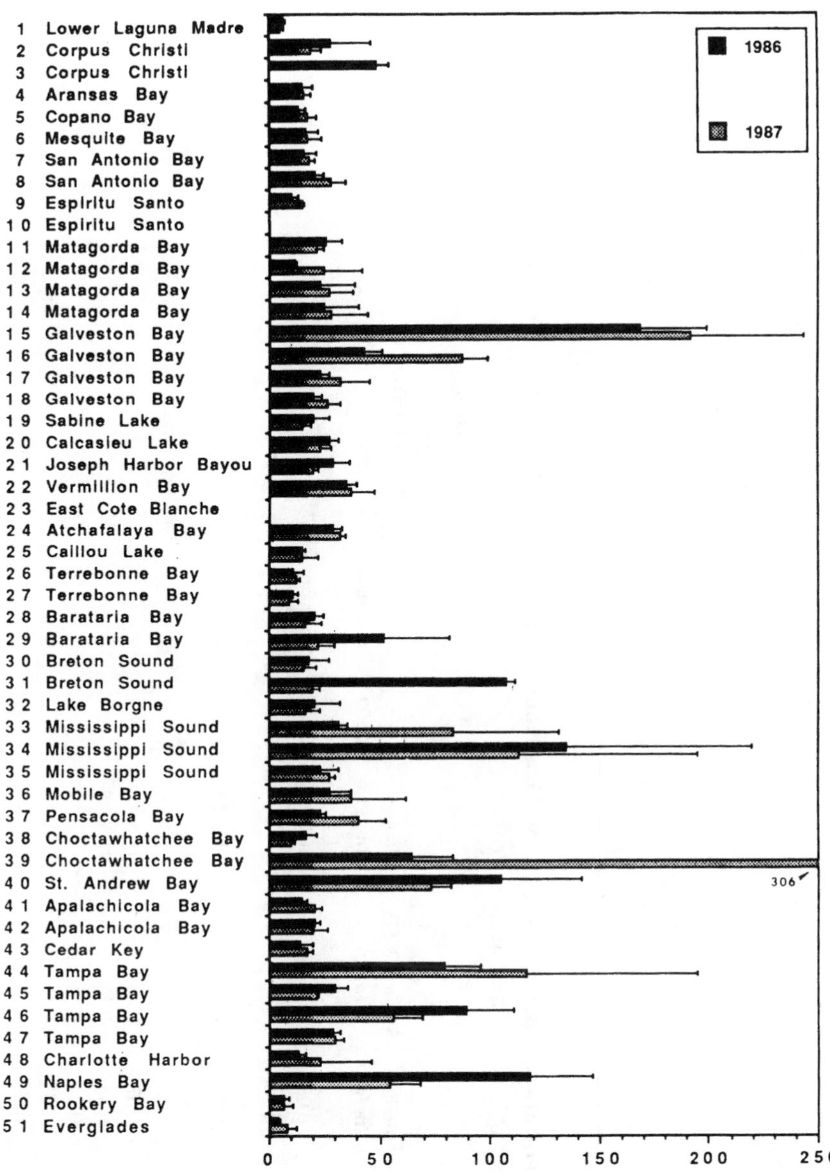

FIGURE 3.4. Average total (non-DDT) pesticide concentrations in oysters from Status and Trends Mussel Watch sampling sites in the Gulf of Mexico. (From Sericano, J. L., Atlas, E. L., Wade, T. L., and Brooks, J. M., *Mar. Environ. Res.*, 29, 161, 1990. With permission.)

TABLE 3.1
Chlorinated Hydrocarbon Concentrations (ng/g dry weight) and Distribution Frequencies (%) in Gulf of Mexico Sediments During 1987 as Recorded by NOAA's Status and Trends Program

	Incidence (%)	Median (ng/g)	Mean ± 1 SD (ng/g)	Range (ng/g)	Concentration (ng/g) Percent Distribution 0.00 →	0.02 →	1.00 →	10.0 →	100 →
HCB	72	0.06	0.25 ± 0.57	<0.02–3.62	28	67	5		
Lindane	19	<0.02	0.07 ± 0.18	<0.02–1.74	81	18	1		
Heptachlor	17	<0.02	0.05 ± 0.12	<0.02–1.14	83	16	1		
Aldrin	19	<0.02	0.08 ± 0.29	<0.02–2.87	81	17	1	1	
Heptachlor epoxide	17	<0.02	0.08 ± 0.35	<0.02–3.82	83	15	1	1	
alpha-Chlordane	72	0.06	1.18 ± 5.08	<0.02–43.5	28	57	12	3	
trans-Nonachlor	70	0.07	0.80 ± 3.80	<0.02–31.4	30	60	8	2	
Dieldrin	43	<0.02	0.36 ± 1.21	<0.02–9.47	57	36	7		
Mirex	32	<0.02	0.18 ± 0.45	<0.02–3.58	68	27	5		
Total (non-DDT) pesticides		0.86	3.05 ± 10.3	<0.02–89.4					
o-p′-DDE	13	<0.02	0.50 ± 3.13	<0.02–29.3	87	10	2	1	
p-p′-DDE	86	0.32	3.71 ± 19.4	<0.02–195	14	59	23	3	1
o-p′-DDD	42	<0.02	3.61 ± 30.1	<0.02–319	58	35	6		1

p-p′-DDD	66	0.27	17.7 ± 183	<0.02–2240	34	43	19	3	1
o-p′-DDT	30	<0.02	0.48 ± 2.46	<0.02–25.7	70	24	5	1	
p-p′-DDT	44	<0.02	5.97 ± 57.2	<0.02–691	56	36	6	1	1
Total DDTs		0.89	32.0 ± 275	<0.02–3270					
di-PCBs			0.58 ± 1.90						
tri-PCBs			2.57 ± 11.0						
tetra-PCBs			11.0 ± 70.0						
penta-PCBs			13.4 ± 86.5						
hexa-PCBs			15.6 ± 99.0						
hepta-PCBs			9.53 ± 54.8						
octa-PCBs			2.61 ± 14.7						
nona-PCBs			0.43 ± 2.92						
Total PCBs		5.40	55.7 ± 328	<0.02–3730					

Source: Sericano, J. L., Atlas, E. L., Wade, T. L., and Brooks, J. M., *Mar. Environ. Res.*, 29, 161, 1990. With permission.

TABLE 3.2
Chlorinated Hydrocarbon Concentrations (ng/g dry weight) and Distribution Frequencies (%) in Gulf of Mexico Oysters During 1987 as Recorded by NOAA's Status and Trends Program

	Incidence	Median	Mean ± 1 STD	Range	Concentration (ng/g)									
					Percent Distribution									
	(%)	(ng/g)	(ng/g)	(ng/g)	0.00	→	0.25	→	1.00	→	10.0	→	100	→
HCB	14	<0.25	0.36 ± 0.45	<0.25–4.33		86		10		4				
Lindane	80	1.20	1.74 ± 1.80	<0.25–9.06		20		24		56				
Heptachlor	23	<0.25	0.54 ± 0.99	<0.25–7.04		77		14		9				
Aldrin	9	<0.25	0.34 ± 0.56	<0.25–6.66		91		7		2				
Heptachlor epoxide	98	2.45	3.30 ± 3.93	<0.25–27.3		2		11		82		5		
alpha-Chlordane	100	6.42	14.1 ± 29.0	0.65–292				1		71		27		1
trans-Nonachlor	99	4.78	11.6 ± 27.7	<0.25–289		1		5		73		20		1
Dieldrin	95	3.71	6.08 ± 7.31	<0.25–51.6		5		10		68		17		
Mirex	38	<0.25	1.38 ± 2.77	<0.25–16.1		62		13		22		3		
Total (non-DDT) pesticides		22.0	39.4 ± 64.9	3.50–623										
o-p′-DDE	24	<0.25	1.33 ± 7.84	<0.25–85.7		76		15		8		1		
p-p′-DDE	100	12.5	29.4 ± 100	0.57–1170				5		38		55		2
o-p′-DDD	73	1.19	9.98 ± 81.3	<0.25–975		27		16		50		6		1

p-p′-DDD	98	8.02	25.0 ± 111	<0.25–1310	2	5	53	37	3
o-p′-DDT	17	<0.25	0.68 ± 1.79	<0.25–18.9	83	3	13	1	
p-p′-DDT	36	<0.25	1.49 ± 2.95	<0.25–25.6	64	8	26	2	
Total DDTs		26.1	67.9 ± 298	3.02–3570					
di-PCBs			1.62 ± 4.70						
tri-PCBs			7.59 ± 11.9						
tetra-PCBs			25.7 ± 54.8						
penta-PCBs			66.2 ± 151						
hexa-PCBs			26.8 ± 48.5						
hepta-PCBs			4.88 ± 7.10						
octa-PCBs			0.98 ± 2.54						
nona-PCBs			0.29 ± 1.19						
Total PCBs		62.3	134 ± 242	3.60–1740					

Source: Sericano, J. L., Atlas, E. L., Wade, T. L., and Brooks, J. M., *Mar. Environ. Res.*, 29, 161, 1990. With permission.

TABLE 3.3
Average Concentrations of Chlorinated Hydrocarbon Residues in Biota and Sediments from the Gulf of Mexico and Adjacent Estuaries

Location	Sample	n	PCBs (ng/g)	DDTs (ng/g)	Dieldrin (ng/g)	Others (ng/g)
Escambia Bay (FL)	Fish	7	11	—	—	—
Aransas Bay (TX)	Fish, crab, oysters	9	—	49	9	—
Gulf of Mexico	Fish, shrimp	46	66	62	—	—
Gulf of Mexico	Plankton	29	95	7	—	—
Gulf of Mexico	Fish	18	33	19	—	—
San Antonio Bay (TX)	Crabs	62	—	16	2.1	—
San Antonio Bay (TX)	Oysters	30	—	25	8.9	—
San Antonio Bay (TX)	Clams	43	—	20	2.9	—
San Antonio Bay (TX)	Shrimp	23	—	2	1.8	—
Mississippi Delta (LA)	Plankton	5	84	1	12	—
Mississippi Delta (LA)	Fish (mesopelagic)	27	25	10	—	—
Estuaries of Texas, Mississippi, Louisiana, Alabama, and Florida	Fish	24	203	18.2	15.2	Toxaphene (200) Ethyl parathion (75) Methyl parathion (47)
St. Louis and Mississippi Bays (MS)	Mollusks, fish	37	—	—	—	Mirex (139)
Gulf of Mexico	Fish, shrimp	27	25	10	—	—
Mexican coastal lagoons	Oysters	9	55	15	0.9	Chlordane (nd) Endrin (nd)
Apalachicola River (FL)[a]	Clams	9	21	29	2	Chlordane (21) Hep. epoxide (0.3)
Mississippi Delta (LA)	Sediments	—	18.7	4.2	—	—
Gulf of Mexico coast	Sediments	—	2	1.3	—	—
Nueces Estuary (TX)	Sediments	—	4.7	1.5	—	HCB (0.11) Lindane (0.03) Chlordane (0.77)
Galveston Bay (TX)	Sediments	—	1.1	0.2	0.1	HCB (0.49)
Apalachicola Bay (FL)	Sediments	56	3.0	3.3	—	—
Apalachicola River (FL)	Sediments	12	1.0	1.7	<0.1	Chlordane (<1)

Note: nd = not detected.

[a] Median concentrations.

Source: Sericano, J. L., Atlas, E. L., Wade, T. L., and Brooks, J. M., *Mar. Environ. Res.*, 29, 161, 1990. With permission.

TABLE 3.4
Average Concentrations and Ranges of Chlorinated Hydrocarbons in Oysters and Sediments from the Gulf of Mexico as Determined by National Programs

Year	N	DDTs in Oysters (ng/g)	PCBs in Oysters (ng/g)	DDTs in Sediments (ng/g)	PCBs in Sediments (ng/g)
1965	58	257 ± 542[a] (<33–4730)	—	—	—
1966	152	346 ± 484[a] (<33–3890)	—	—	—
1967	155	292 ± 428[a] (<33–2790)	—	—	—
1968	136	450 ± 575[a] (<33–6490)	—	—	—
1969	142	284 ± 497[a] (<33–3530)	—	—	—
1970	144	234 ± 301[a] (<33–2200)	—	—	—
1971	140	217 ± 495[a] (<33–2840)	—	—	—
1972	60	162 ± 206[a] (<33–933)	—	—	—
1976	9	18.7 ± 12.4[b] (6.0–42.0)	71.2 ± 104 (<20–336)	—	—
1977	9	11.0 ± 9.10[b] (2.8–28.0)	83.0 ± 87.8 (16–297)	—	—
1984	11			1.83 ± 1.89 (0.1–7.0)	23.0 (nd–34.0)

[a] Recalculated on dry weight basis.
[b] Calculated as DDE.

Source: Sericano, J. L., Atlas, E. L., Wade, T. L., and Brooks, J. M., *Mar. Environ. Res.*, 29, 161, 1990. With permission.

TABLE 3.5
Chlordane-Related Compound Concentrations and Distribution Frequencies in the Gulf of Mexico During the 1986–1990 Period as Determined by the Status and Trends Mussel Watch Program[a]

	Concentration (ng/g)			Concentration (ng/g) Distribution (%)				
	Median	**Mean ± 1 SD**	**Range**	**0.00–<0.25**	**0.25–<1.00**	**1.00–<10.0**	**10.0–<100**	**100+**
1986 (N = 147)								
Heptachlor	<0.25	0.51 ± 0.69	<0.25–4.62	63	29	8		
Heptachlor epoxide	1.87	2.71 ± 3.31	<0.25–24.5	14	12	70	4	
alpha-Chlordane	5.23	10.9 ± 14.4	0.91–96.3		1	72	27	
trans-Nonachlor	4.58	10.0 ± 13.8	0.60–71.9		1	76	23	
ΣChlordane	13.1	24.1 ± 30.3	2.00–175			36	61	3
1987 (N = 143)								
Heptachlor	<0.25	0.54 ± 0.99	<0.25–7.04	77	14	9		
Heptachlor epoxide	2.45	3.30 ± 3.93	<0.25–27.3	2	11	82	5	
alpha-Chlordane	6.42	14.1 ± 29.0	0.65–292		1	71	27	1
trans-Nonachlor	4.78	11.6 ± 27.7	<0.25–289	1	5	73	20	1
ΣChlordane	14.4	29.5 ± 58.5	2.12–590			28	66	6

1988 (N = 132)								
Heptachlor	<0.25	0.49 ± 0.66	<0.25–5.81	73	16	11		
Heptachlor epoxide	1.69	2.44 ± 2.33	<0.25–14.3	3	15	80	2	
alpha-Chlordane	5.80	9.76 ± 10.7	0.40–60.2		2	72	26	
trans-Nonachlor	5.02	8.98 ± 11.5	<0.25–81.6	1	4	70	25	
ΣChlordane	13.9	21.7 ± 22.8	1.29–132			35	64	1
1989 (N = 135)								
Heptachlor	0.33	0.78 ± 1.03	<0.25–8.23	40	40	20		
Heptachlor epoxide	0.60	1.25 ± 1.41	<0.25–9.33	20	41	39		
alpha-Chlordane	3.14	7.00 ± 9.84	<0.25–48.3	1	7	76	16	
trans-Nonachlor	2.33	7.29 ± 14.5	<0.25–99.1	1	16	69	14	
ΣChlordane	7.68	16.3 ± 25.0	1.37–159			61	37	2
1990 (N = 138)								
Heptachlor	0.72	1.34 ± 1.81	<0.25–15.2	30	28	41	1	
Heptachlor epoxide	1.20	2.63 ± 4.53	<0.25–29.9	27	19	50	4	
alpha-Chlordane	4.16	5.81 ± 5.66	<0.25–36.3	5	6	72	17	
trans-Nonachlor	3.13	5.55 ± 6.49	<0.25–29.8	4	17	62	17	
ΣChlordane	10.7	15.3 ± 14.0	<1.00–69.4		4	42	54	

[a] Concentrations on a dry weight basis.

Source: Sericano, J. L., Wade, T. L., Brooks, J. M., Atlas, E. L., Fay, R. R., and Wilkinson, D. L., *Environ. Pollut.*, 82, 23, 1993. With permission.

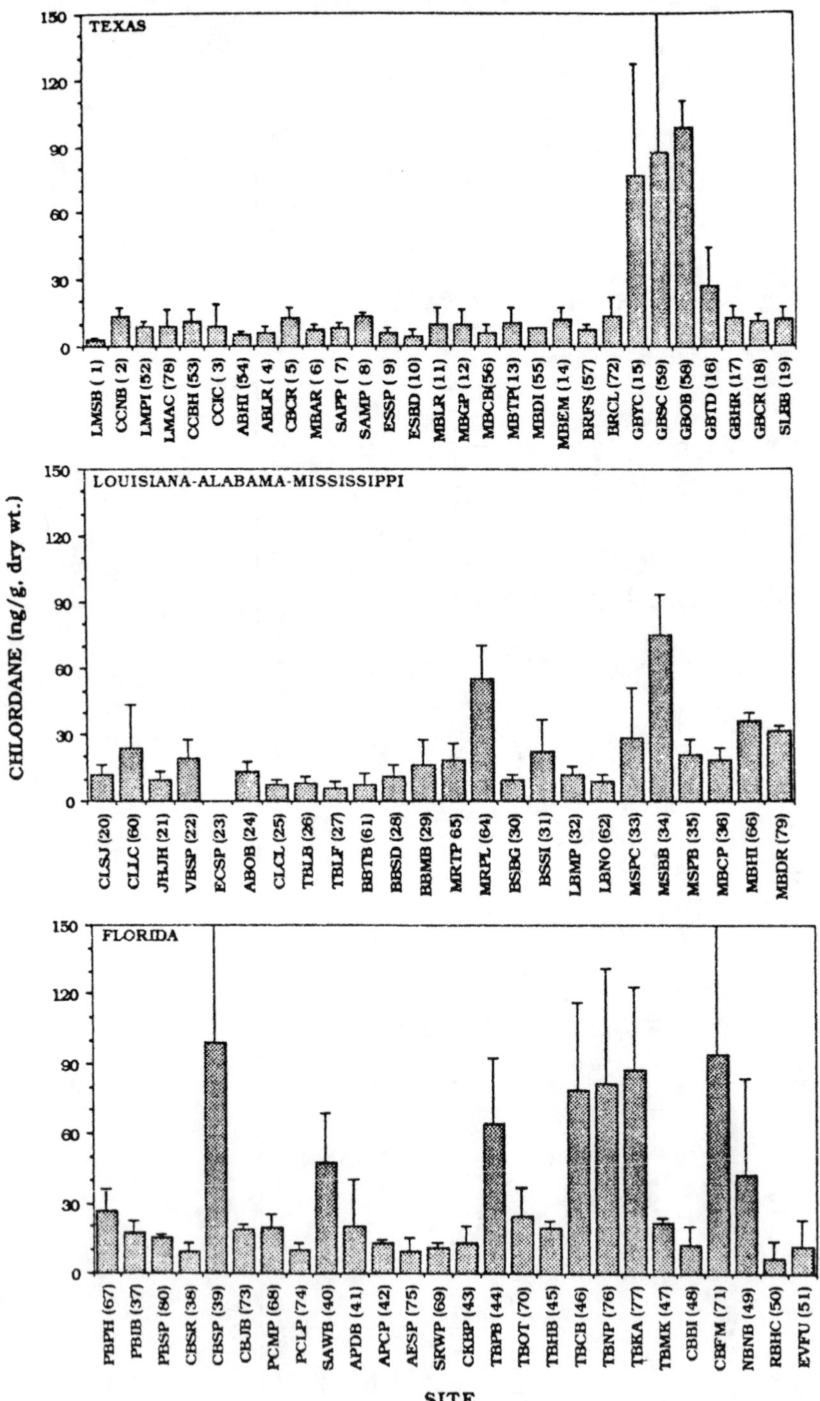

FIGURE 3.5. Average total chlordane concentrations (±1 standard deviation) in oyster samples collected from the Gulf of Mexico by the Status and Trends Mussel Watch Program during 1986 to 1990. (From Sericano, J. L., Wade, T. L., Brooks, J. M., Atlas, E. L., Fay, R. R., and Wilkinson, D. L., *Environ. Pollut.*, 82, 23, 1993. With permission.)

APPENDIX 4. PACIFIC COAST SURVEYS

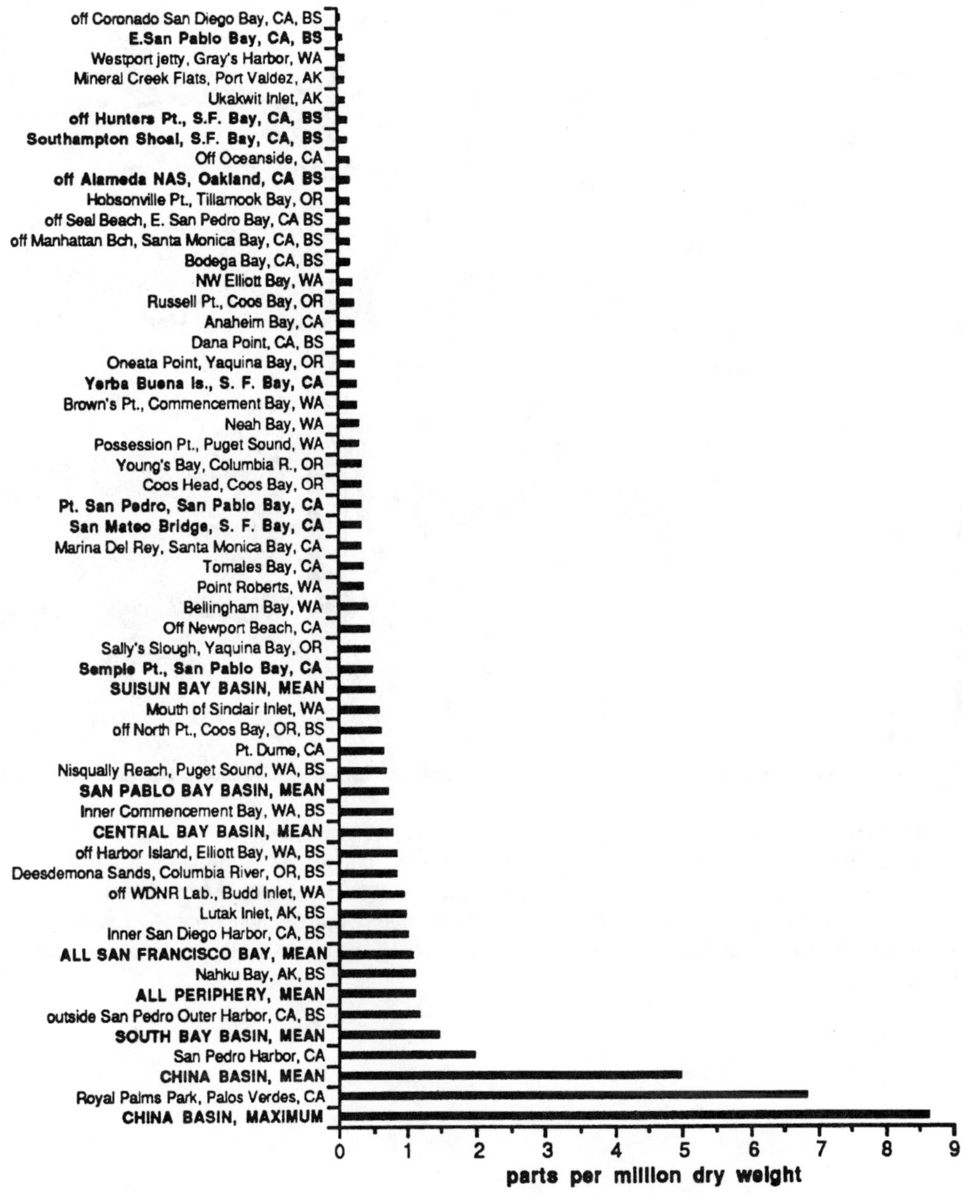

FIGURE 4.1. Comparison of cadmium concentrations in surface sediments of San Francisco Bay based on data collected by the Status and Trends Benthic Surveillance and Mussel Watch projects. Benthic surveillance sites are indicated by a BS after the site name. All other sites are in lower-case print. (From Long, E., Macdonald, D., Matta, M. B., VanNess, K., Buchman, M., and Harris, H., *Status and Trends in Concentrations of Contaminants and Measures of Biological Stress in San Francisco Bay,* NOAA Tech. Mem. NOS OMA 41, National Oceanic and Atmospheric Administration, Seattle, WA, 1988.)

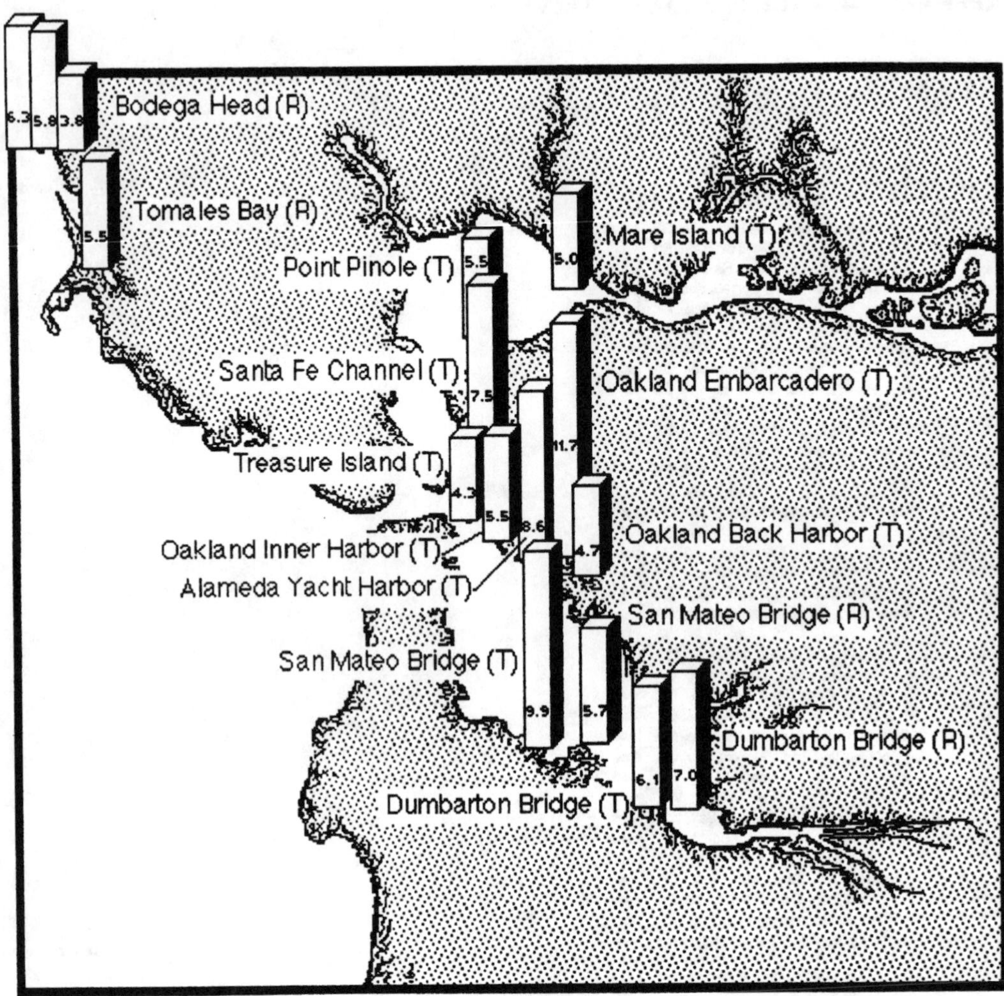

FIGURE 4.2. Cadmium concentrations in transplanted (T) coastal mussels, *Mytilus californianus,* and resident (R) bay mussels, *Mytilus edulis,* sampled from 1985 to 1986 by the California Mussel Watch Program in San Francisco Bay and neighboring sites. (From Long, E., Macdonald, D., Matta, M. B., VanNess, K., Buchman, M., and Harris, H., *Status and Trends in Concentrations of Contaminants and Measures of Biological Stress in San Francisco Bay,* NOAA Tech. Mem. NOS OMA 41, National Oceanic and Atmospheric Administration, Seattle, WA, 1988.)

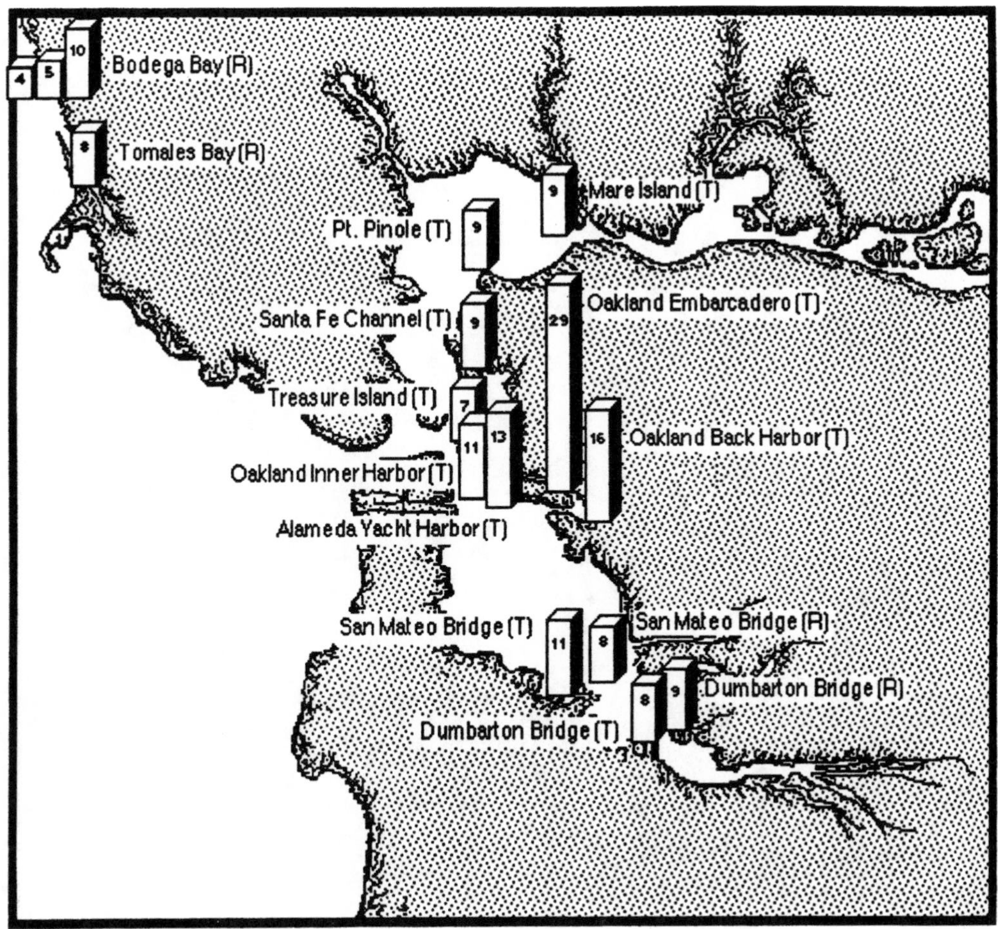

FIGURE 4.3. Copper concentrations in transplanted (T) coastal mussels, *Mytilus californianus,* and resident (R) bay mussels, *Mytilus edulis,* sampled from 1985 to 1986 by the California Mussel Watch Program and the National Status and Trends Program in San Francisco Bay and neighboring sites. (From Long, E., Macdonald, D., Matta, M. B., VanNess, K., Buchman, M., and Harris, H., *Status and Trends in Concentrations of Contaminants and Measures of Biological Stress in San Francisco Bay,* NOAA Tech. Mem. NOS OMA 41, National Oceanic and Atmospheric Administration, Seattle, WA, 1988.)

8 Radioactive Waste

INTRODUCTION

Radioactivity denotes the process by which a radionuclide undergoes spontaneous disintegration (decay) of its unstable nucleus with the emission of one or more radiations and the formation of a daughter nuclide.[1] Radiation emitted from a radioactive substance can be in the form of particles (i.e., alpha, beta, and neutron particles) or electromagnetic waves (i.e., gamma rays and X-rays). These particles and rays are sufficiently energized that they produce positively and negatively charged ion pairs when penetrating matter and, thus, are referred to as ionizing radiation. Because ionizing radiation can severely damage biological tissue, it poses a major hazard to living organisms.

Alpha particles consist of two protons and two neutrons. These relatively slow moving, positively charged particles have a large mass and are intensely ionizing when passing through biological tissue. However, they possess low penetrating power.

Beta particles have greater penetrating capacity than alpha particles. Formed during the conversion of a neutron to a proton, these small charged particles are ejected from the nuclei of radioactive atoms. Negatively charged beta particles are termed negatrons, and positively charged beta particles, positrons. Negatrons are more frequently emitted from radionuclides than positrons.

Neutron particles can be generated naturally in the atmosphere via cosmic ray interaction, as well as anthropogenically during fission reactions in nuclear reactors and at sites of nuclear detonations. These particles, comprised of an electron and a proton, have a much greater range than either alpha or beta radiation. Their great kinetic energy and penetrating capacity are linked to their large mass and chargeless state.

The emission of gamma rays commonly accompanies the release of alpha and beta particles during the decay of many radionuclides. Beta-emitters usually also release energy in the form of gamma photons. While gamma rays are released from the nuclei of radioactive atoms, X-rays arise from the electron shells. Both gamma rays and X-rays are strongly ionizing to organismal tissue through which they pass, having very short wavelengths and great penetrating power.

SOURCES OF IONIZING RADIATION

As conveyed above, radionuclides originate from both anthropogenic and natural background sources. Anthropogenic sources of ionizing radiation in the sea are mainly associated with nuclear power generation, past nuclear weapons testing, military related projects, and direct disposal of radioactive wastes. Natural background radiation is ascribable to cosmogenic and primordial radionuclides. Cosmogenic radionuclides form by the interaction of primary cosmic rays with matter in the atmosphere and on the surface of the Earth. Most of these radionuclides exist in minute quantities, the two exceptions being tritium (^{3}H) and carbon-14 (^{14}C). Terrestrial gamma rays from naturally occurring radionuclides also contribute to this background radiation. Primordial radionuclides derive from internal radioactive sources in the Earth that were generated at the time of formation of the planet. Detectable primordial radionuclides of significance include ^{40}K, ^{238}U, and ^{232}Th, all with a half-life greater than 10^{9} years. Primordial radionuclides occur in the sediments and throughout the water column of estuaries and oceans. Among this group of radionuclides, ^{40}K

constitutes the dominant, naturally occurring source of internal radiation in estuarine and marine organisms.

Anthropogenic Sources of Radioactivity

In estuarine and coastal marine environments, a wide array of organisms accumulates radionuclides from various stages of the nuclear fuel cycle. Agricultural, industrial, medical, and scientific uses of radioisotopes may also contribute to the pool of artificial radionuclides in these environments that affect biotic communities. Radioactive wastes are produced at several stages in the nuclear fuel cycle, notably uranium mining, milling, conversion, enrichment, fuel fabrication, power plant operation, and reprocessing. The wastes occur in solid, liquid, or gaseous form and are classified according to the concentration and potential hazard of radioactive material they contain. In addition to differing in physical form, radioactive wastes can be distinguished by their chemical form and, consequently, their potential environmental impact, as well as the nature of the radiation they emit. Effective management of radioactive waste must take into account the types of ionizing radiation — and the associated energy levels — emitted by radionuclides.

Naturally Occurring Background Radioactivity

Estuarine and marine environments, apart from being affected by anthropogenic sources of radioactivity, are unavoidably exposed to naturally occurring background radioactivity, which is measured in the International System of Units (i.e., Becquerel [Bq], Gray [Gy], and Sievert [Sv]). As noted previously, natural background radiation is due to cosmogenic radionuclides, terrestrial gamma rays from naturally occurring radionuclides, and internal radioactivity, principally potassium-40 (^{40}K).[2] The radioactivity from all natural radionuclides in seawater amounts to about 750 dpm/l (97% derived from ^{40}K). Cosmic rays supply a dose rate of about 4×10^{-8} Sv/hr at the water surface, which declines to approximately 5×10^{-9} Sv/hr at 20-cm depth. The cosmic dose rate drops even further at greater depths, being negligible below 100 m. The total background radioactivity in surface seawater (~12.6 Bq/l) is less than that in marine sands (185-370 Bq/kg) and muds (740-1110 Bq/kg). Estimates of annual doses received by marine organisms from natural sources of radiation are typically less than 5 mGy/yr.[3]

TYPES OF RADIOACTIVE WASTE

Six categories of radioactive waste are recognized: (1) high-level wastes, (2) transuranic wastes, (3) low-level wastes, (4) uranium and mill tailings, (5) decontamination and decommissioning wastes from nuclear reactors, and (6) gaseous effluents. The U.S. Interagency Review Group on Nuclear Waste Management[4,5] defines these six categories as follows:

1. *High-level wastes:* These wastes are either fuel assemblies that are discarded after having served their useful life in a nuclear reactor (spent fuel) or the portion of the wastes generated in reprocessing that contain virtually all of the fission products and most of the actinides not separated out during reprocessing. The wastes are being considered for disposal in geologic repositories or by other technical options designed to provide long-term isolation of the wastes from the biosphere.
2. *Transuranic wastes:* These wastes are produced primarily from the reprocessing of defense spent reactor fuels, the fabrication of plutonium to produce nuclear weapons, and, if it should occur, plutonium fuel fabrication for use in nuclear power reactors. Transuranic wastes contain low levels of radioactivity but varying amounts of long-lived elements above uranium in the periodic table of elements, mainly plutonium. They are currently defined as materials containing >370 Bq/g of transuranic activity.

3. *Low-level wastes:* These wastes contain <370 Bq/g of transuranic contaminants. Although low-level wastes require little or no shielding, they have low, but potentially hazardous concentrations of radionuclides and, consequently, require management. Low-level wastes are generated in almost all activities involving radioactive materials and are presently being disposed of by shallow land burial.
4. *Uranium mine and mill tailings:* These wastes are the residues from uranium mining and milling operations. They are hazardous because they contain low concentrations of radioactive materials which, although naturally occurring, contain long-lived radionuclides. The tailings, with a consistency similar to sand, are generated in large volumes (about 10^{10} kg/yr in the U.S.) and are presently stored in waste piles at the site of mining and milling operations. A program is underway to either immobilize or bury uranium mine and mill tailings to prevent them from being dispersed by wind or water erosion.
5. *Decontamination and decommissioning wastes:* As defense and civilian reactors and other nuclear facilities reach the end of their productive lifetimes, parts of them will have to be handled as either high- or low-level wastes and disposed of accordingly. Decontamination and decommissioning activities will generate significant quantities of wastes in the future.
6. *Gaseous effluents:* These wastes are produced in many defense and commercial nuclear facilities, such as reactors, fuel fabrication facilities, uranium enrichment plants, and weapons manufacturing facilities. They are released into the atmosphere in a controlled manner after passing through successive stages of filtration and mixing with air where they are diluted and dispersed.

Owing to the 1972 London Dumping Convention — the only major international agreement covering ocean disposal of radioactive wastes — high-level radioactive wastes are not dumped at sea. However, the inadvertent sinking of nuclear submarines and airplane crashes carrying nuclear weapons have accidently delivered high-level radioactive wastes to the sea in past years.[4] High-level radioactive waste contains per tonne of material more than 37,000 TBq tritium, 37 TBq beta and alpha-emitters, 3.7 TBq strontium-90 and cesium-137, or 0.037 TBq alpha-emitters with half-lives over 50 years.[7] Intermediate-level radioactive waste has not been dumped at sea since 1982. Only low-level radioactive waste was dumped in marine waters after 1982, being packaged in steel drums or concrete containers and disposed of at approved sites.[8] This waste was derived largely from nuclear power plants, as well as from scientific and medical work.

Aside from releases of low-level radioactive waste, major point sources of radionuclides in shallow waters are associated with fuel-reprocessing facilities. The coastal discharges at Windscale and Dounreay in England and La Hague in France provide examples. Surface currents have transported a substantial quantity of radiological releases from these facilities through the North Sea into the Norwegian Sea and North Atlantic Ocean.[9]

Between 1946 and 1970, approximately 107,000 containers of low-level radioactive waste were dumped by the U.S. at four sites including: (1) the Farallon Island site in the Pacific Ocean west of San Francisco, (2) the 2800-m site in the northwest Atlantic Ocean off New Jersey, (3) the 3800-m site in the northwest Atlantic off New Jersey, and (4) the Massachusetts Bay site. An estimated total activity in these containers at the time of packing amounted to about 4.3×10^{15} Bq. Approximately 4000 canisters of low-level radioactive waste were disposed of in Massachusetts Bay during this 25-year period.[6]

The radioactive waste disposal process at sea is carefully regulated and monitored, as demonstrated by the Northeast Atlantic Dumpsite (NEADS) located at approximately 46°N latitude, 17°W longitude. In existence since 1967, the NEADS site lies at a mean depth of 4.4 km, with seafloor sediment consisting of 350- to 370-m thick marl facies. Although the contents of radioactive waste dumped at this site are expected to corrode and leach through their containment at some time in the future, the delay should provide enough time to ensure the loss of radioactivity via decay of shorter-lived radionuclides. The slow release of the pollutants should also result in great dilution. Theoretical models predict minimal, if any, effect of the radioactive waste on marine organisms in and around this deep-sea dumpsite. Exposure of marine organisms at the NEADS site is expected to be at or near the natural background level of radiation and well below the dose rates at which harmful effects are manifested.[7] Despite the predicted minimal impact of deep-sea radioactive dumpsites in general, political pressure has halted the program of radioactive waste disposal in the deep sea.[8]

Large areas of the deep seabed have been assessed as potential disposal sites for high-level radioactive waste. This is so because they are (1) far removed from seismically and tectonically active lithospheric plate boundaries, (2) far from active or young volcanoes, (3) comprised of thick layers of very uniform fine-grained clays, and (4) devoid of natural resources likely to be exploited in the foreseeable future. The sedimentary records of these sites infer tens of millions of years of slow, uninterrupted deposition of fine-grained clay, which supports predictions of their future stability. The geologic and oceanographic processes governing the deposition of these sediments are well understood. In addition, the properties of the sediments (e.g., permeability, ion-retardation, mechanical strength) indicate that they can act as a primary barrier to the escape of buried radionuclides.[9]

REGULATORY CONTROLS

The oceanic input of radioactive waste from human activity has varied substantially since 1960 due to reductions in fallout from nuclear bomb detonations after the Second Nuclear Test Ban Treaty in 1963, the increase in the number of nuclear power plants on line after 1970, and accelerated usage of radionuclides in agriculture, industry, and medicine after 1975. The acceleration of nuclear power generation has had a particularly dramatic impact during the last 25 years. For example, the number of nuclear power plants worldwide grew from 66 to about 430 between 1970 and 1996. In the U.S. alone, 118 commercial nuclear power plants were operating by 1991.[10] These facilities became a greater source of anthropogenic radioactivity in estuaries and oceans than nuclear detonations after termination of atmospheric nuclear weapons testing in 1980.

Several regulatory controls have been implemented to control radioactive wastes in the sea. The 1972 International Convention on the Prevention of Marine Pollution by Dumping of Wastes and Other Matter (also known as the London Dumping Convention) and the U.S. Marine Protection, Research, and Sanctuaries Act of 1972 (commonly called the Ocean Dumping Act) prohibit the dumping of high-level wastes in the sea and have done so since 1972. International control of radioactive wastes in the sea improved considerably after enactment of the London Dumping Convention in 1975.[11] To enforce the provisions of the London Dumping Convention, contracting parties are required to promulgate national legislation. Substances prohibited from dumping at sea are listed in Annex I of the London Dumping Convention. Annex II contains a list of substances that may be dumped with special care. Annex III specifies the criteria considered in the issuance of permits for the dumping of matter in the ocean. The International Atomic Energy Agency (IAEA) has been designated as the competent international authority to the London Dumping Convention. Not only must the provisions of Annex III of the London Dumping Convention be satisfied for the dumping of radioactive wastes at sea, but recommendations of the IAEA also must be met.[12]

The disposal of low-level radioactive waste in the U.S. is regulated by the Low-Level Radioactive Waste Policy Act of 1980 and its Amendments (1985). The Nuclear Waste Policy Act of 1982 provides a framework for resolving many of the management problems associated with low- and high-level radioactive waste in the U.S. Aside from radioactive waste legislation, a number of waste management strategies — dilution and dispersion, isolation and containment, and isotopic dilution with stable isotopes — have been adapted to minimize the environmental impact of radioactive waste in the U.S. The packaging and dumping of radioactive wastes in the deep sea exemplify two waste management strategies. For instance, the use of drums to contain the waste and their placement in the deep-sea environment represent the practice of isolation and containment. If the radioactive waste escapes from the drums on the sea floor, it will be subject to dilution and dispersion.

In estuaries, which most commonly receive artificial radionuclides in low-level waste discharged in liquid effluents from nuclear power plants, the preferred management option is one of dilution and dispersion. The release of some radioactive materials into estuarine and marine environments appears to be the inevitable consequence of the use of nuclear power. It is absolutely necessary to strictly regulate the amount of this low-level radioactivity discharged into these environments.

Practices deemed to be critical in the handling of high-level radioactive waste (isolation and containment) may not be practical nor necessary as a strategy in managing certain other types of radioactive waste, including the low-level effluents discharged by nuclear power plants. In fact, the disposal of certain categories of radioactive waste offers some advantages over any storage approach.[13] In June 1990, the U.S. Nuclear Regulatory Commission even approved a policy allowing low-level radioactive waste (e.g., clothing, equipment, and parts from nuclear power plants) to be dumped in ordinary landfills and recycled into consumer products (e.g., frying pans and jewelry).

While isolation and containment, together with dilution and dispersion, are methods of dealing with radioactive wastes removed from their source, isotopic dilution serves as a novel approach of controlling waste toxicity (radiation dose) at the source, thus avoiding further treatment. According to this practice, certain radioactive wastes can be maintained below the maximum permissible concentrations by adding stable isotopes of the same chemical forms and equilibrating them sufficiently before the wastes are discharged to the environment. This management scheme may prove to be the most economical means of controlling certain radioactive wastes.[14] Many technical questions remain, however, regarding the best waste disposal strategy and design needed to protect marine systems.

The International Commission on Radiological Protection (ICRP) has developed basic guidelines of radiological protection applicable to better management of low-level radioactive waste disposal in the marine environment. The ICRP is concerned primarily with the radiological protection of human life rather than other living systems, although it views the level of safety required for the protection of human beings as likely to be more than adequate to protect other species of organisms. This protection is achieved through the application of the ICRP system of dose limitiation as follows:[15]

1. No practice shall be adopted unless its introduction produces a positive net benefit.
2. All exposures shall be kept as low as reasonably achievable, economic and social factors being taken into account.
3. The dose equivalent to individuals shall not exceed the limits recommended for the appropriate circumstances by the ICRP.

BEHAVIOR OF RADIONUCLIDES IN ESTUARINE AND MARINE ENVIRONMENTS

Complex interactions of physical, chemical, and biological factors act to disperse, dilute, or concentrate radioactive substances in estuarine and marine environments. Radionuclides behave chemically the same as the stable form of the element and similar to other elements within the same column of the periodic table of the elements. For example, radionuclides such as ^{45}Ca, ^{90}Sr, ^{140}Ba, ^{226}Ra, and ^{45}Ca behave like calcium, whereas ^{40}K, ^{86}Rb, and ^{137}Cs behave like potassium. Consequently, ^{90}Sr concentrates largely in shells, exoskeletons, or bone, and ^{137}Cs in the soft tissue of an organism's body.[2]

As radionuclides enter estuarine or marine waters in an effluent, they are subject to turbulent and molecular diffusion, with dispersion lowering their concentration away from the effluent source. Radioactive particles, like many other types of contaminants, tend to attach to particulate materials, with the strength of the attachment dependent on various types of physical and chemical forces related to the properties of the radionuclides and the particulate surfaces. The interaction of radionuclides with particles in an estuary or open ocean can be described most succinctly in terms of simple ion exchange or adsorption equalibria.[16] Deposition of the radionuclides is enhanced by chemical adsorption or exchange, precipitation scavenging, impaction, and gravitational settling. The sea floor serves as the ultimate repository for most radionuclides in estuarine and coastal marine systems.[17-19]

Bottom sediments play an important role in the contamination of biological systems. Because of the accumulation of radionuclides in seafloor sediments of shallow water systems, the manipulation and processing of radionuclide-sorbed organic detritus by detritus feeders facilitate the recycling of radioactive substances through biotic compartments. The mineralization of radionuclide-sorbed detritus by microbes releases radioactive substances to bottom sediments or interstitial waters. The bioturbating activity of benthic fauna, together with turbulence from bottom currents or storms, can roil and resuspend the detritus, enabling radionuclides to remobilize to other areas of the system where they can reenter food chains. In contrast, radionuclides enter the organic detritus pool after the death of plants and animals that concentrated radionuclides in their tissues when alive.

Aside from the uptake of radionuclides by organisms from bottom sediments, radionuclides enter biotic compartments by organismal uptake directly from seawater or from other organisms via ingestion. Marine algae and other autotrophs obtain radionuclides directly from seawater as well as sediments. Some nuclides may even be assimilated from surface-deposited material that enters plant tissue through stomates or epidermal tissues. In grazing food webs, herbivores accumulate radioactive material largely from consumption of primary producers. Omnivores and carnivores, in turn, obtain radionuclides by consuming herbivores and other prey that contain the contaminants, and they assimilate some of the substances directly from seawater or sediments. The propensity of these upper-trophic-level organisms to accumulate radioactive substances from environmental compartments is related to their behavioral patterns, physical attributes, and other inherent characteristics.

The flow of ^{137}Cs through marine systems provides an example. This radionuclide tends to move from the water compartment to sediments, where it is available to detritivores and bottom feeders. Some of the ^{137}Cs is also accumulated by bottom-dwelling plants and phytoplankton and subsequently moves through the food chain, first to herbivorous consumers, then to primary consumers (e.g., carnivorous invertebrates and small fish), and finally to secondary consumers (i.e., large fish). The biological half-life is weeks to months. Cesium concentration factors for marine algae, mollusks, crustaceans, and fish typically average 10, 10, 50, and 30, respectively.[14]

Smaller organisms (e.g., phytoplankton and zooplankton), which have large surface area to volume ratios, accumulate radionuclides relatively quickly. The degree of uptake by larger organisms (e.g., macroalgae, macroinvertebrates, and fish), while typically less than that of smaller forms, nevertheless can still be significant in those individuals that cannot effectively regulate radionuclide concentrations in their tissues. For many marine species, the uptake of radionuclides is proportional, or nearly proportional, to the elemental concentration in seawater.[20]

Bioaccumulation of radionuclides by estuarine and marine organisms appears to be greatest in areas nearby nuclear fuel processing plants and industrial facilities producing nuclear weapons.[21,22] When the radionuclide concentration is assessed along food chains, little evidence exists for biomagnification at higher trophic levels. Exceptions can be found, however, such as ^{137}Cs, which preferentially accumulates in fish.[23]

RADIATION EFFECTS ON ESTUARINE AND MARINE ORGANISMS

Radiation exposure of estuarine and marine organisms can lead to a range of observable effects depending on the total dose, dose rate, type of radiation, and exposure period. These effects may be manifested as genetic changes, physiological changes (e.g., alteration of the hemopoietic and reproductive systems), aberrant growth and development, cancer induction, and death.[14] Alterations in genetic material of the cells are among the most serious effects of radiation exposure since such alterations cause heritable mutations, most of which are deleterious. While the chance of damage and degree of damage increase proportionally to the radiation dose, any dose of ionizing radiation, no matter how small, can induce mutation.[24]

Laboratory investigations of radiation effects on marine organisms have concentrated on changes induced by single, acute, relatively large doses, often culminating with the death of the organisms. In natural situations, however, these organisms are subject to chronic exposures of relatively low radiation doses, where the effects are likely to be sublethal.[25] In this latter case, the dose rates are not uniform, and the total dose results from both internal irradiation (i.e., from absorption and ingestion) and external irradiation.[2] Chronic exposures do not necessarily lead to ill effects in organisms because tissue repair processes may be adequate. In effect, competition exists between damage to organismal tissue and repair at low dose rates. The probability of tissue damage rises substantially at higher radiation doses. Chronic exposures of <10 mGy are unlikely to generate measurable deleterious changes in marine populations or communities.[26]

The ranges of acute lethal doses of ionizing radiation to estuarine and marine organisms vary widely among taxonomic groups. For example, mollusks are relatively insensitive, with acute lethal doses of about 100 to 1000 Gy. Fishes are more sensitive than mollusks; their acute lethal doses range from approximately 10 to 50 Gy. Mammals, with acute lethal exposures of 2 to 13 Gy, are the most sensitive group. The lethal doses of ionizing radiation for bacteria vs. mammals differ by nearly three orders of magnitude. As noted by Talmage and Meyers-Schone,[2] direct mortality is not significant for any taxonomic group until a dose of ~2 Gy is reached. An entire community can be disrupted at acute lethal exposures of 10 Gy or more. However, these exposure levels reflect the effects of unusual events, such as catastrophic nuclear accidents or nuclear warfare.[1]

REFERENCES

1. Whicker, F. W. and Schultz, V., *Radioecology: Nuclear Energy and the Environment*, Vol. 1, CRC Press, Boca Raton, FL, 1982.
2. Talmage, S. S. and Meyers-Schone, L., Nuclear and thermal, in *Handbook of Ecotoxicology*, Hoffman, D. J., Rattner, B. A., Burton, G. A., Jr., and Cairns, J., Jr. (Eds.), Lewis Publishers, Boca Raton, FL, 1995, 469.
3. International Atomic Energy Agency, *Effects of Ionizing Radiation on Aquatic Organisms and Ecosystems,* Tech. Rep. Ser. No. 172, International Atomic Energy Agency, Vienna, 1976.
4. U.S. Interagency Review Group on Nuclear Waste Management, Report to the President by IRG, TID-29442, March 1979, National Technical Information Service, U.S. Department of Commerce, Washington, D.C., 1979.
5. U.S. Interagency Review Group on Nuclear Waste Management, Subgroup Report on Alternative Technology Strategies for the Isolation of Nuclear Waste, TID-28318, October 1979, National Technical Information Service, U.S. Department of Commerce, Washington, D.C., 1979.
6. Park, P. K., Kester, D. R., Duedall, I. W., and Ketchum, B. H., Radioactive wastes in the ocean: an overview, in *Wastes in the Ocean*, Vol. 3, *Radioactive Wastes and the Ocean*, Park, P. K., Kester, D. R., Duedall, I. W., and Ketchum, B. H. (Eds.), John Wiley & Sons, New York, 1983, 3.
7. Clark, R. B., *Marine Pollution*, 3rd ed., Clarendon Press, Oxford, 1992.
8. Gage, J. D. and Tyler, P. A., *Deep-Sea Biology: A Natural History of Organisms at the Deep-Sea Floor*, Cambridge University Press, Cambridge, 1991.
9. Heath, G. R., Hollister, C. D., Anderson, D. R., and Leinen, M., Why consider subseabed disposal of high-level nuclear wastes?, in *Wastes in the Ocean*, Vol. 3, *Radioactive Wastes and the Ocean*, Park, P. K., Kester, D. R., Duedall, I. W., and Ketchum, B. H. (Eds.), John Wiley & Sons, New York, 1983, 303.
10. U.S. Nuclear Regulatory Commission, Generic Environmental Impact Statement for License Renewel of Nuclear Plants, NUREG-1437, Vol. 1, Washington, D.C., 1991.
11. Hagen, A. A., History of low-level radioactive waste disposal in the sea, in *Wastes in the Ocean*, Vol. 3, *Radioactive Wastes and the Ocean*, Park, P. K., Kester, D. R., Duedall, I. W., and Ketchum, B. H. (Eds.), John Wiley & Sons, New York, 1983, 47.

12. Schell, W. R. and Nevissi, A. E., Radionuclides at the Hudson Canyon disposal site, in *Wastes in the Ocean*, Vol. 3, *Radioactive Wastes and the Ocean*, Park, P. K., Kester, D. R., Duedall, I. W., and Ketchum, B. H. (Eds.), John Wiley & Sons, New York, 1983, 183.
13. Templeton, W. L., Disposal of low-level radioactive wastes in the ocean, in *Oceanic Processes in Marine Pollution*, Vol. 3, *Marine Waste Management: Science and Policy*, Camp, M. A. and Park, P. K. (Eds.), Robert E. Krieger Publishing, Malabar, FL, 1989, 75.
14. Preston, A., Deep-sea disposal of radioactive wastes, in *Wastes in the Ocean*, Vol. 3, *Radioactive Wastes and the Ocean*, Park, P. K., Kester, D. R., Duedall, I. W., and Ketchum, B. H. (Eds.), John Wiley & Sons, New York, 1983, 107.
15. International Commission on Radiological Protection, Recommendation of the International Commission on Radiological Protection, Publication No. 26, International Commission on Radiological Protection, Pergamon Press, New York, 1977.
16. Edgington, D. N. and Nelson, D. M., The chemical behavior of long-lived radionuclides in the marine environment, in *Behavior of Long-Lived Radionuclides Associated with Deep-Sea Disposal of Radioactive Wastes: Report of a Co-Ordinated Research Program Organized by the International Atomic Energy Agency,* Tech. Rept., International Atomic Energy Agency, Vienna, 1986, 41.
17. Donoghue, J. F., Bricker, O. P., and Olsen, C. R., Particle-borne radionuclides as tracers for sediment in the Susquehanna River and Chesapeake Bay, *Est. Coastal Shelf Sci.,* 29, 341, 1989.
18. Hamilton, E., Radionuclides and large particles in estuarine sediments, *Mar. Pollut. Bull.,* 20, 603, 1989.
19. Teksoz, G., Yetis, U., Tuncel, G., and Balkas, T. I., Pollution chronology of the Golden Horn sediments, *Mar. Pollut. Bull.,* 22, 447, 1991.
20. Fowler, S. W., Biological transfer and transport processes, in *Pollutant Transfer and Transport in the Sea*, Kullenberg, G. (Ed.), CRC Press, Boca Raton, FL, 1982, 1.
21. Eisenbud, M., *Environmental Radioactivity from Natural, Industrial, and Military Sources*, 3rd ed., Academic Press, New York, 1987.
22. Langford, T. E., *Electricity Generation and the Ecology of Natural Waters*, Liverpool University Press, England, 1983.
23. Pentreath, R. J., Radionuclides in marine fish, Oceanogr. Mar. Biol. Annu. Rev., 15, 365, 1977.
24. McConnaughey, B. H. and Zottoli, R., *Introduction to Marine Biology*, C. V. Mosby, St. Louis, 1983.
25. Knowles, J. F. and Greenwood, L. N., The effects of chronic irradiation on the reproductive performance of *Ophryotrocha diadema* (Polychaeta, Dorvilleidae), *Mar. Environ. Res.,* 38, 207, 1994.
26. National Council on Radiation Protection and Measurements, *Effects of Ionizing Radiation on Aquatic Organisms,* NCRP Rept. No. 109, National Council on Radiation Protection and Measurements, Bethesda, MD, 1991.

APPENDIX 1. NATURAL RADIONUCLIDE SOURCES AND CONCENTRATIONS

TABLE 1.1
Characteristics of Common Nuclear Radiations[a]

Radiation	Rest Mass	Charge	Typical Energy Range	Path Length (Order of Magnitude) Air	Solid	General Comments
α	4.00 amu	2+	4–10 MeV	5–10 cm	25–40 μm	Identical to ionized He nucleus
β (negatron)	5.48×10^4 amu 0.51 meV	–	0–4 MeV	0–1 m	0–1 cm	Identical to electron
Positron (β positive)	5.48×10^4 amu 0.51 meV	+	—	0–1 m	0–1 cm	Identical to electron except for charge
Neutron	1.0086 amu 939.55 meV	0	0–15 MeV	0–100 m	0–100 cm	Free half-life: 16 min
γ (e.m. photon)[a]	—	0	10 keV–3 MeV	0.1–10 m^4	1 mm–1 m	Photons from nuclear transitions

[a] Exponential attenuation in the case of electromagnetic radiation.

Source: Eichholz, G. G., *Environmental Aspects of Nuclear Power,* Ann Arbor Science, Ann Arbor, MI, 1976. With permission.

TABLE 1.2
Oceanic Radiocarbon Inventories[a]

Source	Inventory (RCU)
Cosmogenic radiocarbon	
Fairhall et al.	1.94×10^4
Killough and Emanuel	2.00×10^4
Recommended value	1.97×10^4
Bomb-carbon (as of ~1972)	
Stuiver et al. (extrapolated)	310
Broecker et al.	303
Recommended value	303

Note: See original source for particular studies.

[a] Inventories are normalized ("recorrected") to $\delta^{13}C = 0$.

Source: Lassey, K. R., Manning, M. R., and O'Brien, B. J., *Rev. Aquat. Sci.,* 3, 117, 1990. With permission.

TABLE 1.3
Natural Levels of Radioactivity in Surface Sea Water

Radionuclide	Concentration (Bq/l)
Potassium-40	11.84
Tritium (^3H)	0.022–0.11
Rubidium-87	1.07
Uranium-234	0.05
Uranium-238	0.04
Carbon-14	0.007
Radium-228	$(0.0037–0.37) \times 10^{-2}$
Lead-210	$(0.037–0.25) \times 10^{-2}$
Uranium-235	0.18×10^{-2}
Radium-226	$(0.15–0.17) \times 10^{-2}$
Polonium-210	$(0.022–0.15) \times 10^{-2}$
Radon-222	0.07×10^{-2}
Thorium-228	$(0.007–0.11) \times 10^{-3}$
Thorium-230	$(0.022–0.05) \times 10^{-4}$
Thorium-232	$(0.004–0.29) \times 10^{-4}$

Source: Clark, R. B., *Marine Pollution,* 3rd ed., Clarendon Press, Oxford, 1992. With permission.

TABLE 1.4
Concentrations (× 10 Bq/l) of ^{210}Po and ^{210}Pb in Sea Water

Sample (m)	^{210}Po	^{210}Pb
Surface water		
≤20	0.19–3.7	<0.37–5.0
Deepwater		
500–1000	1.4–4.9	0.6–3.6
1000–2000	1.2–4.1	0.8–3.4
2000–3000	0.8–5.3	1.1–5.0
3000–4000	2.0–3.6	2.7–4.6
>4000	1.0–5.2	2.7–3.7

Source: Woodhead, D. S. and Pentreath, R. J., in *Wastes in the Ocean,* Vol. 3, *Radioactive Wastes and the Ocean,* Park, P. K., Kester, D. R., Duedall, I. W., Ketchum, B. H. (Eds.), John Wiley & Sons, New York, 1983, 134. With permission.

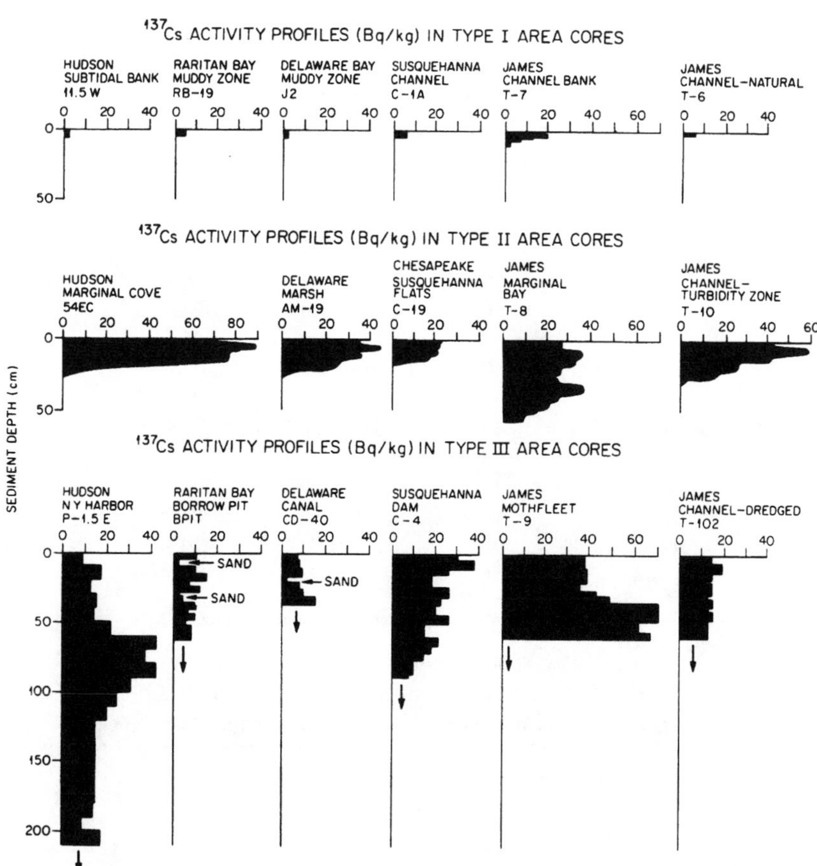

FIGURE 1.1. Cs profiles in estuarine sediments. Type I is typical of areas where the sediment surface is presently at or near a state of equilibrium with its respective wave, current, and flow regime. Type II is typical of marginal cove and mudflat areas where the sediment surface is approaching its equilibrium depth. Type III is typical of areas where the sediment surface is far below its equilibrium depth (such as forest areas on deltas) or in areas that have been affected by human activities such as dredging or construction. (From Olsen, C. R., Larsen, I. L., Mulholland, P. J., Von Damm, K. L., Grebmeier, J. M., Schaffner, L. C., Diaz, R. J., and Nichols, M. M., *Estuaries,* 16, 683, 1993. With permission.)

TABLE 1.5
Concentrations of ^{137}Cs and $\delta^{13}C$ in Bottom Sediments Collected Along a Salinity Gradient of the Savannah River Estuary

Station	Salinity (%)	^{137}Cs[a] (Bq/kg)	$\delta^{13}C$ (%)
R-1	0	37.0 ± 0.6	–26.9
R-2	0	62.6 ± 0.7	–27.2
E-1	<5	12.4 ± 0.3	–25.6
E-2	7	13.2 ± 0.3	–22.2
E-3	10.9	16.0 ± 1.1	–22.4
E-4	13	13.7 ± 0.4	–20.9
E-5	13.9	7.6 ± 0.6	–20.7
E-6	14.3	0.9 ± 0.2	–20.8
E-7	24	0.6 ± 0.1	–22.0
M-1	>30	<0.4	–20.8
M-8	>30	<0.4	–20.4

[a] Statistical counting errors are expressed as one sigma.

Source: Olsen, C. R., Larsen, I. L., Mulholland, P. J., Von Damm, K. L., Grebmeier, J. M., Schaffner, L. C., Diaz, R. J., and Nichols, M. M., *Estuaries,* 16, 683, 1993. With permission.

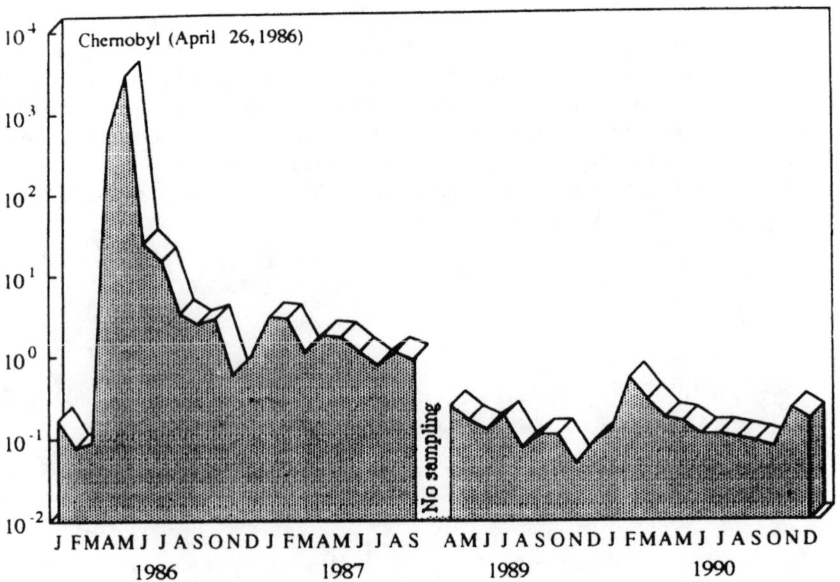

FIGURE 1.2. Monthly depositions of ^{137}Cs (Bq/m^2) detected in the Padan area, Adriatic Sea, between 1985 and 1990. (From Marzano, F. N. and Triulzi, C., *Mar. Pollut. Bull.,* 28, 244, 1994. With permission.)

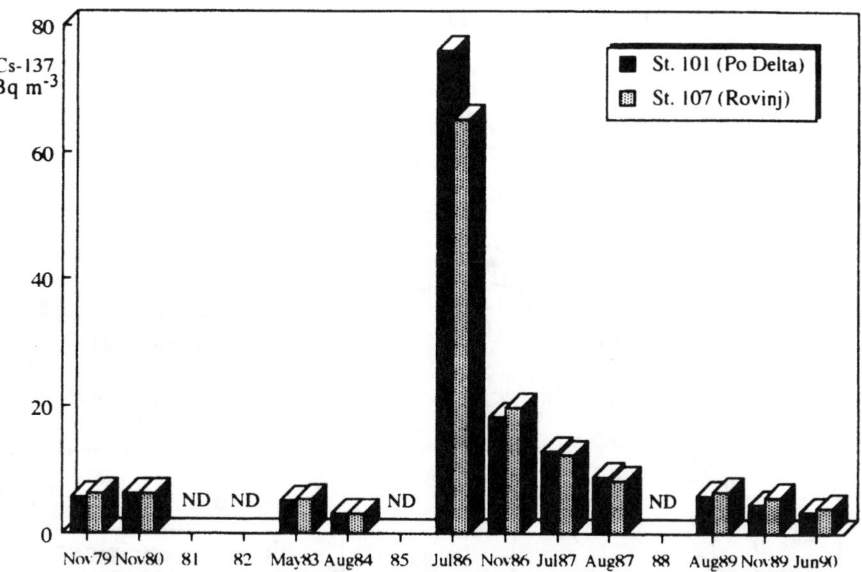

FIGURE 1.3. Temporal trends of ^{137}Cs concentrations (Bq/m^3) detected in seawater samples in the Po Delta and Rovinj areas, Adriatic Sea, from 1979 to 1990. (From Marzano, F. N. and Triulzi, C., *Mar. Pollut. Bull.*, 28, 244, 1994. With permission.)

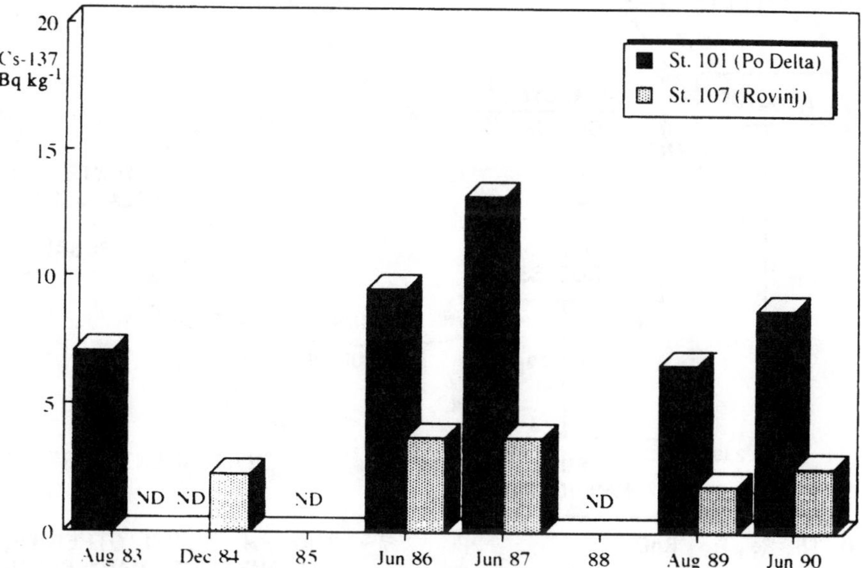

FIGURE 1.4. Temporal trends of ^{137}Cs concentrations (Bq/kg dry weight) detected in sediment samples of the Adriatic Sea system from 1979 to 1990. (From Marzano, F. N. and Triulzi, C., *Mar. Pollut. Bull.*, 28, 244, 1994. With permission.)

APPENDIX 2. ANTHROPOGENIC RADIONUCLIDE SOURCES AND CONCENTRATIONS

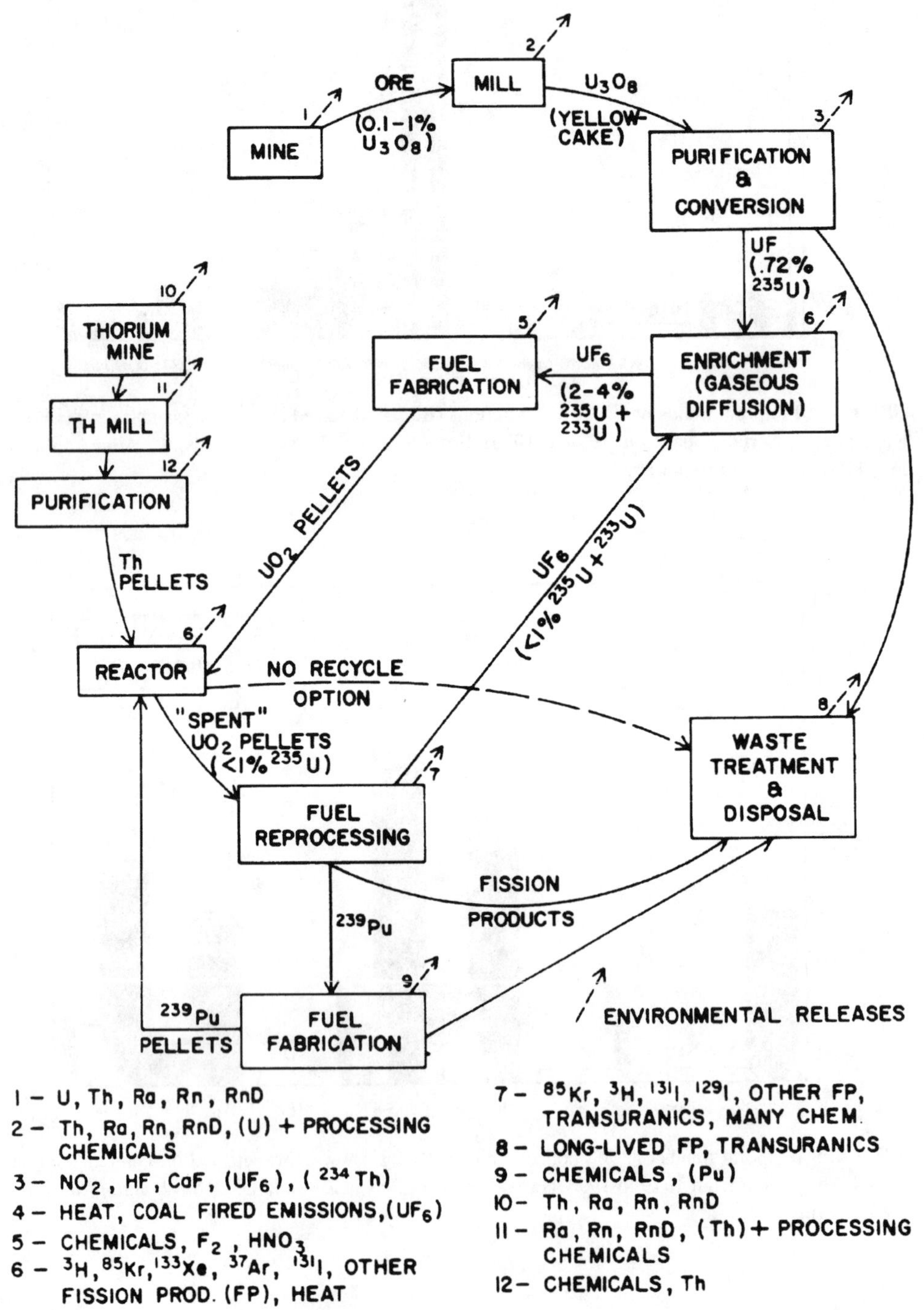

FIGURE 2.1. Basic elements and options within the nuclear fuel cycle, illustrating radioactive releases to the environment. (From Whicker, F. W. and Schultz, V., *Radioecology: Nuclear Energy and the Environment*, Vol. 1, CRC Press, Boca Raton, FL, 1982. With permission.)

TABLE 2.1
Radionuclides Generated by Nuclear Power Plants that Are of Potential Environmental Concern

Radionuclide	Half-Life	Radiation	Source	Element Analogue
^{3}H	12.3 years	β	Fission product, neutron activation	H
^{14}C	5568 years	β	Neutron activation	C
^{24}Na	15 hours	β, γ	Neutron activation	Na
^{32}P	14 days	β	Neutron activation	P
^{35}S	87 days	β	Neutron activation	S
^{41}Ar	110 minutes	β, γ	Neutron activation	—
^{45}Ca	164 days	β	Neutron activation	Ca
^{54}Mn	291 days	γ	Neutron activation	Mn
^{55}Fe	2.6 years	γ	Neutron activation	Fe
^{59}Fe	45 days	β, γ	Neutron activation	Fe
^{57}Co	270 days	γ	Neutron activation	Co
^{58}Co	71 days	$β^+$, γ	Neutron activation	Co
^{60}Co	5.2 years	β, γ	Neutron activation	Co
^{65}Zn	245 days	$β^+$, γ	Neutron activation	Zn
^{85}Kr	10 years	β, γ	Fission product	—
^{89}Sr	51 days	β	Fission product	Ca
^{90}Sr	28 years	β	Fission product	Ca
^{91}Y	58 days	β, γ	Fission product	—
^{95}Zr	65 days	β, γ	Fission product	—
^{103}Ru	40 days	β, γ	Fission product	—
^{106}Ru	1.0 years	β, γ	Fission product	—
^{129}I	1.7×10^7 years	β, γ	Fission product	I
^{131}I	8.1 days	β, γ	Fission product	I
^{134}Cs	2 years	β, γ	Fission product	K
^{137}Cs	27 years	β, γ	Fission product	K
^{140}Ba	12.8 days	β, γ	Fission product	Ca
^{143}Ce	33 hours	β, γ	Fission product	—
^{144}Ce	285 days	β, γ	Fission product	—
^{147}Nd	11 days	β, γ	Fission product	—
^{239}Pu	24,360 years	α, γ	Neutron activation	—
^{239}Np	2.3 days	β, γ	Neutron activation	—
^{241}Am	4770 years	α, γ	Neutron activation	—
^{242}Cm	163 days	α, γ	Neutron activation	—

Source: Whicker, F. W. and Schultz, V., *Radioecology: Nuclear Energy and the Environment,* Vol. 1, CRC Press, Boca Raton, FL, 1982. With permission.

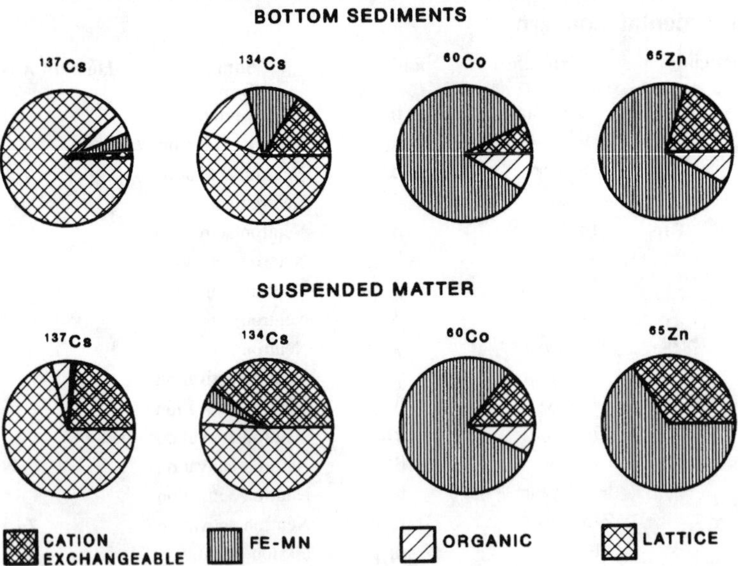

FIGURE 2.2. Radionuclide partitioning among different particulate phases in suspended matter and sediment samples collected in the Susquehanna River near the Peach Bottom Nuclear Power Station. (From Olsen, C. R., Larsen, I. L., Mulholland, P. J., Von Damm, K. L., Grebmeier, J. M., Schaffner, L. C., Diaz, R. J., and Nichols, M. M., *Estuaries,* 16, 683, 1993. With permission.)

TABLE 2.2
Bomb-Carbon Inventories According to Broecker et al.[a] for Mid-Latitude Atlantic and Pacific Oceans[b]

	Atlantic Ocean		Pacific Ocean	
Latitude Band	Col. Inventory (10^{12} atoms/m^2)	Inventory (RCU)	Col. Inventory (10^{12} atoms/m^2)	Inventory (RCU)
50°N–10°N	122	36.5	93	49.4
10°N–10°S	30	3.8	50	20.1
10°S–50°S	86	24.0	94	54.7
50°N–50°S	91	64.3	82	124.2

[a] Broecker, W. S. et al., *J. Geophys. Res.,* 90, 6953, 1985.

[b] All inventories estimated are as of 1972, and all are subject to 5% upward adjustment to compensate for an implicit fractionation correction. RCU, radiocarbon units.

Source: Lassey, K. R., Manning, M. R., and O'Brien, B. J., *Rev. Aquat. Sci.,* 3, 117, 1990. With permission.

TABLE 2.3
Bomb-Carbon Summary for Individual Ocean Regions Showing Either Excess or Deficient Inventories Relative to the Amount Expected if the CO_2 Invasion Rate Were Uniform Over the Ocean[a]

Latitude Band	Water Area (10^{12} m^2)	Inventory (RCU)	Input (RCU)	Inventory-Input[b] (RCU)
	Atlantic Ocean (I = 22.3 mol/m²/year)			
80°N to 40°S	18.6	26.6	19.7	+6.9
40°N to 20°N	15.8	23.0	14.5	+8.5
20°N to 20°S	26.7	10.8	23.1	–12.3
20°S to 45°S	18.4	18.5	14.2	+4.3
45°S to 80°S	15.1	5.2	12.6	–7.4
80°N to 80°S	94.6	84.1	84.1	0.0
	Indian Ocean (I = 19.4 mol/m²/year)			
25°N to 15°S	27.0	13.2	24.1	–10.9
15°S to 45°S	29.8	40.5	20.3	+20.2
45°S to 70°S	20.7	15.9	25.2	–9.3
25°N to 70°S	77.5	69.6	69.6	0.0
	Pacific Ocean (I = 19.2 mol/m²/year)			
65°N to 40°N	15.1	7.3	13.4	–6.1
40°N to 15°N	35.0	35.9	27.8	+8.1
15°N to 10°S	50.0	23.7	39.1	–15.4
10°S to 55°S	63.0	61.3	43.6	+17.7
55°S to 80°S	13.8	6.6	10.9	–4.3
65°N to 80°S	176.9	134.8	134.8	0.0
	World Ocean (I = 20.1 mol/m²/year)			
	349.0	288.5	288.5	0.0

[a] All inventories are estimated as of 1972 and are subject to 5% upward adjustment to compensate for implicit fractionation correction. RCU, radiocarbon units.

[b] The mean CO_2 invasion rate, I, is adjusted for each ocean and for the "World Ocean" so that ocean-average input matches inventory. This requires that, like the inventories, the invasion rate estimates are also subject to 5% upward adjustment, which when applied become I = 23.5, 20.4, 20.2, and 21.1 mol/m²/year for the Atlantic, Indian, Pacific, and World Oceans, respectively.

Source: Lassey, K. R., Manning, M. R., and O'Brien, B. J., *Rev. Aquat. Sci.*, 3, 117, 1990. With permission.

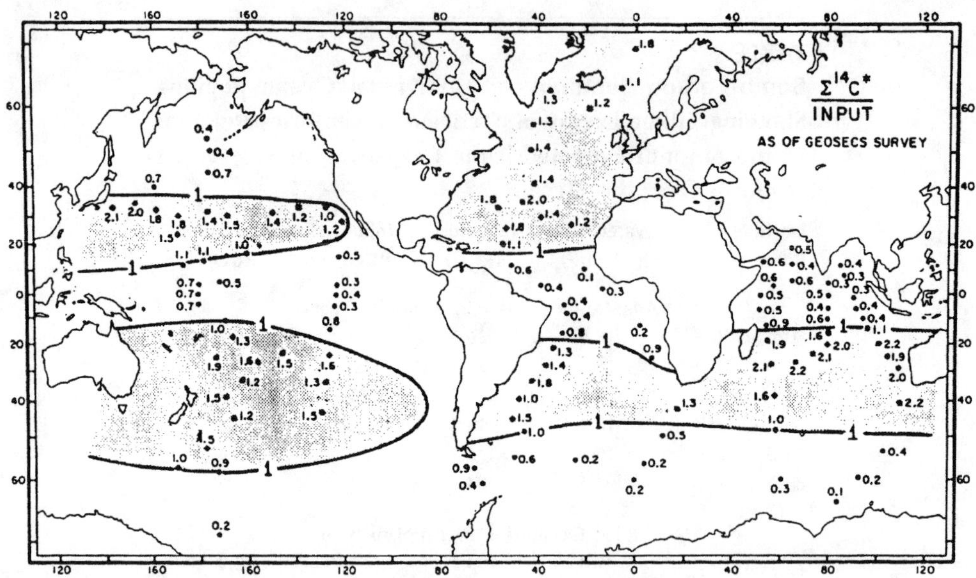

FIGURE 2.3. Map showing the ratio of water column inventory to net input of bomb radiocarbon for the stations occupied during the GESECS program. Regions with excess inventory over input are shaded. The ratio averaged over each of the three main oceans is unity. (From Broecker, W. S. et al., *J. Geophys. Res.*, 90, 6953, 1985. With permission.)

TABLE 2.4
Estimated Number of Nuclear Explosions 16 July 1945 to 5 August 1963 (Signing of the Partial Test Ban Treaty)

	U.S.		U.S.S.R.		U.K.		France		
Year	A	U	A	U	A	U	A	U	**Total**
1945	3	0							3
1946	2[a]	0							2
1947	0	0							0
1948	3	0							3
1949	0	0	1	0					1
1950	0	0	0	0					0
1951	15	1	2	0					18
1952	10	0	0	0	1	0			11
1953	11	0	4	0	2	0			17
1954	6	0	7	0	0	0			13
1955	17[a]	1	5[a]	0	0	0			23
1956	18	0	9	0	6	0			33
1957	27	5	15[a]	0	7	0			54
1958	62[b]	15	29	0	5	0			111
1949–58, exact years not available			18						18
1959	0	0	0	0	0	0			0[d]
1960	0	0	0	0	0	0	3	0	3[d]
1961	0	10	50[a]	1[c]	0	0	1	1	63[d]
1962	39[a]	57	43	1[c]	0	2	0	1	143
1 Jan to 5 Aug 1963	4	25	0	0	0	0	0	2	31
Total	217	114	183[e] (214)[f]	2[c]	21	2	4	4	547 (576)[f]

Note: A = atmospheric; U = underground.

[a] One of these tests was carried out under water.

[b] Two of these tests were carried out under water.

[c] Soviet information released in September 1990 did not confirm whether these were underground or atmospheric tests.

[d] The U.K., the U.S., and the former U.S.S.R. observed a moratorium on testing, November 1958 to September 1961.

[e] The total figure for Soviet atmospheric tests includes the 18 additional tests conducted in the period 1948–1958, the exact years for which are not available.

[f] The totals in brackets include the (probably atmospheric) explosions revealed by Soviet authorities in September 1990, the exact years for which have still not been announced. See *SIPRI Yearbook 1991,* p. 41. If the two tests in 1961 and 1962 (see note c) were atmospheric tests, this figure should read 216, under the column for atmospheric tests.

Source: Ferm, R., *World Armaments and Disarmament,* Appendix 6A, *SIPRI Yearbook 1993,* Oxford University Press, Oxford, 1993.

TABLE 2.5
Estimated Number of Nuclear Explosions 6 August 1963 to 31 December 1992

	U.S.[a]		U.S.S.R.		U.K.[a]		France		China		India		
Year	**A**	**U**	**A**	**U**	**A**	**U**	**A**	**U**	**A**	**U**	**A**	**U**	**Total**
1963	0	15	0	0	0	0	0	1					16
1964	0	38	0	6	0	1	0	3	1	0			49
1965	0	36	0	10	0	1	0	4	1	0			52
1966	0	43	0	15	0	0	6	1	3	0			68
1967	0	34	0	17	0	0	3	0	2	0			56
1968	0	45[b]	0	15	0	0	5	0	1	0			66
1969	0	38	0	16	0	0	0	0	1	1			56
1970	0	35	0	17	0	0	8	0	1	0			61
1971	0	17	0	19	0	0	6	0	1	0			43
1972	0	18	0	22	0	0	3	0	2	0			45
1973	0	16[c]	0	14	0	0	5	0	1	0			36
1974	0	14	0	18	0	1	8	0	1	0	0	1	43
1975	0	20	0	15	0	0	0	2	0	1	0	0	38
1976	0	18	0	17	0	1	0	4	3	1	0	0	44
1977	0	19	0	18	0	0	0	8[d]	1	0	0	0	46
1978	0	17	0	27	0	2	0	8	2	1	0	0	57
1979	0	15	0	29	0	1	0	9	1	0	0	0	55
1980	0	14	0	21	0	3	0	13	1	0	0	0	52
1981	0	16	0	22	0	1	0	12	0	0	0	0	51
1982	0	18	0	32	0	1	0	6	0	1	0	0	58
1983	0	17	0	27	0	1	0	9	0	2	0	0	56
1984	0	17	0	29	0	2	0	8	0	2	0	0	58
1985	0	17	0	9[e]	0	1	0	8	0	0	0	0	35
1986	0	14	0	0[e]	0	1	0	8	0	0	0	0	23
1987	0	14	0	23	0	1	0	8	0	1	0	0	47
1988	0	14	0	17	0	0	0	8	0	1	0	0	40
1989	0	11	0	7	0	1	0	8	0	0	0	0	27
1990	0	8	0	1	0	1	0	6	0	2	0	0	18
1991	0	7	0	0	0	1	0	6	0	0	0	0	14
1992	0	6	0	0	0	0	0	0	0	2	0	0	8
Total	0	611	0	463 (500)[f]	0	21	44	140	23	15	0	1	1318 (1355)[f]

Note: A = atmospheric; U = underground.

[a] One of these tests was carried out under water.

[b] Five devices used simultaneously in the same test are counted here as one explosion.

[c] Three devices used simultaneously in the same test are counted here as one explosion.

[d] Two of these tests may have been conducted in 1975 or 1976.

[e] The U.S.S.R. observed a unilateral moratorium on testing, August 1985 to February 1987.

[f] The totals in brackets include the explosions revealed by the Soviet authorities in September 1990, the exact years of which have still not been announced. See *SIPRI Yearbook 1991*, p. 41.

Source: Ferm, R., *World Armaments and Disarmament*, Appendix 6A, *SIPRI Yearbook 1993*, Oxford University Press, Oxford, 1993. With permission.

TABLE 2.6
Estimated Total Number of Nuclear Explosions 16 July 1945 to 31 December 1992

U.S.[a]	U.S.S.R.[b]	U.K.[a]	France	China	India	Total
942	648 (715)	44	192	38	1	1865 (1931)[b]

[a] All British tests from 1962 have been conducted jointly with the U.S. at the Nevada Test Site. Therefore, the number of U.S. tests is actually higher than indicated here.

[b] The figures in brackets for the former Soviet Union include additional tests announced by the Soviet authorities in September 1990 for the period 1949 to 1990. See *SIPRI Yearbook 1991,* p. 41.

Source: Ferm, R., *World Armaments and Disarmament,* Appendix 6A, *SIPRI Yearbook 1993,* Oxford University Press, Oxford, 1993. With permission.

TABLE 2.7
Average and Range of Radioactivity in the North Atlantic Ocean in the Early 1970s Due to Radioactive Fallout From Nuclear Weapons Tests

	Radioactivity (pCi/l)	
Radioisotope	**Average**	**Range**
Tritium (^3H)	48	(31–74)
^{137}Cs	0.21	(0.03–0.80)
^{90}Sr	0.13	(0.02–0.50)
^{14}C	0.02	(0.01–0.04)
^{239}Pu		(0.0003–0.0012)

Source: Clark, R. B., *Marine Pollution,* 2nd ed., Clarendon Press, Oxford, 1989. With permission.

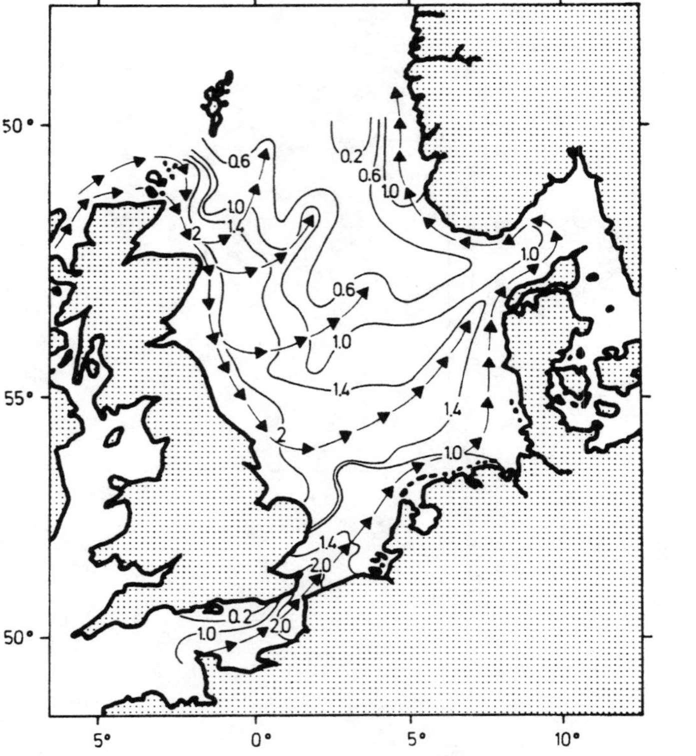

FIGURE 2.4. ^{137}Cs distribution in the North Sea resulting from waste discharge of the French processing plant La Hague at Cherbourg and from the British processing plant Windscale on the Irish Sea. (From Gerlach, S. A., *Marine Pollution: Diagnosis and Therapy*, Springer-Verlag, Berlin, 1981. With permission.)

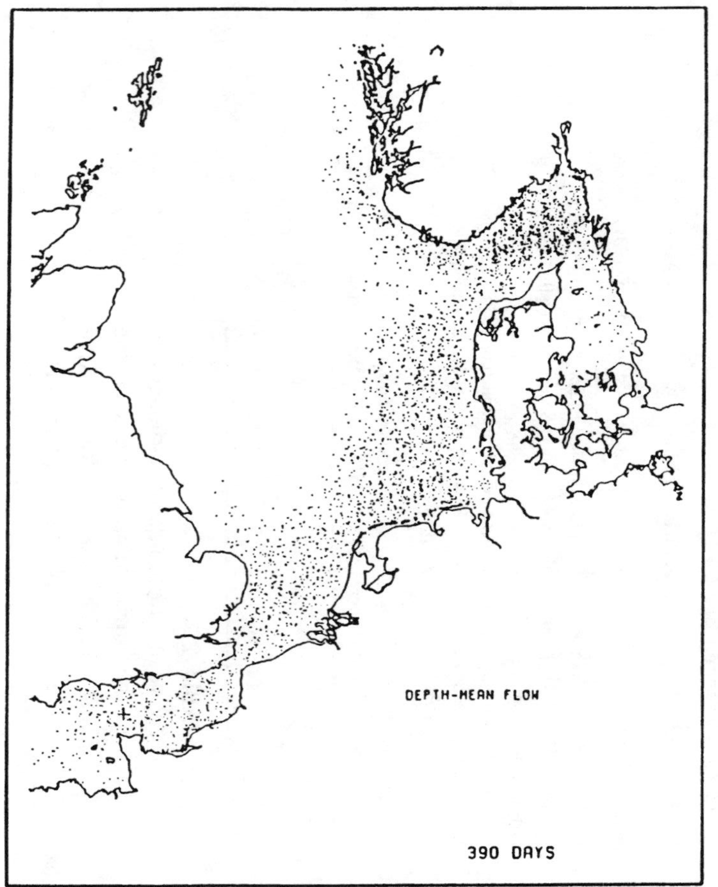

FIGURE 2.5. Particle pattern for continuous discharge (one particle per day) at the nuclear reprocessing plant at Cap de la Hague for a particle "age" (time after release) of 390 days; period: 1971 to 1981. (From Backhaus, J. O., in *Modeling Marine Systems*, Vol. 1, Davies, A. M. (Ed.), CRC Press, Boca Raton, FL, 1990, 132. With permission.)

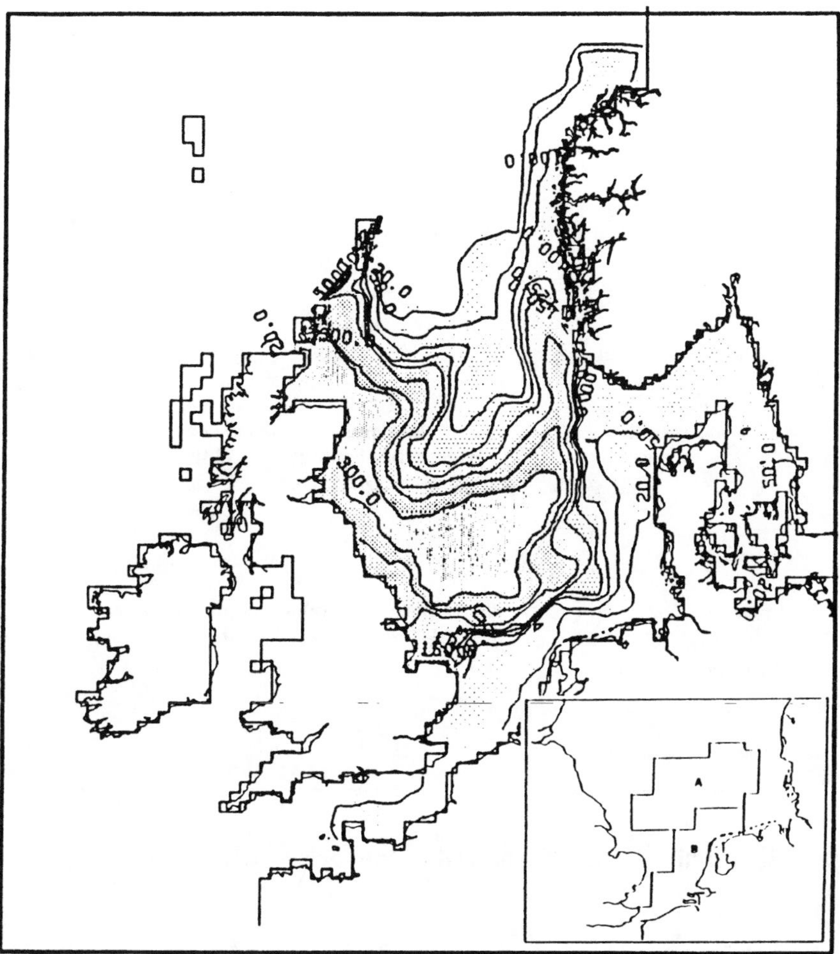

FIGURE 2.6. Spatial distribution of simulated concentrations (Bq/m^3) of Cs in the North Sea for February 1978. (From Backhaus, J. O., in *Modeling Marine Systems,* Vol. 1, Davies, A. M. (Ed.), CRC Press, Boca Raton, FL, 1990, 124. With permission.)

APPENDIX 3. RADIOACTIVE WASTE

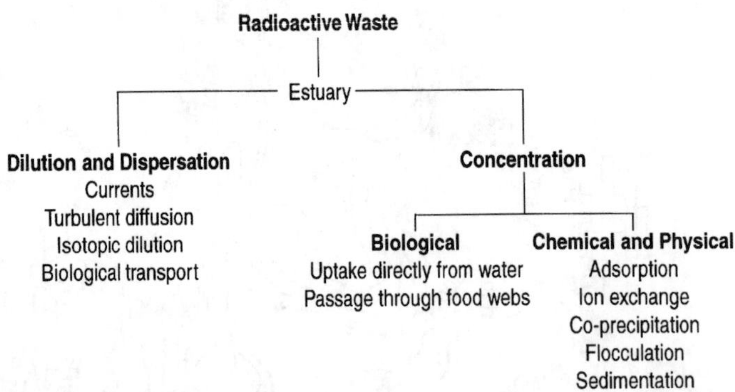

FIGURE 3.1. Processes tending to concentrate, disperse, and dilute radioactive waste in estuaries. (From Langford, T. E., *Electricity Generation and the Ecology of Natural Waters,* Liverpool University Press, England, 1983. With permission.)

TABLE 3.1
Inventory of Radioactive Waste Discharged in the Ocean (units = Bq)

	$^{239,240}Pu$	^{137}Cs	^{90}Sr	^{14}C	^{3}H
Total worldwide fallout by early 1970s	1.2×10^{16}	6.2×10^{17}	4.3×10^{17}	2.2×10^{17}	1.1×10^{20}
North Atlantic Ocean, early 1970s	2.3×10^{15}	1.2×10^{17}	8.5×10^{16}		2.4×10^{19}
Windscale discharge (1957–1978)	5.2×10^{14}	3.1×10^{16}	4.8×10^{15}		1.4×10^{16}
	Total α-Emitters		**Total β- and α-Emitters (Other than ^{3}H)**		
The NEA dumpsite (1967–1979)	3.1×10^{15}		9.5×10^{15}		9.7×10^{15}

Source: Needler, G. T. and Templeton, W. L., *Oceanus,* 24, 60, 1981. With permission.

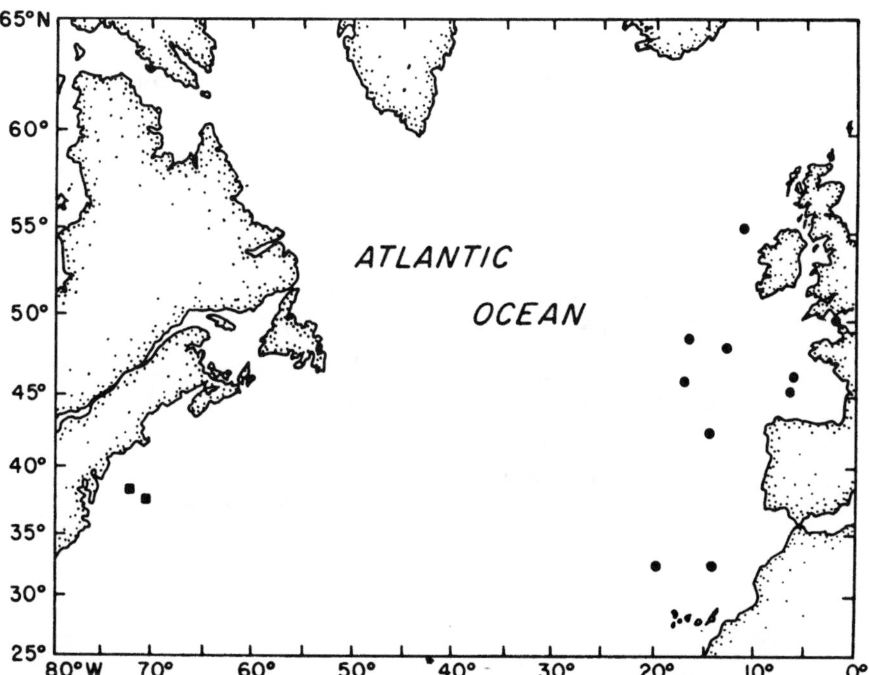

FIGURE 3.2. Radioactive solid waste disposal sites in the North Atlantic Ocean, 1946–1979. Darkened circles = European sites; darkened squares = U.S. sites. (Modified from Preston, A., in *Wastes in the Ocean,* Vol. 3, *Radioactive Wastes and the Ocean,* Park, P. K., Kester, D. R., Duedall, I. W., and Ketchum, B. H. (Eds.), John Wiley & Sons, New York, 1983, 107.)

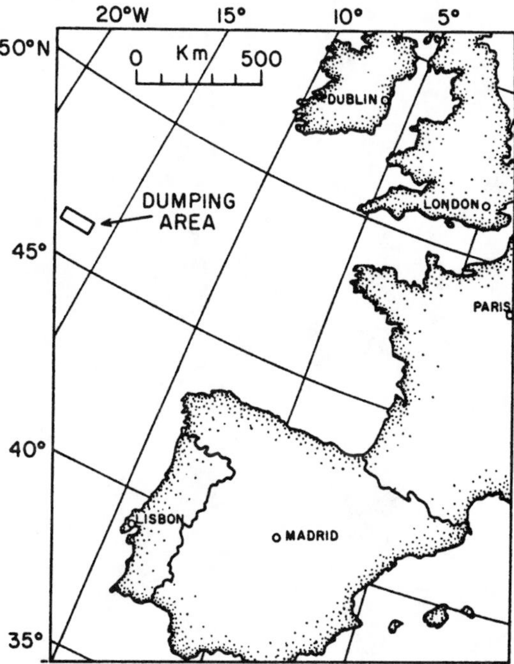

FIGURE 3.3. Location of the Northeast Atlantic low-level radioactive waste dumpsite area in the North Atlantic Ocean. (From Organization for Economic Cooperation and Development-Nuclear Energy Agency, Review of the Continued Suitability of the Dumping Site for Radioactive Waste in the Northeast Atlantic, Organization for Economic Cooperation and Development-Nuclear Energy Agency, Paris, 1980.)

TABLE 3.2
Activity at 12 Radioactive Solid Waste Dumpsites in the North Atlantic Ocean, 1949 to 1979

	Dumped Activity (Bq)	
	α	β–γ
Average per site[a]	3.7×10^{13}	2.6×10^{15}
Range per site[b]	$(1.1 \text{ to } 2.8) \times 10^{17}$	$(1.9 \text{ to } 3.7) \times 10^{17}$
Maximum Bq (per site per year)	5.2×10^{13}	3.5×10^{15}
Total for all 12 sites	4.8×10^{14}	3.1×10^{16}

[a] The sites used for more than 4 years.

[b] Range of years the sites were used was 1 to 14.

Source: Preston, A., in *Wastes in the Ocean,* Vol. 3, *Radioactive Wastes and the Ocean,* Park, P. K., Kester, D. R., Duedall, I. W., and Ketchum, B. H. (Eds.), John Wiley & Sons, New York, 1983, 107. With permission.

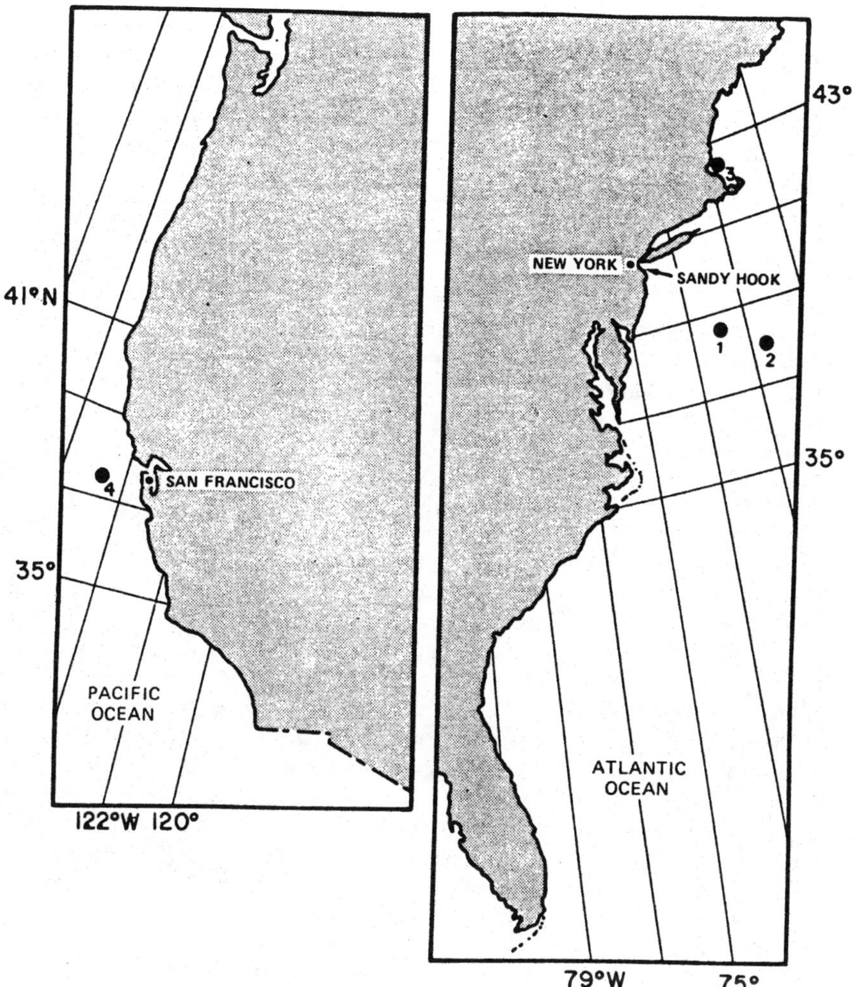

FIGURE 3.4. Major inactive low-level radioactive waste dumpsites of the U.S. (1) 2800-m site, (2) 3800-m site, (3) Massachusetts Bay site, and (4) Farallon Island site. (From U.S. General Accounting Office, *Hazards of Post Low-Level Radioactive Waste Ocean Dumping Have Been Overemphasized,* Report EMD-82–9, U.S. General Accounting Office, Washington, D.C., 1981.)

TABLE 3.3
U.S. Dumpsites Which Have Received Packaged Radioactive Wastes

Distance from Land (km)	Central Coordinates	Depth (m)[a]	Activity (Bq)[b]	Number of Containers	Years Licensed
Pacific Ocean					
80.5 W of San Francisco	37°38′N:123°08′W	896–1700	5.4×10^{14}	47,500	1946–1970
32 NE of Honolulu	21°28′N:157°25′W	3500	3×10^{10}	39	1959–1960
483 N of Midway Island	34°58′N:174°52′W	5490	5.2×10^{11}	7	1960
56 SW of Port Hueneme	33°40′N:119°35′W	1830–1940	4×10^{12}	3114	1946–1962
370 W of California-Oregon border	42°12′N:129°31′W	3294	3.5×10^{10}	26	1955–1958
56 W of California-Oregon border	42°04′N:125°01′W	4099	3×10^{9}	4	1960
306 NW of California-Oregon border	43°52′N:127°44′W	2928	3×10^{9}	2	1960
1600 WSW of Los Angeles	30°43′N:139°05′W	3660–4560	3.5×10^{10}	26	1955–1958
1290 SW of San Francisco	28°47′N:135°00′W	3477	4×10^{10}	29	1955–1960
322 W of San Diego	32°00′N:121°30′W	2210–3660	1.3×10^{12}	4415	1959–1962
1290 WNW of San Francisco	40°07′N:135°24′W	1830–1990	8×10^{9}	29	1960
564 NW of Cape Flattery	50°56′N:136°03′W	3294	3.6×10^{12}	1197	1958–1966
564 NW of Cape Flattery	52°25′N:140°12′W	3294	1×10^{12}	163	1962–1969
North Pacific	51°30′N:136°31′W	—	2×10^{10}	38	1946–1962
North Pacific	52°05′N:140°00′W	—	2×10^{10}	41	1946–1962
North Pacific	47°00′N:138°54′W	—	3.6×10^{12}	361	1946–1962
Atlantic Ocean					
Massachusetts Bay	42°25′N:70°35′W	92	9.0×10^{13}	4008	1952–1959
129 E of Cape Henry	36°56′N:74°23′W	1830–1967	3.2×10^{12}	843	1949–1967
225 SE of Sandy Hook	38°30′N:72°06′W	1830–2800	2.8×10^{15}	14,301	1951–1956 1959–1962
354 SE of Sandy Hook	37°50′N:70°35′W	1830–3800	7.8×10^{13}	14,500	1957–1959
350 E of Charleston	31°32′N:76°30′W	915–3660	2.3×10^{10}	119	1955–1962
Central Atlantic	36°20′N:43°49′N 45°00′W	3660–5289	1.8×10^{13}	432	1959–1960
Gulf of Mexico					
274 from New Orleans	27°14′N:89°33′W	1930	4×10^{11}	1	1958
402 from Appalachicola	25°40′N:85°17′W	3111	7×10^{7}	78	1955–1957

[a] Absence of data indicated by dash denotes data not available.
[b] Activity given in estimated quantities at time of disposal.

Source: Mattson, R. J., Statement before the Subcommittee on Oceanography of the Committee on Merchant Marine and Fisheries, 20 November 1980, U.S. House of Representatives, Washington, D.C.

TABLE 3.4
Radioactive Wastes Dumped at the Northeastern Atlantic Dumpsite, 1967 to 1980

Year	Gross Weight (tons)	Approximate Radioactivity (Bq)[a]		
		α	β–γ	^{3}H
1967	10,900	9×10^{12}	2.81×10^{14}	—
1968	1700[b]	1.1×10^{13} [b]	7.0×10^{13} [b]	—
1969	9180	1.9×10^{13}	8.55×10^{14}	—
1970	1700[b]	1.1×10^{13} [b]	7.0×10^{13} [b]	—
1971	3970	2.3×10^{13}	4.15×10^{14} [c]	—
1972	4130	2.5×10^{13}	8.0×10^{14} [c]	—
1973	4350	2.7×10^{13}	4.67×10^{14} [c]	—
1974	2270	1.6×10^{13}	—	3.7×10^{15}
1975	4460	2.9×10^{13}	1.13×10^{15}	1.10×10^{15}
1976	6770	3.3×10^{13}	1.20×10^{15}	7.8×10^{14}
1977	5600	3.5×10^{13}	1.34×10^{15}	1.18×10^{15}
1978	8040	4.1×10^{13}	1.59×10^{15}	1.56×10^{15}
1979	5400	5.2×10^{13}	1.52×10^{15}	1.57×10^{15}
1980	8400	7.0×10^{13}	4.67×10^{15}	3.87×10^{15}

Note: 1967 to 1979 data from Organization for Economic Cooperation and Development-Nuclear Energy Agency, Review of the Continued Suitability of the Dumping Site for Radioactive Waste in the Northeast Atlantic, Organization for Economic Cooperation and Development-Nuclear Energy Agency, Paris, 1980. 1980 data from Inter-Governmental Maritime Consultative Organization, Report of the Fifth Consultative Meeting of Contracting Parties to the Convention on the Prevention of Marine Pollution by Dumping of Wastes and Other Matter, Inter-Governmental Maritime Consultative Organization, London, 1980.

[a] Absence of data indicated by a dash denotes no information available.
[b] Assumed U.K. yearly average based on 5-year period.
[c] Includes ^{3}H.

TABLE 3.5
Summary of Radioactive Ocean-Dumping Operations, 1946 to 1979

Responsible Authority and Dates	Location	Details
United States (1946–1970)	Farallon Islands; Massachusetts Bay; off mid-Atlantic states	4.3×10^{15} Bq total in about 107,000 drums; mainly at these four Atlantic and Pacific Ocean sites
Japan (1955–1969)	Shallow, nearshore sites about 40 km from mouth of Tokyo Bay	
United Kingdom (pre-1967)	Deep areas of North Atlantic, southwest of England	1.8×10^{15} Bq (other western European countries also dumped during this period)
Republic of Korea (1968–1973)	Territorial sea	
Nuclear Energy Agency and its members (post-1967)	Northeastern Atlantic dumpsite	3.1×10^{15} Bq in 1979; total dumping in excess of 1.9×10^{16} Bq

Source: Modified from Finn, D. P., in *Wastes in the Ocean,* Vol. 3, *Radioactive Wastes and the Ocean,* Park, P. K., Kester, D. R., Duedall, I. W., and Ketchum, B. H. (Eds.), John Wiley & Sons, New York, 1983, 65.

TABLE 3.6
Sources and Concentrations of Different Types of Radioactive Waste

Low-Level Waste

Low-level wastes are produced in large volumes by all the various medical, industrial, scientific, and military applications of radioactivity. They include contaminated solutions and solids, protective, cleaning and decontamination materials, laboratory ware, and other equipment. They also include gases and liquids operationally discharged from power stations and other facilities. It has been estimated that during the period 1980 to 2000 about 3.6×10^6 m^3 of such wastes will be produced, some of which is sufficiently low in activity to be directly discharged into the environment, either with or without prior dilution or chemical treatment. Typical maximum activity concentrations in low-level waste are 4 GBq t^{-1} (alpha) and 12 GBq t^{-1} (beta and gamma). Much low level waste is currently disposed of by shallow burial in landfill sites, often with the co-disposal of other, nonradioactive, controlled wastes or by discharge into surface waters in rivers, lakes, estuaries, or coastal seas or by discharge into the atmosphere. If the environmental biophysicochemical behavior of the radionuclides in question and their possible pathways back to man are well understood then it is possible to make reliable estimates of the likely resultant doses of radioactivity to those most exposed in the population. If these doses are suitably low then the disposal methods may be deemed to be acceptable. However, at the present time there are many uncertainties in the understanding of such behaviour, pathways, and doses; nevertheless, the very large volumes of low level waste being generated will, through lack of economically and environmentally viable alternatives, continue to be disposed of in these relatively uncontrolled ways.

Intermediate-Level Waste

Intermediate-level wastes are sufficiently active to prevent their direct discharge into the environment, with maximum specific activities of typically 2×10^{12} Bq m^{-3} (α) and 2×10^{-14} Bq m^{-3} (β and γ). They comprise much of the solid and liquid wastes generated during fuel reprocessing, residues from power station effluent plants, and wastes produced by the decommissioning of nuclear facilities. Very large quantities (~2000 t y^{-1}) of intermediate- and low-level wastes have been disposed of by dumping in deep ocean waters in the NE Atlantic. Although some authorities still consider this method of disposal to be the best practicable environmental option for these categories of waste, it is no longer practiced. Intermediate-level waste produced in the U.K. is now stored on land, mainly at Sellafield, awaiting further policy decisions.

High-Level Waste

High-level wastes mainly consist of spent fuel and its residues and very active liquids generated during fuel reprocessing. Typical maximum activities are 4×10^{14} Bq m^{-3} (α) and 8×10^{16} Bq m^{-3} β and γ). At present such wastes generated in the U.K. are stored at Sellafield in storage ponds where it is proposed they will be vitrified prior to further storage (to allow the decay of shorter lived nuclides) and finally disposal in deep repositories. No such repository yet exists, but deep mines and boreholes on land and sea as well as other more exotic solutions including extraterrestrial disposal have all been proposed.

Source: Hewitt, C. N., in *Pollution: Causes, Effects, and Control,* 2nd ed., Harrison, R. M. (Ed.), Royal Society of Chemistry, Cambridge, 1990, 343. With permission.

APPENDIX 4. RADIONUCLIDES IN ESTUARINE AND MARINE ORGANISMS

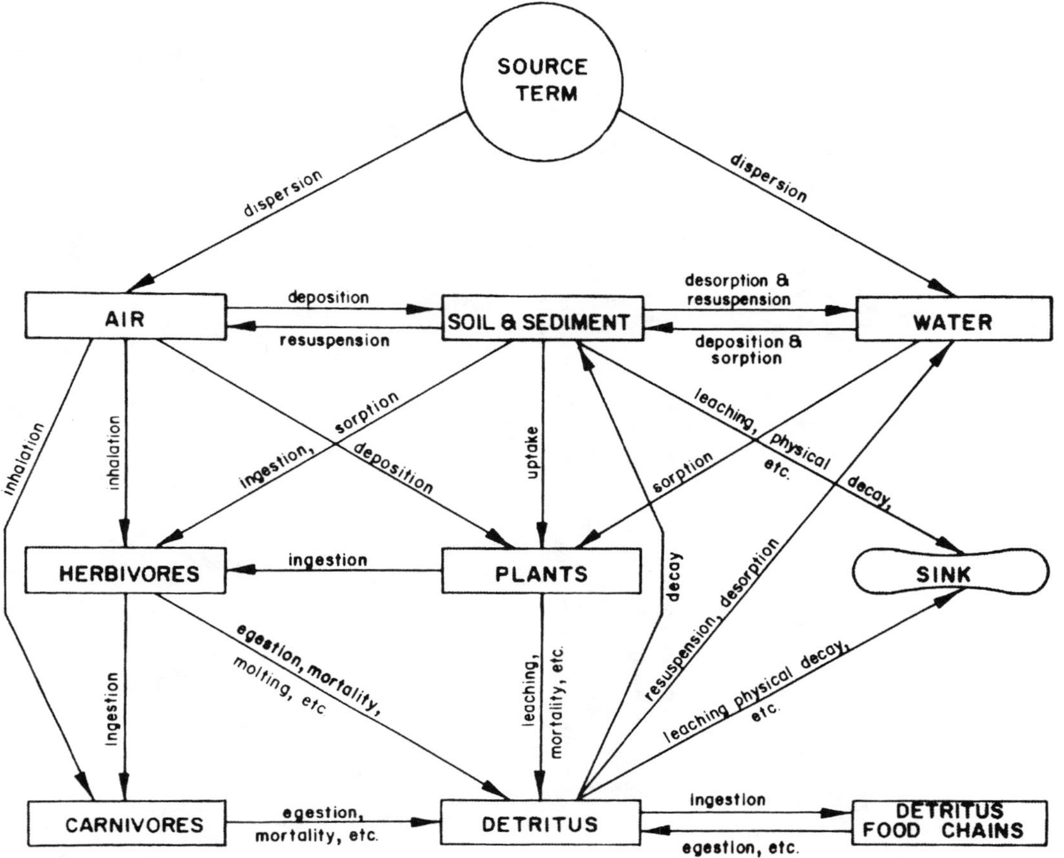

FIGURE 4.1. General transport processes operating on radionuclides in ecosystems. Boxes define ecosystem components, and arrows depict the flow of the materials through functional processes. (From Whicker, F. W. and Schultz, V., *Radioecology: Nuclear Energy and the Environment,* Vol. 1, CRC Press, Boca Raton, FL, 1982. With permission.)

TABLE 4.1
Some Fission Product Radionuclides of Potential Biological Importance

Radionuclide	Fission Yield (%)[a]	Radiation	Half-Life	Important Element Analogs
^{3}H	0.01	β	12 years	H
^{85}Kr	0.29	β, γ	10 years	—
^{90}Sr	5.77	β	28 years	Ca
^{89}Sr	4.79	β	51 days	Ca
^{137}Cs	6.15	β, γ	27 years	K
^{131}I	3.1	β, γ	8.1 days	I
^{129}I	0.9	β, γ	1.7×10^{7} years	I
^{144}Ce[b]	6.0	β, γ	285 days	—
^{103}Ru[b]	3.0	β, γ	40 days	—
^{106}Ru[b]	0.38	β, γ	1.0 years	—
^{95}Zr[b]	6.2	β, γ	65 days	—
^{140}Ba[b]	6.32	β, γ	12.8 days	Ca
^{91}Y	5.4	β, γ	58 days	—
^{141}Ce[b]	5.7	β, γ	33 hours	—
^{147}Nd[b]	2.7	β, γ	11 days	—

[a] Based upon thermal neutron fission of ^{235}U.

[b] Decay to radioactive daughters.

Source: Whicker, F. W. and Schultz, V., *Radioecology: Nuclear Energy and the Environment,* Vol. 1, CRC Press, Boca Raton, FL, 1982. With permission.

TABLE 4.2
Estimates of Annual Doses (mrad/year) Received by Marine Organisms From Natural Sources of Radiation

Source	Taxonomic Group	Marine (20-m depth)
Cosmic	Phytoplankton	4.4
	Zooplankton	4.4
	Mollusca	4.4
	Crustacea	4.4
	Fish	4.4
Water	Phytoplankton	3.5
	Zooplankton	1.8
	Mollusca	0.9
	Crustacea	0.9
	Fish	0.9
Sediment (β^+)	Phytoplankton	0
	Zooplankton	0
	Mollusca	27–324
	Crustacea	27–324
	Fish	0–324
Internal	Phytoplankton	17–64
	Zooplankton	23–138
	Mollusca	65–131
	Crustacea	69–188
	Fish	24–37
Sum of natural sources	Phytoplankton	25–72
	Zooplankton	29–168
	Mollusca	97–460
	Crustacea	101–517
	Fish	29–366

Source: Whicker, F. W. and Schultz, V., *Radioecology: Nuclear Energy and the Environment,* Vol. 1, CRC Press, Boca Raton, FL, 1982. With permission.

TABLE 4.3
Comparison of Annual Doses Received by Marine and Freshwater Organisms from Natural Sources of Radiation (mGy/year)

Freshwater Organisms

	Phytoplankton	Zooplankton	Mollusca	Crustacea	Fish
Cosmic	0.24	0.24	0.19	0.19	0.19–0.24
Water	6.4×10^{-4}–0.54	9×10^{-5}–7.4×10^{-2}	4×10^{-5}–3.1×10^{-2}	4×10^{-5}–3.1×10^{-2}	4×10^{-5}–6.1×10^{-2}
Sediments ($\beta + \gamma$)	0	0	0.27–3.2	0.27–3.2	0–3.2
Internal	—	—	—	—	0.32–0.42
Total	0.24–0.78	0.24–0.31	0.46–3.5	0.46–3.5	0.51–4.0

Marine Organisms

	Phytoplankton	Zooplankton	Mollusca	Crustacea	Fish
Cosmic	4.4×10^{-2}	4.4×10^{-2}	4.4×10^{-2}	4.4×10^{-2}	4.4×10^{-2}
Water	3.5×10^{-2}	1.8×10^{-2}	9×10^{-3}	9×10^{-3}	9×10^{-3}
Sediments ($\beta + \gamma$)	0	0	0.27–3.2	0.27–3.2	0–3.2
Internal	0.17–0.64	0.23–1.4	0.65–1.3	0.69–1.9	0.24–0.37
Total	0.24–0.72	0.29–1.7	0.97–4.6	1.0–5.2	0.29–3.7

Source: International Atomic Energy Agency, *Effects of Ionizing Radiation on Aquatic Organisms and Ecosystems,* Tech. Rep. Ser. No. 172, International Atomic Energy Agency, Vienna, 1976.

TABLE 4.4
Concentrations (×10^{-3} Bq/g) of Po in Deep Water and Comparable Shallow Water or Coastal Organisms

Sample	Deep Water	Shallow or Coastal Water
Teleost fish		
Muscle	0.27–8.62	0.06–7.29
Liver	6.36–195	10.4–481
Bone	1.52–4.85	0.12–19.6
Gonad	1.44–41	—
Elasmobranch fish		
Muscle	0.11–0.48	0.19–2.29
Liver	0.1–0.84	4.11–17.9
Cartilage	1.52–3.66	0.55–1.71
Gonad	0.09–0.27	—
Amphipods		
Gill	37.6	—
Viscera and muscle	19.6	—
Exoskeleton	7.6	—
Whole body	—	58.8
Sea cucumbers (eviscerated)	13.3	—

Source: Woodhead, D. S. and Pentreath, R. J., in *Wastes in the Ocean,* Vol. 3, *Radioactive Wastes and the Ocean,* Park, P. K., Kester, D. R., Duedall, I. W., and Ketchum, B. H. (Eds.), John Wiley & Sons, New York, 1983, 133. With permission.

TABLE 4.5
Radioactivity (Bq/kg wet weight) in Rattail Fish Taken in the Northeastern Atlantic Ocean

Sample Number	Sample Location and Identification	Sample Description	$^{239,240}Pu$	^{137}Cs	^{241}Am	^{55}Fe
Control Site						
6093	52°30.26′N–17°43.62′W Station 91-MAFF *Cirolana*-June 1980 Depth-4046 m	Flesh	0.0014 ± 0.0002	0.178 ± 0.005	0.00047 ± 0.0001	0.95 ± 0.02
6094	Same	Bone	0.007 ± 0.003	0.11 ± 0.03	0.0025 ± 0.0013	1.9 ± 0.1
Dumpsite						
6096	45°59.55′N–17°23.20′W Station 139-MAFF *Cirolana*-June 1980 Depth-4729m	Flesh	0.00070 ± 0.00015	0.175 ± 0.007	0.00030 ± 0.00013	1.60 ± 0.03
6097	Same	Bone	0.0013 ± 0.0005	0.18 ± 0.02	0.0005 ± 0.0007	3.22 ± 0.07

Note: Samples were obtained in June 1980, and sample analyses were carried out by H. Livingston, Woods Hole Oceanographic Institution, in April–May 1982. Since $t_{1/2}$ of ^{55}Fe is 2.7 y, its radioactivity is corrected to 1 June 1980. No ^{242}Cm, nor ^{244}Cm, was detectable, while ^{238}Pu measurements yielded the standard deviations as large as the measured values (not given). Sources of radionuclides: $^{239,240}Pu$ and ^{241}Am originate from nuclear fuel; ^{137}Cs is a fission product; and ^{55}Fe is an activation product.

Source: Park, P. K., Kester, D. R., Duedall, I. W., and Ketchum, B. H., in *Wastes in the Ocean,* Vol. 3, *Radioactive Wastes and the Ocean,* Park, P. K., Kester, D. R., Duedall, I. W., and Ketchum, B. H. (Eds.), John Wiley & Sons, New York, 1983, 481. With permission.

TABLE 4.6
Dose-Equivalent Rates (×10^{-6} Sv/h) to Certain Marine Organisms from ^{40}K and ^{210}Po Accumulated in Tissue

Nuclide	Sample	Deep Water	Shallow or Coastal Water
^{40}K	Fish, whole body	0.025	0.025
^{210}Po	Teleost fish		
	Muscle	0.017–0.54	0.0039–0.39
	Liver	0.40–12	0.65–30
	Bone	0.095–0.30	0.0074–1.2
	Gonad	0.090–2.6	—
	Elasmobranch fish		
	Muscle	0.0069–0.030	0.012–0.14
	Liver	0.0062–0.053	0.26–1.1
	Cartilage	0.095–0.23	0.034–0.11
	Gonad	0.0055–0.017	—
	Amphipods		
	Gill	2.3	—
	Viscera and muscle	1.2	—
	Exoskeleton	0.47	—
	Whole body	—	3.7
	Sea cucumbers (minus viscera)	0.83	—

Source: Woodhead, D. S. and Pentreath, R. J., in *Wastes in the Ocean,* Vol. 3, *Radioactive Wastes and the Ocean,* Park, P. K., Kester, D. R., Duedall, I. W., and Ketchum, B. H. (Eds.), John Wiley & Sons, New York, 1983, 133. With permission.

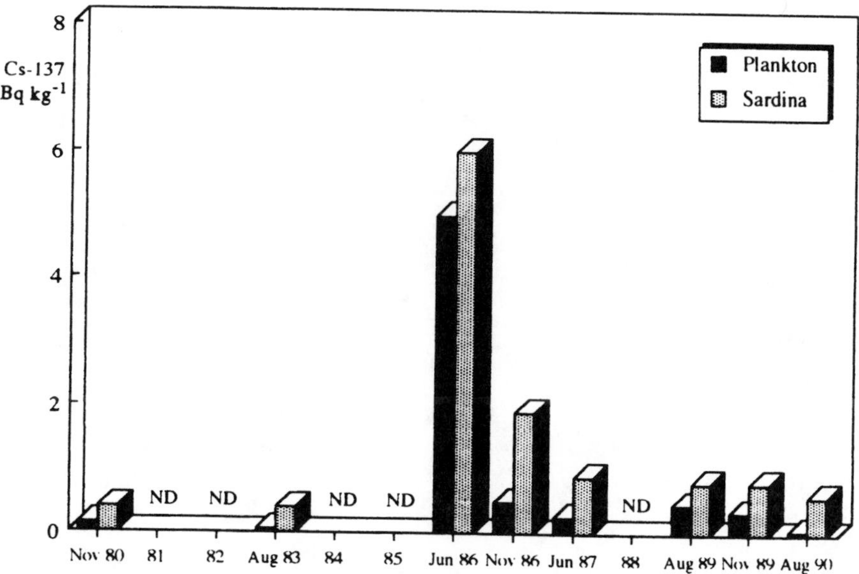

FIGURE 4.2. Temporal trends (1980 to 1990) of ^{137}Cs concentrations (Bq/kg wet weight) detected in samples of plankton and *Sardina pilchardus* collected near the Po Delta, Adriatic Sea. (From Marzano, F. N. and Triulzi, C., *Mar. Pollut. Bull.*, 28, 244, 1994. With permission.)

TABLE 4.7
Concentration Factors of ^{137}Cs Measured in Plankton and Soft Bottom Benthos of the Adriatic Sea During 1989 and 1990

Sample	CF (Bq/kg wet weight per Bq/l)
Transects facing the Po River delta (101 and 201)	
Mixed plankton	29–152
Poriphera sp.	140
Chlamys opercularis	100–229
Pinna pectinata	113–170
Ophiura ophiura	143
Astropecten penthacantus	162
Trieste, May 1986	
Plankton	48
Mytilus edulis	23
Patella coerulea	100
Recommended by IAEA (1985)	
Zooplankton	30
Mollusca	30

Source: Marzano, F. N. and Triulzi, C., *Mar. Pollut. Bull.*, 28, 244, 1994. With permission.

TABLE 4.8
Plutonium Concentrations (Bq/kg wet weight) in Marine Invertebrates of the Ravenglass Estuary Along the Irish Sea for the Years 1984 and 1986

	Date Collected	Concentration	
		$^{239/240}Pu$	^{238}Pu
Polychaetes			
Arenicola marina	10/1/84	72.4	16.3
Arenicola marina	4/2/86	60.5	13.8
Nereis diversicolor	10/1/84	48.6	11.0
Nereis diversicolor	8/27/86	25.6	6.1
Mollusks			
Hydrobia ulvae	6/2/86	107.2	27.6
Littorina littorea	5/5/86	49.3	11.0
Cerastoderma edule	10/27/86	19.9	4.7
Mytilus edulis	11/28/84	15.5	4.0
Mytilus edulis	4/2/86	6.0	1.4
Macoma balthica and *Tellina tenuis*	11/1/84	35.9	8.3
	10/27/86	20.1	5.1
Scrobularia plana	5/5/86	89.3	22.5
Crustaceans			
Corophium volutator	9/12/84	77.9	18.9
Corophium volutator	9/26/86	16.9	3.9
Carcinus maenas	7/30/86	10.2	2.4
Crangon crangon (R. Irt)	8/27/86	6.8	1.6
Crangon crangon (R. Esk)	8/28/86	1.8	0.4

Source: Lowe, V. P. W., *Environ. Pollut.*, 73, 1, 1991. With permission.

TABLE 4.9
Gamma-Emitting Radionuclides in Marine Invertebrates at Ravenglass (Bq/kg wet weight), Pre- and Post-Chernobyl Deposition (May 2–5, 1986)

	Date Collected	^{40}K	^{60}Co	^{95}Zr	^{95}Nb	^{103}Ru	^{106}Ru	^{110}Ag	^{125}Sb	^{131}I	^{134}Cs	^{137}Cs	^{144}Ce	^{154}Eu	^{241}Am
Polychaetes															
Arenicola marina	10/1/84	307.1	20.9	101.6	322.6	13.6	1123.4	—	—	—	8.2	321.3	—	10.3	171.0
Arenicola marina	4/2/86	206.8	8.8	—	4.1	—	105.3	—	—	—	1.6	132.5	—	—	93.5
Arenicola marina	5/5/86	161.9	7.0	—	9.2	407.5	167.5	11.2	—	776.0	63.6	224.7	—	—	—
Nereis diversicolor	10/1/84	148.6	60.5	309.7	815.6	33.3	3717.8	—	—	—	—	358.3	—	—	—
Nereis diversicolor	8/27/86	22.7	4.3	1.9	3.6	4.5	73.9	2.6	—	—	1.7	41.4	2.5	—	37.7
Mollusks															
Hydrobia ulvae	6/2/86	36.9	16.0	—	—	213.5	325.4	77.1	—	—	13.8	139.4	—	—	—
Littorina littorea	5/5/86	61.4	16.3	—	45.1	2453.1	891.7	77.3	29.0	—	388.5	836.2	—	—	—
Cerastoderma edule	10/27/86	19.9	29.9	—	9.0	—	138.0	12.6	—	—	—	33.4	—	—	75.8
Mytilus edulis	11/28/84	49.9	—	25.7	77.3	4.7	421.8	—	—	—	—	27.3	—	—	—
Mytilus edulis	4/2/86	58.8	3.0	—	—	—	94.4	—	—	—	—	8.0	—	—	—
Mytilus edulis	5/5/86	88.4	—	—	—	168.7	123.6	13.3	—	—	10.5	30.6	—	—	—
Macoma balthica and *Tellina tenuis*	11/1/84	39.7	3.8	43.2	106.8	3.8	631.9	—	—	—	—	34.9	—	—	—
	10/27/86	—	—	—	4.0	4.1	96.1	—	—	—	—	23.5	—	—	39.3
Scrobicularia plana	5/5/86	56.6	32.5	—	25.0	292.3	1648.4	18.4	—	—	36.5	141.7	—	—	—
Crustaceans															
Corophium volutator	9/12/84	—	—	104.8	267.6	—	635.4	—	—	—	—	133.2	—	—	—
Corophium volutator	9/26/86	—	—	—	9.6	—	49.8	—	—	—	—	11.7	—	—	29.4
Carcinus maenas	7/30/86	0.8	5.5	—	—	4.2	32.4	16.0	—	—	1.4	17.5	—	—	13.6
Crangon crangon (R. Irt)	8/27/86	30.2	1.3	—	—	—	26.9	7.0	—	—	0.9	21.4	—	—	8.2
Crangon crangon (R. Esk)	8/28/86	64.0	—	—	—	—	—	13.9	—	—	—	27.9	—	—	1.8

Note: — = not detected.

Source: Lowe, V. P. W., *Environ. Pollut.*, 73, 1, 1991. With permission.

TABLE 4.10
Mean Radiocesium and Plutonium Concentrations (Bq/kg wet weight) in Tissues of Birds Collected During the Period 1980 to 1984

		^{137}Cs		^{134}Cs		$^{239/240}$Pu		^{238}Pu	
Species	Locality	Pectoral Muscle (N)	Liver (N)	Pectoral Muscle (N)	Liver (N)	Pectoral Muscle (N)	Liver (N)	Pectoral Muscle (N)	Liver (N)
Greylag goose (*Anser anser*)	Ravenglass	57.7 (1)	27.9 (1)	— (1)	— (1)	0.1 (1)	13.3 (1)	0.03 (1)	3.2 (1)
Pinkfoot goose (*Anser brachyrhynchus*)	Caerlaverock	1.2 ± 0.1 (2)	— (2)	— (2)	— (2)	0.01 ± 0.01 (2)	0.07 ± 0.01 (2)	<0.01 ± 0.01 (2)	<0.04 ± 0.001 (2)
Shelduck	Ravenglass	294.9 ± 55.6 (3)	214.9 ± 61.6 (3)	17.4 (1)	— (3)	0.67 ± 0.33 (3)	12.3 ± 1.9 (2)	0.15 ± 0.06 (3)	2.7 (1)
(*Tadorna tadorna*)	Flookburgh	148.4 ± 106.6 (3)	109.3 ± 70.9 (2)	11.9 (1)	— (3)	0.04 ± 0.03 (2)	0.19 ± 0.23 (3)	0.01 ± 0.01 (2)	0.05 ± 0.04 (3)
Wigeon	Ravenglass	157.8 ± 93.7 (4)	99.7 ± 36.2 (4)	— (4)	— (4)	0.56 ± 0.37 (4)	8.08 ± 5.53 (4)	0.16 ± 0.12 (3)	2.44 ± 1.25 (3)
(*Anas penelope*)	Caerlaverock	2.1 ± 1.6 (4)	— (4)	— (4)	— (4)	0.05 ± 0.04 (4)	0.58 ± 0.53 (3)	<0.01 ± 0.004 (4)	0.08 ± 0.02 (3)
	Flookburgh	27.1 ± 11.5 (2)	83.0 ± 125.5 (18)	— (2)	— (18)	0.01 ± 0.01 (2)	0.44 ± 0.31 (15)	<0.01 ± 0.003 (2)	0.12 ± 0.05 (15)
Mallard	Ravenglass	166.7 ± 5.0 (2)	126.1 ± 23.5 (2)	87 (1)	— (2)	— (1)	3.39 (1)	— (1)	1.11 (1)
(*Anas platyrhynchos*)	Flookburgh	(0)	280.3 ± 238.4 (2)	(0)	— (3)	(0)	0.56 (1)	(0)	0.12 (1)
Merganser (*Mergus serrator*)	Flookburgh	143.8 (1)	250.7 (1)	8.1 (1)	13.1 (1)	0.02 (1)	<0.04 (1)	<0.01 (1)	<0.04 (1)
Moorhen[a] (*Gallinula chloropus*)	Braystones	115.9 (1)	76.4 (1)	55.9 (1)	43.0 (1)	0.05 (1)	0.26 (1)	<0.02 (1)	<0.10 (1)
Black-headed gull (*Larus ridibundus*)	Ravenglass	13.8 ± 14.1 (8)	11.1 ± 9.1 (9)	— (8)	— (9)	0.32 ± 0.77 (9)	0.54 ± 0.67 (8)	<0.03 ± 0.03 (8)	0.17 ± 0.19 (8)
(whole chicks)	Ravenglass		24.9 (7)		0.8 (7)		0.47 (7)		0.09 (7)
	Norfolk	6.2 (6)	3.1 (6)	2.1 (6)	1.5 (6)	0.06 (6)	0.09 (6)	<0.01 (6)	<0.05 (6)
Great black-backed gull (*Larus marinus*)	Ravenglass	158.0 (1)	163.2 (1)	— (1)	— (1)	0.11 (1)	5.32 (1)	— (1)	1.11 (1)
Lesser black-backed gull (*L. fuscus*)	Ravenglass	9.0 (1)	— (1)	— (1)	— (1)	not analyzed	not analyzed	not analyzed	not analyzed
Herring gull (*L. argentatus*)	Flookburgh	155.8 ± 146.1 (2)	87.8 ± 74.8 (2)	8.2 ± 7.6 (2)	— (2)	0.12 ± 0.14 (2)	0.15 ± 0.04 (2)	0.02 ± 0.02 (2)	0.04 ± 0.01 (2)
Curlew	Ravenglass	140.3 ± 138.2 (3)	104.3 ± 120.2 (3)	— (3)	— (4)	0.34 ± 0.34 (4)	2.4 ± 2.5 (3)	0.09 ± 0.09 (4)	0.14 ± 0.12 (3)
(*Numenius arquata*)	Flookburgh	49.3 ± 69.0 (2)	98.9 ± 86.0 (5)	— (2)	— (6)	0.031 (1)	0.07 ± 0.01 (6)	<0.018 (1)	<0.05 ± 0.03 (5)

TABLE 4.10 (continued)
Mean Radiocesium and Plutonium Concentrations (Bq/kg wet weight) in Tissues of Birds Collected During the Period 1980 to 1984

		^{137}Cs		^{134}Cs		$^{239/240}Pu$		^{238}Pu	
Species	**Locality**	**Pectoral Muscle (N)**	**Liver (N)**	**Pectoral Muscle (N)**	**Liver (N)**	**Pectoral Muscle (N)**	**Liver (N)**	**Pectoral Muscle (N)**	**Liver (N)**
Bar-tailed godwit (*Limosa lapponica*)	Ravenglass	478.1 (1)	510.0 (1)	— (1)	— (1)	0.03 (1)	0.91 (1)	<0.02 (1)	0.17 (1)
Oystercatcher	Ravenglass	612.8 ± 34.0 (2)	463.1 ± 67.2 (2)	— (2)	— (2)	0.53 ± 0.3 (2)	4.1 ± 2.2 (2)	0.2 (1)	1.8 (1)
(*Haematopus ostralegus*)	Flookburgh	22.2 (1)	19.5 (1)	— (1)	— (1)	0.04 (1)	0.09 (1)	<0.01 (1)	0.04 (1)
Carrion crow	Ravenglass	161.7 ± 158.4 (4)	130.5 ± 80.1 (3)	14.5 ± 11.7 (2)	— (4)	0.17 ± 0.14 (3)	0.7 ± 0.6 (2)	<0.04 ± 0.03 (2)	<0.08 (1)
(*Corvus corone*)	Flookburgh	17.4 ± 4.2 (3)	8.0 ± 1.6 (2)	— (3)	— (3)	not analyzed	not analyzed	not analyzed	not analyzed
	Penton (OS Grid Ref. NY 454777)	— (2)	— (2)	— (2)	— (2)	not analyzed	not analyzed	not analyzed	not analyzed

[a] Collected 3 June 1986 (post-Chernobyl).

Source: Lowe, V. P. W., *Environ. Pollut.*, 73, 1, 1991. With permission.

TABLE 4.11
Levels of ^{134}Cs and ^{137}Cs in Bq/l (Milk) and Bq/kg (Muscle and Liver) from North Rona (Northeast Atlantic) Seals (Levels of ^{40}K Are Included for Reference)

Seal Ref.	Sample	^{137}Cs	^{134}Cs	^{40}K
A6	Milk	2.0		102.9
A6	Milk	1.66		89.1
A7	Milk	1.37		229.2
A9	Milk	3.9	0.4	90.1
B2	Milk	2.5		85.0
J2	Milk	4.64		181.9
J2	Milk	3.63		84.7
L5	Milk	3.3		106.0
L5	Milk	4.33		96.9
N0	Milk	2.3		76.5
N0	Milk	2.56		212.0
N5	Milk	3.36		147.9
N5	Milk	1.98		90.3
N7	Milk	1.13		90.5
R4	Milk	2.65		105.0
R5	Milk	4.8	0.7	94.5
R9	Milk	2.6		80.1
V2	Milk	3.72		215.9
V2	Milk	1.96		92.9
Mean values		2.91	0.55	121.1
Pup1	Muscle	27.5	2.78	171.1
Pup1	Liver	14.59	1.3	156.7
Female	Muscle	14.3	1.65	178.8
Mean values		18.8	1.91	168.9

Source: Anderson, S. S., Livens, F. R., and Singleton, D. L., *Mar. Pollut. Bull.*, 21, 343, 1990. With permission.

TABLE 4.12
Levels of ^{134}Cs and ^{137}Cs in Bq/l (Milk) and Bq/kg (Muscle and Liver) From Isle of Man (North Sea) Seals (Levels of ^{40}K Are Included for Reference)

Seal. Ref.	Sample	^{137}Cs	^{134}Cs	^{40}K
C0	Milk	2.56		94.6
F2	Milk	2.82		146.3
C1	Milk	1.44		87.5
C2	Milk	4.5		86.9
C3	Milk	2.38		84.0
C4	Milk	3.9		66.9
D0	Milk	3.06		88.2
D1	Milk	1.53		93.7
D2	Milk	3.3		113.0
D3	Milk	3.27		84.6
D4	Milk	3.6		89.7
Mean values		2.94		94.1
Pup2	Muscle	11.08	1.14	157.4
	Liver	6.36	0.64	116.1
Mean values		8.72	0.89	136.8

Source: Anderson, S. S., Livens, F. R., and Singleton, D. L., *Mar. Pollut. Bull.*, 21, 343, 1990. With permission.

TABLE 4.13
Levels of Actinides in Bq/l (Milk) or Bq/kg (Muscle and Liver) in Seal Tissues From the Northeast Atlantic Ocean and North Sea

	Seal Ref.	$^{239,240}Pu$	^{238}Pu	^{241}Am
Milk	R5	<0.3	<0.3	<0.3
Milk	R9	<0.2	<0.2	<0.2
Milk	B2	<0.2	<0.2	<0.2
Milk	A9	<0.2	<0.2	<0.3
Milk	D2	<0.3	<0.3	<0.2
Milk	C2	<0.2	<0.2	<0.3
Milk	C4	<0.3	<0.3	<0.3
Milk	D4	<0.2	<0.2	<0.2
Muscle and liver				
Rona pup muscle		2.25 ± 0.31	0.55 ± 0.17	<0.3
Rona pup liver		3.52 ± 0.38	1.18 ± 0.21	<0.3
Rona adult muscle		<0.5	<0.5	<0.3
May pup muscle		<0.5	<0.5	<0.3
May pup liver		<0.5	<0.5	<0.3

Source: Anderson, S. S., Livens, F. R., and Singleton, D. L., *Mar. Pollut. Bull.*, 21, 343, 1990. With permission.

TABLE 4.14
Concentration Factors (C.F.) for Different Classes of Marine Organisms

Element	Group	C.F. Range	Mean C.F.
Cs	Plants	17–240	51
	Mollusca	3–28	15
	Crustacea	0.5–26	18
	Fish	5–244	48
Sr	Plants	0.2–82	21
	Mollusca	0.1–10	1.7
	Crustacea	0.1–1.1	0.6
	Fish	0.1–1.5	0.43
Mn	Plants	2000–20,000	5230
	Mollusca	170–150,000	22,080
	Crustacea	600–7500	2270
	Fish	35–1800	363
Co	Plants	60–1400	553
	Mollusca	1–210	166
	Crustacea	300–4000	1700
	Fish	20–5000	650
Zn	Plants	80–2500	900
	Mollusca	2100–33,000	47,000
	Crustacea	1700–15,000	5300
	Fish	280–15,500	3400
Fe	Plants	300–6000	2260
	Mollusca	1000–13,000	7600
	Crustacea	1000–4000	2000
	Fish	600–3000	1800
I	Plants	30–6800	1065
	Mollusca	20–20,000	5010
	Crustacea	20–48	31
	Fish	3–15	10
Ce	Plants	120–4500	1610
	Mollusca	100–350	240
	Crustacea	5–220	88
	Fish	0.3–538	99
K	Plants	4–31	13
	Mollusca	3.5–10	8
	Crustacea	8–19	12
	Fish	6.7–34	16
Ca	Plants	1.8–31	10
	Mollusca	0.2–112	16.5
	Crustacea	0.5–250	40
	Fish	0.5–7.6	1.9
Cu	Plants	—	1000
	Mollusca	—	286
	Fish	0.1–5	2.55
Mo	Plants	12–42	23
	Mollusca	11–27	17
	Crustacea	8.9–17.3	13
	Fish	7.6–23.8	17

TABLE 4.14 (continued)
Concentration Factors (C.F.) for Different Classes of Marine Organisms

Element	Group	C.F. Range	Mean C.F.
Mn	Plants	15–2000	448
	Mollusca	1–3.6	2.2
	Crustacea	1–100	38
	Fish	0.4–26	6.6
Ar-Nb	Plants	170–2900	1119
	Mollusca	8–165	81
	Crustacea	1–100	51
	Fish	0.05–247	86

Source: Eisenbud, M., *Environmental Radioactivity,* 3rd ed., Academic Press, New York, 1987. With permission.

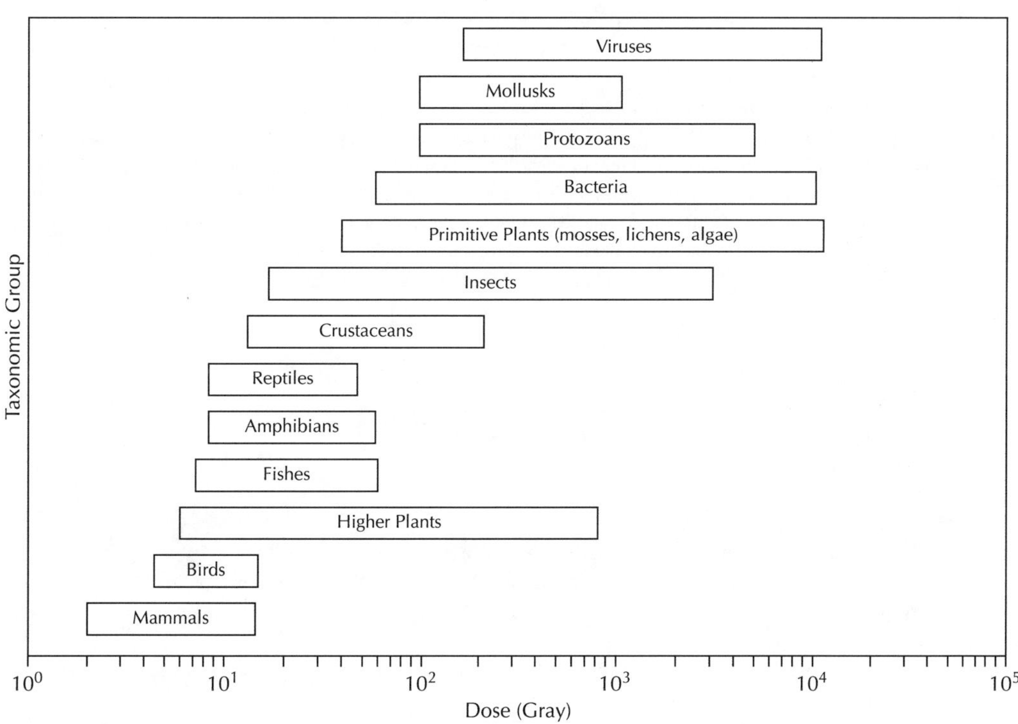

FIGURE 4.3. Ranges of acute lethal radiation doses to various taxonomic groups. (From Whicker, F. W. and Schultz, V., *Radioecology: Nuclear Energy and the Environment,* Vol. II, CRC Press, Boca Raton, FL, 1982. With permission.)

TABLE 4.15
Substances Controlled by the London Dumping Convention

Black List

1. Organohalogen compounds
2. Mercury and mercury compounds
3. Cadmium and cadmium compounds
4. Persistent plastics and other persistent synthetic materials, for example, netting and ropes, which may float or may remain in suspension in the sea in such a manner as to interfere materially with fishing, navigation, or other legitimate uses of the sea.
5. Crude oil, fuel oil, heavy diesel oil, lubricating oils, hydraulic fluids, and mixtures containing any of these, taken on board for the purpose of dumping.
6. High-level radioactive wastes or other high-level radioactive matter, defined on public health, biological, or other grounds, by the competent international body in this field, at present the International Atomic Energy Agency, as unsuitable for dumping at sea.
7. Materials in whatever form (such as solids, liquids, semi-liquids, gases, or in a living state) produced for biological and chemical warfare.
8. The preceding paragraphs of this annex do not apply to substances which are rapidly rendered harmless by physical, chemical, or biological processes in the sea provided they do not: (i) make edible marine organisms unpalatable, or (ii) endanger human health or that of domestic animals. The consultative procedure provided for under Article XIV should be followed by a Party if there is doubt about the harmlessness of the substance.
9. This Annex does not apply to wastes or other materials (such as sewage sludges and dredged spoils) containing the matters referred to in paragraphs 1 to 5 above as trace contaminants. Such wastes shall be subject to the provisions of Annexes II and III as appropriate.
10. Paragraphs 1 and 5 of the Annex do not apply to the disposal of wastes or other matter referred to in these paragraphs by means of incineration at sea. Incineration of such wastes or other matter at sea requires a prior special permit. In the issue of special permits for incineration the Contracting Parties shall apply the Regulations for the Control of Incineration of Wastes and Other Matter at Sea set forth in the Addendum to this Annex (which shall constitute an integral part of this Annex) and take full account of the Technical Guidelines on the Control of Incineration of Wastes and Other Matter at Sea adopted by the Contracting Parties in consultation.

Grey List

The following substances and materials require special permits, issued only according to the articles of the LDC.

A. Wastes containing significant amounts of the matters listed below:
   - Arsenic, lead, copper, zinc, and their compounds
   - Organosilicon compounds
   - Cyanides
   - Fluorides
   - Pesticides and their byproducts not covered in Annex I (Black List)

B. In the issue of permits for the dumping of large quantities of acids and alkalis, consideration shall be given to the possible presence in such wastes of the substances listed in paragraph A, and to beryllium, chromium, nickel, vanadium, and their compounds.

C. Containers, scrap metal, and other bulky wastes liable to sink to the sea bottom which may present a serious obstacle to fishing or navigation.

D. Radioactive wastes or other radioactive matter not included in Annex I (Black List). In the issue of permits for the dumping of this matter, the contracting parties should take full account of the recommendations of the competent international body in this field, at present the International Atomic Energy Agency.

E. In the issue of special permits for the inceration of substances and materials listed in this Annex, the Contracting Parties shall apply the Regulations for the Control of Incineration of Wastes and Other Matter at Sea set forth in the addendum to Annex I (Black List) and take full account of the Technical Guidelines on the Control of Incineration of Wastes and Other Matter at Sea adopted by the Contracting Parties in consultation, to the extent specified in these Regulations and Guidelines.

Source: Final Act of the London Dumping Convention, Annex I and II, Office of the London Dumping Convention, International Maritime Organization, London, 1972.

9 Dredging and Dredged-Spoil Disposal

INTRODUCTION

The U.S. Army Corps of Engineers (ACOE) maintains navigable waterways of the U.S. by dredging large volumes of sediment each year. According to Engler and Mathis,[1] annual amounts of sediment dredged from these waterways average about 290×10^6 m^3 in maintenance projects and approximately 78×10^6 m^3 in new projects. A substantial fraction of this dredged material is dumped in mounds at sea, although a considerable quantity also may be used as landfill for shoreline modification, wetland restoration, sanitary cover, and agricultural soil replenishment. Another important use of dredged spoils over the years has been in the creation of some 2000 estuarine islands which provide habitat for waterfowl and other organisms. During the past several decades, the trend has been less use of dredged material as fill for development because of concerns about the quality of the sediments. The most extensive use of dredged spoils today is in beach nourishment and restoration projects to protect shorelines. Since a large volume of sediment is removed from harbors, ports, and approach channels in proximity to industrial and urban centers, the dredged spoils often are contaminated by various types of pollutants, such as heavy metals, organochlorine compounds, PAHs, petroleum hydrocarbons, and other substances. To properly regulate and strictly limit adverse ecological impacts of ocean dumping of dredged materials, Section 103 of the Marine Protection, Research, and Sanctuaries Act of 1972 specifies that all dredged material targeted for dumping into ocean waters must be evaluated cooperatively by the ACOE and the U.S. Environmental Protection Agency (USEPA). Hence, all three phases of dredged material — solid, liquid, and suspended particulates — are carefully assessed for every proposed disposal operation to minimize adverse environmental effects.

While most investigations of dredging and dredged-spoil disposal operations focus on environmental impacts, they often ignore or downplay their ecological benefits. Dredging can adversely affect estuarine and marine environments in three principal ways: (1) by destroying bottom habitats, (2) by impairing water quality, and (3) by directly or indirectly impacting organisms, most notably those inhabiting the benthos.[2,3] Beneficial effects of dredging include: (1) increased nutrients that can enhance productivity of a system, (2) improved circulation in estuaries and embayments, (3) increased recreational and commercial usage of a waterbody, and (4) sediment supply for beach nourishment, landfill projects, and soil improvement.[4]

Dredging and dredged-spoil disposal may exert severe stresses on biotic communities. The most acute biotic impact of dredging is the physical removal of bottom-dwelling organisms from their natural habitat and subsequent translocation to spoil disposal sites. When the spoils and organisms are dumped at subaerial locations, mortality approaches 100%. Subaqueous disposal usually causes less mortality, although mechanical damage by the dredge or smothering by the sediment can injure or kill large numbers of organisms. Benthic communities recover slowly from dredging effects through recolonization and succession of populations on the altered benthic habitat.[5] However, the period of recovery varies greatly depending on the characteristics of the seafloor sediment, hydrography of the dredged site, types of organisms inhabiting the pre-dredged site, and other factors.

Dredged spoils comprise the largest single source of solid waste dumped at sea in U.S. and foreign waters. Most dredged-spoil dumpsites are located close to shore (10 to 50 km), which minimizes transportation costs. The suitability of dredged sediments for ocean disposal has become more strictly regulated during the past 25 years. Dredged spoils in the U.S. are assessed by means of liquid-phase, suspended particulate, and solid-phase bioassays, together with chemical analyses of the liquid phase. In addition, the USEPA requires a thorough physical, chemical, and biological assessment of all ocean disposal sites prior to their designation as final and acceptable dumpsites. For sediment posing high environmental risk due to contaminants, disposal methods typically stress containment. The management strategy for subaqueous disposal of the most contaminated dredged material entails the confinement of the spoils in a low-energy depression followed by covering (capping) with 1 to 2 m of clean sediment. The cap effectively immobilizes the contaminants. Dumpsites located in areas of low biological productivity, low-energy hydraulic regimes, and insignificant storm currents should have a low potential for creating adverse environmental conditions.[1] Open-water disposal of contaminated spoils may be the alternative of choice when the degree of contamination does not contribute to any ecological harm of an area. However, for most contaminated spoils, the disposal alternative usually chosen is that which provides some type of material containment.

DREDGING

Types of Dredges

There are two categories of dredges: mechanical devices and hydraulic devices. Mechanical dredges lift sediment from the sea floor and transport it to a disposal site. This category includes dipper, bucket, and ladder dredges. Hydraulic dredges first loosen sediment by means of agitation with water jets or cutterheads, thereby producing a slurry that can be pumped to a disposal area or stored in hoppers for later dumping at a designated location. Agitation, hopper, suction, and cutterhead dredges are types of hydraulic devices. These dredges are more efficient than mechanical systems.

The following descriptions of mechanical and hydraulic dredges are derived from Stickney[3] and Machemehl:[6]

Dipper dredge. This dredge, basically comprised of a power shovel mounted on a barge, is commonly employed to remove fractured rock or loose boulders, handling up to about 10 m^3 of hard material per cycle. Three large diameter pipes or spuds are activated to maintain stability of the barge during dredging. The discharge of dredged material occurs within reach of the dipper boom. Dipper dredges typically work in waters shallower than 20 m.

Bucket dredge. Most commonly used on small-scale dredging projects, a bucket dredge is essentially a crane mounted on a barge. Spuds or anchor lines provide stability for the barge during dredging operations. Depending on the type of material dredged, three types of buckets can be mounted (i.e., clamshell, orange peel, dragline). Dredged sediment is dumped within the length of the boom.

Ladder dredge. Unlike the dipper dredge, the ladder dredge consists essentially of an endless chain of buckets mounted on a barge. To maintain position during dredging, the barge utilizes side cables. The capacity of each bucket amounts to about 1 to 2 m^3 of dredged material which is dumped into chutes or onto belts and discharged over the side of the vessel. This dredge generally operates in waters less than 30 m.

Agitation dredge. The concept of the agitation dredge dates back more than 2000 years, making this device the oldest of all dredges. Incorporating aspects of both mechanical and hydraulic dredging systems, this type of dredging device consists of an object dragged along the sea floor in the presence of a water current which transports suspended sediment away from the dredged site.

Hopper dredge. This dredge, which often works in waters over 20 m, does not discharge spoils overboard via chutes or belts like the ladder dredge, but retains the spoils in a hopper for subsequent disposal at a designated site. In effect, the hopper dredge is a self-propelled vessel designed to dredge, load, and temporarily store spoils for hauling to designated disposal locations. The capacity of the hopper can exceed 10,000 m^3.

Hydraulic pipeline dredge. The most commonly used dredge is the hydraulic pipeline dredge, accounting for more than two thirds of all sediment dredged from the coastal zone.[7] This dredge can rapidly excavate substantial volumes of seafloor material, ranging from light silts to heavy rock. It employs a cutter head to loosen the sediment, with the dredged material typically pumped through floating pipes to disposal sites remote from the dredge itself. The hydraulic pipeline dredge operates in waters up to 20 m.

Dredging Effects

The most direct effect of dredging is the destruction of the benthic habitat. Sediments are removed from the sea floor, and some particle resuspension occurs due to the action of the dredge. Benthic organisms are excavated along with the dredged sediment. This action commonly results in mass mortality, thereby disrupting the entire benthic community until recolonization of the dredged site takes place. The recolonization process may be protracted, with recovery typically requiring 1 or more years to complete. Opportunistic pioneering fauna, such as the polychaete worms *Capitella* spp., initially occupy the site and are later supplanted by equilibrium assemblages of organisms in a successional sequence.[5,8] Species originally inhabiting the site usually return to the disturbed area. However, the rate of recolonization varies considerably depending on the geographical location, sediment composition, and type of organisms. Thus, Stickney and Perlmutter[9] reported that recolonization of a muddy bottom by benthic infauna occurred within days of dredging. This rapid rate of recolonization was largely attributed to the type of organisms inhabiting the area, principally short-lived species with relatively motile stages in their life cycles.

While the greatest immediate impacts on the benthic community and habitat are ascribable to sediment removal from the seabed, other adverse effects are attributable to redeposition of suspended sediment subsequent to dredging, increased turbidity at the dredging site, and the release of toxic substances concomitant with the roiling of sediments during the dredging process. Plants may be particularly susceptible to increased turbidity and redeposition of suspended particulates. When turbidity in the vicinity of the dredged site is high and light attenuation sufficiently reduced as a consequence, primary production by phytoplankton and benthic flora can decrease abruptly even beyond the immediate limits of the dredged site. However, these effects are ephemeral. For instance, phytoplankton primary production typically returns to pre-dredged levels shortly after dredging is terminated. Benthic macroflora appear to be less resilient, especially in areas where dredging strips most plant biomass and sediment redeposition and turbidity are elevated. For example, Godcharles[10] observed no recolonization of seagrass beds (i.e., *Thalassia testudinum* and *Syringodium filiforme*) for at least 1 year following hydraulic clam dredging in Florida.

In some cases, the release of nutrients from bottom sediments during dredging can enhance primary production, although this process may promote eutrophic conditions in those systems already characterized by high concentrations of nitrogen and phosphorus in the water column.[11] Dredged sediment often contains large concentrations of nutrient elements. For example, at the site of overboard dredging in Chesapeake Bay, Biggs[12] found that nitrogen and phosphorus levels increased 50 to 100 times. Windom[13] similarly documented increased ammonia at dredging sites in Charleston Harbor, SC, Cape Fear River, NC, and Terry Creek, GA; however, these elevated ammonia concentrations later declined due to plant uptake. Release of ammonia from sediments stimulates phytoplankton production and is followed by higher levels of pH, dissolved oxygen, and biological oxygen demand in estuarine waters.

Because bottom sediments in harbors and coastal waterways near highly populated and industrialized centers commonly are contaminated with heavy metals, chlorinated hydrocarbons, petroleum hydrocarbons, and other chemical compounds, dredging in these areas often releases the pollutants to the water column. Hence, Windom[2] recorded slightly higher mercury levels during dredging operations in Georgia estuaries. Hall[14] likewise detailed higher concentrations of heavy metals (i.e., cadmium, chromium, copper, iron, nickel, manganese, and lead) in the water column associated with episodes of dredging. However, observations at other estuarine dredging sites have revealed initial decreases in concentrations of heavy metals near the dredge, followed by gradual increases in concentrations.[13] These changes reflect the early removal of metals from the water column due to their sorption onto newly suspended particles. Later, the sediments release some of their heavy metal load, raising the metal concentration in the water.

Lunsford et al.,[15] analyzing the uptake of the organochlorine insecticide kepone in the wedge clam (*Rangia cuneata*) from the James River estuary, registered higher residue levels during dredging of contaminated sediments. Processes such as dissolution, diagenesis, resuspension, and microbial degradation influence the flux and residence times of contaminants released with sediments into the water column during dredging. As a result, it also affects their distribution and fate, especially in shallow estuaries and embayments.

At times, dredging violates water quality standards. For instance, hopper dredging in Chesapeake Bay has been shown to violate water quality standards of turbidity for recreation, aquatic life, and shellfishing established in Maryland in the near-field zone less than 300 m behind the dredge. Nichols et al.[16] differentiated two separate turbidity plumes associated with hopper dredging in the estuary: (1) a near-bottom plume produced by the agitation, cutting, and turbulence of the draghead as it moves through bottom sediments; and (2) an upper plume in near-surface and mid-depth water generated by the discharge of hopper overflow slurries. They discerned suspended sediment concentrations in the near-field zone ranging from 840 to 7200 mg/l, or 50 to 400 times above the normal background level, which persisted for less than 3 min at a fixed point. In the far field (>300 m behind the dredge), the amount of suspended sediment during the period of overflow discharge (40 to 60 min) averaged 81 mg/l in the upper water layer, or approximately 5 times the normal background. During an entire dredge-loading cycle (1.5 to 2.0 h), near-bottom water contained an average suspended sediment concentration of 137 mg/l, or 8 times the normal background.

Benthic communities in Chesapeake Bay appear to be minimally impacted by high suspended sediment loads in the near-field zone at dredging sites. Macrobenthic assemblages also exhibit few effects of the redeposition of the suspended material. There may be several explanations for these observations: (1) benthic communities are dominated by populations with high motilities, flexible life-history strategies, and short lifespans; (2) the rate of sediment redeposition from dredging plumes is not excessive for the survival of the biota; and (3) the dredged sediments contain little chemical contamination that is hazardous to the benthos.[17]

In contrast to potential impacts on benthic communities due to dredging, several beneficial effects also have been linked to this activity. For example, the deepening of waterways in estuaries often substantially improves circulation, resulting in more even distribution of temperature, salinity, dissolved oxygen, and other physical-chemical factors which may enable organisms to spread and colonize new habitat area. Finfish and shellfish migrations (e.g., flounders and shrimp) may be facilitated by improved circulation and increased depths of waterways. In addition, for some species the greater water depths can create important sanctuaries from low or high temperatures during winter and summer months, respectively. They also may generate additional spawning grounds for some species.

DREDGED-SPOIL DISPOSAL SITES

Because the bottom sediments of rivers, harbors, embayments, and estuaries are sinks for many chemical contaminants, as alluded to above, the disposal of these sediments subsequent to dredging is a major concern. Dredged material removed from these areas generally is high in organic matter and clay, both of which are biologically and chemically active. Sediment properties influencing the reaction of dredged material with contaminants include the amount and type of clay, organic matter content, amount and type of potentially reactive iron and manganese, and the oxidation-reduction, pH, and salinity status of the sediment.[1] When contaminated dredged sediment is dumped in marine waters, toxicity and food-chain transfers often are anticipated, particularly in biologically productive areas.[18] However, most documented chemical releases from subaqueous disposal sites are nutrients rather than highly toxic contaminants, and the principal impacts at marine disposal sites are coupled to physical rather than geochemical effects.[1] Nevertheless, the trend in recent years favors land disposal over subaqueous disposal of dredged spoils, owing to potential contaminant impacts on marine food chains and environments.

Land-based disposal alternatives, like open-water disposal, offer no panacea for the dumping of dredged spoils. Land disposal sites typically lie near highly productive nearshore areas. Prior to about 1960, these sites often were located on coastal marshes and wetlands, which constitute important nursery and breeding grounds for marine organisms and highly productive zones for estuarine and nearshore regions. Subaerial disposal in these habitats commonly changes the geochemistry of the dredged material, resulting in enhanced release potential of some of the chemical constitutents. Despite these problems, wetland disposal of dredged spoils has remained the only practical method in many cases.

Two types of sites are considered in open-water, dredged-spoil disposal: retentive and dispersive sites.[19] Low-energy hydrodynamic environments generally typify retentive sites, which are selected to ensure that the spoils remain confined to a localized area. However, when a low-energy hydrodynamic environment is lacking, an inerodible cap may be placed over the spoils to prevent their movement and to preclude the mobilization of contaminants from the dredged material into the water column. In spite of the design effectiveness of a sheltered site, both approaches can fail. Major storms roil bottom sediments even in the most protected estuaries and embayments, and the construction of an impenetrable protective cap over disposed sediments may not be achievable. Regardless, careful monitoring of capped disposal sites in ocean waters for extended periods of time has demonstrated conclusively that capping is technically feasible and successful under normal tidal and wave conditions.[20,21] Laboratory investigations have shown that a 50-cm cap of clay, silt, and sand precludes the transfer of contaminants from dredged spoils into biota inhabiting overlying waters, even when large numbers of bioturbating infauna inhabit the sediments.[21] Cap materials consisting mainly of clay and silt afford greater protection against contaminant transfer than those comprised principally of sand.

Where advection is high and the expectation exists for significant dispersal of dredged spoils, dumpsites are usually designated as dispersive sites. Selection of dispersive sites is largely predicated on the hydrodynamic energy regime of an area. However, other factors also must be evaluated in the selection process, notably the volume of dredged spoils, the method of disposal, bathymetry, and character of the sea floor. Field observations, laboratory tests, and numerical modeling studies are performed in the design of disposal management plans for both retentive and dispersive open-water disposal sites in order to minimize environmental impacts.[22] For example, a dispersive condition is maintained at the Alcatraz disposal site in San Francisco Bay due to constraints on dredged mud characteristics developed from laboratory tests on erosion rates as well as results obtained from numerical modeling of the dumping process. A numerical model also has been utilized in Corpus Christi Bay to determine how much dredged spoil returns to the navigation channel and to devise a location for disposal that will minimize that return. A model has likewise been employed in Puget Sound to ensure that most of the dredged spoils remain at the disposal site.[23]

Large volumes of dredged spoils are released at open-water sites after being transported by a hopper or barge, or discharged directly overboard of a sidecaster dredge. The effectiveness of sidecasting depends on the occurrence of cross-currents of sufficient strength to transport and disperse the sediment away from the dredged site. Numerical models, once again, have proven to be extremely useful in simulating the pattern of deposition resulting from sidecasting operations.

Upon release from a hopper or barge at open-water locations, dredged material descends as a dense, fluid-like jet through the water column. Three distinct transport phases or stages define the behavior of the released material.[24] Some of the dredged material consists of solid blocks or clods of very dense cohesive sediment. Site water is entrained in the jet as the sediment descends rapidly toward the sea floor. A fraction of it remains in the upper part of the water column and cannot be accounted for in a mass balance. Currents transport this low density material away from the disposal site. Since some of the suspended material may be contaminated, the potential degradation of water quality in clean areas removed from the impacted disposal location is a cause of concern. The descending jet and its core of cohesive material tend to collapse, most frequently when they strike the sea floor and much less commonly when the jet encounters a layer in the water column of equal or greater density than the jet. The principal adverse effects of dredged-spoil disposal at open-water sites on the overall water quality of an estuary or coastal area are mainly confined to the location of the descending jet of material.

The transport of contaminated sediment away from the descending jet of disposed material does not appear to be great. Mass balance calculations of dredged material released from hoppers and barges at open-water sites, for instance, indicate that typically less than 5% of the volume of sediment discharged from these vessels disperses away from the disposal site. Field measurements of the loss of dredged sediment released from hoppers or barges in Long Island Sound, New York Bight, San Francisco Bay, and Duwamish Waterway amount to 1, 3.7, 1 to 5, and 2 to 4%, respectively.[24] Hence, there is little potential for large-scale contamination of clean bottom sediments away from the immediate area of the dumpsite.

The U.S. Army Corps of Engineers has conducted considerable dredged-material research. This comprehensive research program has yielded substantial amounts of data on the environmental effects of dredging and dredged-spoil disposal operations.[25] In the case of dredge-spoil dumping, considerable research is now directed toward environmental impacts, either directly or indirectly, on the structure and function of biotic communities. In this respect, dredged-spoil disposal may be of even greater concern than dredging, with high levels of suspended solids, sediment buildup, and oxygen depletion at disposal sites generally creating adverse conditions for biotic communities. In particular, the presence of biostimulants and toxins, often chemically or physically sorbed within the sediment matrix, may pose a long-term water quality problem for the general area while directly threatening faunal and floral populations at the impacted site.

EVALUATIVE REQUIREMENTS

Section 103 of the Marine Protection, Research, and Sanctuaries Act of 1972 (MPRSA), Public Law 92-532, requires that all proposed operations involving the transportation and dumping of dredged material into ocean waters must be evaluated to determine their potential environmental impact. This is accomplished through the Secretary of the Army, using criteria developed by the Administrator of the USEPA. The U.S. ACOE is the permitting authority for dredged material, subject to USEPA review, as specified in Section 103 of the MPRSA. Environmental evaluations must be in accordance with applicable criteria published in Title 40, Code of Federal Regulations, Parts 220-228 (40 CFR 220-228). Proposed ocean disposal of dredged material also must comply with the permitting and dredging regulations found in Title 33 CFR, Parts 320-330 and 335-338.[26] All ocean dumping evaluation must consider the environmental impact; the general compatibility of dredged spoils with sediments at the disposal site; the need for ocean dumping, with a thorough review of alternatives; the impact on aesthetics, recreation, and economics; materials prohibited from disposal by international treaty; and the impacts on other uses of the oceans. The evaluations in ACOE regulations 33 CFR 209.120 and 33 CFR 209.145 must be applied.[1,27]

Dredged spoils that are not chemically contaminated must be evaluated only in terms of their compatibility with the disposal site. Other dredged material must be thoroughly tested for potential environmental impacts, not only for the presence of possible chemical contaminants, but also for possible biological effects. Three procedures may be applied, including: (1) elutriate testing, (2) bulk or total sediment analysis, or (3) bioassay tests (liquid-phase, suspended particulate phase, and solid-phase bioassays). The elutriate test measures the quantity of a substance exchanged between the sediment and the aqueous phase during dredged-spoil disposal and provides a basis for categorizing dredged spoils as polluted or unpolluted.

Bulk or total sediment analysis provides a basis for estimating mass loads of wastes to estuarine and marine environments and, therefore, a basis for assessing the amount of substances that possibly can alter the biology and chemistry of a disposal site. It places numerical limits on a dry weight basis on several pollution parameters (i.e., volatile solids, chemical oxygen demand, oil and grease, Kjeldahl nitrogen, mercury, lead, and zinc). If any one of these parameters exceeds the specified limits, the dredged spoils are considered to be polluted and unacceptable for open-water disposal. However, studies have shown that bulk sediment analyses do not predict short- or long-term release of contaminants and that no relationship exists between bulk sediment concentration and bioaccumulation. Hence, bulk sediment analyses do not provide an adequate assessment of water quality effects nor any level of environmental protection.[1]

Bioassay tests yield an operational measure of the toxicity of dredged material on organisms. Three bioassay tests are conducted: liquid-phase, suspended particulate phase, and solid-phase bioassays. These tests indicate that solid-phase material generally accounts for the greatest potential impact on benthic organisms at a disposal site.

Because of the long-term organism-sediment interaction taking place subsequent to dredged-spoil disposal, the solid-phase material must be evaluated for long-term bioaccumulation. Bioaccumulation tests are conducted to ensure that chemical constituents in dredged spoils do not exceed concentrations that will result in the accumulation of toxic substances in the human food chain.[18] Alternatively, any contaminants in the dredged material can be isolated from areas of biological activity by capping them with clean sediments.[1]

IMPACTS OF DREDGED MATERIAL DISPOSAL

A wide range of physical, chemical, and biological impacts may arise from dredged-spoil disposal. Chief among the physical effects of dredged-spoil disposal are increased turbidity, burial of benthic organisms, and changes in grain size or sediment type at the dumpsite.[28] Changes in water column turbidity associated with dredged-spoil disposal in open-water environments tend to be relatively short lived and to have minimal biological impact.[29] Of greater concern is the burial or smothering of organisms in a dredged-spoil mound. While many benthic organisms may not survive loading of the dredged material, some mobile species have the capability to migrate vertically through the sediments to recover from the overburden.[1] In addition, recolonization often proceeds rapidly with the cessation of dumping.

Increases in trace contaminants (e.g., PCBs, DDT, petroleum hydrocarbons, etc.) may have less of an effect on biotic communities than changes in redox potential (Eh) and pH accompanying sediment disposal. Changes in Eh and pH can have a significant effect on the chemical behavior of metals in marine systems as recounted by Kester et al.[28] Some metals are mobilized as Eh and pH levels change, becoming biologically available to various organisms.

Perhaps the greatest impact on benthic communities occurs when dredged spoils have sediment characteristics much different than those of seafloor sediments at the dumpsite. The most dramatic impacts arise when sandy dredged material is placed onto a muddy substrate or when muddy dredged material is dumped onto a sandy substrate. Such changes in sediment type frequently result in acute perturbations in benthic communities commonly manifested as reductions in population abundances and species diversity.[30]

Land-based disposal alternatives can produce more environmental problems than open-ocean disposal alternatives. For example, the intertidal or upland disposal of coarse-textured contaminated dredged spoils with low content of organic matter may present a high environmental risk because of the likelihood of contaminant leaching into surface water or groundwater supplies. A high potential also exists at these sites for long-term mobilization and leaching of toxic materials from noncalcareous dredged spoils containing large concentrations of reactive iron and total or pyritic sulfide. Intertidal disposal of dredged spoils, in particular, may pose a greater environmental risk than subaqueous disposal because: (1) the greater hydraulic energy conditions at some intertidal sites contribute to erosion and dispersion of solids, (2) important and sensitive benthic and aquatic habitats usually are associated with nearshore areas, and (3) plants have the potential to assimilate certain contaminants and cycle them into wetland ecosystems.[1,31]

HABITAT CREATION, PROTECTION, AND RESTORATION

Dredged-material disposal has been used most effectively in recent years for the creation, protection, and restoration of habitats. The ACOE has used dredged spoils to construct 23 underwater berms for storm wave attenuation, bottom topographic relief, and fisheries habitat improvement.[32] Coastal revitalization often entails the use of uncontaminated dredged spoils as landfill for shoreline nourishment or land improvement, such as in beach replenishment, storm protection, and sediment stabilization projects. Important alternate uses of dredged spoils include the creation of oyster beds, fishing reefs, and clam flats.[33,34] Another valuable usage of dredged material is in elevating shallow water vegetated habitats. This application has proven to be successful in seagrass recolonization and restoration projects.[32] Habitat creation, protection, and restoration in the coastal zone have been facilitated by the development of innovative dredging and dredged-material placement technologies (e.g., multihead pipe heads, perforated pipes, flexible pipes, diffusers).[35]

In conclusion, dredged spoils, when environmentally compatible, have been used in salt marsh creation, spoil-island development, beach nourishment, and substrate enhancement. The utilization of dredged material in habitat creation serves as an attractive alternative to the more commonly practiced subaerial or open-ocean disposal of this material. The impacts of dredged-spoil disposal on wetlands and open-water sites in many areas of the world underscore the value of these alternate uses of dredged material.

REFERENCES

1. Engler, R. M. and Mathis, D. B., Dredged-material disposal strategies, in *Oceanic Processes in Marine Pollution*, Vol. 3, *Marine Waste Management: Science and Policy*, Camp, M. A. and Park, P. K. (Eds.), Robert E. Krieger Publishing, Malabar, FL, 1989, 53.
2. Windom, H. L., Environmental aspects of dredging in estuaries, *J. Waterways, Harbors, Coastal Eng. Div. SCE*, 98(WW4), 475, 1972.
3. Stickney, R. R., *Estuarine Ecology of the Southeastern United States and Gulf of Mexico*, Texas A & M University Press, College Station, TX, 1984.
4. U.S. Army Corps of Engineers, *Final Environmental Statement: Operation and Maintenance of the New Jersey Intracoastal Waterway and Manasquan, Barnegat, Absecon, and Cold Spring Inlets, New Jersey*, Unpubl. Tech. Rept., U.S. Army Corps of Engineers, Philadelphia, 1975.
5. Rhoads, D. C., McCall, P. L., and Yingst, J. Y., Disturbance and production on the estuarine sea floor, *Am. Sci.*, 66, 577, 1978.
6. Machemehl, J. L., Mechanics of dredging and filling, in *Proceedings of the Seminar on Planning and Engineering in the Coastal Zone*, Coastal Plains Center for Marine Development Service, Washington, D.C., 1972.
7. Kennish, M. J., *Ecology of Estuaries: Anthropogenic Effects*, CRC Press, Boca Raton, FL, 1992.

8. Rhoads, D. C. and Boyer, L. F., The effects of marine benthos on physical properties of sediments: a successional perspective, in *Animal-Sediment Relations: The Biogenic Alteration of Sediments*, McCall, P. L. and Tevesz, M. J. S. (Eds.), Plenum Press, New York, 1982, 3.
9. Stickney, R. R. and Perlmutter, D., Impact of intracoastal waterway dredging on a mud bottom benthos community, *Biol. Conserv.*, 7, 211, 1975.
10. Godcharles, M. F., *A Study of the Effects of a Commercial Hydraulic Clam Dredge on Benthic Communities in Estuarine Areas,* Marine Research Laboratory Tech. Ser. No. 64, Florida Department of Natural Resources, St. Petersburg, 1971.
11. Jaworski, N. A., Retrospective study of the water quality issues of the upper Potomac estuary, *Rev. Aquat. Sci.*, 3, 11, 1990.
12. Biggs, R. B., Environmental effects of overboard spoil disposal, *J. Sanitary Eng. Div.*, 94 (SA3), 477, 1968.
13. Windom, H. L., Water-quality aspects of dredging and dredge-spoil disposal in estuarine environments, in *Estuarine Research*, Vol. 2, *Geology and Engineering*, Cronin, L. E. (Ed.), Academic Press, New York, 1975, 559.
14. Hall, L. A., The effects of dredging and relamation on metal levels in water and sediments from an estuarine of Trinidad, West Indies, *Environ. Pollut.*, 56, 189, 1989.
15. Lunsford, C. A., Weinstein, M. P., and Scott, L., Uptake of kepone by the estuarine bivalve, *Rangia cuneata*, during the dredging of contaminated sediments in the James River, Virginia, *Water Res.*, 21, 411, 1987.
16. Nichols, M., Diaz, R. J., and Schaffner, L. C., Effects of hopper dredging and sediment dispersion, Chesapeake Bay, *Environ. Geol. Water. Sci.*, 15, 31, 1990.
17. Schaffner, L. C., Diaz, R. J., Olsen, C. R., and Larsen, I. L., Faunal characteristics and sediment accumulation processes in the James River estuary, Virginia, *Est. Coastal Shelf Sci.*, 25, 211, 1987.
18. Kamlet, K. S., Dredged-material ocean dumping: perspectives on legal and environmental impacts, in *Dredged-Material Disposal in the Ocean*, Vol. 2, *Wastes in the Ocean*, Kester, D. R., Ketchum, B. H., Duedall, I. W., and Park, P. K. (Eds.), John Wiley & Sons, New York, 1983, 29.
19. McAnally, W. H. and Adamec, S. A., Jr., *Designing Open Water Disposal for Dredged Muddy Sediments,* Unpubl. Tech. Rept., U.S. Army Engineer Waterways Experiment Station, Vicksburg, MS, 1987.
20. O'Connor, J. M. and O'Connor, S. G., *Evaluation of the 1980 Capping Operations at the Experimental Mud Dump Site,* New York Bight Apex Tech. Rep. D-83-3, U.S. Army Engineer Waterways Experiment Station, Vicksburg, MS, 1983.
21. Parker, J. H. and Valente, R. M., *Long-Term Sand Cap Stability: New York Dredged-Material Disposal Site,* Contract Rept. CERC-88-2, U.S. Army Engineer Waterways Experiment Station, Vicksburg, MS, 1988.
22. Brannon, J. M., Hoeppel, R. E., and Gunnison, D., Capping contaminated dredged material, *Mar. Pollut. Bull.,* 18, 175, 1987.
23. McAnally, W. H., Jr. and Adamec, S. A., Jr., Designing open water disposal for dredged muddy sediments, in *Dynamics of Turbid Coastal Environments*, Uncles, R. J. (Ed.), *Cont. Shelf Res.*, 7, 1445, 1987.
24. Truitt, C. L., Dredged-material behavior during open-water disposal, *J. Coastal Res.,* 4, 489, 1988.
25. Kirby, C. J., Keeley, J. W., and Harrison, J., An overview of the technical aspects of the Corps of Engineers national dredged-material research program, in *Estuarine Research*, Vol. 2, *Geology and Engineering*, Cronin, L. E. (Ed.), Academic Press, New York, 1975, 523.
26. U.S. Environmental Protection Agency and U.S. Department of the Army, *Draft Ecological Evaluation of Proposed Discharge of Dredged Material into Ocean Waters,* Tech. Rep., EPA Contract No. 68-C8-0105, Washington, D.C., 1990.
27. U.S. Environmental Protection Agency — Corps of Engineers Technical Committee on Criteria for Dredged and Fill Material, *Ecological Evaluation of Proposed Discharge of Dredged Material into Ocean Waters: Implementation Manual for Section 103 of Public Law 92-532 (Marine Protection, Research, and Sanctuaries Act of 1972),* U.S. Army Engineer Waterways Experiment Station, Vicksburg, MS, 1977.
28. Kester, D. R., Ketchum, B. H., Duedall, I. W., and Park, P. K., The problem of dredged-material disposal, in *Dredged-Material Disposal in the Ocean*, Vol. 2, *Wastes in the Ocean*, Kester, D. R., Ketchum, B. H., Duedall, I. W., and Park, P. K. (Eds.), John Wiley & Sons, New York, 1983, 3.

29. Stern, E. M and Stickle, W. B., *Effects of Turbidity and Suspended Material in Aquatic Environments,* Literature Review, Tech. Rep. D-78-21 (NTIS No. AD-A056035), U.S. Army Engineer Waterways Experiment Station, Vicksburg, MS, 1978.
30. Wright, T. D., *Aquatic Dredged Material Disposal Impacts,* Tech. Rep. DS-78-1 (NTIS No. AD-A060250), U.S. Army Engineer Waterways Experiment Station, Vicksburg, MS, 1978.
31. Gambrell, R. P., Kahlid, R. A., and Patrick, W. H., Jr., *Disposal Alternatives for Contaminated Material as a Management Tool to Minimize Adverse Environmental Effects,* Synthesis Report DS-78-8 (NTIS No. AD-A073158), U.S. Army Engineer Waterways Experiment Station, Vicksburg, MS, 1978.
32. National Research Council, *Restoring and Protecting Marine Habitat: The Role of Engineering and Technology,* National Academy Press, Washington, D.C., 1994.
33. Landin, M. C., *Dredged Material: A Recognized Resource,* Permanent International Association of Navigation Congresses (PIANC), Brussels, Belgium, 1991, 112.
34. Landin, M. C., Beneficial uses of dredged material projects: how, when, where, and what to monitor, and why it matters, in *Proceedings of Ports 92*, American Society of Civil Engineers, New York, 1992, 142.
35. Herbich, J. B., *Handbook of Dredging Engineering*, McGraw-Hill, New York, 1992.

APPENDIX 1. DREDGING

TABLE 1.1
Types of Dredging Devices and Their Relationship to Sediment Type and Disposal Method

Dredge Type	Sediment Type	Disposal Conveyance
Mechanical devices		
Dipper dredge	Blasted rock	Vessel
Bucket dredge	Coarse-grain size	Vessel
Ladder dredge	Fine-grain size	Vessel
Agitation dredge	Mud, clay	Prevailing current
Hydraulic devices		
Agitation dredge	Mud, clay	Prevailing current
Hopper dredge	Fine-grain size	Vessel
Suction dredge	Soft mud, clay	Pipeline
Cutterhead dredge	Consolidated, coarse-grain size	Pipeline
Dustpan dredge	Sand	Pipeline
Sidecasting dredge	Fine-grain size	Short pipe

Source: Kester, D. R., Ketchum, B. H., Duedall, I. W., and Park, P. K., in *Wastes in the Ocean,* Vol. 2, *Dredged Material Disposal in the Ocean,* Kester, D. R., Ketchum, B. H., Duedall, I. W., and Park, P. K. (Eds.), John Wiley & Sons, New York, 1983, 3. With permission.

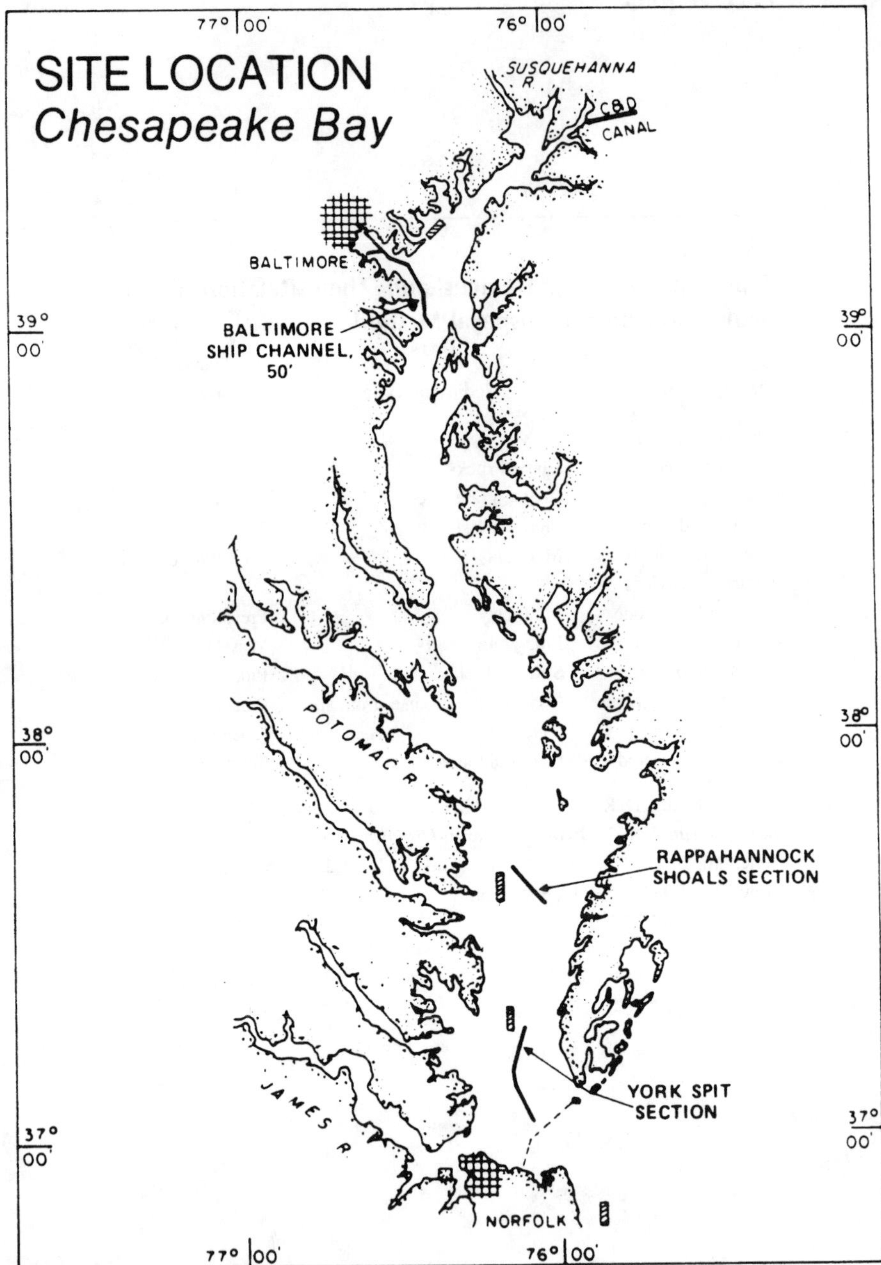

FIGURE 1.1. Map of Chesapeake Bay showing the location of the Baltimore shipping channel, dredged-spoil disposal sites (cross-hatched), and the benthic study area at Rappahannock Shoals in the central bay. (From Nichols, M., Diaz, R. J., and Schaffner, L. C., *Environ. Geol. Water Sci.*, 15, 31, 1990. With permission.)

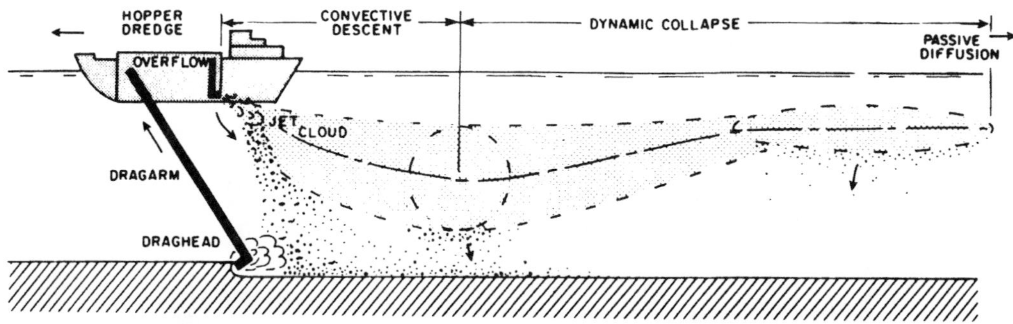

FIGURE 1.2. Hopper dredging operations in Chesapeake Bay showing an upper turbidity plume produced by overflow discharge and a near-bottom turbidity plume produced by draghead agitation and settling of particulates from the upper plume. The conceptual model depicts three transport phases for hopper overflow discharge: convective descent, dynamic collapse, and passive diffusion. (From Nichols, M., Diaz, R. J., and Schaffner, L. C., *Environ. Geol. Water Sci.,* 15, 31, 1990. With permission.)

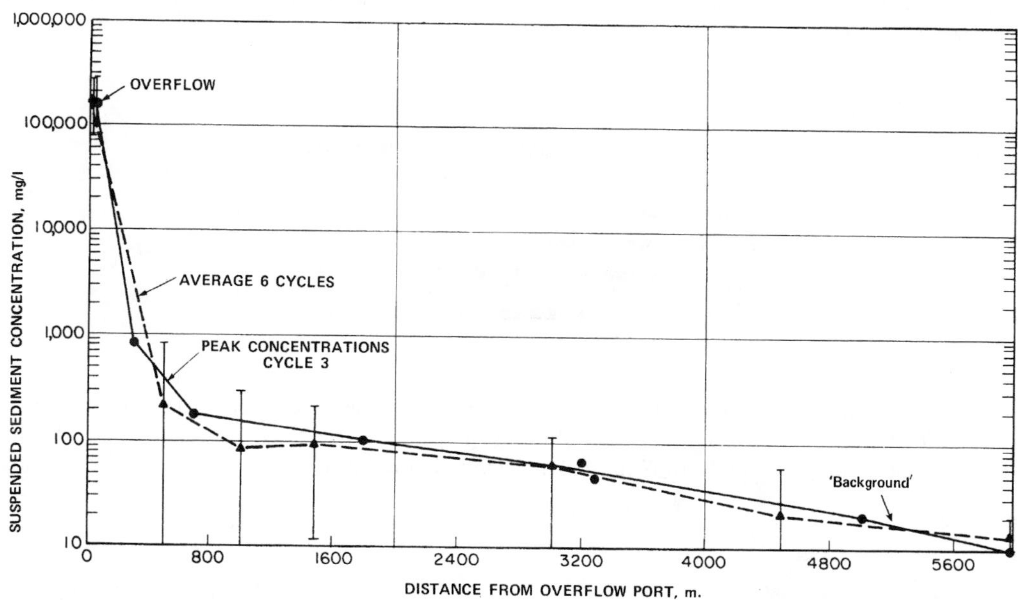

FIGURE 1.3. Dispersion curves for average peak suspended sediment concentrations at mid-depth (7 m) along the plume axis over six dredge cycles and for one dredge cycle (number 3) during hopper dredging operations in Chesapeake Bay. (From Nichols, M., Diaz, R. J., and Schaffner, L. C., *Environ. Geol. Water Sci.,* 15, 31, 1990. With permission.)

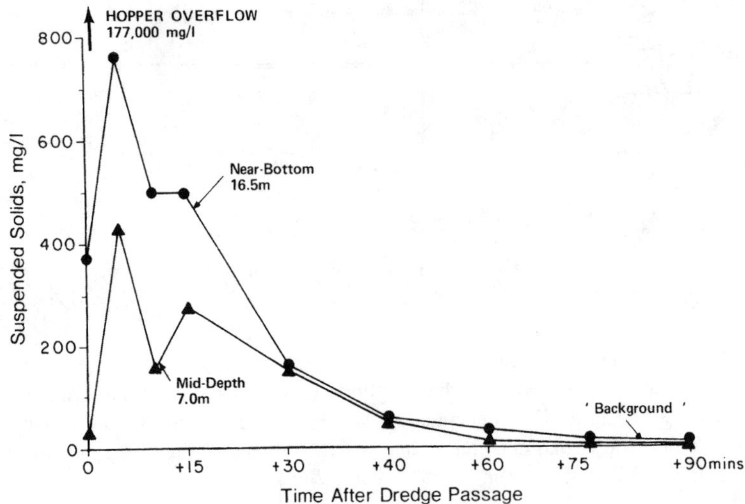

FIGURE 1.4. Temporal variations of average suspended sediment concentration in Chesapeake Bay at three fixed points in a plume to 90 minutes after passage of a hopper dredge. (From Nichols, M., Diaz, R. J., and Schaffner, L. C., *Environ. Geol. Water Sci.,* 15, 31, 1990. With permission.)

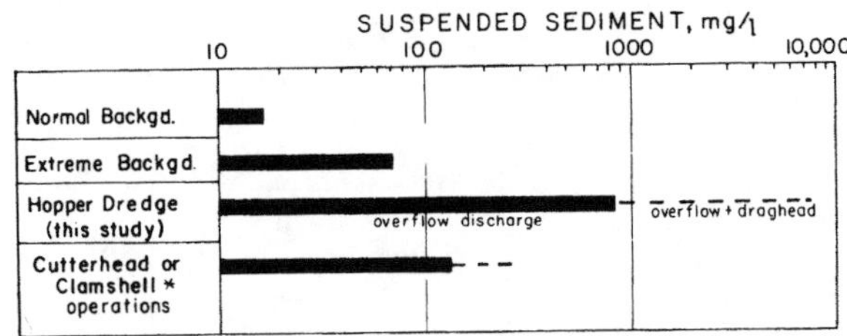

FIGURE 1.5. Comparison of suspended sediment concentration ranges for different background conditions in central Chesapeake Bay and different types of dredging operations. (From Nichols, M., Diaz, R. J., and Schaffner, L. C., *Environ. Geol. Water Sci.,* 15, 31, 1990. With permission.)

TABLE 1.2
Suspended Solids Levels Reported from Bucket Dredging Operations

Location	Suspended Solids Field Characteristics
San Francisco Bay, CA	Nearfield concentrations of total suspended sediments were 21–282 mg/l. Suspended sediment concentrations in the water column 50 m downstream from the dredge were generally less than 200 mg/l and averaged 30–90 mg/l relative to background concentrations outside the plume of approximately 40 mg/l. The visible plume was about 300 m long at the surface and approximately 450 m long at a bottom depth of 10 m.
Lower Thames River Estuary, CT	Maximum suspended sediment concentrations of 68, 110, and 168 mg/l at the surface, mid-depth (3 m), and near bottom (10 m), respectively, were noted within 100 m downstream. These maximum concentrations decreased very rapidly to the background levels of 5 mg/l within 300 m at the surface and 500 m near the bottom. Suspended sediment concentrations adjacent to the dredge were 200–400 mg/l and approached background within approximately 700 m. Major perturbations were confined within 300 m of the dredge.
New Haven Harbor, CT	Suspended sediment plume (defined by transmissometer readings) was a well defined, small-scale feature extending over a distance of approximately 1000 m downstream.
Patapsco River, MD	Suspended sediment concentrations 22 m downstream from the dredging operation were 30 mg/l at near bottom depths of 10 m relative to background water column concentrations of approximately 10 mg/l or less.
Japan	Maximum suspended sediment concentrations 7 m downstream from the dredging operation ranged from 150–300 mg/l (defined using turbidity measurements) relative to background levels of less than 30 mg/l. These levels decreased by 50% at a distance of 23 m. Turbidity near the surface was generally lower than levels at mid-depth or near the bottom.
St. Johns River, FL	Sediment resuspension caused by bucket dredges showed that the plume downstream of a typical bucket operation may extend approximately 1000 ft (300 m) at the surface and 1500 ft (450 m) near the bottom. The average suspended sediment concentrations of all samples collected within 800 ft (240 m) of the dredge along upper-water-column and near-bottom transects were approximately 106 and 134 mg/l, respectively. A comparison of suspended sediment concentrations from open and enclosed bucket dredge operations showed considerable reductions in suspended sediment concentrations in the upper water column (>50%) but increases in concentrations in the lower water column (>50%) due to "shock" waves created by the closed bucket.
Thames River Estuary, CT	The composition of material suspended by the dredge indicates that variations are similar to those produced by local storm events. Storm events affect a significantly larger area and display a higher frequency of occurrence than that characterizing typical dredging schedules. Both storm events and dredges increase particulate organic carbon concentrations and bias the material composition in favor of the inorganic components.

Source: Houston, L. J., LaSalle, M. W., and Lunz, J. D., in *Estuarine Research in the 1980s,* Smith, C. L. (Ed.), State University of New York Press, Stony Brook, NY, 1992. With permission.

TABLE 1.3
Comparison of Dissolved Oxygen Levels Baywide and in the Vicinity of an Operating Bucket Dredge in Haverstraw Bay

				DO Levels at Dredge Site			
	Baywide Ag. DO Levels[a]			Downstream		Upstream	
Date	NVC	NTC	SH	SR	SW	SR	SW
8/20	6.80	6.20	8.03				
8/27	6.50	6.60	6.87				
9/3	7.03	7.12	7.47		Pre-dredging		
9/11	6.13	6.16	6.21	6.30	6.00	6.30	5.60
9/17	6.53	6.43	6.84	6.50	6.70	6.40	6.20
9/22	5.63	5.79	6.19	5.60	—	5.60	—
9/30	6.37	6.21	6.54	7.40	—	7.30	—
10/8	7.83	7.96	8.10	7.90	—	8.20	—
10/13	7.97	7.89	8.29		Post-dredging		
10/23	8.37	8.40	8.76				

Note: DO = dissolved oxygen; NVC = navigation channel (32 ft MLW); NTC = natural river channel (23–28 ft MLW); SH = shallows (under 10 ft MLW); SR = sunrise sample; SW = slack water sample (not taken if within 1 hour of sunrise sample).

[a] Values based on average from three baywide transects.

Source: Houston, L. J., LaSalle, M. W., and Lunz, J. D., in *Estuarine Research in the 1980s,* Smith, C. L. (Ed.), State University of New York Press, Stony Brook, NY, 1992, 82. With permission.

TABLE 1.4
Average Oxygen Saturation Levels for Three Weekly Transects Sampled Before, During, and After Dredging of Haverstraw Bay

		Mean (SD) % Oxygen Saturation			
Transect	**Station**	**Before**	**During**	**After**	**Significance**[a]
16	WS	86.0 (4.8)	76.1 (4.1)	85.9 (0.4)	D:B&A[b]
	WNC	80.2 (1.3)	74.1 (4.6)	83.7 (3.3)	D:A
	MIDC	77.9 (1.5)	74.2 (7.3)	85.2 (5.3)	none
	ENC	79.9 (1.3)	73.8 (5.9)	85.2 (2.6)	D:A
	ES	87.8 (9.9)	77.3 (5.7)	83.3 (6.6)	none
14	WS	82.8 (2.0)	74.8 (5.8)	82.9 (2.3)	none
	WNC	78.6 (7.8)	74.2 (4.8)	81.6 (1.1)	none
	MIDC	76.0 (6.9)	73.5 (6.7)	81.1 (0.4)	none
	ENC	82.7 (2.8)	73.3 (7.6)	81.2 (1.7)	none
	ES	82.3 (1.1)	77.7 (7.2)	86.9 (0.3)	none
	FES	104.9 (28.9)	78.4 (8.3)	86.1 (2.2)	none
12	WS	93.4 (13.6)	72.8 (10.1)	83.8 (0.8)	D:B
	WNC	84.3 (4.2)	72.1 (7.8)	80.6 (2.5)	none
	MIDC	75.9 (9.8)	73.3 (8.5)	76.8 (8.1)	none
	ENC	84.0 (1.6)	72.6 (7.3)	71.8 (16.6)	none
	ES	92.5 (9.7)	81.3 (6.1)	73.2 (14.8)	none

Note: Based on nonparametric Tukey test where Q(0.05,3) = 2.394: D:B&A = mean oxygen saturation levels during dredging are significantly different from mean levels before and after dredging. D:A = mean oxygen saturation level during dredging is significantly different from mean level after dredging. D:B = mean oxygen saturation level during dredging is significantly different from mean level before dredging.

[a] Significant Kruskal-Wallis test, H(0.05, 5, 3, 2) = 5.25
[b] D = during; B = before; A = after.

Source: Houston, L. J., LaSalle, M. W., and Lunz, J. D., in *Estuarine Research in the 1980s,* Smith, C. L. (Ed.), State University of New York Press, Stony Brook, NY, 1992, 82. With permission.

TABLE 1.5
Average Optical Turbidity for Three Transects Sampled Before, During, and After Dredging of Haverstraw Bay

Transect	Station	Mean (SD) Optical Turbidity (in NTU)			Significance[a]
		Before	During	After	
16	WS	4.5 (4.5)	9.5 (9.5)	12.5 (0.7)	B:A[b]
	WNC	4.7 (2.4)	9.8 (1.8)	15.3 (0.4)	B:A
	MIDC	5.8 (1.6)	9.9 (2.5)	14.0 (7.1)	B:A&D
	ENC	5.6 (2.9)	8.6 (8.6)	16.5 (10.7)	none
	ES	3.8 (1.5)	8.7 (3.5)	8.3 (0.4)	none
14	WS	4.5 (1.6)	9.0 (1.7)	9.9 (1.6)	B:A&D
	WNC	4.8 (3.4)	10.6 (4.7)	17.2 (12.4)	none
	MIDC	6.2 (3.6)	11.6 (5.4)	14.5 (4.9)	none
	ENC	4.2 (1.3)	8.7 (3.3)	13.1 (6.9)	B:A
	ES	3.6 (1.3)	7.8 (3.6)	10.5 (3.5)	none
	FES	4.4 (0.7)	7.2 (3.0)	8.3 (1.0)	none
12	WS	3.8 (0.5)	7.0 (2.3)	9.4 (0.9)	B:A
	WNC	4.6 (0.5)	8.6 (1.0)	18.1 (12.6)	B:A
	MIDC	7.0 (3.6)	10.9 (7.1)	19.0 (12.7)	none
	ENC	5.1 (3.0)	8.7 (3.1)	14.5 (4.9)	none
	ES	4.1 (2.2)	7.0 (2.7)	10.0 (1.4)	none

Note: Based on nonparametric Tukey test where Q(0.05,3) = 2.394: B:A = before dredging mean turbidity significantly different than after dredging turbidity level. B:A&D = before dredging mean turbidity significantly different than both dredging and after dredging turbidity levels.

[a] Significant Kruskal-Wallis Test, H(0.05,5,3,2) = 5.25.
[b] D = during; B = before; A = after.

Source: Houston, L. J., LaSalle, M. W., and Lunz, J. D., in *Estuarine Research in the 1980s,* Smith, C. L. (Ed.), State University of New York Press, Stony Brook, NY, 1992, 82. With permission.

TABLE 1.6
Average Weekly Sunrise (SR) and Slack Water (SL) Bottom Dissolved Oxygen Levels 90 m Upstream and Downstream of a Bucket Dredge Operation in Haverstraw Bay

	Downstream Station				Upstream Station				Control Station			
	All Days		Dredging Days		All Days		Dredging Days		All Days		Dredging Days	
Week	SR	SL	SR	SL	SR	SL	SR	SL	SR	SL	SR	SL
1	6.53	6.40	6.80	6.60	6.60	6.13	6.60	6.20	6.65	6.40	6.60	6.50
(N)[a]	(4)	(3)	(1)	(1)	(4)	(3)	(1)	(1)	(4)	(3)	(1)	(1)
2	6.24	6.02	—	6.40	6.13	6.05	—	6.30	6.43	6.16	—	6.40
(N)	(7)	(5)	(0)	(1)	(7)	(5)	(0)	(1)	(7)	(5)	(0)	(1)
3	6.13	6.10	6.40	6.10	6.22	6.05	6.65	6.05	6.19	6.23	6.85	6.23
(N)	(6)	(4)	(2)	(4)	(6)	(4)	(2)	(4)	(6)	(4)	(2)	(4)
4	7.31	7.33	7.40	7.43	7.37	7.35	7.41	7.43	7.37	7.20	7.50	7.37
(N)	(6)	(4)	(1)	(3)	(6)	(4)	(1)	(3)	(6)	(4)	(1)	(3)
5	8.38	8.15	8.57	8.15	8.40	8.10	8.60	8.10	8.73	8.15	9.00	8.15
(N)	(4)	(2)	(3)	(2)	(4)	(2)	(3)	(2)	(4)	(2)	(3)	(2)

Note: If slack water was within 1 hour of sunrise, only one set of samples was taken that day. Dredging began on a Thursday (9/10/87) and ended on a Friday (10/9/87).

[a] N = number of daily samples.

Source: Houston, L. J., LaSalle, M. W., and Lunz, J. D., in *Estuarine Research in the 1980s,* Smith, C. L. (Ed.), State University of New York Press, Stony Brook, NY, 1992, 82. With permission.

APPENDIX 2. DREDGED-SPOIL DISPOSAL

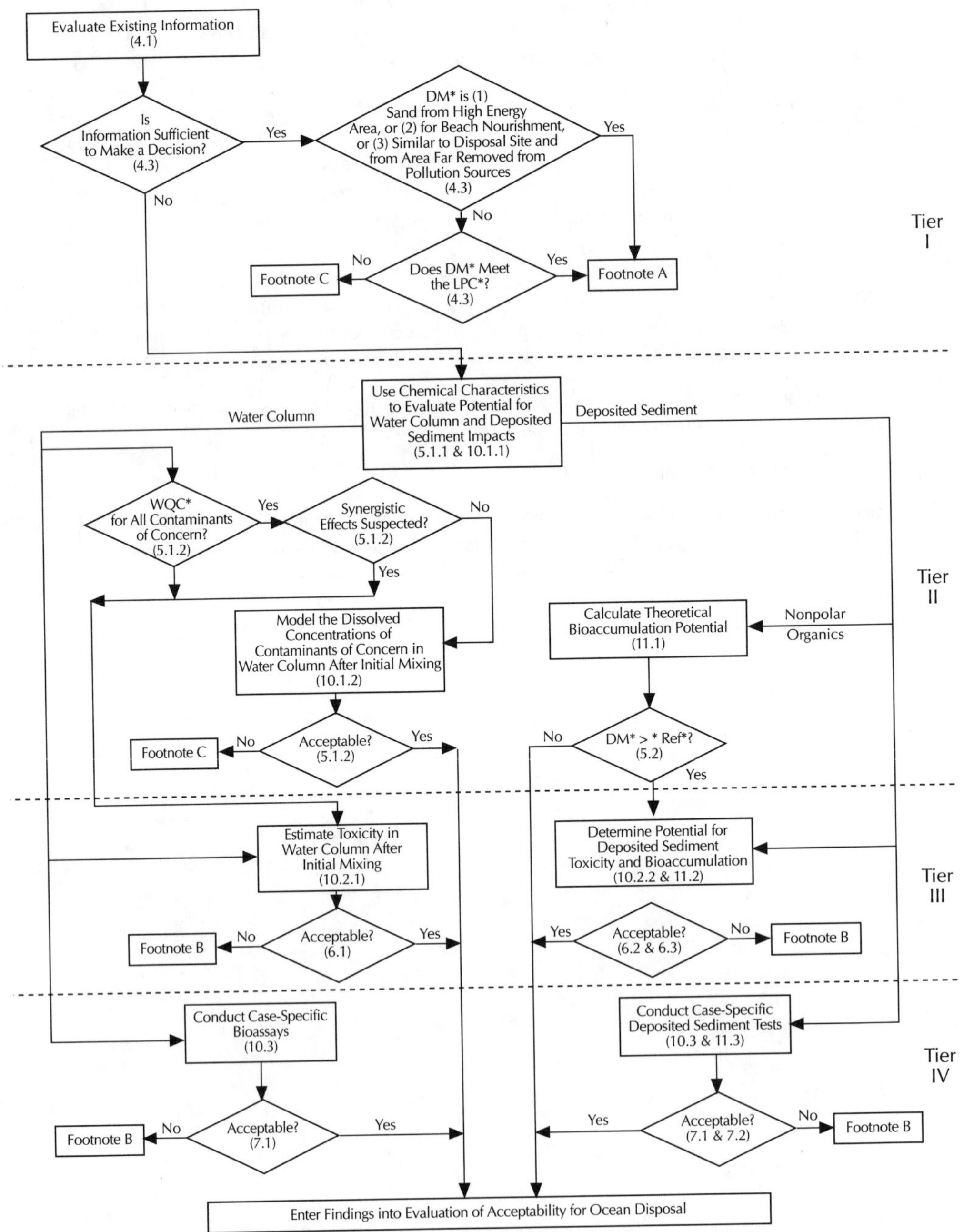

FIGURE 2.1. Tiered approach to evaluating potential impacts of ocean disposal of dredged material. (From U.S. Environmental Protection Agency and U.S. Department of the Army, *Ecological Evaluation of Proposed Discharge of Dredged Material into Ocean Waters,* Tech. Rep., EPA Contract No. 68-C8–0105, U.S. Government Printing Office, Washington, D.C., 1990.)

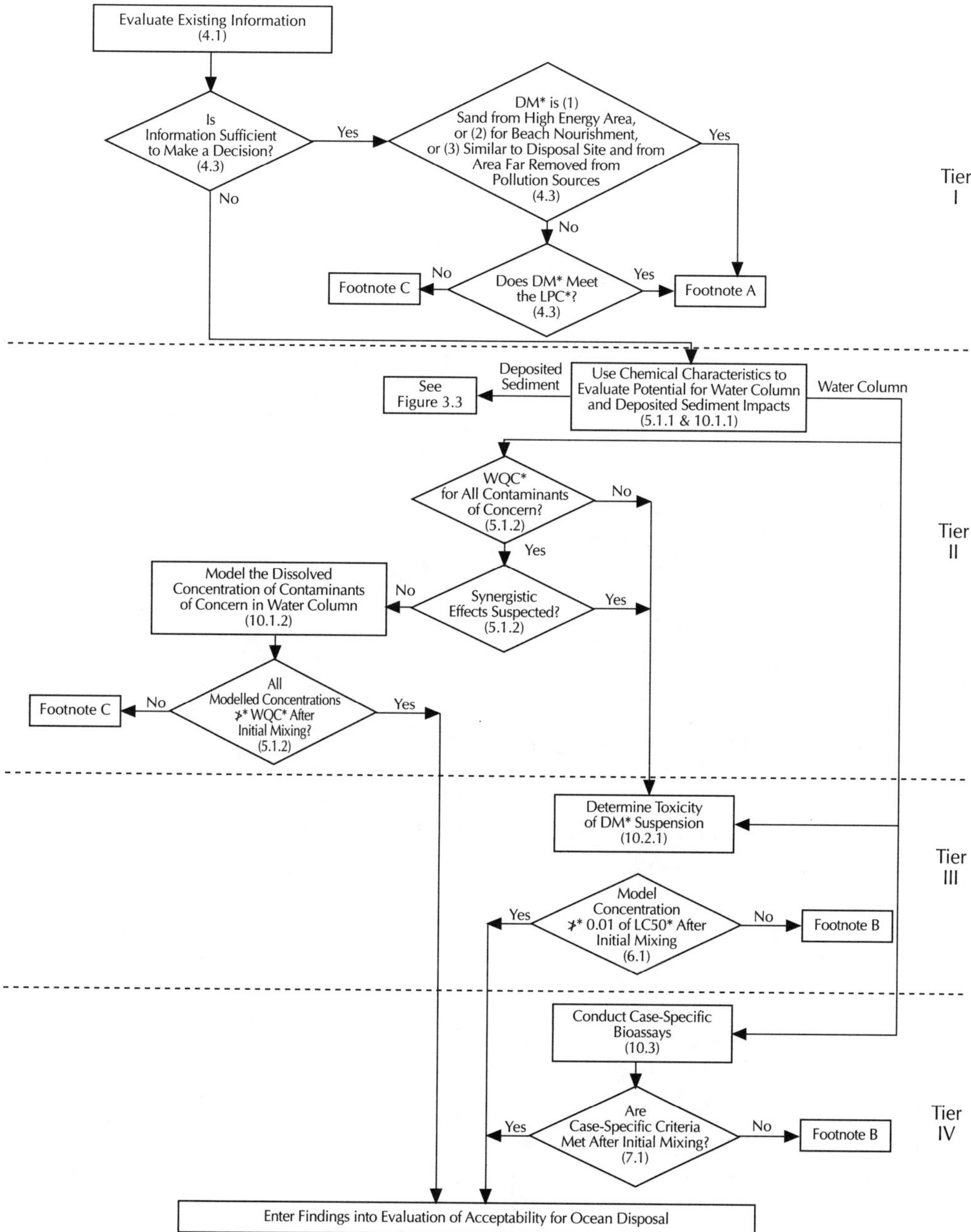

FIGURE 2.2. Tiered approach to evaluating potential water column impacts of dredged material. (From U.S. Environmental Protection Agency and U.S. Department of the Army, *Ecological Evaluation of Proposed Discharge of Dredged Material into Ocean Waters,* Tech. Rep., EPA Contract No. 68-C8–0105, U.S. Government Printing Office, Washington, D.C., 1990.)

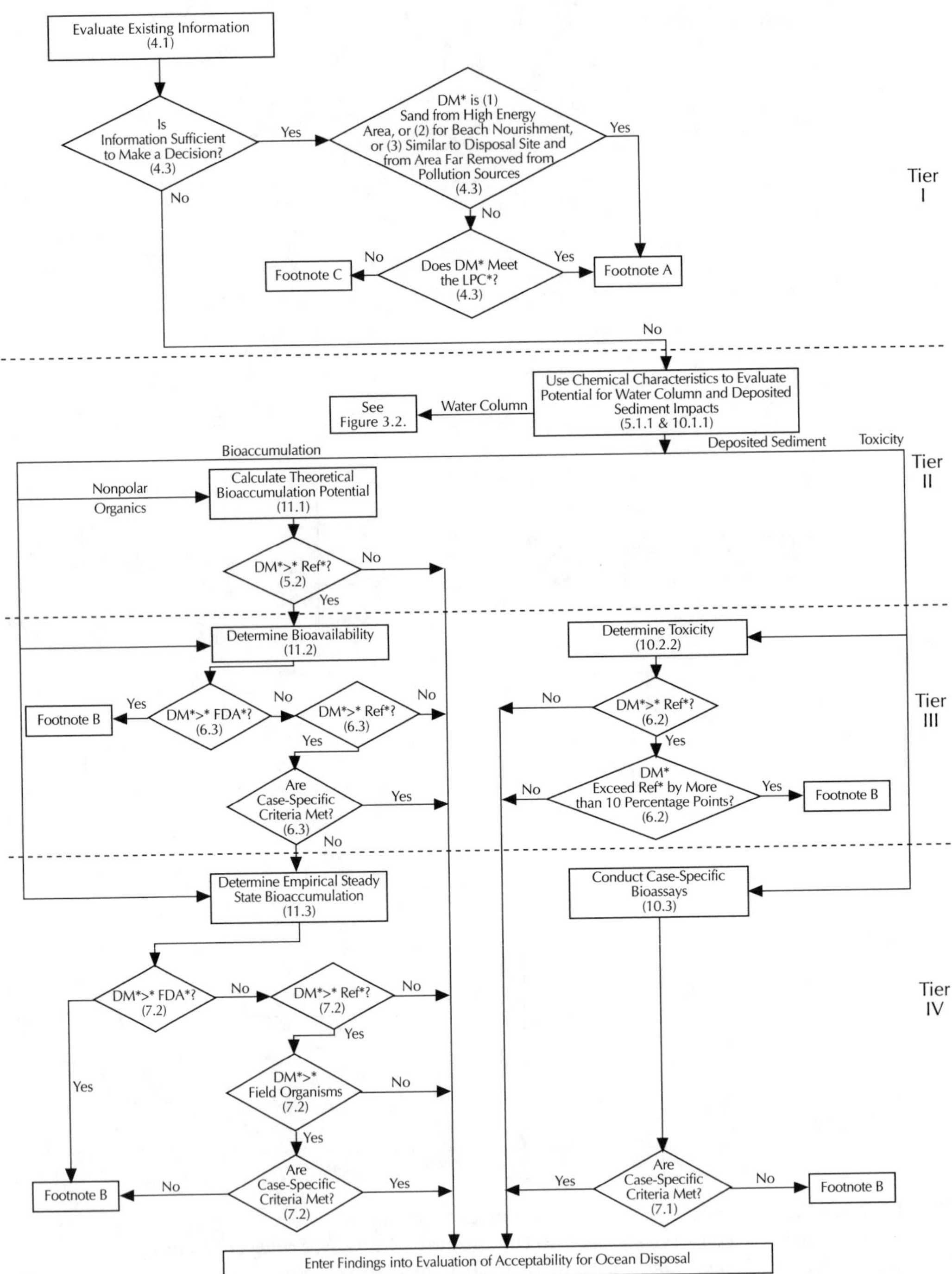

FIGURE 2.3. Tiered approach to evaluating potential benthic impacts of deposited dredged material. (From U.S. Environmental Protection Agency and U.S. Department of the Army, *Ecological Evaluation of Proposed Discharge of Dredged Material into Ocean Waters,* Tech. Rep., EPA Contract No. 68-C8–0105, U.S. Government Printing Office, Washington, D.C., 1990.)

KEY TO FIGURES 2.1, 2.2, AND 2.3.

* Key To Nomenclature

DM – Dredged material
LPC – It is necessary to demonstrate, using existing data comparable to the data that would be generated if testing were conducted, that the limiting permissible concentration (LPC) would not be exceeded.
$>$ – Is statistically greater than
$\ngtr$ – Is not statistically greater than
WQC – Applicable Marine Water Quality Criteria
LC50 – Acutely toxic concentration, i.e., lethal concentration to 50% of test organisms
Ref – Reference material
FDA – Food and Drug Administration Action Levels for Poisonous or Deleterious Substances in Fish and Shellfish for Human Consumption

Notes:

A. Ocean disposal at a designated site is acceptable if all other requirements of the regulations are satisfied.

B. Ocean disposal without management action is not acceptable.

C. Ocean disposal is not acceptable.

TABLE 2.1
Priority Pollutants and 301(H) Pesticides of Concern in Dredged Material Disposal Operations

Structural Compound Class	PP[a]	Pollutant
Phenols	65	Phenol
	34	2,4-Dimethylphenol
Substituted phenols	21	2,4,6-Trichlorophenol
	22	Para-chloro-meta-cresol
	24	2-Chlorophenol
	31	2,4-Dichlorophenol
	57	2-Nitrophenol
	58	4-Nitrophenol
	59	2,4-Dinitrophenol
	60	4,6-Dinitro-o-cresol
	64	Pentachlorophenol
Organonitrogen compounds	5	Benzidine
	28	3,3′-Dichlorobenzidine
	35	2,4-Dinitrotoluene
	36	2,6-Dinitrotoluene
	37	1,2-Diphenylhydrazine
	56	Nitrobenzene
	61	*N*-nitrosodimethylamine
	62	*N*-nitrosodiphenylamine
	63	*N*-nitrosodipropylamine
Low molecular weight polynuclear aromatic hydrocarbons (PAH)	1	Acenaphthene
	55	Naphthalene
	77	Acenaphthylene
	78	Anthracene
	81	Phenanthrene
	80	Fluorene
High molecular weight PAH	39	Fluoranthene
Phthalates	66	Bis(2-ethylhexy)phthalate
	67	Butyl benzyl phthalate
	68	Di-*n*-butyl phthalate
	69	Di-*n*-octyl phthalate
	70	Diethyl phthalate
	71	Dimethyl phthalate
Polychlorinated biphenyls (PCB) as arochlors	106	PCB-1242
	107	PCB-1254
	108	PCB-1221
	109	PCB-1232
	110	PCB-1248
	111	PCB-1260
	112	PCB-1016
Miscellaneous oxygenated compounds	129	TCDD (dioxin)
	54	Isophorone
Pesticides	89	Aldrin
	90	Dieldrin
	91	Chlordane
	92	DDT[b]
	95	Endosulfan[c]
	98	Endrin
	99	Endrin aldehyde
	100	Heptachlor
	101	Heptachlor epoxide
	102	Alpha-hexachlorocyclohexane
	103	Beta-hexachlorocyclohexane
	104	Delta-hexachlorocyclohexane
	105	Gamma-hexachlorocyclohexane
	113	Toxaphene

	72	Benzo(a)anthracene
	73	Benzo(a)pyrene
	74	Benzo(b)fluoranthene
	75	Benzo(k)fluoranthene
	76	Chrysene
	79	Benzo(ghi)perylene
	82	Dibenzo(a,h)anthracene
	83	Ideno(1,2,3-cd)pyrene
	84	Pyrene
Chlorinated aromatic hydrocarbons	8	1,2,4-Trichlorobenzene
	9	Hexachlorobenzene
	20	2-Chloronaphthalene
	25	1,2-Dichlorobenzene
	26	1,3-Dichlorobenzene
	27	1,4-Dichlorobenzene
Chlorinated aliphatic hydrocarbons	52	Hexachlorobutadiene
	12	Hexachloroethane
	53	Hexachlorocyclopentadiene
Halogenated ethers	18	Bis(2-chloroethyl)ether
	40	4-Chlorophenyl ether
	41	4-Bromophenyl ether
	42	Bis(2-chloroisopropyl)ether
	43	Bis(2-chlorethoxy)methane
Volatile halogenated alkenes	29	1,1-Dichlorethylene
	30	1,2-*trans*-Dichlorethylene
	33	*trans*-1,3-Dichloropropene
	33	*cis*-1,3-Dichloropropene
	85	Tetrachlorethene
	87	Trichlorethene
	88	Vinyl Chloride
Volatile aromatic hydrocarbons	4	Benzene
	38	Ethylbenzene
	86	Toluene
	—	Mirex[d]
	—	Methoxychlor[d]
	—	Parathion[e]
	—	Malathion[e]
	—	Guthion[e]
	—	Demeton[e]
Volatile halogenated alkanes	6	Tetrachloromethane
	10	1,2-Dichloroethane
	11	1,1,1-Trichloroethane
	13	1,1-Dichloroethane
	14	1,1,2-Trichloroethane
	15	1,1,2,2-Tetrachloroethane
	16	Chloroethane
	23	Chloroform
	32	1,2-Dichloropropane
	44	Dichloromethane
	45	Chloromethane
	46	Bromomethane
	47	Bromoform
	48	Dichlorobromoethane
	49	Fluorotrichloromethane (removed)
	50	Dichlorodifluoromethane (removed)
	51	Chlorodibromomethane

TABLE 2.1 (continued)
Priority Pollutants and 301(H) Pesticides of Concern in Dredged Material Disposal Operations

Structural Compound Class	PP[a]	Pollutant	Structural Compound Class	PP[a]	Pollutant
Volatile chlorinated aromatic hydrocarbons	7	Chlorobenzene			
Volatile unsaturated carbonyl compounds	2	Acrolein			
	3	Acrylonitrile			
Volatile ethers	19	2-Chlorethylvinylether			
		Bis(chloromethyl)ether (removed)			
Metals	114	Antimony			
	115	Arsenic			
	117	Beryllium			
	118	Cadmium			
	119	Chromium			
	120	Copper			
	122	Lead			
	123	Mercury			
	124	Nickel			
	125	Selenium			
	126	Silver			
	127	Thallium			
	128	Zinc			
Miscellaneous	121	Cyanide			
	116	Asbestos			

[a] PP = priority pollutant.
[b] Includes DDT, DDD, and DDE.
[c] Includes *alpha*-endosulfan, *beta*-endosulfan, and endosulfan sulfate.
[d] Chlorinated 301(H) pesticides that are not on the priority pollutant list.
[e] Organophosphorus 301(H) pesticides that are not on the priority pollutant list.

Source: U.S. Environmental Protection Agency and U.S. Department of the Army, *Ecological Evaluation of Proposed Discharge of Dredged Material into Ocean Waters,* Tech. Rep., EPA Contract No. 68-C8–0105, U.S. Government Printing Office, Washington, D.C., 1990.

TABLE 2.2
Summary of Field Studies of Dredged Material Behavior During Open-Water Disposal

	Site Characteristics		Dredging/Disposal Characteristics					
Site	Water Depth (m)	Bottom Currents (cm/s)	Dredged Sediment	Dredge Type	Disposal Type	Typical Volume (cu/m)	Monitoring Technique/Device	Sediment in Upper Water Column (% of Original)
Long Island Sound	18–20	6–30	Silt-clay	Clamshell	Scow	900–2300	Transmissometer	1
Carquinez[a]	14	9–24	Silt-clay	Trailing suction hopper	Hopper	1000	Transmissometer and gravimetric	1–5
Ashtabula (Lake Erie)	15–18	0–21	Sandy silt	Trailing suction hopper	Hopper	690	Transmissometer and gravimetric	1[b]
New York Bight	26	6–24	Marine silt	Trailing suction hopper	Hopper	6000	Transmissometer and gravimetric	1[b]
Saybrook (Long Island Sound)	52	21–70	Marine silt	Clamshell	Scow	1100	Transmissometer and gravimetric	1[b]
Elliott Bay	67	0–21	Sandy silt	Clamshell	Scow	380–535	Transmissometer and gravimetric	1[b]
Rochester (Lake Ontario)	117–45	0–21	Riverine silt	Trailing suction hopper	Hopper	690	Transmissometer and gravimetric	1[b]
New York Bight	15–25	N/R	Silt-clay	Clamshell	Scow	1375–3000	Mass balance	3.7
Duwamish Waterway	20–21	6	Silt-clay	Clamshell	Scow	840	Gravimetric and mass balance	2–4

[a] Limited data at two additional sites included.
[b] Synthesis of all sites reported.

Source: Truitt, C. L., *J. Coastal Res.*, 4, 489, 1988. With permission.

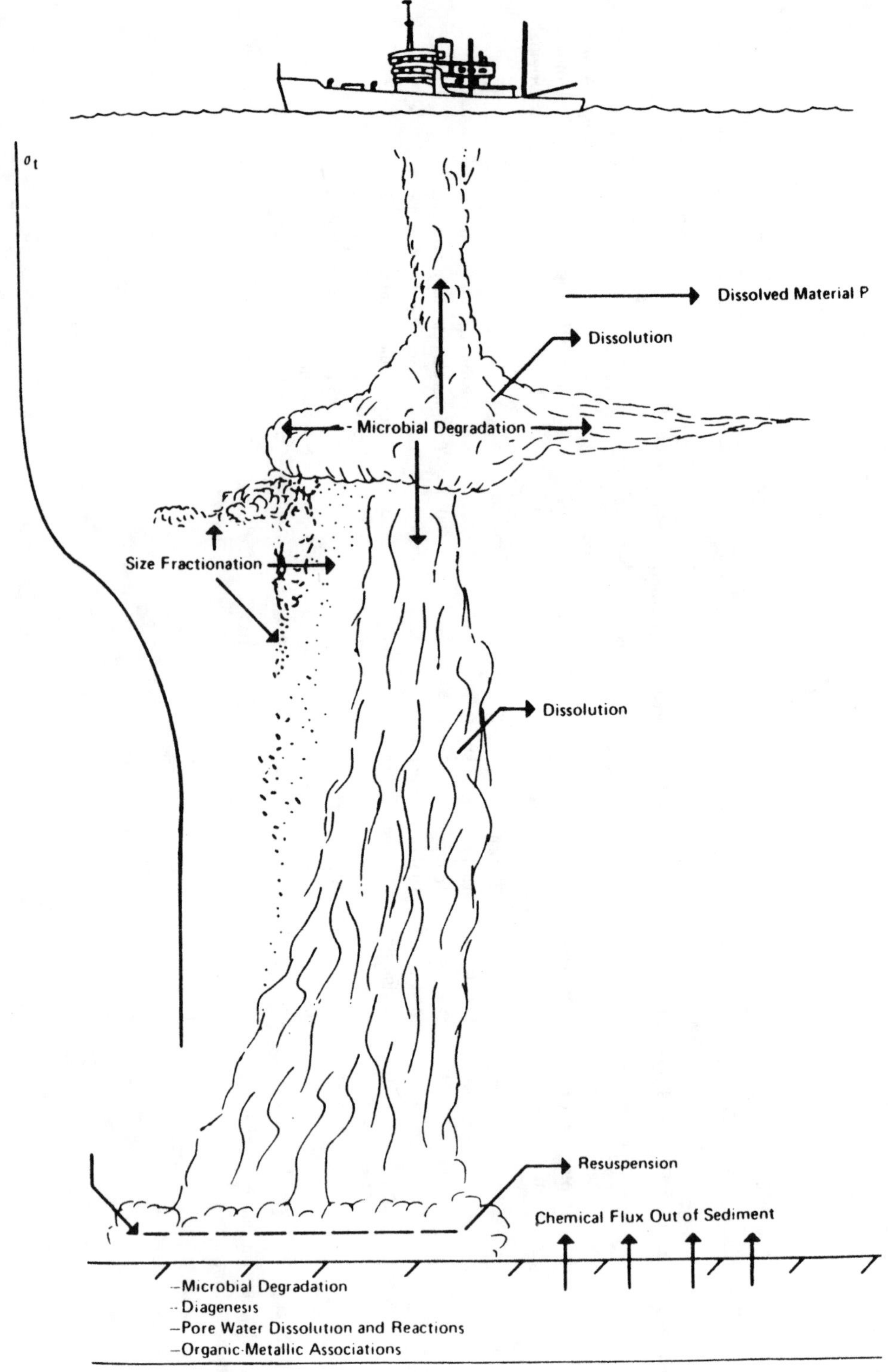

FIGURE 2.4. The processes influencing the distribution and fate of organic pollutants associated with dredged material disposal. (From U.S. Army Corps of Engineers, Dredged Material Research Program, Vicksburg, MS, 1973–1978).

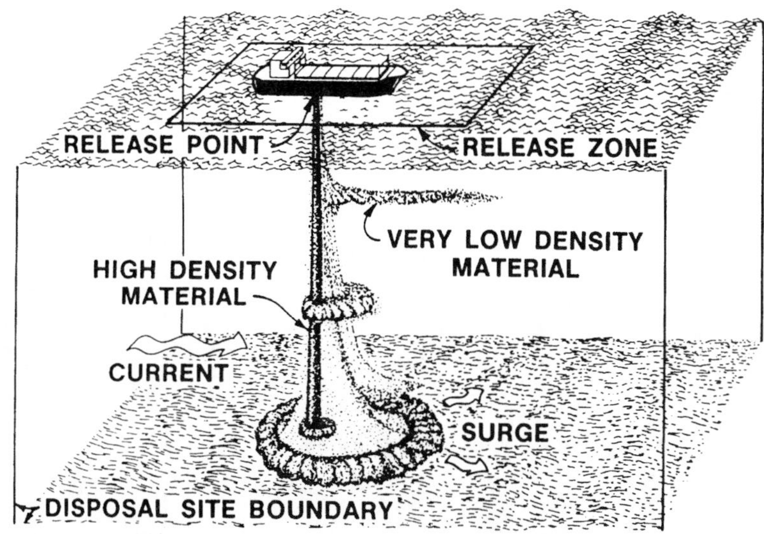

FIGURE 2.5. Sediment transport processes operating during open-water dredge disposal. (From Pequegnat, W. E., Pequegnat, L. H., James, B. M., Kennedy, E. A., Fay, R. R., and Fredericks, A. D., *Procedural Guide for Designation Surveys of Ocean Dredged-Material Disposal Sites,* Tech. Rep. EL-81–1, U.S. Army Corps of Engineers, Washington, D.C., 1981.)

FIGURE 2.6. Idealized profile of the Duwamish Waterway capping mound for dredged material disposal, Seattle, WA. (From Brannon, J. M. and Poindexter-Rollings, M. E., *Sci. Total Environ.,* 91, 115, 1990. With permission.)

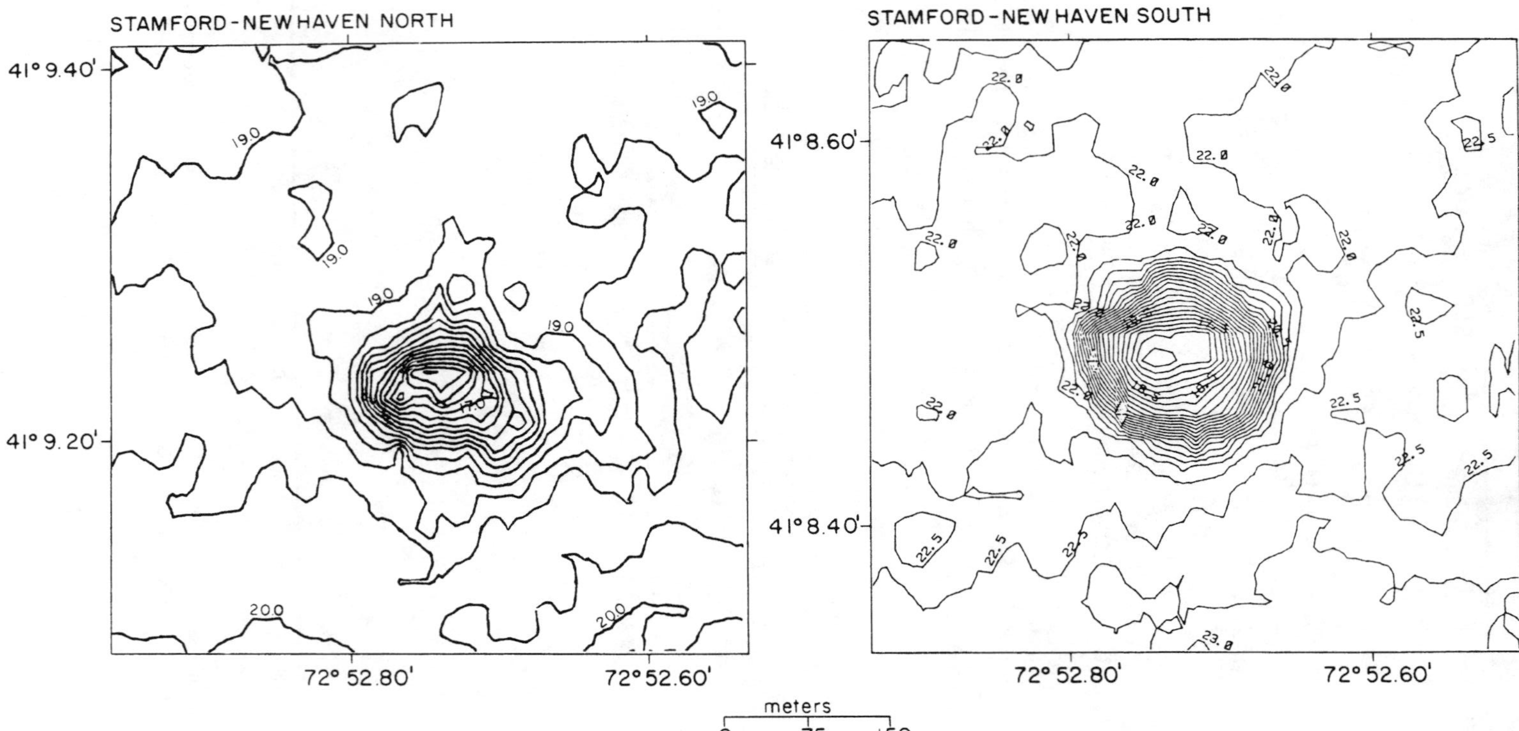

FIGURE 2.7. High precision bathymetric surveys at the Stamford-New Haven north and south dredged-spoil disposal sites after dumping of cap material. (From Morton, R. W., in *Wastes in the Ocean*, Vol. 2, *Dredged Material Disposal in the Ocean*, Kester, D. R., Ketchum, B. H., Duedall, I. W., and Park, P. K. (Eds.), John Wiley & Sons, New York, 1983, 99. With permission.)

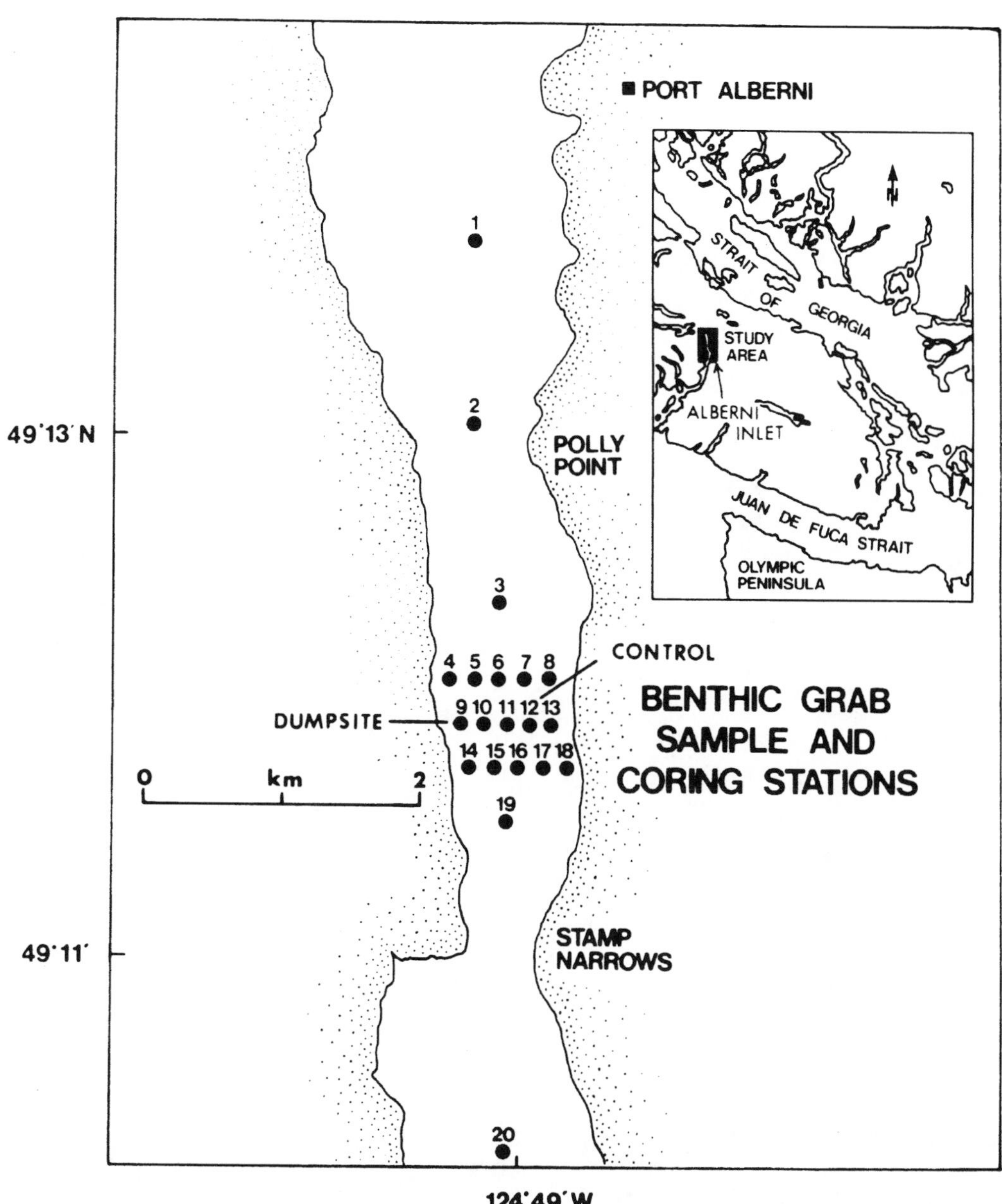

FIGURE 2.8. Dredged material disposal sites in Alberni Inlet. (From Levings, C. D., Anderson, E. P., and O'Connell, G. W., in *Wastes in the Ocean,* Vol. 6, *Nearshore Waste Disposal,* Ketchum, B. H., Capuzzo, J. M., Burt, W. V., Duedall, I. W., Park, P. K., and Kester, D. R. (Eds.), John Wiley & Sons, New York, 1985, 131. With permission.)

10 Effects of Electric Generating Stations

INTRODUCTION

Large open-cycle, once-through cooling electric generating stations (>500 MW) are sited on estuaries or other coastal systems because they require substantial quantities of condenser cooling water. These power stations, whether fossil-fueled (i.e., coal, oil, or gas-fired) or nuclear, discharge tremendous amounts of heat to the aquatic environment during operation. For example, fossil-fueled power plants have an average thermal efficiency of about 38%, and nuclear power plants, approximately 33%. Hence, more than 60% of the heat energy produced by these facilities is rejected to the environment. The average heat rejection of fossil-fueled power plants exceeds 5500 Btu/kWh of electricity produced, and that of nuclear power plants is about 50% greater.[1,2] A 1000 MW nuclear power plant with an open-cycle, once-through cooling system requires 3 to 5 billion l/day of cooling water. Such a unit discharges thermal effluent with an average temperature of approximately 15°C above ambient.[3]

Rejected heat from large power plants generally results in significant calefaction of receiving waters that often affects biological communities. The most acute effects of heated discharges typically arise in near-field regions in close proximity to the facilities, although subtle effects also may be detected in far-field regions where temperature increases in some areas are no more than 1 or 2°C. Changes in species composition, species diversity, population density, and primary production frequently are observed in near-field regions. The physiological and behaviorial responses of organisms affected by thermal discharges commonly involve: (1) an alteration of metabolic rates leading to a diminution of growth, and (2) behavioral adjustment manifested in avoidance or attraction reactions to the heated effluent. Large numbers of finfish occasionally experience heat- or cold-shock mortality within thermal plume areas.

Aside from the thermal loading of estuarine and coastal waters, other impacts of power plants result from chemical releases, as well as impingement of organisms on intake screens and entrainment of organisms through plant condensers.[4-6] The most notable chemical impacts are associated with power plant biocides, particularly the chlorine used to prevent and remove buildup of bacterial slime and other fouling organisms in condenser tubes.[7,8] Biocide releases have been implicated periodically in mortality of organisms in receiving waters and hence are closely regulated. There are several documented cases of greening in shellfish owing to the release of copper from power plants.[9] While low-level radioactive waste is routinely discharged from nuclear power plants to receiving waters, the concentration of this material is too low to cause hazardous conditions for aquatic organisms or man.

Finfish and other nonentrainable organisms are frequently impinged on intake screens and occasionally on trash racks during plant operation. Smaller organisms, such as the early life history stages of fish and invertebrates (i.e., eggs and larvae), pass through the condenser system and are exposed to thermal, chemical, and mechanical stresses that lower their survival rates. Both entrainment and impingement can reduce the productivity of a system due to the loss of large numbers of organisms at the plant site.

THERMAL DISCHARGES

Initial concerns regarding the impacts of power plant cooling systems on the aquatic environment centered on thermal loading. Some of the earliest investigations on the biological effects of heated discharges were conducted during the 1950s and 1960s when power plants tended to be much smaller than those constructed during the 1970s and 1980s. For example, in 1955 electric generating stations averaged about 35 MW in size, with the largest facility being approximately 300 MW.[2] Larger power plants were built during the 1960s, and by the 1970s the size of most electric generating facilities amounted to 500 MW or more. It was common for power plants constructed in the 1980s to exceed 1000 MW. Nuclear power plant construction accelerated dramatically after 1970, with the number of units operating worldwide rising from 66 in 1970 to about 430 today. Of the 118 nuclear power plants currently operating in the U.S., 74 use once-through cooling systems. Of these units, 19 discharge directly to estuaries or coastal marine waters. Fewer units (N = 44) employ closed-cycle cooling, mainly mechanical or natural draft cooling towers.[10] There are many more fossil-fueled power plants than nuclear units on-line.

Thermal loading associated with cooling water discharges interferes with many important biological functions such as photosynthesis, respiration, enzyme activity, feeding, and reproduction. At elevated temperatures, the metabolic rate of faunal inhabitants increases while the dissolved oxygen content of the water decreases. For instance, in estuaries dissolved oxygen levels decline from 14.6 mg/l at 0°C to 6.6 mg/l at 40°C. Thus, while the oxygen demand of the biota rises at higher temperatures, less dissolved oxygen is available. Under extreme conditions, many individuals may die suddenly, whereas others become more susceptible to disease and chemical toxins, resulting in delayed mortality. As plant and animal remains accumulate on the estuarine bottom, bacterial decomposition accelerates, further diminishing dissolved oxygen levels of receiving waters. Elevated water temperatures, together with organic loading and increased bacterial respiration in the summer, may promote severe anoxia or hypoxia that can decimate benthic populations.

Thermal effluent also alters other properties of receiving waters, such as density, viscosity, surface tension, and nitrogen solubility, all of which decrease with rising water temperature. Changes in density can have a significant effect on water column stratification and local hydrographic conditions. Lower viscosity coupled to waste heat discharges may cause an increase in settling rates of particles, thereby affecting sedimentation rates in near-field regions. Some fish experience nitrogen embolism when the concentration of dissolved nitrogen changes rapidly in the discharge zone. At elevated temperatures, vapor pressure also rises, which fosters higher evaporation rates in power plant outfalls.

The combination of heated waters and biocides in the discharge zone reduces primary production at some sites. Within the phytoplankton community, less desirable and more heat-tolerant blue-green algae tend to replace diatoms, green algae, and other more favorable forms, often culminating in noxious blooms that place large biological oxygen demands on receiving waters. Blue-greens gradually replace diatoms and green algae as water temperatures rise from about 10 to 38°C. Extreme temperatures of discharge waters, together with scouring of the bottom habitat at some plant outfalls, locally decimate or eliminate benthic populations.

Although elevated outfall temperatures may not be lethal, they can induce potentially deleterious sublethal effects such as a decrease in reproduction or hatching success of eggs, a retardation in larval development, and an alteration in respiratory rates. Some populations have a much greater risk of death when the temperature exceeds some upper threshold that may not be much greater than the optimal temperature for the species. Several mechanisms may cause thermal death of higher organisms: the failure of smooth muscle peristalsis, denaturation of protein in the cells, increased lactic acid in the blood, and oxygen deficit due to increased respiratory activity.[11] Most of these aforementioned harmful effects can be averted by acclimation of the organisms.

Fish have the ability to thermoregulate behaviorally; consequently, many species will avoid or will be attracted to a thermal plume depending on the season of the year and other factors. The thermal preference and temperature tolerance of a species strongly influence the movements of individuals in an outfall area.[12,13] Direct kills of fish from heat stress are not common because the individual perceives the thermal characteristics of a plume to be incompatible with its preferred temperature range and, therefore, moves away from the area of stress. However, once a species has adapted to a thermally impacted area, abrupt shutdowns of a power plant can generate rapid temperature reductions that may be lethal to the organism. Thus, in temperate estuaries, cold shock mortality of fish populations in near-field regions of a power plant is not unusual.

Thermal plumes usually do not affect fish migration. Fish typically avoid areas of maximum temperature during migration by swimming around or under thermal plumes.[14] Temperatures as high as 33 to 35°C have not created a barrier to movements of some fish species.[15]

Heated effluents that enter shallow rivers or enclosed estuaries often alter primary and secondary production, community respiration, species composition, biomass, and nutrient dynamics in near-field regions, owing to modifications of physical, chemical, physiological, and behavioral phenomena.[11,16] Changes in the distribution of organisms are common, with the replacement of heat-sensitive species of algae, invertebrates, and fish by more temperature-resistant forms occurring in proximity to the discharge site. However, no long-term problems in biotic communities have been uncovered by extensive monitoring of far-field regions.[10] Thermal impacts are most likely to arise in areas of shallow, enclosed, and/or poorly mixed estuaries and in systems vulnerable to the siting of clusters of power plants that strain available water resources. Examples of concentrations of power plants on receiving waters include a series of electric generating stations on the Hudson River (i.e., Bowline, Lovett, Indian Point, Roseton, Danskammer, and Albany) and a series of stations sited along the subestuaries of the Chesapeake Bay estuarine system.

BIOCIDES

Chlorine is used in coastal power plants to control biofouling of the cooling water circuit. Because micro- and macrofouling of condenser tube walls lower the heat transfer efficiency of the condenser, accelerate the fluid frictional resistance, and increase the corrosion of metals, chlorine treatment of cooling waters is used to remove the buildup of bacterial slime and other biofouling organisms on the condenser circuits. Although chlorine destroys bacterial slimes very efficiently when added at regular intervals to plant cooling water, it does not control macrofouling (e.g., mussels or clams) as well. Mussels and other large shell-bearing organisms respond to intermittent chlorine dosing by closing their shells until the biocide concentration dissipates. However, chlorine rapidly kills microorganisms directly and hydrolyzes the extracellular polymers that hold biofilms together, thereby preventing their accumulation on heat exchanger surfaces.[17]

Nontarget organisms are also susceptible to power plant chlorination. Estuarine and marine organisms entrained in the coolant (i.e., phytoplankton, zooplankton, meroplankton, and ichthyoplankton), as well as organisms inhabiting outfall waters, tend to be adversely affected. Hence, the chlorine dose administered at the plant — typically in liquid, gaseous, or hypochlorite form[5] — must be strictly regulated to minimize the concentration of the biocide released to receiving waters, while still maintaining effective biofouling control in the plant. Effluent guidelines established by the U.S. Environmental Protection Agency do not allow the concentration of total residual chlorine to exceed 0.2 mg/l, and chlorine discharge from any power plant unit is permitted for no more than 2 hours a day.[18]

A byproduct of power plant chlorination is the propensity of chlorine to form toxic residual organic compounds (chloramines) which can be hazardous to many aquatic organisms. In addition, some of the organochlorine compounds that form can be highly refractory. The toxicity and fate of many of these persistent chlorinated compounds in estuarine and marine environments have not been determined.

In terms of toxicity of chlorine to freshwater and marine organisms, results of laboratory bioassay investigations indicate that the freshwater acute toxicity threshold runs from a chlorine concentration-time point of 1.0 mg/l-1 min to 0.0015 mg/l-7500 min. Similarly for marine organisms, the marine acute toxicity threshold runs from a chlorine concentration-time point of 0.15 mg/l-1 min to 0.02 mg/l-100 min.[19] The chronic toxicity data suggest that relatively low concentrations of chlorine may elicit sublethal responses, including physiological, biochemical, or behavioral (e.g., avoidance) changes in both freshwater and marine forms.[20,21] Based on the site-specific assessment of power plant chlorination in the marine environment, chronic toxicity thresholds for saltwater organisms amount to approximately 0.3 mg/l for short exposure periods and 0.02 mg/l for long exposure periods.[19]

Phytoplankton populations are affected significantly by chlorination, although recovery appears to be rapid. For example, chlorination (0.1 to 1.2 mg/l) at the Millstone Point Nuclear Power Plant on Long Island Sound resulted in a 79% reduction of photosynthetic activity in entrained water.[22] Marine phytoplankton photosynthesis decreased by 70 to 80% in the thermal plume of the San Onofre nuclear power stations in California, where chlorine levels ranged from 0.02 to 0.04 mg/l.[23] Phytoplankton populations are highly resilient to chlorine exposure, however, because of their short generation times.[24]

Meroplankton and ichthyoplankton are likewise sensitive to chlorine exposure. For instance, Capuzzo[25] and Capuzzo et al.[26] found decreased growth and metabolic activity in larval lobsters (*Homarus americanus*) subjected to prolonged sublethal exposure of chlorine or chloramine. After 30- to 60-min exposures of these larvae to 1.0 and 0.1 mg/l concentrations of chlorine, acute increases in respiration were noted. The lobster larvae were even more sensitive to chloramine.

The tolerance of ichthyoplankton to chlorine appears to be age related. Studies of the eggs and larvae of blueback herring (*Alosa aestivalis*), striped bass (*Morone saxatilis*), and white perch (*Morone americana*) show that newly laid eggs are less tolerant to chlorine than older eggs. However, older larvae of these species tend to be less tolerant of the biocide than younger larvae.[27] The eggs of plaice (*Pleuronectes platessa*) and sole (*Solea solea*) are more tolerant of chlorine than the larvae of these species.[20] Later stages of some species, such as postlarvae of Atlantic herring *Clupea harengus* L.,[28] exhibit considerable sensitivity to chlorine as well.

Chlorine exposure can be detrimental to many juvenile and adult benthic invertebrates and finfish. The toxicity of chlorine to macrobenthic invertebrates is a complicated function of exposure period, temperature, and other factors.[19,20] In general, chlorine begins to be lethally effective on marine fish at concentrations of about 0.01 mg/l,[19] although temperature, the presence of other compounds (e.g., nitrogenous substances), and the physiological condition of the animal appear to affect their tolerance.[5,20] Chlorine avoidance thresholds in estuarine fish populations range from approximately 0.04 to 0.15 mg/l.[29] When exposed to sufficiently high chlorine levels, juvenile and adult fish eventually die from anoxia as the biocide attacks the gills, resulting in the oxidation of hemoglobin to methemoglobin.[20]

IMPINGEMENT

Numerous fish and macroinvertebrates are impinged on intake water screening devices of power plants each year, culminating in sizable mortalities. For example, more than 7 million fish were impinged on intake screens of the P. H. Robinson Power Plant in Galveston Bay, TX, over a 1-year period from 1969 to 1970. At the Millstone Point Nuclear Power Station in Connecticut, impingement mortality of Atlantic menhaden (*Brevoortia tyrannus*) exceeded 2 million individuals in the late summer and early fall of 1971.[11] From September 1975 through August 1977, approximately 13 million fish and macroinvertebrates were impinged at the Oyster Creek Nuclear Generating Station in New Jersey.[30] Holmes[31] reported weekly impingement estimates of up to 60,000 fish at the Fawley Generating Station in England. Although the absolute numbers of estuarine and coastal marine organisms impinged on power plant intake screens range into the millions over an annual cycle at some large power plants, the actual impact of these losses on resident populations is difficult to ascertain.

Many factors influence impingement rates. Apart from intake flow velocities, type of intake screens, and configuration of the intake structure, the behavior and physiology of the organisms themselves also play prominent roles. Seasonal variations in impingement, which can be substantial, are affected by migration of populations, changes in water temperature, and swimming performance. The condition of an organism is clearly important in its ability to withstand currents and hence impingement on intake screens. The swimming performance of a fish is contingent upon numerous factors, such as its developmental stage and nutritional condition, as well as the occurrence of parasite infestations, the presence of predators or prey, tides, time of day, and season.

The physical location and characteristics of intake structures may be more important than intake velocity in determining screen kill at many plants.[32] The volume of cooling water drawn into some facilities has been shown to correlate with the severity of impingement.[33] The presence of fish in the intake zone increases the probability of impingement.[34] For power plants sited on estuaries, seasonal fluctuations of finfish impingement primarily reflect the seasonal migration of populations in the estuary, which brings individuals into and out of intake areas, as well as the magnitude of temperature changes, which control the ability of individuals to withstand intake currents. These effects are evident at the Oyster Creek Nuclear Generating Station where seasonal temperature changes are substantial and maximum impingement of fish on intake screens occurs in the Spring and Fall when many species migrate into (Spring) or emigrate out of (Fall) Barnegat Bay.[30]

Many studies have been conducted on the effects of intake screening and spray wash systems on impingement mortality. Power plants generally utilize drum or band-type intake screens with apertures of 5 to 9 mm. Debris and organisms entrapped on conventional intake screens are removed by pressurized spray, while the screens rotate vertically in a screen well. As the spray washes the organisms off the screens, they collect in sluiceways or channels which return them to natural waterways. The debris is dumped in containers and transported to licensed disposal sites.

Improvements in intake screening devices during the past 25 years have significantly reduced impingement mortality. For instance, the retrofitting of Ristroph intake screens in 1974 onto the Surry Power Station, located on the James River, VA, greatly reduced impingement mortality. Results of an 18-month study showed that more than 93% of impinged fish survived when Ristroph screens were used at the station.[35] After Ristroph screens were retrofitted onto the Oyster Creek Nuclear Generating Station in 1978, mortality of impinged fish decreased by 24%.[36] These Ristroph screens consist of a continuously rotating traveling design modified with a low pressure spray wash and fish recovery and return system. The Ristroph screens at the Oyster Creek Nuclear Generating Station contain water-tight fish buckets which collect impinged organisms washed from the screens and return them to the plant discharge canal via a sluiceway.

Other modifications of intake systems which have been implemented to mitigate impingement mortality rely heavily on behavioral barriers and guidance systems. Most of these modifications are designed to attract or repel fish from intake structures by incorporating deflection methods and exclusion structures. The main deflection techniques developed to ameliorate impingement include electrical barriers, air bubble curtains, artificial lights, acoustic barriers, louvers, and velocity caps.[37,38] Among these methods, louvers and velocity caps appear to afford the greatest success in directing fish away from intake screens and guiding them safely into areas for removal or bypass.[37] In regard to exclusion devices, fine screens or clinker bunds may be effective, although they clog easily and consequently require much backwashing and general maintenance. In addition to these modifications, the use of horizontal rather than conventional vertical traveling screens may offer great potential for guiding fish from plant intakes. Perforated pipe intakes, wedgewire screens, rotating disc screens, open setting screens, fine mesh screens, radial well intakes, porous-like filters, and rapid filter beds also hold promise in impingement mitigation.[39]

ENTRAINMENT

The absolute number of organisms lost in large power plants due to entrainment mortality is usually extremely high. It is not unusual for annual entrainment estimates at large power plants located on estuaries to exceed 1 billion individuals. While not all of the eggs, larvae, microinvertebrates, and small juvenile fish passing through the cooling water condenser systems of power plants die, large percentages of them are injured or destroyed by mechanical and hydraulic shocks (e.g., pressure changes, abrasion of particles, impacts against piping, and turbulence), thermal stresses, and chemical toxicity (e.g., chlorination of cooling water). Mechanical damage appears to be of overriding importance at many sites.

Biotic factors must also be considered in entrainment mortality. For example, the seasonal densities of entrainable organisms in impacted systems are closely correlated with observed mortalities. Furthermore, the sizes, life stages, and relative susceptibility to injury of the entrained species play a critical role in the observed mortality of estuarine organisms in power plants.

Those organisms which survive passage through the plant cooling water systems may experience lower survivorship when secondarily entrained in outfall waters. Sublethal exposure to mechanical, thermal, and chemical shocks in power plants reduces the ability of the entrained organisms to avoid predators in discharge plumes.[40] Delayed mortality of the entrained organisms occurs in outfall waters, further depleting the number of viable eggs and larvae in receiving waters. The cropping or predatory effect of power plants has been examined extensively using predictive models to assess potential impacts on populations and communities in near- and far-field regions. Schubel and Marcy[41] provide a detailed review of the biological effects of power plant entrainment.

Entrainment mortality estimates are available on most coastal power plants, and several examples follow. The Connecticut Yankee Nuclear Power Plant on the Connecticut River killed an estimated 179 million fish larvae during 1969 and 1970. Over a 5-day period in 1972, the Millstone Point Nuclear Power Plant killed an estimated 150 million fingerling Atlantic menhaden.[11] The estimated number of hard clam (*Mercenaria mercenaria*) larvae lost at the Oyster Creek Nuclear Generating Station from September 1975 through August 1976 amounted to 1.14×10^{10}. Somewhat lower mortality figures were registered on bay anchovy (*Anchoa mitchilli*) larvae and juveniles (1.61×10^9) entrained and killed at the station during this time period.[30] Entrainment data collected at the Pilgrim Nuclear Power Station on Cape Cod Bay, MA, from 1983 to 1988 indicated that phytoplankton survival ranged from 48 to 98% after in-plant passage, and zooplankton survival, 95 to 100%. In 1986, the estimated number of fish eggs and larvae entrained at this power plant equalled 1.70×10^9 and 2.75×10^9, respectively.[42] While these numbers are unequivocally large, they must be compared to the many billions of eggs and larvae in far-field regions not impacted by entrainment. Accurate assessment impacts of entrainment mortality on entire systems clearly require evaluation of a host of complicating factors.

ASSESSMENT IMPACTS

Considerable controversy over the years has surrounded the assessment of impingement and entrainment impacts on resident populations and communities. Mathematical population models have been formulated to assess these impacts.[43-46] Major limitations on the use of the models include inaccurate estimates of population sizes and the inability to accurately account for marked fluctuations in reproductive success and survival of organisms from year to year.[5,43] In addition, biological compensation, a basic tenet of fisheries management, may operate to mollify or eliminate impingement and entrainment effects.[40] For example, high egg and larval mortalities can be compensated by increases in fecundity and survival of resident populations, together with predator mortality, unless the populations diminish to very low levels.[5,47]

Among the most important abiotic factors affecting impingement and entrainment are the siting and cooling water system design of the power plant and the volume and ambient condition of the cooling water used. Biotic factors of significance relate to the abundance, survival, ecological roles, and reproductive strategies of the affected organisms.[48] Much variation in impingement and entrainment impact may arise from a multitude of factors such as seasonal fluctuations, organismal migrations, climatic changes, tides, storms, and unscheduled changes in plant operations.[5]

Results of power plant impact assessments during the last two decades reveal that, while local effects are evident at many sites, there have been no documented cases of long-term, system-wide problems ascribable to a single unit. Potential environmental impacts have been mitigated by adapting new siting considerations, modifying intake structures, reducing cooling water flow, and employing more cooling towers.[10] New engineering designs of systems have greatly allayed many long-term environmental problems associated with power plant operation.

REFERENCES

1. U.S. Environmental Protection Agency, *Development Document for Effluent Limitations, Guidelines, and New Source Performance Standards for the Steam Electric Power Generating Point Source Category,* Tech. Rep. EPA 440/1-74029a, U.S. Government Printing Office, Washington, D.C., 1974.
2. Tebo, L. B., Jr. and Little, J. A., Permitting procedures for thermal discharges, in *Factors Affecting Power Plant Waste Heat Utilization*, Gross L. B. (Ed.), Pergamon Press, New York, 1980, 110.
3. Sorge, E. V., The status of thermal discharges east of the Mississippi River, *Chesapeake Sci.*, 10, 131, 1969.
4. Hocutt, C. H., Stauffer, J. R., Jr., Edinger, J. E., Hall, L. W., Jr., and Morgan, R. P., II (Eds.), *Power Plants: Effects on Fish and Shellfish Behavior*, Academic Press, New York, 1980, 103.
5. Langford, T. E., *Electricity Generation and the Ecology of Natural Waters*, Liverpool University Press, England, 1983.
6. Kennish, M. J., *Ecology of Estuaries: Anthropogenic Effects*, CRC Press, Boca Raton, FL, 1992.
7. Mattice, J. S. and Zittel, H. E., Site-specific evaluation of power plant chlorination, *J. Water Pollut. Control Fed.*, 48, 2284, 1976.
8. Jolley, R. L., Gorchev, H., and Hamiltion, D. H., Jr. (Eds.), *Water Chlorination: Environmental Impact and Health Effects*, Vol. 2, Ann Arbor Science Publishers, Ann Arbor, MI, 1978.
9. Abbe, G. R. and Sanders, J. G., Condenser replacement in a power plant: copper uptake and incorporation in the American oyster, *Crassostrea virginica*, *Mar. Environ. Res.*, 19, 93, 1986.
10. Talmage, S. S. and Meyers-Schone, L., Nuclear and thermal, in *Handbook of Ecotoxicology*, Hoffman, D. J., Rattner, B. A., Burton, G. A., Jr., and Cairns, J., Jr. (Eds.), Lewis Publishers, Boca Raton, FL, 1995, 469.
11. Hall, C. A. S., Howarth, R., Moore, B., III, and Vorosmarty, C. J., Environmental impacts of industrial energy systems in the coastal zone, *Ann. Rev. Energy*, 3, 395, 1978.
12. Coutant, C. C., Temperature selection by fish — a factor in power-plant impact assessments, in *Environmental Effects of Cooling Systems at Nuclear Power Plants*, International Atomic Energy Agency, Vienna, 1975, 575.
13. Olla, B. L., Studholme, A. L., and Bejda, A. J., Behavior of juvenile bluefish *Pomatomus saltatrix* in vertical thermal gradients: influence of season, temperature acclimation, and food, *Mar. Ecol. Prog. Ser.*, 23, 165, 1985.
14. Moss, J. L., Boonyaratpalin, S., and Shelton, W. L., Movement of three species of fishes past a thermally influenced area in the Coosa River, Alabama, in *Energy and Environmental Stress in Aquatic Systems*, Thorp, J. H. and Gibbons, J. W. (Eds.), CONF-771114, DOE Symposium Series 48, National Technical Information Service, Springfield, VA, 1978, 534.
15. Wrenn, W. B., Temperature preference and movement of fish in relation to a long, heated discharge channel, in *Thermal Ecology II*, Esch, G. W. and McFarlane, R. W. (Eds.), National Technical Information Service, Springfield, VA, 1976, 191.
16. Day, J. W., Jr., Hall, C. A. S., Kemp, W. M., and Yanez-Arancibia, A., *Estuarine Ecology*, John Wiley & Sons, New York, 1989.
17. Characklis, W. G., Bryers, J. D., Trulear, M. G., and Zelver, N., Biofouling film development and its effects on energy losses: a laboratory study, in *Condenser Biofouling Control: Symposium Proceedings*, Ann Arbor Science Publishers, Ann Arbor, MI, 1980, 49.
18. Chow, W., Condenser biofouling control: the state-of-the-art, in *Proceedings: Condenser Biofouling Control State-of-the-Art Symposium*, Chow, W. and Massalli, Y. G. (Eds.), Tech. Rep., Electric Power Research Institute, Palo Alto, CA, 1985, 1-1.
19. Mattice, J. S. and Zittel, H. E., Site-specific evaluation of power plant chlorination, *J. Water Pollut. Control Fed.*, 48, 2284, 1976.
20. Morgan, R. P., II and Carpenter, E. J., Biocides, in *Power Plant Entrainment: A Biological Assessment*, Schubel, J. R. and Marcy, B. C., Jr. (Eds.), Academic Press, New York, 1978, 95.

21. Seegert, G., Bogardus, R. B., and Horvatk, F., *Review of the Mattice and Zittel Paper. Site-Specific Evaluation of Power Plant Chlorination Project,* Tech. Rep., Edison Electric Institute, Washington, D.C., 1978.
22. Carpenter, E. J., Peck, B. B., and Anderson, S. J., Cooling water chlorination and productivity of entrained phytoplankton, *Mar. Biol.,* 16, 37, 1972.
23. Eppley, R. W., Ringer, E. H., and Williams, P. M., Chlorine reactions with seawater constituents and the inhibition of photosynthesis of natural marine phytoplankton, *Est. Coastal Mar. Sci.,* 4, 147, 1976.
24. Goldman, J. C. and Davidson, J. A., Physical model of marine phytoplankton chlorination at coastal power plants, *Environ. Sci. Technol.,* 11, 908, 1977.
25. Capuzzo, J. M., The effects of free chlorine and chloramine on growth and respiration rates of larval lobsters (*Homarus americanus*), *Water Res.,* 11, 1021, 1977.
26. Capuzzo, J. M., Lawrence, S. A., and Davidson, J. A., Combined toxicity of free chlorine, chloramine, and temperature to stage I larvae of the American lobster, *Homarus americanus*, *Water Res.,* 10, 1093, 1976.
27. Morgan, R. P., II and Prince, R. D., Chlorine toxicity to eggs and larvae of five Chesapeake Bay fishes, *Trans. Am. Fish. Soc.,* 106, 380, 1977.
28. Dempsey, C. H., The exposure of herring postlarvae to chlorine in coastal power stations, *Mar. Environ. Res.,* 20, 279, 1986.
29. Middaugh, D. P., Couch, J. A., and Crane, A. M., Response of early life history stages of the striped bass, *Morone saxatilis*, to chlorination, *Chesapeake Sci.,* 18, 141, 1977.
30. Jersey Central Power and Light Company, Oyster Creek and Forked River Nuclear Generating Stations 316(a) and (b) Demonstration, Vol. 1, Tech. Rep., Jersey Central Power and Light Company, Morristown, NJ, 1978.
31. Holmes, R. H. A., Fish and Weed on Fawley Power Station Screens, Central Electric Research Laboratory (C.E.R.L.) Lab. Note RD/L/N 129/75, Leatherhead, Surry, England, 1975.
32. Boreman, J., *Impacts of Power Plant Intake Velocities on Fish, Topical Brief No. 1, Fish and Wildlife Resources and Electric Power Generation,* Fish and Wildlife Service, U.S. Department of Interior, Washington, D.C.
33. Clark, J. R. and Brownwell, W., *Electric Power Plants in the Coastal Zone: Environmental Issues*, American Littoral Society Special Publ. No. 7, Highlands, NJ, 1972.
34. Merriman, D. and Thorpe, L. M., Introduction, in *The Connecticut River Ecology Study: The Impact of a Nuclear Power Plant*, Merriman, D. and Thorpe, L. M. (Eds.), American Fisheries Society Monograph No. 1, Lawrence, KS, 1976, 1.
35. White, J. C. and Brehmer, M. L., Eighteen-month evaluation of the Ristroph traveling fish screens, in *Proceedings of the Third National Workshop on Entrainment and Impingement: 316(b) Research and Compliance*, Jensen, L. D. (Ed.), Ecological Analysts, Melville, NY, 1976, 241.
36. Kennish, M. J., Roche, M. B., and Tatham, T. R., Anthropogenic effects on aquatic communities, in *Ecology of Barnegat Bay, New Jersey*, Kennish, M. J. and Lutz, R. A. (Eds.), Springer-Verlag, New York, 1984, 318.
37. Sharma, R. K., Fish Protection at Water Diversions and Intakes: A Bibliography of Published and Unpublished References, Tech. Rep. ANL.ESP-1, Argonne National Laboratory, Argonne, IL, 1973.
38. Hocutt, C. H., Behavioral barriers and guidance systems, in *Power Plants: Effects on Fish and Shellfish Behavior*, Hocutt, C. H., Stauffer, J. R., Jr., Edinger, J. E., Hall, L. W., Jr., and Morgan, R. P., II (Eds.), Academic Press, New York, 1980, 183.
39. Hocutt, C. H. and Edinger, J. E., Fish behavior in flow fields, in *Power Plants: Effects on Fish and Shellfish Behavior*, Hocutt, C. H., Stauffer, J. R., Jr., Edinger, J. E., Hall, L. W., Jr., and Morgan, R. P., II (Eds.), Academic Press, New York, 1980, 143.
40. O'Connor, S. G. and McErlean, A. J., The effects of power plants on productivity of the nekton, in *Estuarine Research*, Vol. 1, *Chemistry, Biology, and the Estuarine System*, Cronin, L. E. (Ed.), Academic Press, New York, 1975, 494.
41. Schubel, J. R. and Marcy, B. C., Jr. (Eds.), *Power Plant Entrainment: A Biological Assessment*, Academic Press, New York, 1978.
42. Boston Edison Company, Marine Ecology Studies Related to Operation of Pilgrim Station, Semi-Annual Rept. No. 33, Boston Edison Company, Braintree, MA, 1989.
43. Van Winkle, W. (Ed.), *Proceedings of the Conference Assessing the Effects of Power-Plant-Induced Mortality on Fish Population*, Pergamon Press, New York, 1977.

44. Murarka, I. P., Validation and Software Documentation of the ANL Fish Impingement Model, Tech. Rept. ANL/ES-62, Argonne National Laboratory, Argonne, IL, 1978.
45. Swartzmann, G. L., *Comparison of Simulation Models Used in Assessing the Effects of Power-Plant-Induced Mortality on Fish Populations,* U.S. Nuclear Regulatory Commission Rep. No. NUREG/CR-0474, Washington, D.C., 1978.
46. Ogawa, H., Modeling of power plant impacts on fish populations, *Environ. Mgmt.,* 3, 321, 1979.
47. Goodyear, C. P., Assessing the impact of power plant mortality on the compensatory reserve of fish populations, in *Proceedings of the Conference Assessing the Effects of Power-Plant-Induced Mortality on Fish Population*, Van Winkle, W. (Ed.), Pergamon Press, New York, 1977, 198.
48. Marcy, B. C., Jr., Kranz, V. R., and Barr, R. P., Ecological and behavioral characteristics on fish eggs and young influencing their entrainment, in *Power Plants: Effects on Fish and Shellfish Behavior*, Hocutt, C. H., Stauffer, J. R., Jr., Edinger, J. E., Hall, L. W., Jr., and Morgan, R. P., II (Eds.), Academic Press, New York, 1980, 29.

APPENDIX 1. POWER PLANT DESIGNS

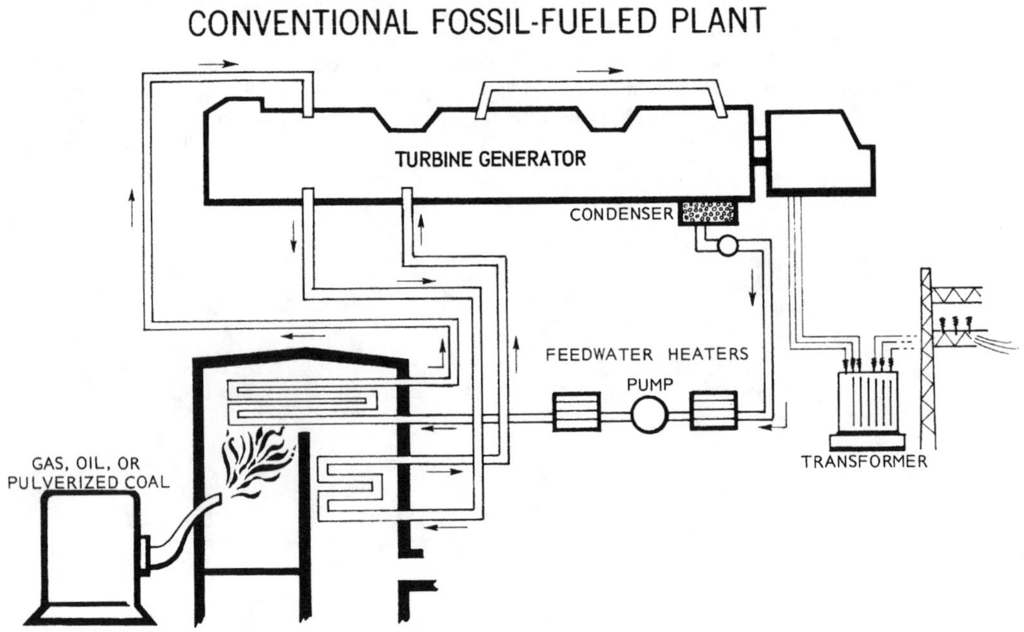

FIGURE 1.1. Simplified schematic diagram of a conventional fossil-fueled power plant. (From the Pennsylvania Department of Education, Harrisburg, PA.)

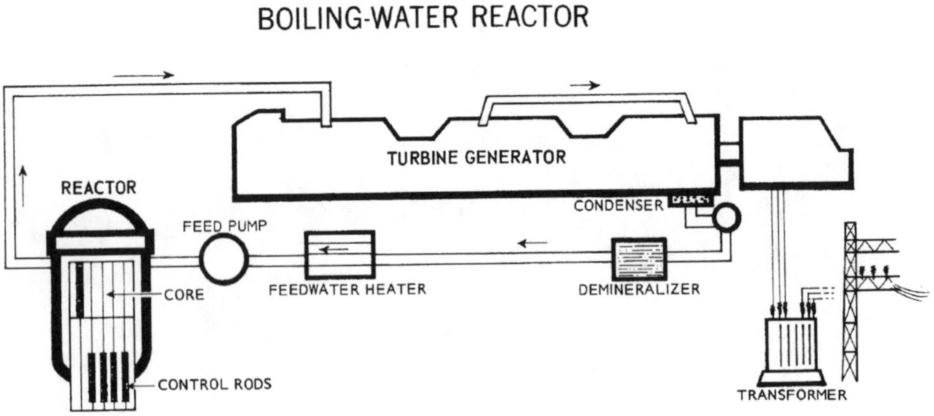

FIGURE 1.2. Simplified schematic diagram of a nuclear power plant consisting of a single boiling-water reactor. (From the Pennsylvania Department of Education, Harrisburg, PA.)

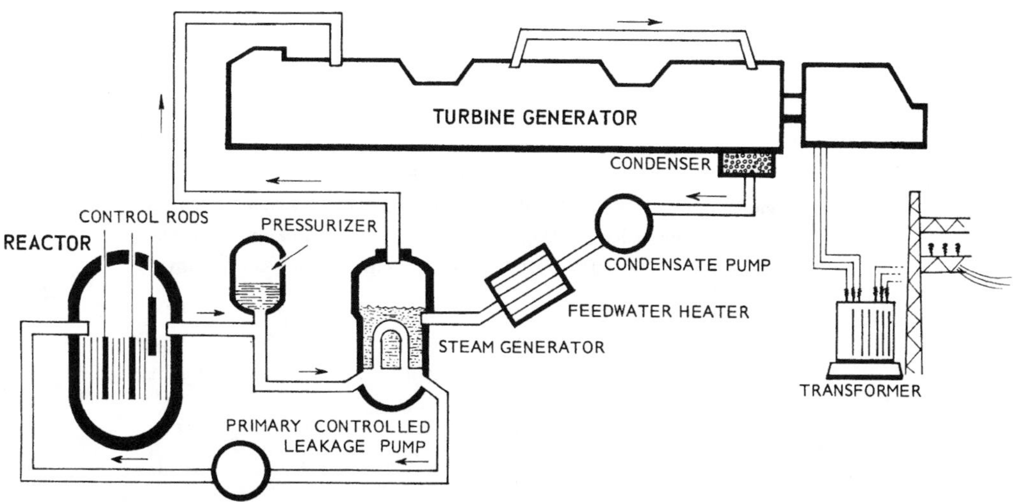

FIGURE 1.3. Simplified schematic diagram of a nuclear power plant consisting of a pressurized water reactor. (From the Pennsylvania Department of Education, Harrisburg, PA.)

TABLE 1.1
Estimated Cooling Water Requirements for a 1000-MWe Steam-Electric Plant Operating at Full Load

	Type of Plant			
	Fossil		Nuclear	
	1980	1990	1980	1990
Plant heat rate,[a] Btu/kWh	9500	9000	10,500	10,000
Condenser flows-cms, for various temperature rises across the condenser				
10°F (5.5°C)	58.9	53.5	82.6	76.7
15°F (8.3°C)	39.3	35.7	55.2	51.2
20°F (11.1°C)	29.4	26.9	41.2	38.5
Consumptive use, cms for various types of cooling				
Once through	0.34	0.28	0.48	0.42
Cooling ponds[b]	0.45	0.40	0.62	0.57
Cooling towers[c]	0.79	0.74	1.13	0.99

[a] For fossil-fueled plants in operation in 1970, a heat rate of 10,000 Btu/kWh and a temperature rise of 13°F (7.2°C) were assumed, except where reported heat rate data were available.

[b] Where appropriate, an additional allowance was made for natural evaporation from the pond surface.

[c] Evaporative towers; includes blowdown and drift.

Source: Eichholz, G. G., *Environmental Aspects of Nuclear Power,* Ann Arbor Science Publishers, Ann Arbor, MI, 1976. With permission.

APPENDIX 2. THERMAL DISCHARGES

TABLE 2.1
Properties of Water as a Function of Temperature

Temp. (°C)	Vapor Pressure (torr)	Viscosity (centipoise)	Density (g/ml)	Surface Tension (dynes/cm)	Oxygen Solubility (mg/l)	Nitrogen Solubility (mg/l)
0	4.579	1.787	0.99984	75.6	14.6	23.1
5	6.543	1.519	0.99997	74.9	12.8	20.4
10	9.209	1.307	0.99970	74.2	11.3	18.1
15	12.788	1.139	0.99910	73.5	10.2	16.3
20	17.535	1.002	0.99820	72.8	9.2	14.9
25	23,756	0.890	0.99704	72.0	8.4	13.7
30	31.824	0.798	0.99565	71.2	7.6	12.7
35	42.175	0.719	0.99406	70.4	7.1	11.6
40	55.324	0.653	0.99224	69.6	6.6	10.8

Source: Eichholz, G. G., *Environmental Aspects of Nuclear Power,* Ann Arbor Science Publishers, Ann Arbor, MI, 1976. With permission.

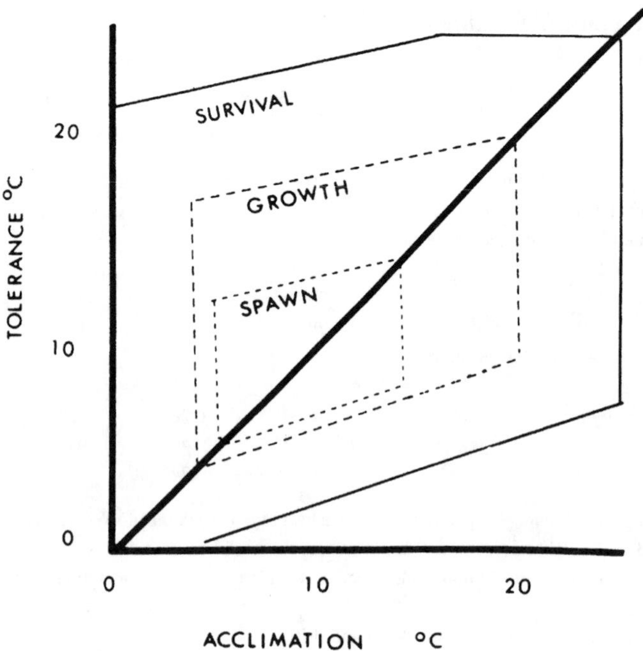

FIGURE 2.1. Temperature tolerance polygon for aquatic organisms. (From Brett, J. R., in *Water Pollution,* Taft, R. A. (Ed.), Sanitation Engineering Center Tech. Rep., unpublished manuscript, 1960, 110. With permission.)

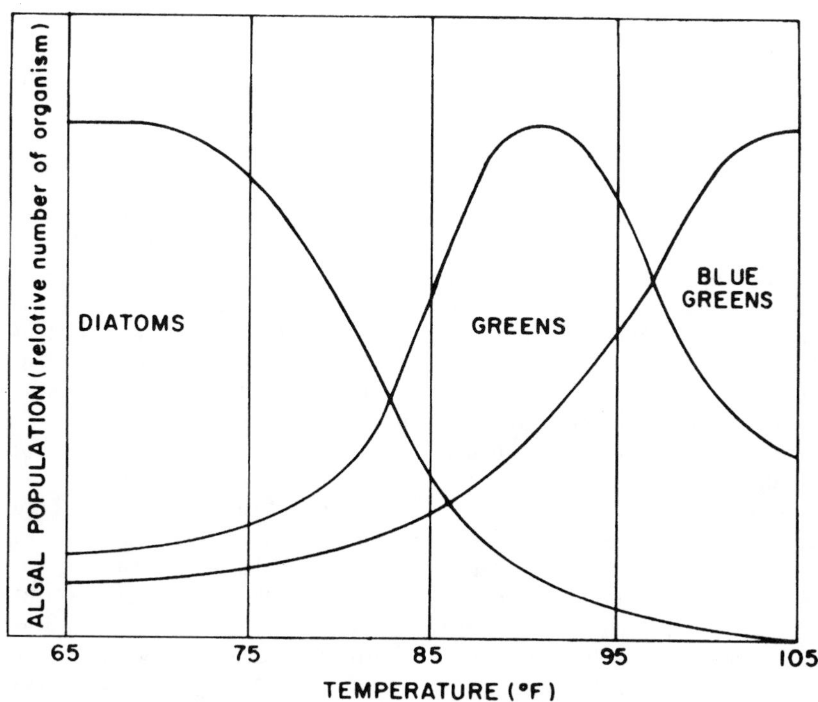

FIGURE 2.2. Population shifts of algae with changes in temperature. (From Eichholz, G. G., *Environmental Aspects of Nuclear Power,* Ann Arbor Science, Ann Arbor, MI, 1976. With permission.)

TABLE 2.2
Lethal Temperature Limits for Adult Marine, Estuarine, and Freshwater Fishes

			Acclimation Temp.[a]			
Species	**°C**	**°F**	**°C**	**°F**	**°C**	**°F**
Alewife	—	—	—	—	26.7–32.2	80.0–90.0
Bass, striped	—	—	6.0–7.5	42.8–45.5	25.0–27.0	77.0–80.0
California killifish	14.0–28.0	57.2–82.4	—	—	32.3–36.5	90.1–97.7
Common silverside	7.0–28.0	44.6–82.4	1.5–8.7	34.8–47.8	22.5–32.5	73.3–90.3
Flounder, winter	21.0–28.0	69.8–82.4	1.0–5.4	33.8–41.6	—	—
	7.0–28.0	44.6–82.4	—	—	22.0–29.0	71.6–84.2
Herring	—	—	-1.0	30.2	19.5–21.2	67.1–70.1
Northern swellfish	14.0–18.0	57.2–82.4	8.4–13.0	47.1–55.4	—	—
	10.0–28.0	50.0–82.4	—	—	28.3–33.0	82.9–90.4
Perch, white	4.4	40.0	—	—	27.8	82.0
Salmon (general)	—	—	0.0	32.0	26.7	80.0
Bass, largemouth	20.0	68.0	5.0	41.0	32.0	89.6
	30.0	86.0	11.0	51.8	34.0	93.2
Bluegill	15.0	59.0	3.0	37.4	31.0	87.8
	30.0	86.0	11.0	51.8	34.0	93.2
Catfish, channel	15.0	59.0	0.0	32.0	30.0	86.0
	25.0	77.0	6.0	42.8	34.0	93.2
Perch, yellow	5.0	41.0	—	—	21.0	69.8
(Winter)	25.0	77.0	4.0	39.2	30.0	86.0
(Summer)	25.0	77.0	9.0	48.2	32.0	89.6
Shad, gizzard	25.0	77.0	11.0	51.8	34.0	93.2
	35.0	95.0	20.0	68.0	37.0	98.6
Shiner, common	5.0	41.0	—	—	27.0	80.6
	25.0	77.0	4.0	39.2	31.0	87.8
	30.0	86.0	8.0	46.4	31.0	87.8
Trout, brook	3.0	37.4	—	—	23.0	73.4
	20.0	68.0	—	—	25.0	77.0

[a] Values are LD_{50} temperature tolerance limits, i.e., water temperatures survived by 50% of the test fish after 1- to 4-day acclimation.

Source: The Federal Water Pollution Control Administration, *Industrial Waste Guide on Thermal Pollution,* Federal Water Pollution Control Administration, Corvallis, OR, 1968.

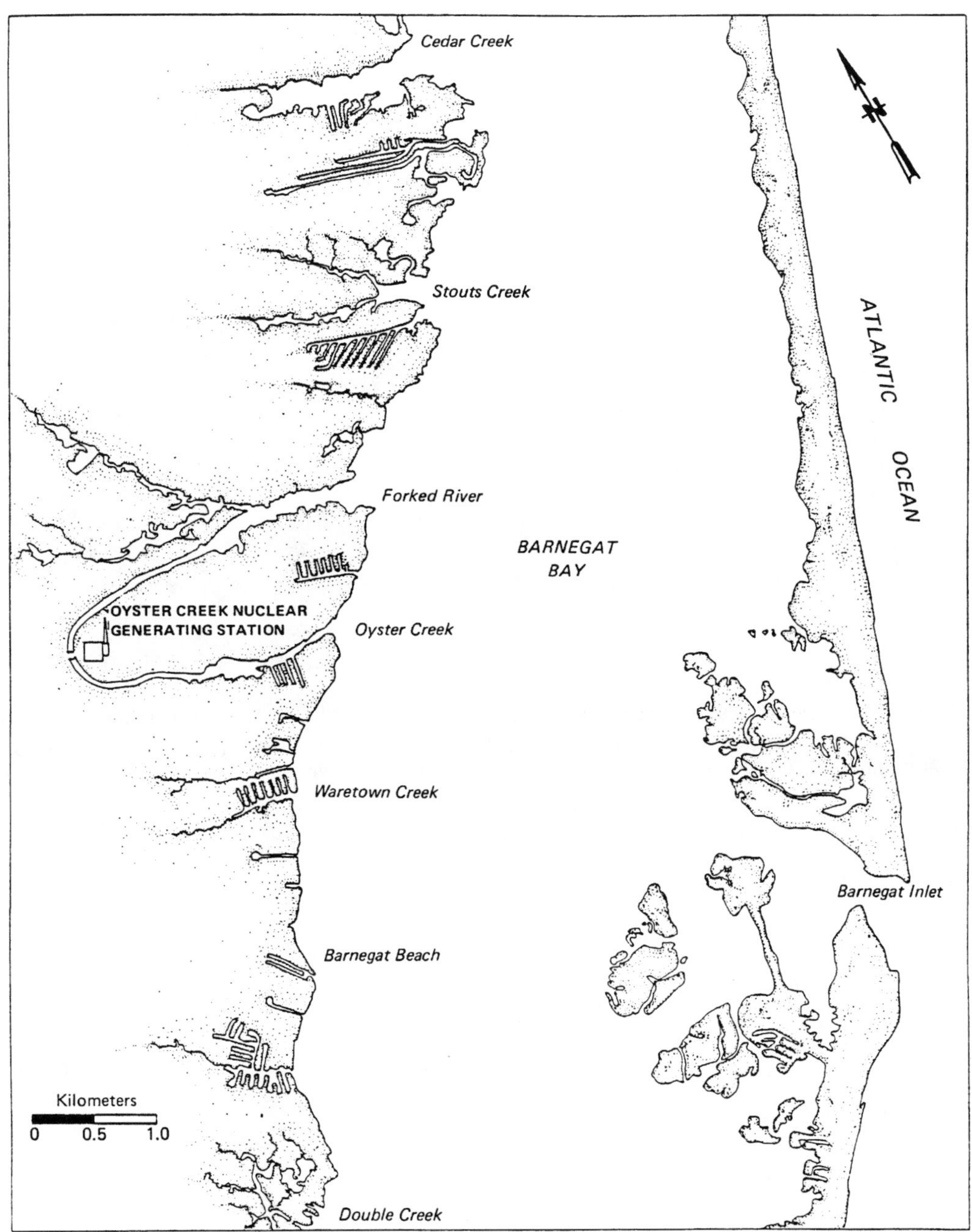

FIGURE 2.3. Location of the Oyster Creek Nuclear Generating Station on Barnegat Bay, NJ, showing the source of intake water (Forked River) for the power plant and the source of discharge water (Oyster Creek) entering the bay. (Modified from Chizmadia, P. A., Kennish, M. J., and Ohori, V. L., in *Ecology of Barnegat Bay, New Jersey,* Kennish, M. J. and R. A. Lutz (Eds.), Springer-Verlag, New York, 1984, 1.)

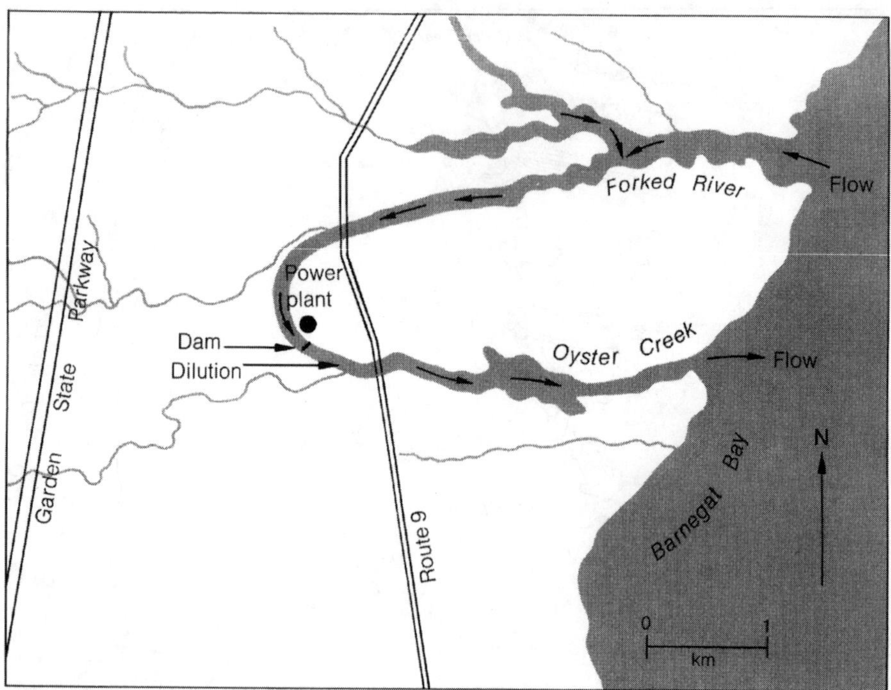

FIGURE 2.4. Flow characteristics at Forked River, Oyster Creek, and adjacent bay localities during operation of the Oyster Creek Nuclear Generating Station. Thermal discharges enter Barnegat Bay via Oyster Creek. (From Kennish, M. J. and Olsson, R. K., *Environ. Geol.,* 1, 41, 1975. With permission.)

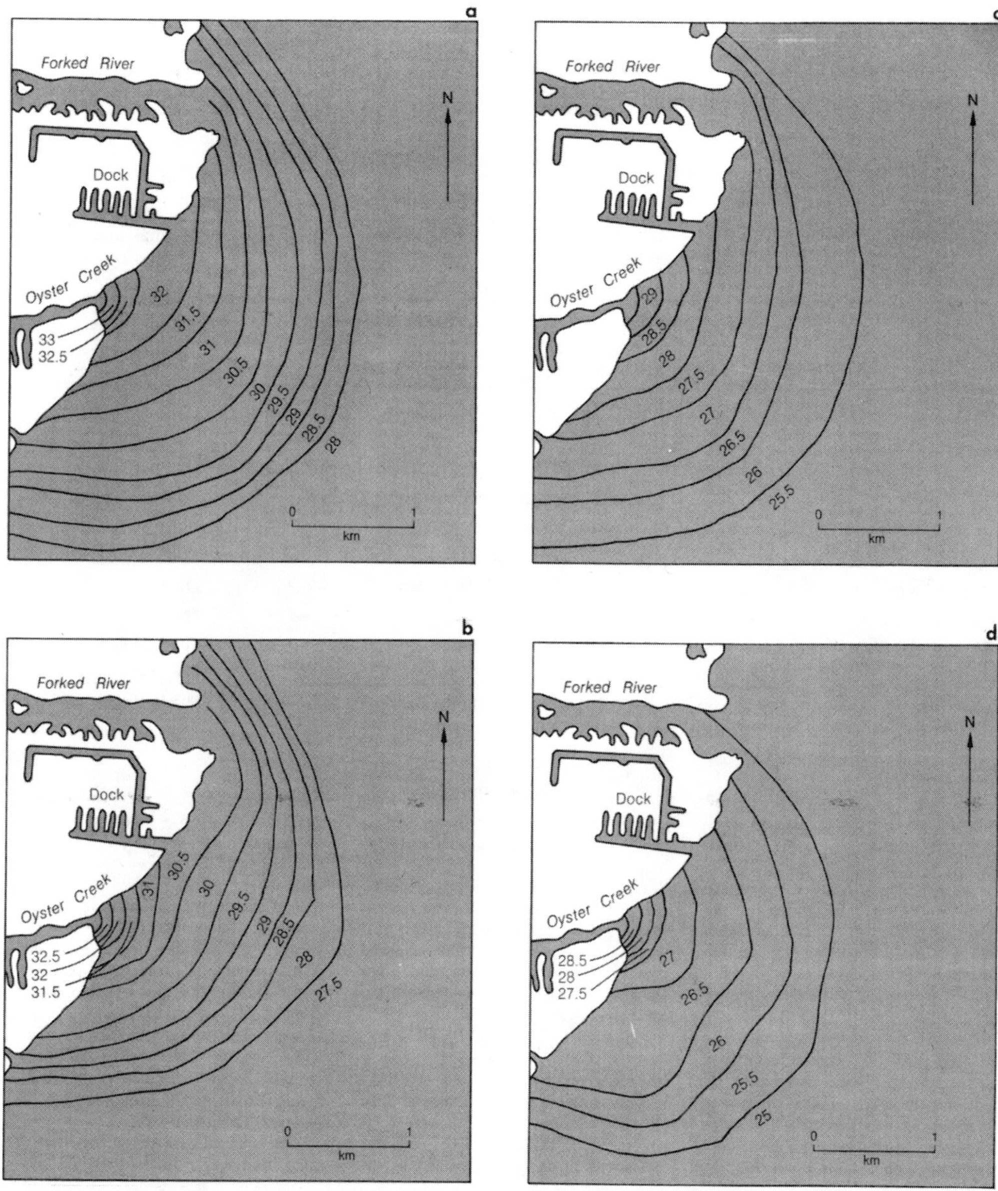

FIGURE 2.5. Thermal plumes entering Barnegat Bay from the Oyster Creek Nuclear Generating Station. (a) Surface temperatures on August 6, 1973. (b) Bottom temperatures on August 6, 1973. (c) Surface temperatures on September 8, 1973. (d) Bottom temperatures on September 8, 1973. (From Kennish, M. J. and Olsson, R. K., *Environ. Geol.,* 1, 41, 1975. With permission.)

TABLE 2.3
Fish Mortality in the Discharge Canal of the Oyster Creek Nuclear Generating Station (OCNGS) and Oyster Creek Caused by Operation of the OCNGS

Date	No.	Species	Size Range (mm)	Probable Cause
1/29/72	100,000–1,000,000	Atlantic menhaden	76–127	Thermal shock
1/5/73	18,000–1,200,000	Atlantic menhaden	102–356	Thermal shock
1/8/73	20	Bay anchovy	—	Thermal shock
2/16/73–2/21/73	Several thousand	Atlantic menhaden	—	Thermal shock
8/9/73	2000–4000	Atlantic menhaden	127–356	Thermal shock
1/7/74	500	Atlantic menhaden	203–280	Chlorine
1/11/74–1/15/74	9900–180,000	Atlantic menhaden	102–356	Thermal shock
	100–3600	Bluefish	228–356	Thermal shock
10/9/74	200	Crevalle jack	—	Thermal shock
2/4/75	100	Atlantic menhaden	—	Thermal shock
	50–100	Bluefish	—	Thermal shock
11/24/75	7–100	Crevalle jack	—	Thermal shock
12/29/75	15–100	Atlantic menhaden	100–250	Thermal shock
	3–200	Bluefish	90–170	Thermal shock
10/21/77	120–200	Blue runner	—	Thermal shock
		Crevalle jack	—	Thermal shock
1/15/79	682	Atlantic menhaden	165–225	Thermal shock
8/2/79	50–100	Striped bass	34–44	Thermal shock
		Northern puffer	Missing parts	
		Goosefish	Missing parts	
		Tautog		
12/17/79	Unknown	Unknown	Unknown	Unknown
12/20/79	12	Bluefish	—	Unknown
	1	Weakfish	—	
	1	Sea robin	—	
	1	Black sea bass	—	
	1	Atlantic menhaden	—	
1/5/80	5483	Atlantic menhaden	240[a]	Thermal shock
	952	Bluefish	295[a]	
	544	Spot	120[a]	
	43	Weakfish	501[a]	
	5	Scup	200[a]	
	1	Butterfish	—	
	1	Northern kingfish	240[a]	
11/22/80	3638	Blue runner	206[a]	Thermal shock
		Crevalle jack	173[a]	
	1038	Bluefish	267[a]	
	17	Smooth dogfish	601[a]	
	3	Ladyfish	298[a]	
	2	Northern kingfish	—	
	1	Gray snapper	118[a]	
	1	American eel	—	
	1	Mojarra	221	

TABLE 2.3 (continued)
Fish Mortality in the Discharge Canal of the Oyster Creek Nuclear Generating Station (OCNGS) and Oyster Creek Caused by Operation of the OCNGS

Date	No.	Species	Size Range (mm)	Probable Cause
12/9/82–12/10/82	166	Crevalle jack	110–204	Thermal shock
	80	Blue runner	171–218	
	76	Bluefish	274–476	
	28	Atlantic needlefish	250–661	
	9	Scup	205–247	
	2	American eel	—	
	1	Conger eel	—	
	1	Northern kingfish	185	
	1	Ladyfish	410	

[a] Mean size.

Source: Kennish, M. J., Roche, M. B., and Tatham, T. R., in *Ecology of Barnegat Bay, New Jersey,* Kennish, M. J. and Lutz, R. A. (Eds.), Springer-Verlag, New York, 1984, 318. With permission.

APPENDIX 3. RADIOACTIVE DISCHARGES

TABLE 3.1
Operating Characteristics at Selected Coastal Nuclear Power Plants

Plant	Gross Generating Capacity (MWe)	Maximum (at Full Power) Condenser ΔT (°C)	Cooling Water Flow (m^3/s)
Diablo Canyon 1 and 2, CA	2190	12	104
Millstone 1 and 2, CT	1522	13	64
Millstone 3, CT	1150	10	61
Pilgrim, MA	655	18	20
San Onofre 1, CA	430	10	20
San Onofre 2 and 3, CA	2200	11	103
St. Lucie 1, FL	850	13	33
St. Lucie 2, FL	850	13	33

Source: Modified from Osterberg, C. L., in *Wastes in the Ocean,* Vol. 4, *Energy Wastes in the Ocean,* Duedall, I. W., Kester, D. R., Park, P. K., and Ketchum, B. H. (Eds.), John Wiley & Sons, New York, 1985, 127.

TABLE 3.2
Oceanic Conditions at Selected Coastal Nuclear Power Plants

Plant	Water Temperature Range (°C)	Average Salinity (‰)	Mean Tidal Range (m)	Typical Currents (cm/s)
Diablo Canyon, CA	10–18	33.5	1.6	18
Millstone, CT	1–24	28.5	0.8	30
Pilgrim, MA	1–17	32	2.8	9
San Onofre, CA	13–22	33.5	1.6	12
St. Lucie, FL	15–29	33	0.8	18

Source: Modified from Osterberg, C. L., Nuclear power wastes and the ocean, in *Wastes in the Ocean,* Vol. 4, *Energy Wastes in the Ocean,* Duedall, I. W., Kester, D. R., Park, P. K., and Ketchum, B. H. (Eds.), John Wiley & Sons, New York, 1985, 127.

TABLE 3.3
Radioactivity (10^9) of Mixed Fission and Activation Products in Liquid Effluent from Selected Nuclear Power Plants

Plant	1970	1971	1972	1973	1974	1975	1976	1977	1978	1979
Millstone 1, CT		730	1900	1200	7300	7400	360	20	6.7	7.8
Pilgrim, MA			56	33	160	300	86	130	65	19
Calvert Cliffs, MD						53	44	130	230	290
Indian Point 1 and 2, NY				81	160	180	180	110	74	72
Indian Point 3, NY									38	15
Millstone 2, CT						0.7	9.6	58	100	180
San Onofre, CA	280	56	1100	590	180	45	270	360	440	410
St. Lucie 1, FL							3	210	100	99

Source: Modified from Osterberg, C. L., in *Wastes in the Ocean,* Vol. 4, *Energy Wastes in the Ocean,* Duedall, I. W., Kester, D. R., Park, P. K., and Ketchum, B. H. (Eds.), John Wiley & Sons, New York, 1985, 127.

TABLE 3.4
Radionuclides in Liquid Effluent from the Calvert Cliffs Nuclear Power Station, MD, in 1980

Radionuclide	Activity (Bq)	Radionuclide	Activity (Bq)
^{3}H	1.8×10^{13}	^{110m}Ag	1.1×10^{9}
^{51}Cr	2.7×10^{10}	^{113}Sn	1.5×10^{8}
^{54}Mn	3.2×10^{9}	^{124}Sb	6.2×10^{8}
^{56}Mn	4.6×10^{7}	^{125}Sb	7.6×10^{9}
^{57}Co	3.0×10^{7}	^{131}I	5.8×10^{9}
^{58}Co	7.4×10^{10}	^{132}I	6.0×10^{7}
^{59}Fe	2.7×10^{8}	^{133}I	2.6×10^{9}
^{60}Co	1.6×10^{10}	^{133}Xe	5.7×10^{11}
^{65}Zn	8.9×10^{8}	^{133m}Xe	3.9×10^{9}
^{85m}Kr	6.6×10^{6}	^{134}Cs	3.9×10^{9}
^{89}Sr	5.2×10^{8}	^{135}I	2.6×10^{8}
^{90}Sr	8.5×10^{8}	^{135}Xe	7.8×10^{9}
^{91}Sr	2.3×10^{7}	^{135m}Xe	1.2×10^{8}
^{95}Nb	1.4×10^{9}	^{137}Cs	6.8×10^{9}
^{95}Zr	7.8×10^{9}	^{140}Ba	2.2×10^{8}
^{97}Zr	2.2×10^{7}	^{140}La	1.2×10^{9}
^{99}Mo	3.2×10^{7}	^{141}Ce	1.0×10^{7}
^{103}Ru	9.2×10^{8}		
^{106}Ru	2.7×10^{7}		

Source: Modified from Osterberg, C. L., in *Wastes in the Ocean,* Vol. 4, *Energy Wastes in the Ocean,* Duedall, I. W., Kester, D. R., Park, P. K., and Ketchum, B. H. (Eds.), John Wiley & Sons, New York, 1985, 127.

TABLE 3.5
Percentages of Soluble and Particulate Radionuclides Discharged by the Oyster Creek Nuclear Generating Station

	Before Discharge		After Discharge	
Radionuclide	Particulate	Soluble	Particulate	Soluble
^{54}Mn	74	26	88	12
^{60}Co	52	48	84	16
^{134}Cs	3	97	17	83
^{137}Cs	3	97	13	87

Note: Uncertainties at the 2-σ confidence level are approximately 10% of the values given.

Source: Blanchard, R. L., in *Effluent and Environmental Radiation Surveillance,* Kelly, J. J. (Ed.), ASTM STP 698, American Society for Testing Materials, Philadelphia, 1980, 139. With permission.

TABLE 3.6
Concentrations (pCi/kg) of Radionuclides in Aquatic Samples at the Haddam Neck and Oyster Creek Nuclear Power Stations

	Haddam Neck			Oyster Creek			
Radionuclide	Fish	Algae	Water[a]	Fish	Clams	Algae	Water[a]
^{3}H	2300	7400	10,000	<250	<250	<250	50
^{54}Mn	<20	560	0.1	10	<20	240	0.39
^{55}Fe	<80	7000	0.6	<80	<100	<60	0.49
^{60}Co	<30	640	0.3	20	190	540	0.81
^{90}Sr	8	28	0.0007	0.3	<10	5.4	0.0011
^{131}I	<20	2300	2	<20	<15	<20	0.12
^{134}Cs	50	1200	0.1	24	<30	<20	2.1
^{137}Cs	110	1900	0.1	36	<20	30	3.5

[a] Water concentrations computed from quantities discharged.

Source: Blanchard, R. L., in *Effluent and Environmental Radiation Surveillance,* Kelly, J. J. (Ed.), ASTM STP 698, American Society for Testing Materials, Philadelphia, 1980, 139. With permission.

TABLE 3.7
Annual Radiation Exposures (mrem/yr) for Aquatic Pathways at the Haddam Neck and Oyster Creek Nuclear Power Stations

Radionuclide	Critical Organ	Haddam Neck Fish[a]	Haddam Neck Water[b]	Oyster Creek Fish[a]	Oyster Creek Clams[c]	Oyster Creek Water[d]
^{3}H	Total body	1.1 (–3)[e]	1.1 (–2)	<1.2 (–4)	<6.1 (–5)	1.2 (–4)
^{54}Mn	GI(LLI)[f]	<2.8 (–3)	3.0 (–5)	1.4 (–3)	<1.4 (–3)	2.8 (–4)
^{55}Fe	Spleen	<1.0 (–3)	1.7 (–5)	<1.0 (–3)	<6.2 (–4)	3.1 (–5)
^{60}Co	GI(LLI)	<8.5 (–3)	1.9 (–4)	5.7 (–3)	2.7 (–2)	1.2 (–3)
^{90}Sr	Bone	7.0 (–2)	1.4 (–5)	2.6 (–3)	<4.4 (–2)	4.8 (–5)
^{131}I	Thyroid	<2.8 (–1)	6.4 (–2)	<2.8 (–1)	<1.1 (–1)	8.5 (–3)
^{134}Cs	Total body	3.7 (–2)	1.7 (–4)	1.8 (–2)	<1.1 (–2)	7.6 (–3)
^{137}Cs	Total body	5.0 (–2)	1.0 (–4)	1.6 (–2)	<4.5 (–3)	7.9 (–3)

[a] Assumes a daily intake of 20 g.
[b] Assumes a daily intake of 1 l and a dilution of 22 in the Connecticut River.
[c] Assumes a daily intake of 10 g.
[d] Assumes a daily intake of 1 l and a dilution of 10 in the bay.
[e] Exponents of 10 are indicated by numbers in parentheses.
[f] GI = gastrointestinal tract; LLI = lower large intestine.

Source: Blanchard, R. L., in *Effluent and Environmental Radiation Surveillance,* Kelly, J. J. (Ed.), ASTM STP 698, American Society for Testing Materials, Philadelphia, 1980, 139. With permission.

TABLE 3.8
Radiation Dose Rates (mrem/yr) to the Critical Organs of Aquatic Samples for Pathways at the Haddam Neck and Oyster Creek Nuclear Generating Stations

Critical Organ	Pathway at Haddam Fish	Pathway at Haddam Water	Pathway at Oyster Creek Fish	Pathway at Oyster Creek Clams	Pathway at Oyster Creek Water
Total body	8.8 (–2)[a]	1.1 (–2)	3.4 (–2)	<1.6 (–2)	1.6 (–2)
GI(LLI)[b]	<1.1 (–2)	2.2 (–4)	7.1 (–3)	2.8 (–2)	1.5 (–3)
Spleen	<1.0 (–3)	1.7 (–5)	<1.0 (–3)	<6.2 (–4)	3.1 (–5)
Bone	7.0 (–2)	1.4 (–5)	2.6 (–3)	<4.4 (–2)	4.8 (–5)
Thyroid	<2.8 (–1)	6.4 (–2)	<2.8 (–1)	<1.1 (–1)	8.5 (–3)

[a] Exponents of 10 are indicated by numbers in parentheses.
[b] GI = gastrointestinal tract; LLI = lower large intestine.

Source: Blanchard, R. L., in *Effluent and Environmental Radiation Surveillance,* Kelly, J. J. (Ed.), ASTM STP 698, American Society for Testing Materials, Philadelphia, 1980, 139. With permission.

APPENDIX 4. BIOCIDES

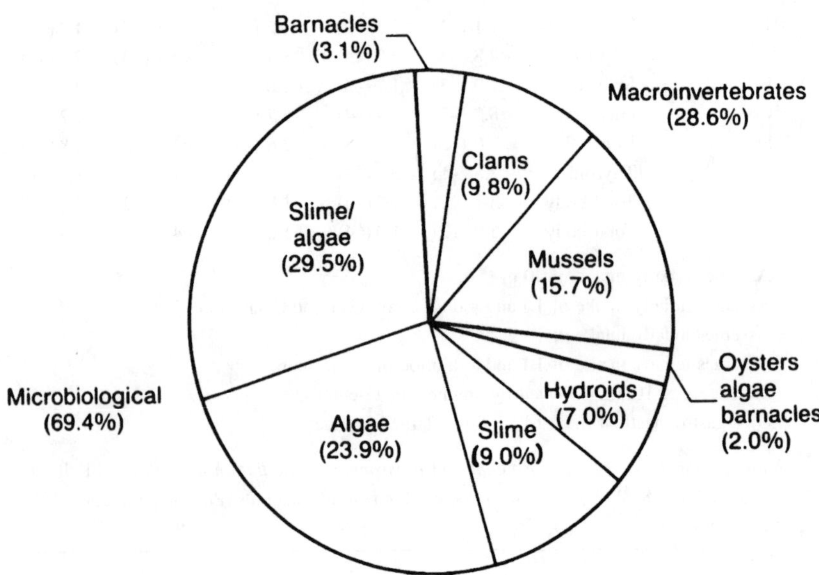

FIGURE 4.1. Biofouling organisms in cooling water systems based on a survey of 365 units. (From Chow, W., in *Proceedings: Condenser Biofouling Control — State-of-the-Art Symposium,* Chow, W. and Massalli, Y. G. (Eds.), Tech. Rep., Electric Power Research Institute, Palo Alto, CA, 1985, 1–1.)

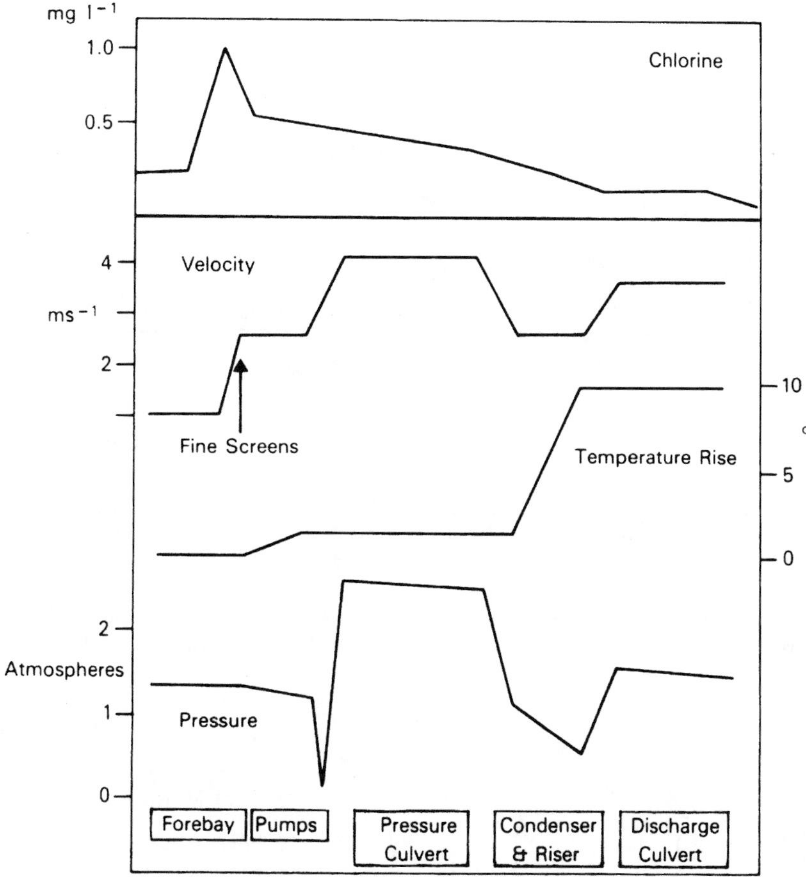

FIGURE 4.2. Typical changes in chlorine as well as velocity, temperature, and pressure in the cooling water system of an electric generating station. (From Langford, T. E., *Electricity Generation and the Ecology of Natural Waters,* Liverpool University Press, England, 1983. With permission.)

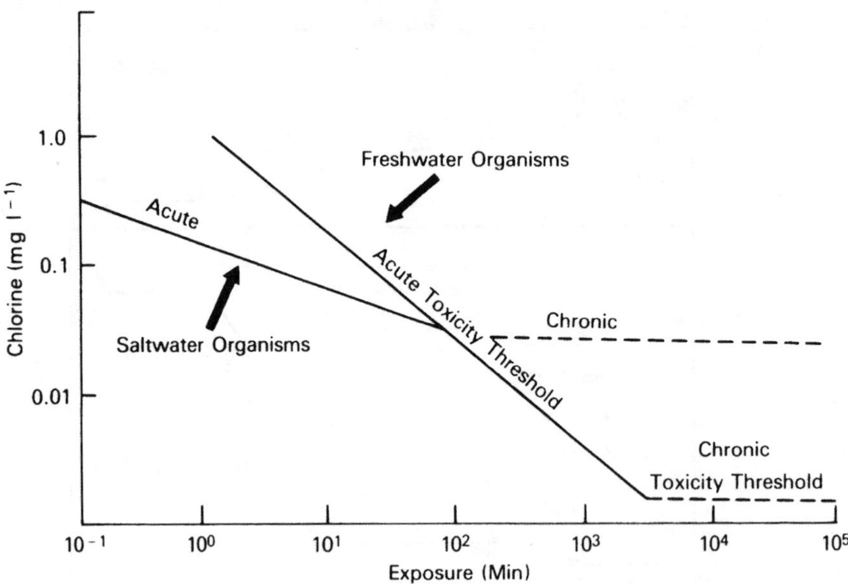

FIGURE 4.3. Chlorine-exposure diagram illustrating acute and chronic toxicity thresholds for marine and freshwater organisms. (From Mattice, J. S. and Zittel, H. E., *J. Water Pollut. Cont.*, 48, 2284, 1976. With permission.)

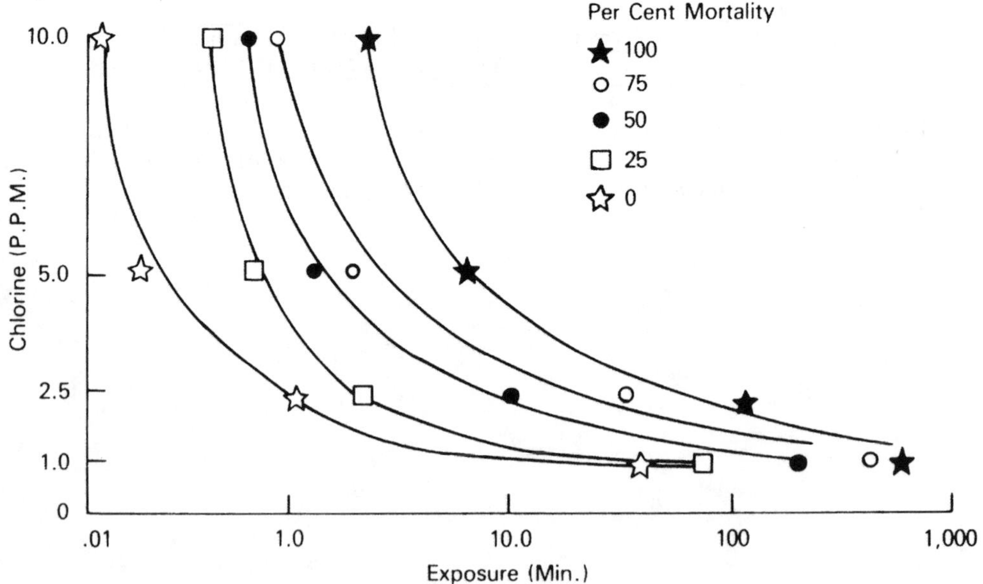

FIGURE 4.4. Mortality data (response isopleths) for the calanoid copepod, *Acartia tonsa*, exposed to chlorine. (From Morgan, R. P., II and Carpenter, E. J., in *Power Plant Entrainment: A Biological Assessment*, Schubel, J. R. and Marcy, B. C. (Eds.), Academic Press, New York, 1978, 95. With permission.)

APPENDIX 5. IMPINGEMENT AND ENTRAINMENT

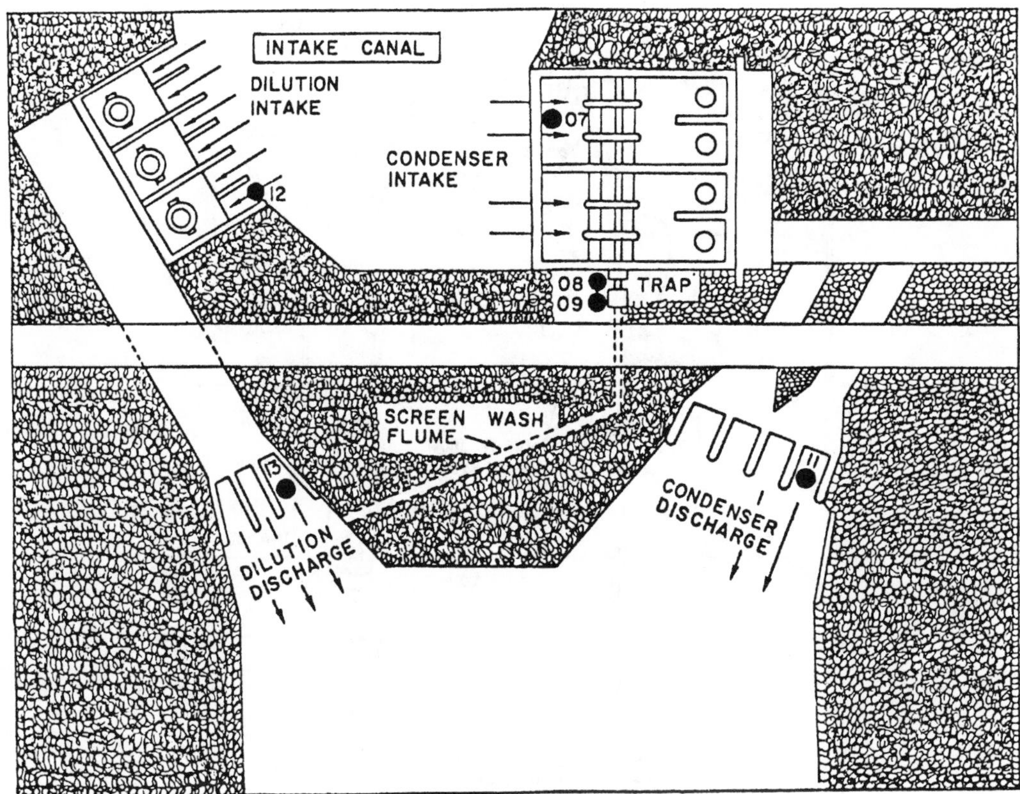

FIGURE 5.1. Configuration of the intake and discharge structures of the Oyster Creek Nuclear Generating Station. (From U.S. Atomic Energy Commission, Final Environmental Statement Related to Operation of the Oyster Creek Nuclear Generating Station, AEC Docket No. 50–219, 1974.)

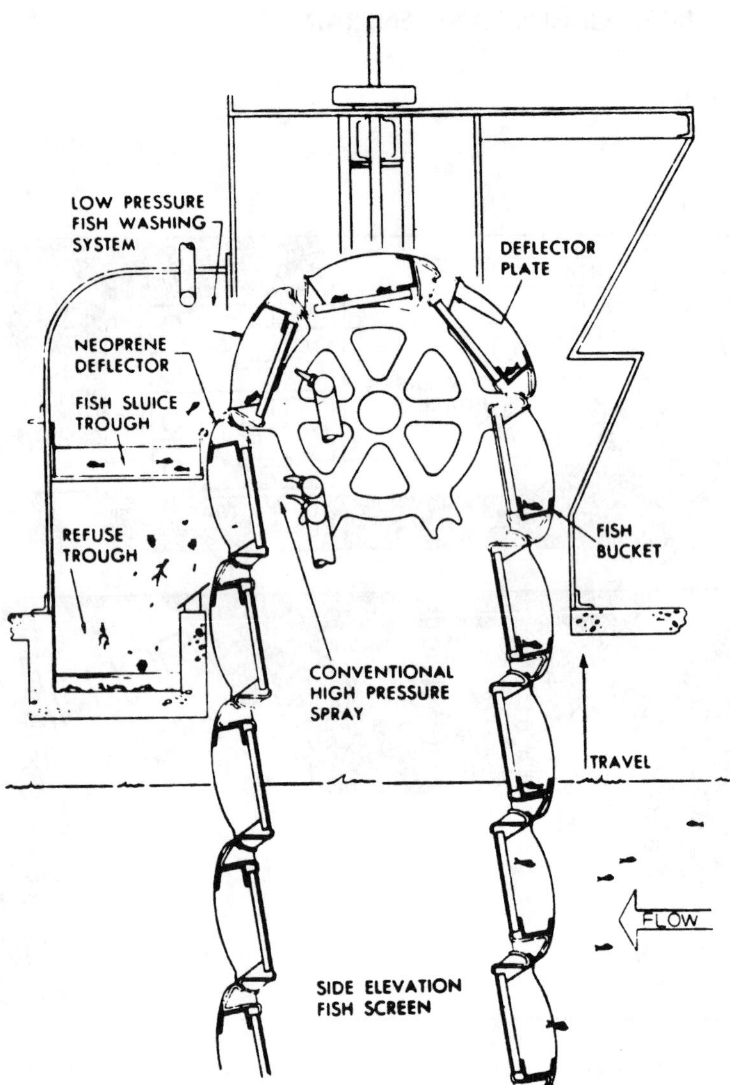

FIGURE 5.2. Ristroph, rotating traveling screen with a fish recovery system. (From Kennish, M. J., Roche, M. B., and Tatham, T. R., in *Ecology of Barnegat Bay, New Jersey,* Kennish, M. J. and Lutz, R. A. (Eds.), Springer-Verlag, New York, 1984, 318. With permission.)

TABLE 5.1
Monthly Estimates of the Number of White Perch Impinged at All Hudson River Power Plants Combined for 1974 and 1975 Year Classes

		Year Class			
		1974		1975	
		Number of Years of Vulnerability		Number of Years of Vulnerability	
Age (years)	Month	2	3	2	3
0	6	0		0	
	7	3486		8898	
	8	14,887		97,910	
	9	26,239		83,980	
	10	112,957		93,888	
	11	245,492		239,150	
	12	607,434		348,596	
	1	415,724		589,206	
	2	270,571		182,891	
	3	139,751		130,261	
	4	609,090		111,820	
	5	91,910		40,151	
1	6	37,242	18,621	27,014	13,507
	7	22,126	11,063	13,835	6918
	8	14,122	7061	6770	3385
	9	19,924	9962	13,791	6896
	10	19,534	9767	25,676	12,838
	11	28,005	14,002	12,552	6276
	12	7803	3902	48,102	24,051
	1	38,078	19,039	143,010	71,515
	2	9293	4646	43,558	21,779
	3	12,444	6222	49,579	24,790
	4	14,103	7052	38,692	19,346
	5	7612	3806	56,365	28,183
2	6		13,057		35,710
	7		6918		8805
	8		3385		12,662
	9		6896		8736
	10		12,838		17,362
	11		6276		19,145
	12		24,051		10,890
	1		71,505		
	2		21,779		
	3		24,790		
	4		19,346		
	5		28,182		

Source: Van Winkle, W., Barnthouse, L. W., Kirk, B. L., and Vaughan, D. S., *Evaluation of Impingement Losses of White Perch at the Indian Point Nuclear Station and Other Hudson River Power Plants,* Tech. Rep. ORNL/NUREG/TM-361, Oak Ridge National Laboratory, Oak Ridge, TN, 1980.

TABLE 5.2
Catch Rates and Seasonality of the Five Dominant Finfish Species Collected in Impingement Samples at the Pilgrim Nuclear Generating Station from January 1976 to December 1980

Species	Rate (h)	Rate[a] (year)	Dominant Season of Occurrence
Atlantic herring	1.62	14,191	Autumn
Rainbow smelt	0.55	4818	Autumn
Atlantic silverside	0.38	3329	Winter/spring
Alewife	0.35	3066	Summer
Cunner	0.10	876	Summer
All fish	3.30	28,908	

[a] Assuming operation of both circulating water pumps 100% of the time.

Source: Lawton, R. P., Anderson, R. D., Brady, P., Sheehan, C., Sides, W., Kouloheras, E., Borgatti, M., and Malkoski, V., in *Observations on the Ecology and Biology of Western Cape Cod Bay, Massachusetts,* Davis, J. D. and Merriman, D. (Eds.), Springer-Verlag, New York, 1984, 191. With permission.

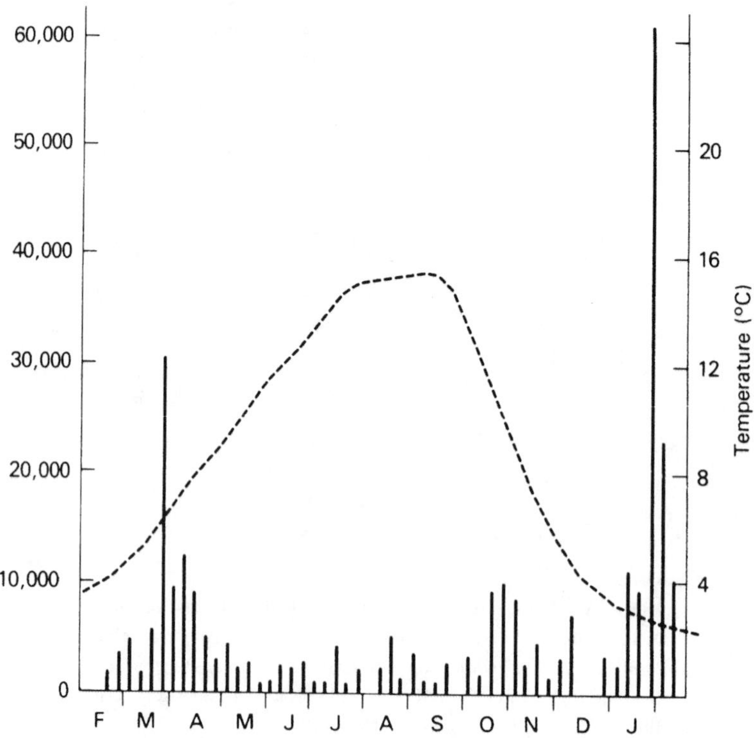

FIGURE 5.3. Weekly impingement of fish on intake screens of the Fawley Generating Station, England, in 1973. Vertical bars depict impingement. Dashed line gives water temperature. (From Langford, T. E., *Electricity Generation and the Ecology of Natural Waters,* Liverpool University Press, England, 1983. With permission.)

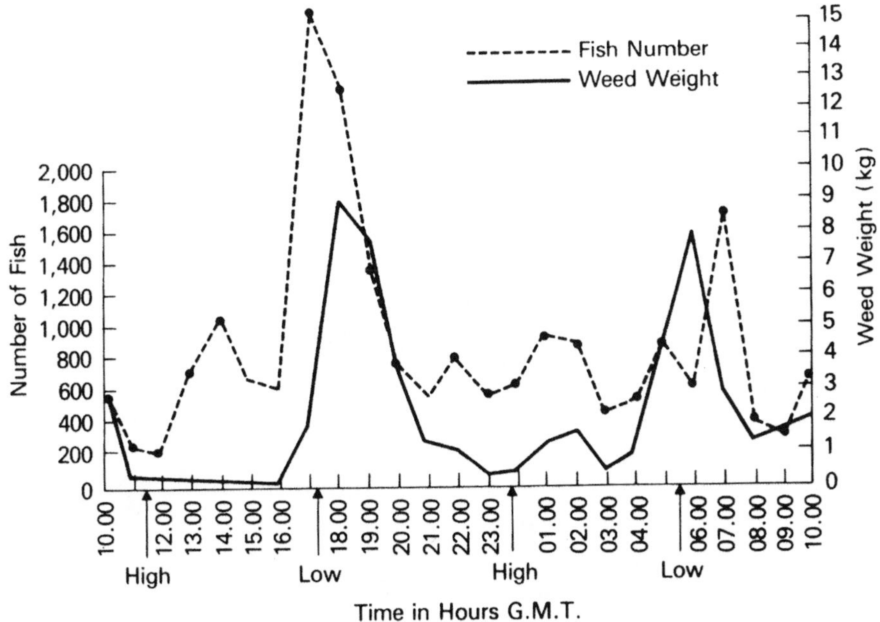

FIGURE 5.4. Hourly impingement of fish and weeds on intake screens of the Fawley Generating Station, England, on November 20 and 21, 1975. (From Langford, T. E., *Electricity Generation and the Ecology of Natural Waters,* Liverpool University Press, England, 1983. With permission.)

TABLE 5.3
Average Yearly Entrainment and Impingement Losses at the Brunswick Steam Electric Plant[a]

Species	Entrainment[b]	Impingement[c]	Impingement[d]
Spot	186,000	724	350
Croaker	123,000	356	235
Shrimp	171,000	675	760
Flounder	7200	21	12
Mullet	5200	34	18
Trout	38,500	169	205
Menhaden	32,500	9744	4000
Anchovy	913,000	2748	1600

[a] Number of organisms × 10^3.

[b] Computed from weekly average plant flows from September 1976 through August 1978 and 5-year average entrainment densities for the same weekly period. These flows are close to the full plant flow.

[c] Two-year averages of measured impingement losses from September 1976 through August 1978.

[d] Five-year averages of measured impingement losses from January 1974 through August 1978.

Source: Lawler, J. P., Hogarth, W. T., Copeland, B. J., Weinstein, M. P., Hodson, H. G., and Chen, H. Y., in *Issues Associated With Impact Assessment, Proc. 5th Natl. Workshop on Entrainment and Impingement,* Jensen, L. D. (Ed.), Ecological Analysts Communications, Sparks, MD, 1981, 159. With permission.

Index

C

D

E

F

G

H

I

J

K

L

M

N

O

P

R

S

T

U

V

W

X

Z